Bruce W. Martin

THE LIBRARY
ST. MARY'S COLLEGE OF MARYLAND
ST. MARY'S CITY, MARYLAND 20686

D1786927

GROUNDWATER MONITORING

GROUNDWATER MONITORING

GUIDELINES AND METHODOLOGY FOR DEVELOPING AND IMPLEMENTING A GROUNDWATER QUALITY MONITORING PROGRAM.

LORNE G. EVERETT

GENERAL ELECTRIC COMPANY
TECHNOLOGY MARKETING OPERATION
120 ERIE BLVD.
SCHENECTADY, N.Y. 12305

Copyright © 1980 by General Electric Company

All rights reserved. No part of this publication may be reproduced, stored in a retrieval system, or transmitted in any form or by any means electronic, mechanical, photo copying, recording or otherwise without the prior written permission of the publisher.

Judgements as to the suitability of information herein for purchaser's purposes are neccessarily the purchaser's responsibility. Although reasonable care has been taken in the preparation of such information, General Electric Company extends no warranties, makes no representations and assumes no responsibility as to the accuracy or suitability of such information for application to the purchaser's intended purposes or for consequences of its use.

80-82885

ISBN 0-931690-14-5

Printed in the United States of America

ACKNOWLEDGMENTS

The material in this book was originally prepared as part of a series of 11 reports by TEMPO, the Center for Advanced Studies of the General Electric Company, under contract to the U.S. Environmental Protection Agency (Contract No. 69-01-0729). Mr. Leslie G. McMillion of the EPA Environmental Monitoring and Support Laboratory, Las Vegas, Nevada provided technical guidance during the course of the study.

Chapter II was written by Dr. David K. Todd, University of California, Berkeley; Dr. Richard M. Tinlin, consultant, Camp Verde, Arizona; Dr. Kenneth D. Schmidt, consultant, Fresno, California; and Dr. Lorne G. Everett, General Electric-TEMPO, Santa Barbara, California. Chapter III was written by Dr. Lorne G. Everett and Dr. Kenneth D. Schmidt, Chapter IV was written by Mr. Norman F. Hampton, consultant, Germantown, Maryland. Chapter V was written by Dr. Donald L. Warner, University of Missouri, Rolla. Chapter VI was originally edited by Dr. Richard M. Tinlin.

I am pleased to acknowledge the assistance, as well as the encouragement, of the EPA and the principal authors toward publishing the work in book form.

LORNE G. EVERETT

Dr. Everett is Manager of the Natural Resource Program for General Electric's Center for Advanced Studies, TEMPO, Santa Barbara, California. His current hydrology interests are related to the design of groundwater quality monitoring programs for coal strip mining, oil shale extraction, uranium mine development, and hazardous waste disposal areas. In addition, he oversees programs relating to minerals and industrial development.

After completing his PhD in Hydrology at the University of Arizona in 1972, Dr. Everett was invited to join the faculty in the Department of Hydrology. Prior to his current position, Dr. Everett was the Manager of TEMPO's Water Resources Program and a principal investigator in developing a national groundwater quality monitoring methodology for the U.S. Environmental Protection Agency. Currently, he oversees a major EPA contract to develop groundwater quality monitoring quidelines for all western coal strip mine operations and for surface and in situ extraction of shale oil. His work has emphasized monitoring in the vadose zone.

Dr. Everett has worked under contract to the U.S. Department of Justice in managing testimony relative to water resource decisions. He has testified before Congress on national legislation relative to water monitoring. In 1978, Dr. Everett was invited by the American Water Resources Association to be the Technical Chairman of a special symposium on water quality monitoring. He is a member of several professional societies and has published extensively. He is the principal author of the book *Establishment of Water Quality Monitoring Program*.

TABLE OF CONTENTS

CHAPTER I - SUMMARY

MONITORING METHODOLOGY 1

MONITORING METHODS AND COSTS 1

DATA MANAGEMENT 2

MONITORING DISPOSAL WELLS 2

ILLUSTRATIVE EXAMPLES 2

SELECTED BIBLIOGRAPHY 3

CHAPTER II - GROUNDWATER MONITORING METHODOLOGY

Section 1 - Introduction 5

BACKGROUND AND SCOPE 5

PURPOSE AND APPROACH 5

Ambient Trend Monitoring 6
Source Monitoring 6
Case Preparation Monitoring 6
Research Monitoring 6

NEEDS AND OBJECTIVES 6

Economic Needs and Objectives 6
EPA Needs and Objectives 8
Monitoring Objectives 10

CONSTRAINTS 11

Section 2 - Groundwater Quality 13

HYDROGEOLOGIC FRAMEWORK 13

Geologic Formations as Aquifers 14
Groundwater Movement 14
Natural Chemical Quality 15

OCCURRENCE OF GROUNDWATER POLLUTION 19

Definition 19
Distribution of Pollutants 19
Mechanisms of Pollution 20
Attenuation of Pollution 28
Distribution of Pollution Underground 29
Evaluation of Pollution Potential 32
Trends in Groundwater Pollution 34

CONSTITUENTS IN POLLUTED GROUNDWATER	34
Quality Categories	34
Effects of Water Use	34
Agricultural Uses	35
Industrial Uses	38
Commercial Uses	39
Domestic Uses	40
SOURCES AND CAUSES OF POLLUTION	42
Agricultural Sources and Causes	42
Irrigation Return Flow	42
Animal Wastes	43
Fertilizers	43
Crop Residues and Dead Animals	43
Pesticide Residues	44
Municipal and Industrial Sources and Causes	44
Surface Disposal of Solid Wastes	44
Surface Disposal of Liquid Wastes	45
Sewer Leakage	48
Tank and Pipeline Leakage	48
Disposal Wells	49
Injection Wells	49
Stockpiles	50
Mining Activities	50
Oilfield Brines	51
Groundwater Basin Management	51
Saline Water Intrusion	51
Aquifers Interchange Through Wells	52
Miscellaneous	54
Spills and Surface Discharges	54
Septic Tanks and Cesspools	54
Highway Deicing	55
QUALITY IN RELATION TO WATER USE	55
Water Quality Standards	55
Drinking Water	56
Irrigation Water	58
Livestock Water	58
Section 3 - Monitoring Methodology	61
CONCEPT OF A MONITORING METHODOLOGY	61
Implementation of a Monitoring Methodology	62
Step 1 - Select Area or Basin for Monitoring	62
Administrative Considerations	62
Physiographic Considerations	62
Priority Considerations	63
Example - Basin Boundary Area	63

Step 2 - Identify Pollution Sources, Causes, and Methods
of Waste Disposal 65
    Municipal 65
    Agricultural 67
    Industrial 68
    Oilfield Wastes 68
    Mining Wastes 69
    Miscellaneous 69
Step 3 - Identify Potential Pollutants 69
    Municipal 72
    Agricultural 73
    Industrial 74
    Oilfield Wastes 75
    Mining Wastes 75
    Miscellaneous 76
Step 4 - Define Groundwater Usage 76
    Municipal Use 76
    Industrial Use 76
    Agricultural Use 77
    Rural Use 77
Step 5 - Define Hydrogeologic Situation 78
Step 6 - Study Existing Groundwater Quality
Step 7 - Evaluate Infiltration Potential of Wastes
at the Land Surface 85
Step 8 - Evaluate Mobility of Pollutants from the Land Surface
to Water Table 87
    Dilution 88
    Filtration 88
    Sorption 89
    Buffering 90
    Chemical Precipitation 90
    Oxidation and Reduction 90
    Volatilization 91
    Biologic Degradation and Assimilation 91
    Radio Active Decay 91
Step 9 - Evaluate Attenuation of Pollutants in the Saturated Zone 92
    Processes Other Than Dilution 92
    Dilution and Related Factors 92
Step 10 - Prioritize Sources and Causes 94
Step 11 - Evaluate Existing Monitoring Programs 96
Step 12 - Establish Alternative Monitoring Approaches 97
    Selection of Portion of System to be Monitored 98
    Selection of Nonsampling Methods 98
    Determination of Required Analyses 99
    Determination of Sampling Frequencies 100
Step 13 - Select and Implement the Monitoring Program 101
Step 14 - Review and Interpret Monitoring Results 102
Step 15 - Summarize and Transmit Monitoring Information .. 102

SELECTED BIBLIOGRAPHY 120

CHAPTER III – GROUNDWATER MONITORING METHODS AND COSTS

Section 1 – Introduction 129

PURPOSE 129

SCOPE 129

HYDROGEOLOGICAL FRAMEWORK 130

GROUNDWATER MONITORING METHODS 131

TYPES OF COST 132

Updating Cost Data 132
Effects of Scale on Costs 135
In-House Versus Out-of-House Costs 135
Accuracy and Costs 136

Section 2 – Monitoring at the Land Surface 137

INTRODUCTION 137

NONSAMPLING METHODS 137

Waste-Load Inventory 137
Calculation of Leaching Potential 137
Pipeline and Tank Tests 138
Testing Artificial Liners for Leakage 139
Aerial Surveillance 140
Notification Procedures 141
Emergency Procedures 147

SAMPLING METHODS 149

Surface Water Bodies 149
Wastewater 150
Solid Wastes 151

Section 3 – Monitoring in the Vadose Zone 152

INTRODUCTION 152

Topsoil 152
Vadose Zone 152
Monitoring Techniques 152
Water Content 152
Water Flow 153
Chemical Changes 153

SOIL SAMPLING AND WELL DRILLING 153

Shallow Wells 154
Deeper Wells 154

DETERMINATION OF WATER CONTENT 154

Neutron Moderation or Moisture Logging 155
Tensiometers 159
Electric Resistance Blocks 160
Calculation of Water Content 160

DETERMINATION OF WATER MOVEMENT 163

Infiltration Across the Soil Surface 163
Movement in Perched Water Tables 163
    Piezometers 163
    Observation Wells 164

WATER MOVEMENT IN THE UNSATURATED STATE 164

Tensiometers 164
Psychrometers 165
Neutron Moisture Logging 166
Moisture Blocks 166
Quantification of Flux in Unsaturated Media 166
    Relating Water Content Changes to Flux 166
    Expressions for Flux and Hydraulic Conductivity 167
    Direct Measurement of Flux 167
Calculation of Approximate Flow Rates 167
Use of Tracers 168

WATER SAMPLING IN THE VADOSE ZONE 168

Sampling in Saturated Regions 168
    Wells and Piezometers 168
    Sampling Tile Drain Outflow 169
    Fiberglass Probes 169
    Multi-Level Sampler 170
    Sampling Chambers 170
Sampling in Unsaturated Regions 170
    Soil Sampling 170
    Suction Cups 170
    Salinity Sensors 174

Section 4 - Monitoring in the Zone of Saturation 175

INTRODUCTION 175

GENERAL MONITORING PROCEDURES 175

Well Inventory and Well Data Collection 175
Geological Framework 175
Water Levels and Groundwater Movement 176
Pump Tests and Aquifer Analysis 177
Water Budget and Modeling 177
Water Temperature and Chemical Quality 178

SPECIFIC MONITORING METHODS 178

Surface Geophysics 178
Well Construction 181
Types of Shallow Wells 181
Costs of Shallow Wells 182
Types of Drilled Wells 182
Costs of Well Drilling 184
Well Casing and Costs 184
Well Screens and Perforated Casing 187
Gravel Packing 187
Well Sealing 189
Well Development 189
Test Pumping 190
Total Well Construction Costs 190
Test Drilling 191
Geological Sampling 192
Borehole Geophysics 194
Water-Level Measurements 196
    Steel Tape 196
    Electric Sounder 197
    Airline 199
    Mechanical Recorder 199
Water Sampling 199
Well Hydraulics 200
Time Changes in Quality 200
Sampling at the Well Discharge 201
Types of Well Sampling 202
Sampling at Open Wells 203
Other Methods of Sampling Groundwater 210

Section 5 - Analysis of Samples 211

INTRODUCTION 211

CUSTODY CONTROL 211

Preparation 211
Sample Collection 212

QUALITY CONTROL 213

SOIL 215

Physical Analyses 215
    Water Content 215
    Bulk Density and Porosity 215
    Particle-Size Distribution 216
    Soil-Water Characteristic Curve 216
    Hydraulic Conductivity 218
    Specific Surface 218
Chemical Analysis 219
    Soluble Ions 219
    Cation Exchange Capacity 221

WATER	221
Determinations	223
Physical	224
Inorganic Chemical	224
Containers	225
Preservation of Water and Waste Samples	226
Sample Collection and Treatment for Groundwaters	226
Selection of Specific Determinations	228
Organic Chemical	232
Containers	232
Sample Preservation	233
Bacteriological	233
Radiological	235
Containers	235
Preservation	235
SELECTED BIBLIOGRAPHY	237

CHAPTER IV - GROUNDWATER DATA MANAGEMENT

Section 1 - Introduction

SCOPE	247

Section 2 - Groundwater Information Management Requirements — 248

GENERAL	248
INFORMATION TO BE MANAGED	249
Station Descriptions	249
Quality Criteria	251
Geologic Data	251
Hydrologic Data	252
Water Quality Parameter Identifiers	253
Water Quality Measurements	255
Temporal Data	255
Information Qualification Data	256
DMA Status Data	256
Information Indexing	257
Summary	257
DATA COLLECTION	259
DATA COMMUNICATIONS	259
DATA STORAGE	261
DATA PROCESSING	263
DATA RETRIEVAL	264

Section 3 - Existing Systems 267

GENERAL 267

STORET 267

WATSTORE 279

NAWDEX 281

Section 4 - Proposed Modifications to Existing System 285

Section 5 - Conclusions and Recommendations 288

LIST OF ABBREVIATIONS AND ACRONYMS 290

SELECTED BIBLIOGRAPHY 292

CHAPTER V - MONITORING DISPOSAL WELLS

Section 1 - Introduction 295

Section 2 - The Subsurface Environment 296

STRATIGRAPHIC GEOLOGY 296

STRUCTURAL GEOLOGY 300

LITHOLOGY 300

FLUIDS 300

Chemistry 300
Viscosity 304
Density 308
Pressure 308
Compressibility 312

MECHANICAL PROPERTIES OF INJECTION AND CONFINING UNITS 313

Porosity 313
Permeability 313
Compressibility 316
Temperature 316
State of Stress 317

HYDRODYNAMICS 318

RESOURCES 318

Section 3 - Acquisition of Subsurface Data 320

PRIOR TO DRILLING 320

DURING WELL CONSTRUCTION AND TESTING 320

Rock Samples 320
Formation Fluids 320
Borehole Geophysical Logs 325
Testing of Injection Units and Confining Beds 327
Drill Stem Testing 331
Injectivity Tests 334

Section 4 - Prediction of Aquifer Response 337

FLOW THEORY 337

REGIONAL FLOW 338

PRESSURE EFFECTS OF INJECTION 339

Multiple Wells 344
Hydrologic Discontinuities 345

RATE AND DIRECTION OF FLUID MOVEMENT 346

HYDRAULIC FRACTURING 350

GENERATION OF EARTHQUAKES 351

Section 5 - Surveillance of Operating Wells 353

INJECTION WELL MONITORING 353

PERIODIC INSPECTION AND TESTING 356

MONITORING WELLS 363

OTHER MONITORING METHODS 366

Section 6 - Conclusions and Recommendations 368

SELECTED BIBLIOGRAPHY 369

CHAPTER VI - ILLUSTRATIVE EXAMPLES

Section 1 - Introduction 375

Section 2 - Groundwater Pollution Case Histories and Evaluation of Monitoring Techniques 377

BRINE DISPOSAL IN ARKANSAS 377

Background	377
Description of Area	377
Location	377
Geologic Setting	377
History	378
Monitoring Methods	379
Objectives	379
Exploration Alternatives	379
Drilling Alternatives	380
Critique of Monitoring Project	381
Results Obtained	381
Recommended Changes	381
PLATING WASTE CONTAMINATION IN LONG ISLAND, NEW YORK	**382**
Background	382
Hydrogeologic Setting	384
Dimensions of Plume	385
Mapping of Plume	385
Steps That Could Have Been Taken	385
Surface Resistivity Survey	385
Comprehensive Chemical Analyses	388
Hydraulic Properties of the Aquifer	388
Contaminant Transport Model	388
What Should Have Been Done	388
Improved Treatment	388
Sampling of Wells and Streams	388
Recovery of Contaminants	388
LANDFILL LEACHATE CONTAMINATION IN MILFORD, CONNECTICUT	**389**
Background	389
Hydrogeologic Setting	389
Dimensions of Plume	389
Mapping of Plume	391
What Could Have Been Done	392
Gas Generation	392
Additional Chemical Determinations	392
Bacteriological Studies	392
Additional Test Wells	393
Infiltration Rates	393
What Should Have Been Done	393
POLLUTION POTENTIAL OF AN OXIDATION POND NEAR TUCSON, ARIZONA	**393**
Map of Nitrate Levels	395
Rationale of Project	396
What Could Have Been Done	398
What Should Have Been Done	399

MULTIPLE-SOURCE NITRATE POLLUTION IN THE FRESNO-CLOVIS,
CALIFORNIA, METROPOLITAN AREA 401

Background 401
Summary of the Monitoring Program 401
    Collection of Additional Water Samples 403
    Site-Specific Data and Interpretation 403
    Special Cases and Assumptions 407
    Institutional Constraints 408
Description of Alternative Monitoring Programs 408
Strengths and Weaknesses of Monitoring Program 409
Description of Optimal Monitoring Program 409

Section 3 - Site-Specific Groundwater Quality Monitoring Examples . . 410

AGRICULTURAL RETURN FLOW 410

Land Surface Monitoring 410
Vadose Zone Monitoring 411
Saturated Zone Monitoring 411

AGRICULTURAL RETURN FLOW EXAMPLE 411

Step 2 - Identify Pollution Sources, Causes,
         and Methods of Waste Disposal 411
Step 3 - Identify Potential Pollutants 412
Step 4 - Define Groundwater Usage 412
Step 5 - Define Hydrogeologic Situation 412
Step 6 - Study Existing Groundwater Quality 412
Step 7 - Evaluate Infiltration Potential of Wastes
         at the Land Surface 413
Step 8 - Evaluate Mobility of Pollutants from the
         Land Surface to the Water Table 413
Step 9 - Evaluate Attenuation of Pollutants
         in the Saturated Zone 413
Step 11 - Evaluate Existing Monitoring Programs 414
Step 12 - Establish Alternative Monitoring Approaches 414
    Land Surface Monitoring 414
    Vadose Zone Monitoring 414
    Saturated Zone Monitoring 414
Step 13 - Select and Implement the Monitoring Program 415
    Land Surface Monitoring 415
    Vadose Zone Monitoring 415
    Saturated Zone Monitoring 417
    Summary 417

SEPTIC TANKS 418

Land Surface Monitoring 418
Vadose Zone Monitoring 419
Saturated Zone Monitoring 419

SEPTIC TANK EXAMPLE 419

Step 2 - Identify Pollution Sources and Causes
    and Methods of Waste Disposal 419
Step 3 - Identify Potential Pollutants 420
Step 4 - Define Groundwater Usage 420
Step 5 - Define Hydrogeologic Situation 420
Step 6 - Study Existing Groundwater Quality 420
Step 7 - Evaluate Infiltration Potentials of Wastes
    at the Land Surface 420
Step 8 - Evaluate Mobility of Pollutants from
    the Land Surface to the Water Table 421
Step 9 - Evaluate Attenuation of Pollutants
    in the Saturated Zone 421
Step 11 - Evaluate Existing Monitoring Programs 421
Step 12 - Establish Alternative Monitoring Approaches 422
  Land Surface Monitoring 422
  Saturated Zone Monitoring 422
Step 13 - Select and Implement the Monitoring Program 422
  Land Surface Monitoring 422
  Vadose Zone Monitoring 422
  Saturated Zone Monitoring 423
  Summary 423

PERCOLATION PONDS AND LINED PONDS 423

Land Surface Monitoring 424
Vadose Zone Monitoring 425
Saturated Zone Monitoring 425

PERCOLATION POND EXAMPLE 425

Step 2 - Identify Pollution Sources, Causes,
    and Methods of Waste Disposal 425
Step 3 - Identify Potential Pollutants 425
Step 4 - Define Groundwater Usage 426
Step 5 - Define Hydrogeologic Situation 426
Step 6 - Study Existing Groundwater Quality 426
Step 7 - Evaluate Infiltration Potential of Wastes
    at the Land Surface 426
Step 8 - Evaluate Mobility of Pollutants from
    the Land Surface to the Water Table 426
Step 9 - Evaluate Attenuation of Pollutants
    in the Saturated Zone 426
Step 11 - Evaluate Existing Monitoring Programs 426
Step 12 - Establish Alternative Monitoring Approaches 427
  Land Surface Monitoring 427
  Vadose Zone Monitoring 427
  Saturated Zone Monitoring 427
Step 13 - Select and Implement the Monitoring Program 428
  Land Surface Monitoring 428
  Vadose Zone Monitoring 428
  Saturated Zone Monitoring 429
  Summary 431

SOLID WASTE LANDFILLS 431

    Land Surface Monitoring 432
    Vadose Zone Monitoring 432
    Saturated Zone Monitoring 433

SOLID WASTE LANDFILL EXAMPLE 433

Step 2 - Identify Pollution Sources, Causes,
         and Methods of Waste Disposal 433
Step 3 - Identify Potential Pollutants 433
Step 4 - Define Groundwater Usage 433
Step 5 - Define Hydrogeologic Situation 433
Step 6 - Study Existing Groundwater Quality 434
Step 7 - Evaluate Infiltration Potential of Wastes
         at the Land Surface 434
Step 8 - Evaluate Mobility of Pollutants from
         the Land Surface to the Water Table 434
Step 9 - Evaluate Attenuation of Pollutants
         in the Saturated Zone 434
Step 11 - Evaluate Existing Monitoring Programs 435
Step 12 - Establish Alternative Monitoring Approaches 435
    Land Surface Monitoring 435
    Saturated Zone Monitoring 435
Step 13 - Select and Implement the Monitoring Program 436
    Land Surface Monitoring 436
    Vadose Zone Monitoring 436
    Saturated Zone Monitoring 436
    Summary 437

SELECTED BIBLIOGRAPHY 438

LIST OF ILLUSTRATIONS

Figure		Page
2-1	Unconfined and confined aquifers (Todd, 1959)	14
2-2	Cumulative curves showing the frequency distribution of various constituents in potable water (modified after Davis and DeWiest, 1966)	16
2-3	A hypothetical drainage basin in a humid region in plan view the distribution of zones of polluted water in the upper part of the zone of saturation. Line AA' shows profile view across basin. At disposal sites 1 and 5 pollutants extend through zone of aeration and cause a polluted zone beneath the water table. Disposal sites 2, 3, 4, and 6 do not pollute the groundwater (modified after LeGrand, 1965b)	20
2-4	Disposal of household wastes through a conventional septic tank system (modified after Miller et al, 1974)	21
2-5	Diagram showing percolation of pollutants from a disposal pit to a water table aquifer (modified after Deutsch, 1963)	22
2-6	Diagram showing the horizontal movement of pollutants beneath a disposal pit as a result of clay lenses in the vadose zone above the water table.	22
2-7	Illustration of a line source of groundwater pollution caused by a leaking sewer. The water table is situated below the sewer and is assumed to be horizontal	23
2-8	Diagram showing pollution of an aquifer by leaching of surface solids (modified after Deutsch, 1963)	23
2-9	Diagram showing how polluted water can be induced to flow from a surface stream to a well (modified after Deutsch, 1963)	24
2-10	Diagram showing floodwater entering a well through an improperly sealed gravel pack (modified after Deutsch, 1963)	24
2-11	Diagram showing movement of pollutants from a recharge well to a nearby pumping well (modified after Deutsch, 1963)	25
2-12	Diagrams showing spread of pollutants injected through wells into water table and artesian aquifers (modified after Deutsch, 1963)	26

2-13	Diagrams showing reversal of aquifer leakage by pumping (Deutsch, 1963)	26
2-14	Diagrams showing aquifer leakage by vertical movement of water through a nonpumping well	27
2-15	Diagrams showing lines of flow of pollutants from a recharge pond above a sloping water table (modified after Deutsch, 1963)	27
2-16	Diagram showing migration of saline water caused by lowering of water levels in a gaining stream	28
2-17	Plan view of a water table aquifer showing the hypothetical areal extent to which specific pollutants of mixed wastes at a disposal site disperse and move to insignificant levels (modified after LeGrand, 1965b)	30
2-18	Types of pollution plumes in the upper part of the zone of saturation (plan view). An "X" marks the core of contamination beneath a waste site and "Z" the point downstream at which some plumes terminate (LeGrand, 1965b)	30
2-19	Changes in plumes and factors causing the changes (modified after LeGrand, 1965a)	32
2-20	Rating chart for pollution potential in unconfined aquifers of unconsolidated alluvial materials (modified after LeGrand, 1964)	33
2-21	Estimated trend of groundwater pollution in the United States during the 20th Century	35
2-22	Water quality cycle – sources and uses of water and effects on water quality (modified after Hassan, 1974)	37
2-23	Agricultural uses of water and their effects on water quality (modified after Hassan, 1974)	38
2-24	Industrial uses of water and their effects on water quality (modified after Hassan, 1974)	39
2-25	Commercial uses of water and their effects on water quality (modified after Hassan, 1974)	40
2-26	Domestic uses of water and their effects on water quality (modified after Hassan, 1974)	41
2-27	Schematic vertical cross section through a coastal aquifer showing freshwater and seawater circulations with a transition zone (Meyer, 1973)	52

2-28	Schematic diagram of upcoming and underlying saline water to a pumping well	53
2-29	Diagram illustrating the relation of water pollution to impairment for a given water use (modified after McGauhey, 1968a)	56
2-30	Boundaries of the Santa Clara-Calleguas groundwater basin, Ventura County, California (California Department of Water Resources, December 1974)	64
2-31	Relation of groundwater pollution plume size and orientation for a pollution source near a river with different groundwater flow directions (Palmquist and Sendlien, 1975).	79
2-32	Groundwater flow system under idealized homogeneous aquifer conditions (modified after Born and Stephenson, 1969)	80
2-33	Idealized pollution plume configuration for various locations of surface pollution sources (modified after Palmquist and Sendlein, 1975)	80
2-34	Flow lines for steady-state condition in an aquifer and positions of a pollution front advancing from a percolation pond (Cole, 1975).	81
2-35	Water table contours in the vicinity of Stockton, California, Fall 1964 (California Department of Water Resources, 1967)	83
2-36	Distribution of saline water in a confined aquifer resulting from an oilfield brine disposal pit in southwestern Arkansas (modified after Fryberger, 1975)	84
2-37	Vertical bar graph of chemical quality expressed in miliequivalents per liter (Hem, 1970)	105
2-38	Vertical bar graph of chemical quality expressed in miliequivalents per liter which also shows hardness as $CaCo_3$ in milligrams per liter (Hem, 1970).	106
2-39	Vertical bar graph of chemical quality expressed in milliequivalents per liter which also shows silica in millimoles per liter (Hem, 1970)	106
2-40	Radial vector diagram of chemical quality expressed in milliequivalents per liter (Hem, 1970).	107
2-41	Pattern diagram of chemical quality expressed in milliequivalents per liter (Hem, 1970)	108

2-42	Circular diagram of chemical quality with subdivisions showing percentages of total milliequivalents per liter (Hem, 1970)	109
2-43	Trilinear diagram of chemical quality expressed in percentages of cations and anions as milliequivalents per liter and represented by two points and a circle (Hem, 1970).	110
2-44	Comprehensive trilinear diagram of chemical quality in which five points represent the portions of cations and anions together with total dissolved solids in milligrams per liter and hardness as $CaCo_3$ in milligrams per liter.	111
2-45	Variation in chloride concentration with time for groundwater at Burlington, Massachusetts (Terry, 1974)	112
2-46	Seasonal variation in nitrate concentration for groundwater at Fresno, California (Schmidt, 1972)	113
2-47	Variation in electrical conductivity of groundwater along the length of the Santa Ana River Basin, California	114
2-48	Variation of total dissolved solids with well depth in a portion of the Santa Clara-Calleguas groundwater basin, Ventura County, California (California Department) . . .	115
2-49	Isosalinity map of groundwater in the Santa Clara-Calleguas groundwater basin, Ventura County, California, as of 1966 (California Department of Water Resources, August 1974)	116
2-50	Vertical cross section showing groundwater pollution movement from waste disposal ponds and control by cooling water recharge ponds and purge wells (Burt, 1972).	117
2-51	Contours of chloride concentration in groundwater surrounding a brine disposal pit in southwestern Arkansas (modified after Fryberger, 1975)	118
2-52	Plume of groundwater pollution from a landfill near Munich, West Germany, shown by lines of chloride concentration (Cole, 1975)	119
3-1	Schematic diagram of groundwater pollution from surface sources.	130
3-2	Small-diameter well costs index and the ENR materials components (after Gibb, 1971).	133

xviii

3-3	Large-capacity well costs index and the ENR construction cost index (modified after Gibb, 1971).	133
3-4	Cost of testing submerged tanks of various volumes, November 1974.	139
3-5	Alerting procedure chart for hazardous material spills (after California Emergency Plan, 1970)	148
3-6	Cost of power-augering 8-inch diameter sample holes to various depths and hole densities, November 1974	155
3-7	Sketch of neutron moisture logger and accessories (after Holmes et al., 1967).	156
3-8	Moisture logs showing growth and dissipation of mounds in the vadose zone (after Wilson and DeCook, 1968).	158
3-9	Cost of neutron moisture logging for various depths and sample space densities, December 1974.	159
3-10	A single manometer tensiometer.	160
3-11	Cost of tensiometers for various depths and sample space densities, December 1974	161
3-12	Cost of multiple electrical resistance blocks and soil moisture meter for various depths and sampling densities, December 1974	162
3-13	Relationship between median grain size and water-storage properties of alluvium from large valleys (Davis and DeWiest, 1966)	162
3-14	Cost of piezometers for various depths and sampling densities, November 1974	165
3-15	Cross section of suction cup assembly and back-filling material (after Parizek and Lane, 1970).	172
3-16	Cross section of porous cup hi/pressure-vacuum soil water sampler	173
3-17	Cost of soil salinity sensors and a salinity bridge for various depths and sampling densities, November 1974.	174
3-18	Cost of shallow, small-diameter monitor wells, October 1974.	182
3-19	Updated (1975) cost of 4-, 5-, and 6-inch wells in sand and gravel (after Gibb, 1971)	191

3-20	Updated (1975) cost of 4-, 5-, and 6-inch wells in consolidated rock (after Gibb, 1971)	193
3-21	Generalized cost of coring and unconsolidated and consolidated formations, October 1974	195
3-22	Cost of electrical conductivity (EC) and gamma logging, September 1974	196
3-23	Cost of steel tapes, September 1974.	198
3-24	Cost of electric sounder and cable, September 1974.	198
3-25	Approximate cost of sampling wells (one round), December 1974.	202
3-26	Cost of cable and flex hose for use with portable submersible pumps, October 1974	208
3-27	Power costs for continuous pumping of 1 hour, 1 day, 1 week for specific pumping heads and specific discharge rates, December 1974.	209
3-28	Costs for determinations of water content and bulk density in soils, September 1974.	215
3-29	Modified Haines Apparatus for obtaining soil-water characteristics curves (after Day et al., 1967).	217
3-30	Soil-water characteristic curve, showing hysteresis. ABC = drying, CDA = wetting (after Day et al., 1967). . . .	218
4-1	User access to groundwater data base	262
4-2	STORET system-station storage format.	272
4-3	STORET system-station type codes.	275
4-4	WATSTORE Water Quality File - data storage format	283
5-1	Generalized columnar section of Cambrian and Ordovician strata in northeastern Illinois (Buschbach, 1964, p. 16)	297
5-2	Isopach map of Mt. Simon Formation in northeastern Illinois (Buschbach, 1964)	298
5-3	Schematic eastwest section of the Eau Claire and equivalent Rome strata (Janssens, 1973, p. 10)	299
5-4	Lithologic ratio map of post-Mt. Simon, pre-Knox rocks (Janssens, 1973, p. 19)	299

5-5	East-west cross section of Paleozoic rocks in the northern Ohio River Valley -- modified after cross sections in American Association of Petroleum Geologists cross section Publication 4, 1966 (Ohio River Valley Water Sanitation Commission, 1973, p. 51).	301
5-6	Map of the Ohio River Basin and vicinity showing some major geologic features. Data modified from published maps (Ohio River Valley Sanitation Commission, 1973, p. 24).	302
5-7	Structure on top of Mt. Simon Formation (Bond, 1972, p. 36).	303
5-8	Isocon map, showing the dissolved solids content in parts per million of water in the upper 100 feet of the Mt. Simon Formation in Illinois	305
5-9	Water viscosity as a function of temperature and salinity (ppm NaCl) (Pirson, 1963, p. 40).	307
5-10	Specific gravity of distilled water as a function of temperature (Pirson, 1963, p. 39).	308
5-11	Specific volume of water as a function of temperature and pressure (Eisenberg and Kauzmann, 1969, p. 186)	309
5-12	Specific gravity of formation waters (D_w) versus total solids in ppm (data for NaCl solutions) (Pirson, 1963, p. 39).	309
5-13	Relation between relative density and dissolved solids content of brines in deep aquifers of the Illinois basin (Bond, 1972).	310
5-14	Hydraulic pressure gradient in a column of water (Katz and Coats, 1968, p. 11)	311
5-15	Compressibility of water (Katz and Coats, 1968, p. 93).	312
5-16	Map showing distribution of the average porosity of the Mt. Simon Formation in Illinois	314
5-17	Reproduction of portfolio map No. 10, American Association of Petroleum Geologists Geothermal Survey of North America (Gould, 1974)	317
5-18	Potentiometric surface of the Mt. Simon Formation In Ohio and vicinity (Clifford, 1973).	319
5-19	Sample log (Moore, 1951).	321

5-20	Fluid passage diagram for a conventional bottom section, drill stem test (Gatlin, 1960).	323
5-21	Schematic illustration of various drill stem test conditions (Kirkpatrick, 1954).	324
5-22	Portion of a Laterlog-gamma ray-neutron log from a deep well in northern Illinois	328
5-23	Portion of a sonic log from a deep well in northern Illinois.	329
5-24	Portion of a temperature log from a deep well	330
5-25a	Normal sequence of events as recorded on the chart during a successful drill stem test (Kirkpatrick, 1954).	331
5-25b	Sequence of events as recorded during a drill stem test when no fluids were produced (Kirkpatrick, 1954).	331
5-26	Example of a plot of data from a drill stem test with dual closed-in periods (Murphy, undated).	332
5-27	Plot of extrapolated pressure from drill stem test data from an injection well in Ohio.	333
5-28	Plot of pressure buildup data from an injectivity test of the Mt. Simon Formation in Ohio.	334
5-29	Plot of recovery data and matching-type curves for an injection test of a well at Mulberry, Florida (Wilson et al., 1973)	336
5-30	Hydrogeology of the lower Floridan aquifer in northwest Florida (Goolsby, 1972).	340
5-31	Generalized north-south geologic section through southern Alabama and northwestern Florida (Goolsby, 1972)	341
5-32	Theoretical potentiometric surface of lower limestone of Floridan aquifer in late 1971 (Goolsby, 1972)	342
5-33	Generalized flow net showing the potential lines and stream lines in the vicinity of an injection well near an impermeable boundary (Ferris et al., 1962).	345
5-34	Theoretical potentiometric surface of lower limestone of Floridan aquifer in late 1971, with flow lines showing the direction of aquifer water and wastewater movement. Solid flow lines show the direction of flow of diverted aquifer water, dashed flow lines show direction of flow of injected wastewater and displaced aquifer water (modified after Goolsby, 1972)	347

5-35	Predicted and probably actual extent of wastewater for a well completed in a carbonate aquifer	349
5-36	Schematic diagram of pressure versus time during hydraulic fracturing (Kehle, 1964).	350
5-37	Schematic diagram of an industrial waste injection well completed in competent sandstone (modified after Warner, 1965)	353
5-38	Pressure history of a well injecting into a carbonate aquifer.	354
5-39	Semilogarithmic plot of two pressure fall-off tests measured in an injection well of the Monsanto Company, Pensacola, Florida (Goolsby, 1971)	355
5-40	Monthly average injection index of two injection wells of the Monsanto Company, Pensacola, Florida (Goolsby, 1971)	356
5-41	Pipe inspection log and photographs of casing pulled after log was run to verify the log (Schlumberger, 1970).	357
5-42	Portion of a casing inspection log run in a wastewater injection well showing possible corrosion in the interval from 1480 to 1510 ft.	358
5-43	Preinjection and postinjection caliper logs from a wastewater injection well at Belle Glade, Florida, showing solution of the limestone aquifer in the 1500- to 1600-ft interval by acidic wastewater (Black, Crow, and Eidsness, 1972).	359
5-44	Borehole televiewer log of a section of casing showing casing perforations, packer seat, and casing collar (Schlumberger, 1970).	360
5-45	Borehole televiewer log showing vertical fractures in the borehole wall of a well in Oklahoma (Zemanek et al., 1970)	361
5-46	Schematic diagram of a cement bond logging tool in a borehole (Grosmangin et al., 1960)	362
5-47	Portions of a cement bond log from an acid wastewater injection well.	363
5-48	Geologic column and construction of a wastewater injection well at Mulberry, Florida, where two aquifers above the injection zone are monitored through the injection well (Wilson et al., 1973)	367

6-1	Location of Polluted areas.	378
6-2	Contours of chloride concentration at bottom of alluvium.	382
6-3	Section A-A' showing brine distribution.	383
6-4	Map showing plating-waste plume, water-table contours, and selected test wells (modified after Perlmutter and Lieber, 1970)	386
6-5	Vertical profiles of hexavalent chromium and cadmium along the center line of the plating-waste plume (modified after Perlmutter and Lieber, 1970).	387
6-6	Three major groundwater environments as delineated by interpretation of resistivity data 15 to 20 feet below land surface, Milford, Connecticut.	390
6-7	Schematic hydraulic profile along section A-A' of Figure 6-6, Milford, Connecticut	391
6-8	Location of pond near Tucson, Arizona.	394
6-9	Nitrate and chloride distribution in wells near the pond site, October 1971	395
6-10	Location of monitoring facilities at the pond site	396
6-11	Map of part of the San Joaquin Valley, California	402
6-12	Chloride concentration contours (mg/ℓ) in groundwater at and downgradient of Fresno sewage treatment plant.	404
6-13	Chloride concentration contours (mg/ℓ) in groundwater east of the Fresno sewage treatment plant	405
6-14	Relation between aquifer penetration and 1970 nitrate for wells in Figarden-Bullard Area	405
6-15	Short-term trends in nitrate during pump test on a large-capacity well in FCMA.	406
6-16	Seasonal trends in nitrate and chloride for a large-capacity well in a septic tank area.	406
6-17	Study areas in the Fresno-Clovis Metropolitan area	407
6-18	Nitrate concentrations in groundwater near Fresno sewage treatment plant	408

LIST OF TABLES

Table		Page
2-1	Relative Abundance of Dissolved Solids in Potable Water	16
2-2	Major Natural Constituents in Groundwater – Their Sources and Effects upon Usability	17
2-3	Classifications of Water	19
2-4	Explanation of Plumes Shown in Figure 2-18	31
2-5	Parameters and Constituents Which May Be Included in Analyses of Groundwater Quality	36
2-6	Principal Sources and Causes of Groundwater Pollution	42
2-7	Normal Range of Mineral Pickup in Domestic Sewage	47
2-8	Drinking Water Standards of the U.S. Public Health Service	57
2-9	Guide for Evaluating the Quality of Water used for Irrigation	59
2-10	Guide for Evaluating the Quality of Water used by Livestock	60
2-11	Major Sources and Causes of Groundwater Pollution and Methods of Waste Disposal	66
2-12	Classification of Major Potential Pollutants in Groundwater	70
2-13	Major Sources of Groundwater Pollution and Types of Pollutants	71
2-14	Inorganic Chemical Pollutants	72
2-15	Analyses of Chemical Quality of Groundwater Presented in Tabular Form	104
3-1	Example of ENR Indexes for September 12, 1974	134
3-2	Example of Eros Data Center Standard Costs – Satellite Products, October 1974	142
3-3	Land-use Classification System for use with Remote-Sensor Data	145
3-4	Total Cost of Aerial Surveillance for Land-Use Mapping, July 1974	146

3-5	Specific Yield Values Assumed for the Southern Part of the Central Valley, California	163
3-6	Earth Resistivity Survey Costs, December 1974	180
3-7	Costs in Dollars for Well Drilling in Unconsolidated Formations, EPA Regions III and IV, October 1974	185
3-8	Costs in Dollars for Well Drilling in Consolidated Formations, EPA Regions III and IV, October 1974	186
3-9	Black Pipe Casing Costs for Wells in Dollars, EPA Region IX, October 1974	188
3-10	PVC Pipe Costs for Wells in Dollars, EPA Region IX, October 1974	188
3-11	Average Cost of Submersible Water Pumps, Found in EPA Region IX, October 1974	204
3-12	Hydrologic Laboratory Tests and Cost Schedule, September 1974	220
3-13	Recommended Sampling and Preservation Techniques for Inorganic Chemical Determinations	227
3-14	Costs of Inorganic Chemical Determinations for Groundwater Pollution, October 1974	229
3-15	Costs of Organic Chemical Determinations for Groundwater Pollution, October 1974	233
3-16	Recommended Sampling and Preservation Techniques for Organic Chemical Determinations	234
3-17	Costs of Bacterial Analysis for Groundwater Pollution, October 1974	235
3-18	Costs of Radiochemical Analysis in Groundwater Pollution, October 1974	236
4-1	Menu of Candidate Water Quality Parameters for Groundwater Monitoring	254
4-2	Summary of Information to be Managed by Groundwater MIS	258
4-3	Existing Environmental Data Management Systems	268
4-4	Computerized Information Indexing Systems	269
4-5	Generalized Data Base Management Packages	270
4-6	STORET Supported Sections of PL 92-500	271

4-7	Established STORET Parameter Codes – Groundwater Specific	277
4-8	Parameters Maintained in Watstore Groundwater Site Inventory File (Baker, 1975)	280
4-9	USGS Numeric Codes for Geologic Age Identification	282
4-10	Proposed Additional STORET Parameter Codes	287
5-1	Typical Description of a Core from the Top of the Mt. Simon Formation in Illinois	304
5-2	Analysis of Water from the Mt. Simon Formation in the Vicinity of Bloomington, Illinois	306
5-3	Table of Equivalency of Permeability Values in Various Units	315
5-4	Laboratory Core Analysis Data from the Mt. Simon Formation in Illinois	322
5-5	Well Logging Methods and Their Applications	325
6-1	Well-Construction Costs for Vadose Zone Monitoring	415
6-2	Costs for Vadose Zone Monitoring	416
6-3	Total Costs for Agricultural Return Flow Monitoring	417
6-4	Cost Summary for Monitoring Septic Tank Pollution	423
6-5	Costs for Monitoring the Vadose Zone	429
6-6	Costs for Saturated Zone Monitoring	431
6-7	Total Costs for Monitoring Industrial Waste Percolation Pond	432
6-8	Monitoring Costs for Solid Waste Landfill	437

CHAPTER I – SUMMARY

The monitoring of groundwater quality is part of a national program to prevent, reduce, and eliminate groundwater pollution. This subject and its many related facets are discussed in Chapters II through VI, most of which are divided into several sections.

MONITORING METHODOLOGY

Chapter II describes the needs, objectives, and constraints of monitoring groundwater quality, and presents a monitoring methodology. Particular emphasis is placed on the problem as viewed by the United States Environmental Protection Agency, given its legislative mandates in the Federal Water Pollution Control Act Amendments of 1972 [PL92-500], and the Safe Drinking Water Act of 1974 [PL93-523]. Section 1 provides background information and describes ambient trend monitoring, source monitoring, case preparation monitoring, and research monitoring. Section 2 provides information on groundwater quality, describes the geologic framework governing the movement of groundwater and natural underground water quality, and discusses the occurrence of groundwater pollution. The constituents in polluted groundwater and the various sources and causes of pollution are reviewed. This section ends with a discussion of groundwater quality in relation to water use. Section 3 develops a methodology for monitoring groundwater quality degradation resulting from man's activities. The methodology is presented in the form of a series of procedural steps arranged in chronological order. By so doing, a straightforward sequence of actions is outlined, which can lead to a groundwater pollution monitoring program in a given area. The methodology forms an integral part of the statewide monitoring programs for Arizona, New Mexico, and Texas. Recent applications of the methodology to develop groundwater quality monitoring guidelines for western coal strip mining and hazardous waste disposal can be found in Everett (1979), Everett and Hulburt (1979), and Everett and Schmidt (1980). Applications of the methodology to the design of monitoring programs for oil shale extraction can be found in Slawson (1979).

MONITORING METHODS AND COSTS

Chapter III describes various groundwater monitoring methods and provides a generalized cost breakdown of the major economic factors for each method. Most of the groundwater-related measuring techniques applicable at the land surface and in top soil, the vadose zone, and the zone of saturation are presented. Each monitoring field is described, referenced, and illustrated. Section 1 gives estimates of itemized capital and operational costs. In addition to reviewing sampling methods utilized in the various zones, Section 2 describes nonsampling techniques that should be considered in a monitoring program. Section 3 presents information on monitoring in the vadose zone. Soil sampling and well drilling techniques are presented, as are techniques to determine water content, water movement, water movement in the unsaturated state, and water sampling in the vadose zone. Monitoring techniques in the vadose zone have been expanded in the report by Wilson (1980). Section 4 summarizes general monitoring procedures and details specific methods of monitoring pollution in the saturated zone. Section 4 discusses the analysis of samples. Procedures are given for custody control, quality control, soil analysis, and water analysis. The material is presented for in-depth reference purposes without recommendations for least cost techniques or approaches.

DATA MANAGEMENT

Chapter IV deals with groundwater data management. The growing concern for subsurface water resources will surely be accompanied by an expanding groundwater data base, a data base that is already quite large. The efficient management of this data base will assure the availability of pertinent information when and where it is needed. This chapter describes the requirements of groundwater data management, surveys some available capabilities that may serve to satisfy these requirements, and shows how these capabilities can be used to manage groundwater data. Section 2 discusses groundwater information management requirements, such as station descriptions, quality criteria, geologic data, hydrologic data, water quality parameter identifiers, water quality measurements, temporal data, information qualification data, designated monitoring agency (DMA) status data, and information indexing. In addition, this section deals with data collection, communications, storage, processing, and retrieval. Section 3 describes existing water information systems such as STORET, WATSTORE, and NAWDEX, and Section 4 discusses proposed modifications to these water information systems. Section 4 presents some conclusions and recommendations.

MONITORING DISPOSAL WELLS

Chapter V provides detailed information on the monitoring of wastewater injection wells. The chapter was written to support Public Law 93-523, the Safe Drinking Water Act and state monitoring programs to control underground injection, which endangers drinking water sources. The chapter discusses the data that characterize the subsurface environment, how these data are obtained, and how they are used to predict and interpret aquifer response. The surveillance of operating injection wells is also treated in detail. Section 2 describes the subsurface environment by reviewing stratigraphic geology, structural geology, lithology, fluids, mechanical properties of injection and confining units, hydrodynamics, and resources. Section 3 describes the subsurface data needed for monitoring cased disposal wells and the methods and tools available for obtaining the data. Section 3 also discusses the acquisition of data from rock samples, formation fluids, borehole geophysical logs, testing of injection and confining units, drill stem testing, and injectivity tests conducted during well construction. To help predict aquifer response, Section 4 discusses flow theory, regional flow, pressure effects of injection, rate and direction of fluid movement, hydraulic fracturing, and generation of earthquakes. The section of surveillance of operating disposal wells, Section 5, discusses injection well monitoring, periodic inspection and testing, monitoring wells, and other monitoring methods. Section 6 presents some conclusions and recommendations.

ILLUSTRATIVE EXAMPLES

Chapter VI contains illustrative examples of the techniques described in the earlier chapters. This chapter is designed to show, by example, site-specific procedures for monitoring various classes of groundwater pollution sources. Section 2 discusses the first five case histories of actual or potential groundwater pollution and the monitoring techniques that were employed, as well as a retrospective view of the techniques and their efficacy. The case histories cover brine disposal in Arkansas; plating waste contamination in Long Island, New York; landfill leachate pollution in Milford, Connecticut; an oxidation pond near Tucson, Arizona; and multiple-source nitrate pollution in the Fresno-Clovis, California metropolitan area. Section 3 presents hypothetical illustrative examples for developing and selecting monitoring alternatives based on a cost comparison and hydrologic judgment. The examples illustrated cover agricultural return flow, septic tanks, percolation ponds, and landfills.

SELECTED BIBLIOGRAPHY

L.G. Everett, *Groundwater Quality Monitoring of Western Coal Strip Mining: Identification and Priority Ranking of Potential Pollution Sources*, EPA-600/7-79-024, January, 1979

L.G. Everett and M.A. Hulburt, *Groundwater Quality Monitoring Designs for Municipal Pollution Sources: Preliminary Designs for Coal Strip mine Impact Assessments*, EPA, in press.

L.G. Everett and K.D. Schmidt, *Establishment of Water Quality Monitoring Programs*, AWRA, March, 1980.

L.G. Wilson, *Monitoring in the Vadose Zone – A review of Technical Elements and Methods*, General Electric Company-TEMPO, GE79TMP-55, April, 1980.

G.C. Slawson, *Groundwater Quality of Monitoring of Western Oil Shale Development: Identification and Priority Ranking of Potential Pollution Sources*, EPA-600/7-79-023, January, 1979.

CHAPTER II

GROUNDWATER MONITORING METHODOLOGY

SECTION 1 – INTRODUCTION

BACKGROUND AND SCOPE

Groundwater serves as a major source of water supply in the United States. Public water supplies for one-third of the nation's 100 largest cities are derived from groundwater. It is estimated that in rural areas 95 percent of the domestic water and over half of the water for livestock and irrigation are obtained from underground resources. Furthermore, most of the day-to-day base flow of the nation's rivers and streams originates from groundwater discharge.

The natural quality of groundwater tends to be degraded by activities of man. Wastes which are not discharged into lakes, streams, or the ocean are deposited on land, and from there may migrate downward to pollute groundwater. The extent of this pollution has grown concomitantly in recent decades with increases in population, agriculture, and industry; however, information as to the magnitude of the problem is meager.

Currently, the U.S. Environmental Protection Agency (EPA) has an investigational program of groundwater pollution underway on a regional basis. From the reports already issued (Fuhriman and Barton, 1971; Scalf et al., 1973; van der Leeden et al., 1973, 1975; Miller et al., 1974), it is apparent that there are literally millions of point sources of pollution in existence.

PURPOSE AND APPROACH

The monitoring methodology described in this book is intended to serve as a set of guidelines for developing and implementing a groundwater quality monitoring program. It should be apparent that factors such as climate, hydrology, population, pollution sources, and water use vary from place to place; therefore, the design of an appropriate monitoring program will also vary accordingly. No one set of guidelines can cover all situations; however, with judgment, the approach presented herein can be extended and interpreted to meet most other situations which will be confronted in the field.

The physical, chemical, and biological mechanisms governing groundwater pollution are reasonably well understood. Yet, applying this knowledge to the many different situations which can result from superimposing a given groundwater pollution source upon particular hydrogeologic environments is difficult. Nonetheless, this document is intended to be a handbook or manual for the development and implementation of a methodology for monitoring groundwater quality. The methodology is expressed in a generalized form so that it can be usefully employed by regional, State, and local water pollution control agencies, and is applicable to all types of groundwater aquifers, areas, and basins. Alternatives in the decision-making process leading to the final monitoring program are considered throughout the methodology.

Monitoring may be defined as a scientifically designed surveillance system of continuing measurements and observations, including evaluation procedures. The EPA is currently involved in establishing, in cooperation with the States, a national groundwater quality monitoring system as part of its legislatively directed program to prevent, reduce, and eliminate groundwater pollution.

Four basic types of monitoring have been defined by the EPA. In terms of groundwater quality, these may be interpreted as follows:

AMBIENT TREND MONITORING
This concerns measurements of groundwater quality and deviations in relation to standards, and involves temporal and spatial trends within a groundwater basin or area.

SOURCE MONITORING
This involves the measurement of effluent quantity and quality for pollution sources which may affect groundwater.

CASE PREPARATION MONITORING
This serves to gather evidence for enforcement actions of past, existing, or anticipated groundwater pollution situations. Implied are carefully documented measurements within a circumscribed area.

RESEARCH MONITORING
This contributes to research investigations on groundwater quality and pollution occurrence and movement.

Of the four above types, the monitoring methodology is directed largely toward source monitoring. Case preparation and research monitoring, while providing valuable data, are clearly specialized needs which do not lend themselves to a national program. Ambient trend monitoring provides background quality information on groundwater resources, such as the ongoing programs of the U.S. Geological Survey. Thus, a national program to protect groundwater quality, relative to those activities of man which pollute groundwater, will focus primarily on measurements relating to pollution sources and methods of waste disposal which contribute to pollution. Furthermore, because it is infeasible to monitor all sources and causes of pollution, the methodology concentrates on identifying the most important sources and methods of disposal. In essence the methodology becomes a resource allocation problem, with the goal of developing a cost-effective monitoring program, which will contribute most to the protection of the nation's groundwaters.

NEEDS AND OBJECTIVES

Growing evidence indicates that the nation's groundwaters (like its other resources such as air and surface water) are becoming excessively polluted. Groundwaters are not being efficiently allocated among their alternative uses. Wastes are an unavoidable byproduct of all man's activities, and must be disposed of somewhere. The substantive questions of optimum production of wastes and their optimum disposal must be addressed. Clearly, the nation's aquifers, like every other sector of the environment, must serve as a repository for the disposal of some of society's wastes. The question, then, is not whether wastes can be placed on and in the ground, but rather where and how much.

ECONOMIC NEEDS AND OBJECTIVES
In order to maximize the value of groundwater resources from society's point of view, it is necessary that they be used in optimal amounts for various appropriate purposes such as irrigation, cooling, drinking, etc, as well as for the disposal of wastes.

The reason that groundwaters become excessively polluted is exactly the same as for the reason that our other environmental resources become excessively polluted; namely, the fact that well-specified (and, therefore, easily enforceable) property rights do not exist in the resource. As Corker's (1971) authoritative legal analysis makes abundantly clear, property rights in groundwater are very poorly specified. As a result, some parties use other parties' groundwater resources, without authorization, as a receptor for their wastes, thereby imposing uncompensated damages on the resource owner. In the absence of any obligation to compensate parties whose groundwater is serving as a waste disposal site, such waste disposal service appears to be free to the polluter (although it is obviously not free from society's point of view). In such a situation, the polluter will utilize such a service until the marginal *private* benefit he obtains by using that service is driven down to zero, even though the marginal social cost may be positive and large. Consequently, the aquifer's waste receptor capabilities are abused, and groundwater becomes excessively polluted. (Crouch et al., 1976).

When well-specified property rights to a resource do not exist, the *market* processes through which our resources are normally allocated break down as an efficient resource allocation mechanism. Therefore, if resources to which well-specified property rights do not exist are to be efficiently allocated among their alternative uses, it is not possible to rely on the normal allocation mechanism provided by market processes. The resource must, instead, be *managed* by some public agency.

As pointed out by the U.S. Council on Environmental Quality (1973, Chapter 3), efficient management of our environmental resources requires that consideration be given to four categories of costs. First, there are *damage costs*. These are costs which are generated directly by a polluting activity. With respect to groundwater resources, one example would be increased physiological damage caused by pollution of drinking water. Another example would be crop losses resulting from salt buildup in an irrigation well. Second, there are *avoidance costs*. These are costs which are incurred by society in order to avoid, or reduce, damage costs. With respect to groundwater resources, one example would be the importation of unpolluted water to replace that previously obtained from a well that has become polluted. Third, there are *abatement costs*. These are costs associated with the reduction of pollution. Such reduction of pollution can be achieved either by controlling the source or by treating the polluted water. With respect to groundwater resources, one example would be the deep injection of noxious effluents, previously disposed of at the land surface, into the atmosphere, surface waters, or landfill, into a safe geologic zone. Fourth, there are *transaction costs*. Transaction costs include the cost of resources allocated to the establishment, and of enforcement of environment-preserving policies and regulations. With respect to groundwater resources, the most important example of a transaction cost, and of special relevance to this study, would be the cost of monitoring groundwater pollution either to generate information on quantity and quality, or to detect violations of, and ensure compliance with, groundwater quality standards.

The essential principle to note is that these four cost categories are, in general, interdependent. Thus, any new groundwater quality policy will probably affect all four categories. For example, if a policy is introduced which is designed to reduce groundwater pollution, it will certainly reduce damage costs but may very well increase abatement, avoidance, and monitoring costs. Usually, each feasible groundwater quality policy will affect these four categories of cost differently. Obviously, the only groundwater quality policy alternative that should be seriously considered for implementation is that set of policies for which the reduction in damage costs exceeds the net increase, if any, in avoidance, abatement, and monitoring costs. To go even further, the most efficient, or optimal, groundwater quality policy among the feasible set of alternatives is that policy which minimizes the sum of the damage costs, avoidance costs, abatement costs, and monitoring costs for a given groundwater pollution situation. Implicitly, the minimization of these costs is equivalent to the maximization of society's income or gross national product.

To illustrate by example, sewer leakage is known to pollute groundwater and may, therefore, impose certain damage costs and avoidance costs. However, with present technology there is no way of controlling this source of pollution that does not impose abatement costs that are far in excess of the reduction in damage and avoidance costs which would be achieved. It follows that the efficient policy is not to monitor and abate this pollution source, but simply to accept the existing level of damage and avoidance costs.

Of course, in many other groundwater pollution situations the reduction in damage costs will exceed the increase in avoidance, abatement, and monitoring costs that bring these reductions about. The objective then becomes that of selecting the avoidance, abatement, and monitoring strategy which generates, at the margin, decreases in damage costs just equal to the increases in the avoidance, abatement, and monitoring costs required by the strategy.

The attainment of this objective will not simply involve the minimization of monitoring costs by the responsible government agency. For example, in any given groundwater pollution situation there may well be several different abatement strategies, each with an associated monitoring requirement, that would achieve the desired level of groundwater quality. If the public agency responsible for selecting among the abatement and monitoring strategies simply chooses that alternative which minimizes its own monitoring costs, this could imply higher private abatement costs with the result that the combined monitoring and abatement costs of that policy would be greater than the combined monitoring and abatement costs of some other strategy. Clearly, this would not be efficient from society's point of view.

EPA NEEDS AND OBJECTIVES

The needs and objectives of the EPA for those data pertaining to groundwater quality, which must be obtained through monitoring programs, are dictated by the mandates and requirements of the Federal Water Pollution Control Act, as amended (Public Law 92-500; 33 USC 1151, et seq.), and the Safe Drinking Water Act (Public Law 93-523; 42 USC 300f, et seq.).

FEDERAL WATER POLLUTION ACT, AS AMENDED. The objective of the Act as amended as stated in Sec. 101 (a) is:

> Sec. 101. (a) The objective of this Act is to restore and maintain the chemical, physical, and biological integrity of the Nation's waters. .

The definition of "pollution" given in the Act indicates clearly the wide spectrum of groundwater quality problems and of the sources and causes of those problems that may need to be considered in designing a groundwater quality monitoring program:

> Sec. 502. (19) The term 'pollution' means the man-made or man-induced alteration of the chemical, physical, biological, and radiological integrity of water.

The Act directs that specific programs be developed to improve and maintain groundwater quality:

> Sec. 102. (a) The *Administrator* (of EPA) *shall*, after careful investigation, and in cooperation with other Federal agencies, State water pollution control agencies, interstate agencies, and the municipalities and industries involved, *prepare or develop comprehensive programs for preventing, reducing, or eliminating the pollution* of the navigable waters and *ground waters and improving the sanitary condition* of surface and *underground waters.* In the development of such comprehensive programs due regard shall be given to the improvements which are necessary to conserve such waters for the protection and propagation of fish and aquatic life and wildlife, recreational purposes, and the withdrawl of such waters for public water supply, agricultural, industrial, and other purposes. For the purpose of this section, the Administrator is authorized to make joint investigations with any such agencies of the condition of any waters in any State or States, and of the discharges of any sewage, industrial wastes, or substances which may adversely affect such waters. [Author's italics.]

Development of cost-effective programs to accomplish this purpose requires an adequate data base that can only be obtained by carefully designed and implemented monitoring programs.

Establishment and conduct of monitoring programs is mandated by the Act:

> Sec. 104 (a) The Administrator (of EPA) shall establish national programs for the prevention, reduction, and elimination of pollution and as part of such programs shall: . . .
>
> (5) in cooperation with the States, and their political subdivisions, and other Federal agencies *establish, equip, and maintain a water quality surveillance system for the purpose of monitoring the quality* of the navigable waters and ground waters and contiguous zone and the oceans. . .and shall report on such quality. . . [Author's italics and insertion.]
>
> Sec. 106 (e) Beginning in fiscal year 1974 the Administrator shall not make any grant under this section (Sec. 106) to any State which has not provided or is not carrying out as a part of its program. . .
>
> (1) the establishment and operation of appropriate devices, methods, systems, and procedures necessary to *monitor*, and to *compile and analyze data on* (including classification according to eutrophic condition), *the quality* of navigable waters and to the extent practicable, ground waters including biological monitoring; and provisions for annually updating such data and including it in the report required under section 305 of this Act; . . . [Author's italics and insertion.]

SAFE DRINKING WATER ACT. This Act, which became effective 16 December 1974, places additional responsibilities of major public health and economic significance upon the Administrator of the EPA to protect the nation's groundwater resources. The following excerpts from the Act indicate the nature and extent of those responsibilities:

> Sec. 1424. (e) If the Administrator (of EPA) determines, on his own initiative or upon petition, that an area has an *aquifer* which is the *sole or principle drinking water source* for the area and *which, if contaminated, would create a significant hazard to public health,* he shall publish notice of that determination in the Federal Register. *After the publication of any such notice, no commitment for Federal financial assistance (through a grant, contract, loan guarantee, or* otherwise) *may be entered into for any project which the Administrator determines may contaminate such aquifer through a recharge zone* so as to create a significant hazard to public health, but a commitment for Federal financial assistance may, if authorized under another provision of law, be entered into to plan or design the project to assure that it will not, so contaminate the aquifer. [Author's italics and insertion.]
>
> Sec. 1442. (a)(1) The Administrator may conduct research, studies, and demonstrations relating...to the provision of a dependably safe supply of drinking water, including...
>
> > (E) *improved methods of protecting underground water sources of* public water systems *from contamination...*
>
> Sec. 1442. (a)(4) The Administrator shall conduct a survey and study of...
>
> (A) *disposal of* waste (including residential waste) *which may endanger* underground water which supplies, or can reasonably be expected to supply, any public water systems, and
>
> (B) means of control of such waste disposal...
>
> Sec. 1442. (a)(5) The Administrator shall carry out a study of methods of underground injection which do not result in the degradation of underground drinking water sources.
>
> (6) The Administrator shall carry out a study of methods of preventing, detecting, and dealing with *surface spills of contaminants* which may *degrade underground water sources* for public water systems...
>
> (8) The Administrator shall carry out a study of the *nature and extent of the impact on underground water* which supplies or can reasonably be expected to supply public water systems of (A) *abandoned injection or extraction wells;* (B) *intensive application of pesticides and fertilizers* in underground water recharge areas; and (C) *ponds, pools, lagoons, pits, or other surface disposal of contaminants in underground water recharge areas.* [Author's italics.]

To carry out these responsibilities effectively, detailed information for many aquifers on groundwater quality, geology, hydrology, and the sources and causes of groundwater quality impairment, will be necessary. Groundwater quality monitoring programs must be designed to provide such data where and when required.

Interstate Compacts. A few interstate compacts – the Federal-interstate compacts for the Delaware River Basin and the Susquehanna River Basin are examples – provide authority for the compact commissions to monitor groundwater quality and to develop and implement plans for quality protection.

State Statutes and Regulations. State laws and implementing regulations vary widely from State to State as regards both groundwater quality monitoring and the development and implementation of plans for quality protection. In California, the State Water Resources Control Board and the nine regional water quality control boards have broad authority and responsibilities in this regard under Division 7 of the Water Code (Porter-Cologne Water Quality Control Act, as amended). Likewise, the Water Code provides with respect to the State Department of Water Resources:

> 229. (Water Code) The Department of Water Resources, either independently or in cooperation with any person or any county, State, Federal, or other agency, to the extent that funds are allocated therefore, *shall investigate conditions of the quality of all waters* (including ground waters) within the State, including saline waters, coastal and inland, as related to *all sources of pollution of whatever nature* and shall report thereon to the Legislature, to the board (State Water Resources Control Board), and to the appropriate regional water quality control board annually, and *may recommend any steps which might be taken to improve or protect the quality of such waters.* The department shall coordinate its investigations fully with the board. [Author's italics and insertions.]

The Water Code of the State of Texas provides similar authorities and responsibilities to the Texas Water Quality Board, and, to some extent, to the Texas Water Development Board.

Some States have not yet progressed as far in monitoring and planning for the improvement and protection of groundwater quality.

MONITORING OBJECTIVES

Simply stated, the objective of a monitoring program should be to collect, manage, and analyze the data on groundwater quality and the sources and causes of groundwater pollution, and the other information – geologic, hydrologic, and economic – necessary to enable the EPA and the State(s) involved to fulfill their statutory responsibilities as regards protection of groundwater quality, that is "...to restore and maintain the chemical, physical, and biological integrity of the nation's waters..."

To fulfill that objective, a specific monitoring program must be developed and implemented for each groundwater basin; such a program will be the end result of the application of the monitoring methodology described in this report. The data to be obtained may be needed for one or more of the following purposes, depending upon the types, extent, and seriousness of quality problems, and the present and future importance of the groundwater system as a source of water supply:

- Provision of background information and quality
- Detection of quality trends
- Identification and assessment of the sources and causes of pollution
- Planning – including formulation, calibration, verification, and use of planning models
- Establishment of water quality standards and effluent limitations
- Formulation of other regulatory control and management actions necessary to protect quality
- Compliance
- Enforcement
- Reporting

For a groundwater basin which is already extensively developed, or in process of development, all these purposes must generally be served, although the relative emphasis will vary.

A wide range of types of data may be needed, depending upon the particular groundwater system involved, its importance, uses, its present and potential quality problems, and the information already available. In addition to water quality data this information should include:

- Geologic characteristics
- Hydrology
- Hydraulic characteristics
- Land use
- Water resource development and use
- Demographic and economic conditions
- Waste generation and disposal

An effective monitoring program must recognize the dynamic nature of groundwater systems as affected by both natural phenomena and man-induced changes. The program must, therefore, be continuing. Its scope and relative emphasis will change over time. The data obtained must be adequate to enable prediction of potential quality problems and for formulation of plans for prevention of groundwater pollution.

CONSTRAINTS

There are several constraints to be considered in the design of a cost-effective monitoring program – institutional, budgetary, legal, social, and personnel.

At the Federal level, there are several agencies in addition to the EPA and U.S. Geological Survey (USGS) which collect groundwater quality data, and which possess information of value to a comprehensive groundwater quality monitoring program. Their monitoring activities are accomplished in support of their statutorily authorized programs, which involve groundwater and its uses in some manner. Other programs of particular significance are those of U.S. Bureau of Reclamation, U.S. Army Corps of Engineers, and U.S. Soil Conservation Service.

There are several State agencies which have statutorily defined authorities and responsibilities with respect to groundwater, and which conduct monitoring activities. Such State entities may include:

- Water resource planning and management agencies
- Water quality control agencies
- Water rights administration
- Department of public health
- Universities
- Soil conservation agencies

The authorities, responsibilities, and activities of the several State agencies may overlap to some extent.

At the regional and local levels of government, counties, municipalities, and special political jurisdictions, such as water supply; sanitary; flood control; and water conservation districts, have interests and responsibilities with respect to groundwater, and many conduct extensive monitoring programs.

The authorities, responsibilities, and programs of the several agencies at each governmental level and the data already available from their monitoring activities must be taken into account in designing a new monitoring system.

The number of geologists, hydrologists, and water quality specialists competent to design and implement a groundwater quality monitoring program and to make effective use of the data generated is limited.

Funds available for monitoring activities are scarce, particularly at the State, regional, and local levels of government. Any increase in budgets will take place gradually, and then only on well-documented justification. As a consequence, the implementation of expanded monitoring programs must, in general, take place gradually.

SECTION 2 – GROUNDWATER QUALITY

HYDROGEOLOGIC FRAMEWORK

GEOLOGIC FORMATIONS AS AQUIFERS

Permeable rock formations which store and transmit significant quantities of water are known as aquifers. A variety of geologic formations can act as aquifers. The most widely developed aquifers in the United States consist of unconsolidated alluvial deposits, chiefly gravels and sands. These geologic features occur as water courses, buried valleys, plains, or intermontane valleys.

Limestone, a consolidated sedimentary rock, serves as an important aquifer when sizable proportions of the original rock have been dissolved and removed. Openings may range from microscopic pores to large solution caverns forming subterranean channels large enough to carry entire streams. Within large openings in limestone aquifers, groundwater can flow rapidly under turbulent conditions. Fractures and faults may allow limestones to serve as significant aquifers. Springs are frequently found in limestone areas.

Volcanic rocks such as basalt flows can form highly permeable aquifers. Flow breccias, porous zones between lava beds, lava tubes, shrinkage cracks, faults, and joints are other permeable zones in volcanic rocks. Most of the largest springs in the United States originate in basalt.

Sandstones and conglomerates represent cemented forms of sand and gravels. Sandstones function best as aquifers where they are only partially cemented or where joints, fractures, or faults are present. Crystalline and metamorphic rocks are usually relatively impermeable; however, groundwater may be developed by shallow wells under fractured or weathered conditions.

The following underground situations include most field conditions. The differences among them are important to waste disposal practices and to management of pollution problems:

- Unconsolidated granular materials extending downward from the ground surface to great depths: Beneath the ground surface in many places are loose granular materials, chiefly clays, silts, and sands, which may represent soil near the ground surface and sedimentary material below. Water and dissolved waste materials move en masse through pores surrounding the mineral particles.

- Unconsolidated granular materials at the ground surface underlain at shallow depths by dense rocks with linear openings: In many places the soils, loose sedimentary materials, or residual weathered materials are only a few feet or a few tens-of-feet thick and are underlain by dense rocks, the only openings in which are joints or solution channels. Water and waste materials move through both types of media.

- Dense rocks at the ground surface: The only movement is through interconnecting joints or solution channels.

The permeability of underground strata may vary by several orders of magnitude, and large variations can often be found over short distances. Some generalizations regarding permeability follow:

- In the vertical direction, the contrast in permeability between loose granular soil and underlying jointed rock can be very large.

- Zones of uniform permeability tend to be associated with particular rock strata.

- Where interbedded sedimentary rocks are flat or gently inclined, the changes of permeability in the vertical direction are frequent and large.

- Changes of permeability in the horizontal direction, although common, are in many cases more gradual than in the vertical direction.

An unconfined aquifer is one where a water table forms the upper surface of the zone of saturation (see the upper aquifer in Figure 2-1). A water table is a surface representing the level of atmospheric pressure within an aquifer. Fluctuations in the water table correspond to changes in the volume of water in storage within an aquifer.

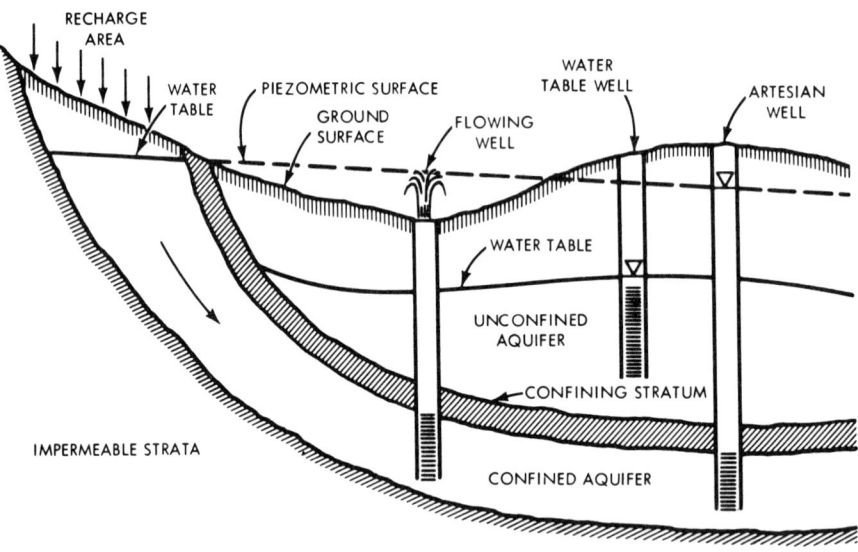

Figure 2-1. Unconfined and confined aquifers (Todd, 1959).

A confined (or artesian) aquifer is one where groundwater is confined, under pressure greater than atmospheric, by overlying less permeable strata. If a well penetrates a confined aquifer, the water level will rise above the bottom of the confining bed, as shown in Figure 2-1. Variations of water levels in wells penetrating confined aquifers result from pressure changes within the aquifer. The piezometric surface of a confined aquifer (see Figure 2-1) is an imaginary surface defining the hydrostatic pressure of the water in the aquifer.

GROUNDWATER MOVEMENT

Most groundwater moves in accordance with Darcy's Law, which states that the velocity is directly proportional to the permeability of the aquifer and the hydraulic gradient. Typical groundwater flow velocities fall in the range of 5 feet per year to 5 feet per day; however, exceptionally higher velocities have been measured in highly permeable outwash glacial deposits, in fractured basalts and granitic rocks, and in cavernous limestones.

Essentially all groundwater is in motion from its source of recharge to its point of discharge. Natural recharge occurs from precipitation and surface water bodies;

artificial recharge results from acts of man, such as by irrigation. Groundwater discharges naturally to springs, surface water bodies, and the ocean. Wells and drains function as manmade discharge points for groundwater.

Pollutants tend to move in the direction of flow of the surrounding groundwater. Within confined aquifers water movement is roughly horizontal because of the negligible vertical movement within the confining strata. There may be in certain situations, however, large vertical head gradients which induce vertical flows from one strata to another. In unconfined aquifers the groundwater also flows generally horizontally beneath a gently sloping water table. In the unsaturated zone above a water table, water usually percolates vertically downward to the water table. Less permeable layers may cause saturated conditions (perched groundwater), and allow horizontal movement.

Water in the saturated zone (below the water table) tends to move faster at relatively shallow depths than that at great depths. Where groundwater occurs to great depths, it is reasonable to consider an upper zone of rapid circulation, a middle zone of delayed circulation, and the lowest zone of very slow water circulation.

NATURAL CHEMICAL QUALITY

All groundwater contains natural chemical constituents in solution. The kinds and amounts of constituents depend upon the geochemical environment, movement, and source of the groundwater. Typically, concentrations of dissolved constituents in groundwater exceed those in surface water. Also, salinities tend to be higher in arid regions and where drainage is poor.

Chemical constituents originate primarily from solution of rock materials. The geologic history of groundwater governs its salinity. Common chemical constituents of groundwater include:

Cations	Anions	Undissociated
Calcium	Carbonate	Silica
Magnesium	Bicarbonate	
Sodium	Sulfate	
Potassium	Chloride	
	Nitrate	

An abundance classification of dissolved solids in fresh water is shown in Table 2-1. Frequently distributions of various constituents are shown in Figure 2-2.

Table 2-2 provides a convenient summary of the major natural chemical constituents in groundwater in terms of their sources and their effects upon the useability of water.

The chemical quality of groundwater is often conveniently described for domestic and industrial use in terms of its salinity and its hardness. Salinity refers to the concentration of total dissolved solids present in the water. Hardness is a measure of the calcium and magnesium content, and is usually expressed as the equivalent of calcium carbonate. Table 2-3 lists classifications of water by salinity and hardness.

Besides solution of rock materials, other natural sources of salinity in groundwater include:

- Water of volcanic origins
- Evapotranspiration through native vegetation
- Airborne salts

Table 2-1

RELATIVE ABUNDANCE OF DISSOLVED SOLIDS
IN POTABLE WATER (Davis and DeWiest, 1966)

Major Constituents (1.0 to 1000 ppm)	Secondary Constituents (0.01 to 10.0 ppm)	Minor Constituents (0.0001 to 0.1 ppm)	Trace Constituents (generally less than 0.001 ppm)
Sodium	Iron	*Antimony	Beryllium
Calcium	Strontium	Aluminum	Bismuth
Magnesium	Potassium	Arsenic	*Cerium
Bicarbonate	Carbonate	Barium	Cesium
Sulfate	Nitrate	Bromide	Gallium
Chloride	Fluoride	*Cadmium	Gold
Silica	Boron	*Chromium	Indium
		Cobalt	Lanthanum
		Copper	*Niobium
		*Germanium	Platinum
		Iodide	Radium
		Lead	*Ruthenium
		Lithium	*Scandium
		Manganese	Silver
		Molybdenum	*Thallium
		Nickel	*Thorium
		Phosphate	Tin
		*Rubidium	*Tungsten
		Selenium	Ytterbium
		*Titanium	*Yttrium
		Uranium	*Zirconium
		Vanadium	
		Zinc	

*These elements occupy an uncertain position in the list.

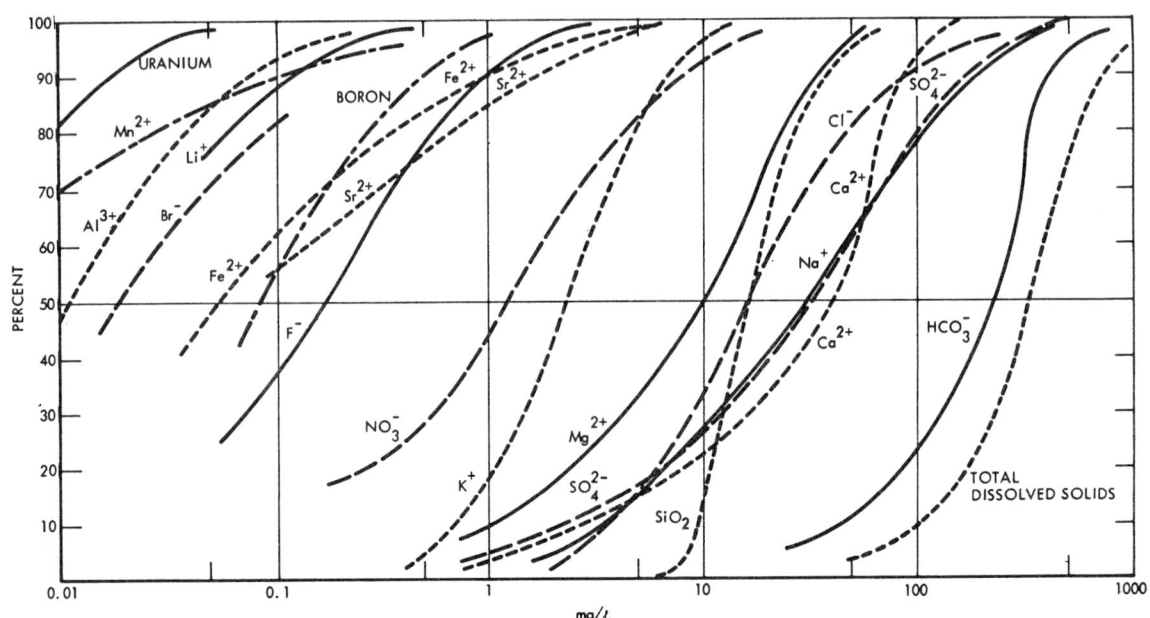

Figure 2-2. Cumulative curves showing the frequency distribution of various constituents in potable water (modified after Davis and DeWiest, 1966).

Table 2-2

...NSTITUENTS IN GROUNDWATER – THEIR
...ON USABILITY (modified after Miller et al., 1974)

...ium and clay minerals, amorphous silica, chert, opal.

...ium and magnesium, silica forms a scale in boilers and on
...ards heat; the scale is difficult to remove. Silica may
... to inhibit corrosion of iron pipes.

...s — amphiboles, ferromagnesian micas, ferrous sulfide (FeS),
...yrite (FeS_2), magnetite (Fe_3O_4) — and
...ides, carbonates, and sulfides or iron clay minerals.

...recipitates after exposure to air; causes turbidity, stains
...d laundry and cooking utensils, and imparts objectionable tastes
...nd drinks. More than 0.2 ppm is objectionable for most

...l water probably comes most often from soils and sediments.
...dimentary rocks and mica biotite and amphibole hornblende
...ge amounts of managanese.

...precipitates upon oxidation; causes undesirable tastes, deposits
...king, stains plumbing fixtures and laundry, and fosters growths
...s, and distribution systems. Most industrial users object to
...ore than 0.2 ppm.

...Mg)
...boles, feldspars, gypsum, pyroxenes, aragonite, calcite, dolomite,

...phiboles, olivine, pyroxenes, dolomite, magnesite, clay minerals.

...nesium combine with bicarbonate, carbonate, sulfate, and silica to
...ng, pipe-clogging scale in boilers and in other heat-exchange
equipment. Calcium and magnesium combine with ions of fatty acid in soaps to form soapsuds. A high concentration of magnesium has a laxative effect, especially on new users of the supply.

Sodium (Na) and Potassium (K)
  Sources: Sodium — Feldspars (albite); clay minerals; evaporites, such as halite (NaCl) and mirabilite ($Na_2SO_4 \cdot 10H_2O$).
  Potassium — Feldspars (orthoclase and microcline), feldspathoids, some micas, clay minerals.
  Effect: More than 50 ppm sodium and potassium in the presence of suspended matter causes foaming, which accelerates scale formation and corrosion in boilers. Sodium and potassium carbonate in recirculating cooling water can cause deterioration of wood in cooling towers. More than 65 ppm of sodium can cause problems in ice manufacture.

Carbonate (CO_3) and Bicarbonate (HCO_3)
  Sources: Bicarbonate — Biosphere; limestone; dolomite.
  Effect: Upon heating, bicarbonate is changed into steam, carbon dioxide, and carbonate. The carbonate combines with alkaline earths — principally calcium and magnesium — to form a crustlike scale of calcium carbonate that retards flows of heat through pipe wells and restricts flow of fluids in pipes. Water containing large amounts of bicarbonate and alkalinity are undesirable in many industries.

(continued)

Table 2-2 (Cont'd.)

MAJOR NATURAL CONSTITUENTS IN GROUNDWATER – THEIR SOURCES AND EFFECTS UPON USABILITY (modified after Miller et al., 1974)

Sulfate (SO_4)
  Sources: Oxidation of sulfide ores; gypsum; anhydrite.
  Effect: Sulfate combines with calcium to form an adherent, heat-retarding scale. More than 250 ppm is objectionable in water in some industries. Water containing about 500 ppm of sulfate tastes bitter; water containing about 1,000 ppm may be cathartic.

Chloride (Cl)
  Sources: Sedimentary rocks and evaporites; ocean tides force salty water upstream in tidal estuaries.
  Effect: Chloride in excess of 100 ppm imparts a salty taste, while concentrations of the order of 1000 or more may cause physiological damage. Food processing industries usually require less than 250 ppm. Some industries — textile processing, paper manufacturing, and synthetic rubber manufacturing — desire less than 100 ppm.

Fluoride (F)
  Sources: Amphiboles (hornblende), apatite, fluorite, mica.
  Effect: Fluoride concentration between 0.6 and 1.7 ppm in drinking water has a beneficial effect on the structure and resistance to decay of children's teeth. Fluoride in excess of 1.5 ppm in some areas causes "mottled enamel" in children's teeth. Fluoride in excess of 6.0 ppm causes pronounced mottling and disfiguration of teeth.

Nitrate (NO_3)
  Sources: Atmosphere; legumes, plant debris, and animal excrement.
  Effect: Water containing large amounts of nitrate (more than 100 ppm) is bitter tasting and may cause physiological distress. Water from shallow wells containing more than 45 ppm has been reported to cuase methemoglobinemia in infants. Small amounts of nitrate help reduce cracking of high-pressure boiler steel.

Dissolved Solids
  Sources: The mineral constituents dissolved in water constitute the dissolved solids.
  Effect: More than 500 ppm is undesirable for drinking and many industrial uses. Less than 300 ppm is desirable for dyeing of textiles and the manufacture of plastics, pulp, paper, and rayon. Dissolved solids cause foaming in steam boilers; the maximum permissible content decreases with increases in operating pressure.

Table 2-3

CLASSIFICATIONS OF WATER
(Davis and DeWiest, 1966)

Based on Concentration of Total Dissolved Solids (TDS)	
Name	Concentration of TDS (parts per million)
Fresh	0-1,000
Brackish	1,000-10,000
Salty	10,000-100,000
Brine	More than 100,000
Based on Hardness	
Name	Hardness as $CaCO_3$ (parts per million)
Soft	0-60
Moderately Hard	61-120
Hard	121-180
Very Hard	More than 180

Within a large body of groundwater the natural chemical composition of the water tends to be relatively consistent, although the concentration of minerals in solution may be variable. However, where geologic differences or multiple sources exist, significant salinity differences occur. Time variations of groundwater quality under natural conditions are minor in comparison with surface water quality changes.

Groundwater under natural conditions tends to increase in salinity with depth. Most of the geologic formations containing groundwater in the United States are underlain by waters varying from brackish to highly saline. Density and permeability differences act to maintain a separation between these waters and the overlying fresh groundwater.

OCCURRENCE OF GROUNDWATER POLLUTION

DEFINITION

Groundwater pollution is the degradation in the natural quality of groundwater produced by acts of man. Pollution may inhibit the use of the water or may create hazards to public health through toxicity or the spread of disease.

This definition follows that appearing in the Federal Water Pollution Control Act of 1972, and quoted in the "Introduction" to this chapter.

DISTRIBUTION OF POLLUTANTS

If it were possible to see zones of groundwater pollution from an aerial vantage point, most would appear so small in relation to the total areas as to be termed scattered points of pollution. Areally extensive sources such as irrigation return flows and seawater intrusion would be identified as nonpoint sources. A line source would result, for example, from recharge of sewage effluent in an ephemeral stream channel.

Figure 2-3 suggests by small dots on a map the distribution of groundwater pollution in a drainage basin. As indicated by the profile view at the top of Figure 2-3, most point sources of pollution will not penetrate through the zone of aeration; this is particularly true where the volume of pollutant is relatively small and where the water table is some distance below ground surface. The profile view of Site 5 in Figure 2-3 shows a pollutant that has reached the water table and has begun to migrate down-gradient with the groundwater flow.

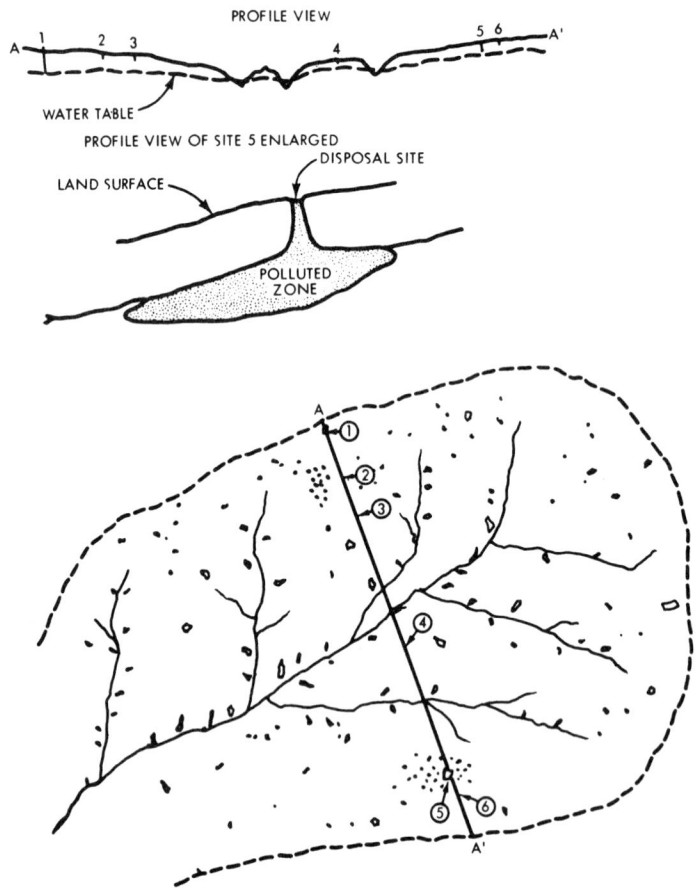

Figure 2-3. A hypothetical drainage basin in a humid region showing in plan view the distribution of zones of polluted water in the upper part of the zone of saturation. Line AA' shows profile view across basin. At disposal sites 1 and 5 pollutants extend through zone of aeration and cause a polluted zone beneath the water table. Disposal sites 2, 3, 4, and 6 do not pollute the groundwater (modified after LeGrand, 1965b).

MECHANISMS OF POLLUTION

Shallow aquifers are usually the most important sources of groundwater for water supply purposes, but the upper portions of these aquifers are also the most susceptible to pollution. The entry of pollutants to shallow aquifers occurs (1) directly through wells, (2) by downward percolation through the zone of aeration, (3) by induced recharge from surface water bodies, (4) by interaquifer flow, and (5) by upconing of deeper saline water due to overpumping of wells.

It should be recognized that the configuration of pollution entry into and movement within the underground is unique for each individual pollution source. Furthermore, because there are many millions of groundwater pollution sources in the United States, it becomes apparent that the possibilities in terms of pollutant movement and distribution are virtually limitless. Notwithstanding this fact, typical flow patterns of groundwater pollutants for a variety of common situations can be described.

The diagrams on the following pages depict some of the frequently occurring pollution geometries. These emphasize vertical cross sections at pollution sources; horizontal pollution movement thereafter is discussed later. Whatever the particular source of pollution may be, these diagrams indicate the hydraulic relationships for a given situation. Where the local hydrogeology is known, paths of probable pollutant movement can be defined. With estimates of permeability, hydraulic gradient, and porosity available, rates of groundwater movement can be ascertained. Rates of pollutant movement are based on groundwater flow rates, chemical interactions with aquifer materials, and changes in water chemistry. Thus, pollutants travel at velocities equal to or less than that of the groundwater.

Figure 2-4 shows the local movement of effluent from a domestic septic tank system. The waste water flows vertically downward to the water table and then is distributed laterally within the groundwater body.

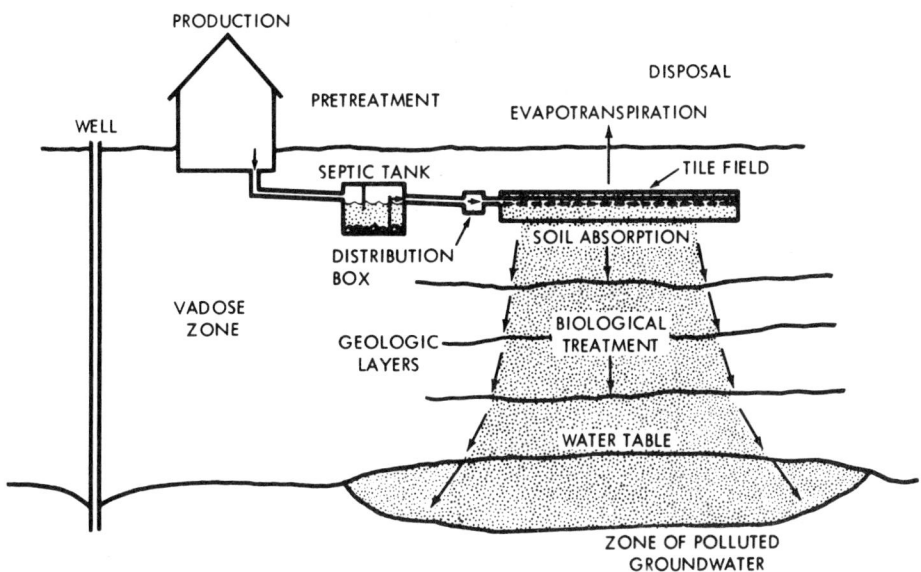

Figure 2-4. Disposal of household wastes through a conventional septic tank system (modified after Miller et al., 1974).

Figure 2-5 illustrates the flow of pollution from a surface source such as a disposal pit, lagoon, or basin. Note that the polluted water flows downward to form a recharge mound at the water table and then moves laterally outward below the water table. If significant layering exists above the water table, horizontal movement may occur in the vadose zone as suggested by Figure 2-6.

Figure 2-7 shows cross-sectional and plan views of groundwater pollution caused by a leaking sewer. The pollution drains downward to the water table and then flows laterally thereafter to form a line source of pollution beneath the sewer.

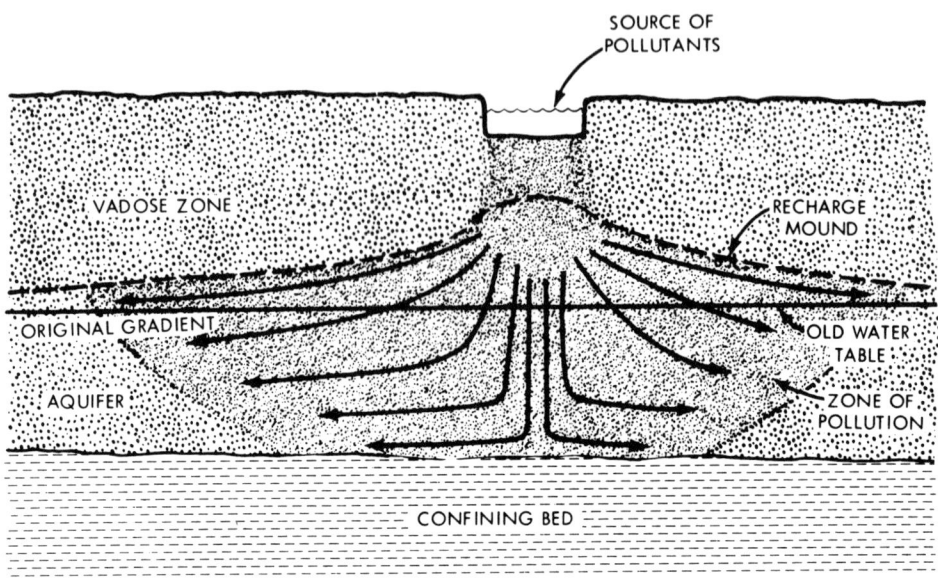

Figure 2-5. Diagram showing percolation of pollutants from a disposal pit to a water table aquifer (modified after Deutsch, 1963).

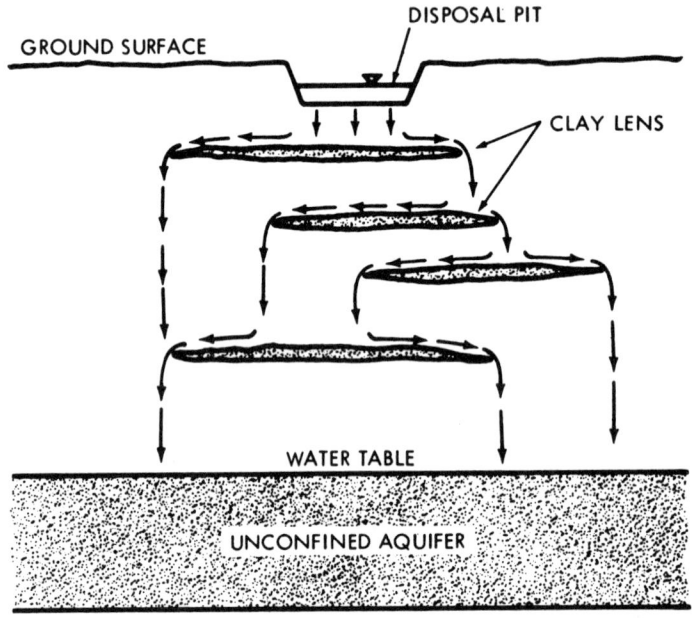

Figure 2-6. Diagram showing the horizontal movement of pollutants beneath a disposal pit as a result of clay lenses in the vadose zone above the water table.

Figure 2-8 indicates how salt leached from a salt stockpile moves downward to the water table, and thereafter, laterally and vertically to a nearby pumping well. Figure 2-9 indicates pollution movement from a surface stream or lake to a nearby pumping well. The drawdown of the water table induces recharge of surface water to groundwater. Many municipal water supply wells are located adjacent to rivers in order to insure continuous water supplies. This is an important groundwater pollution mechanism where rivers are polluted. In arid areas where water tables are relatively deep, polluted surface water can percolate to groundwater regardless of pumping patterns.

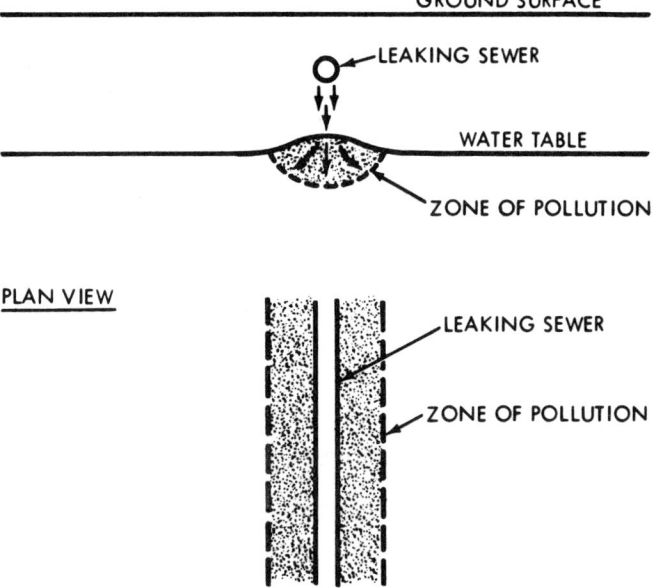

Figure 2-7. Illustration of a line source of ground-water pollution caused by a leaking sewer. The water table is situated below the sewer and is assumed to be horizontal.

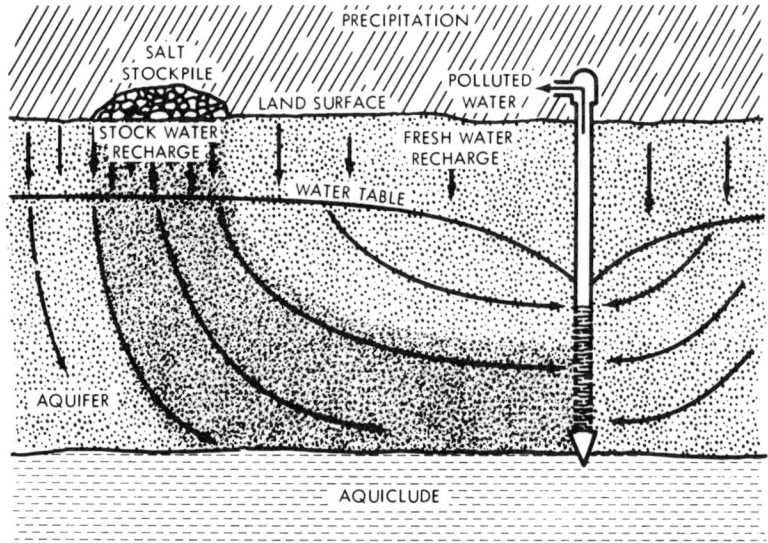

Figure 2-8. Diagram showing pollution of an aquifer by leaching of surface solids (modified after Deutsch, 1963).

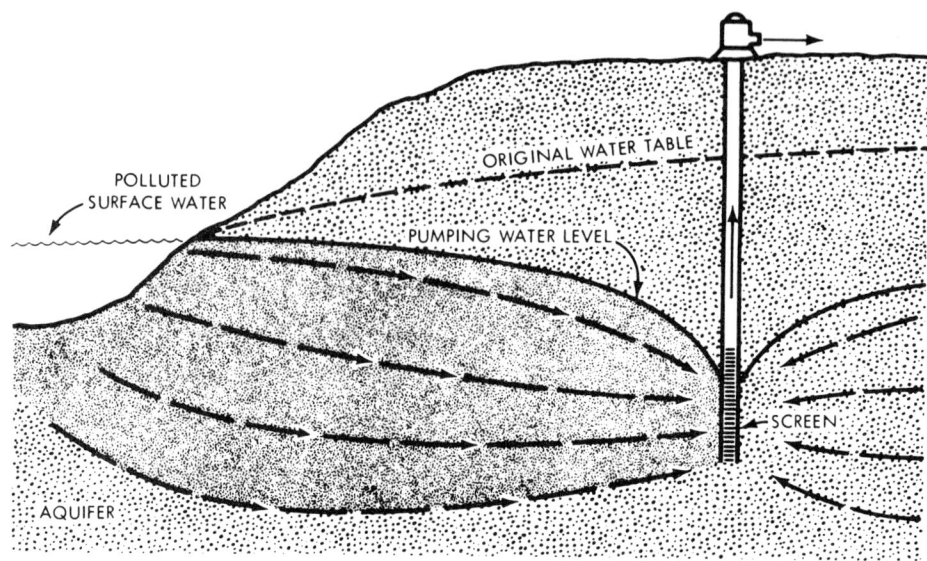

Figure 2-9. Diagram showing how polluted water can be induced to flow from a surface stream to a well (modifed after Deutsch, 1963).

Figure 2-10 suggests how temporary flooding of a well can lead to groundwater pollution. Downward flow of polluted surface water occurs around the well casing if the well has been improperly sealed at the ground surface.

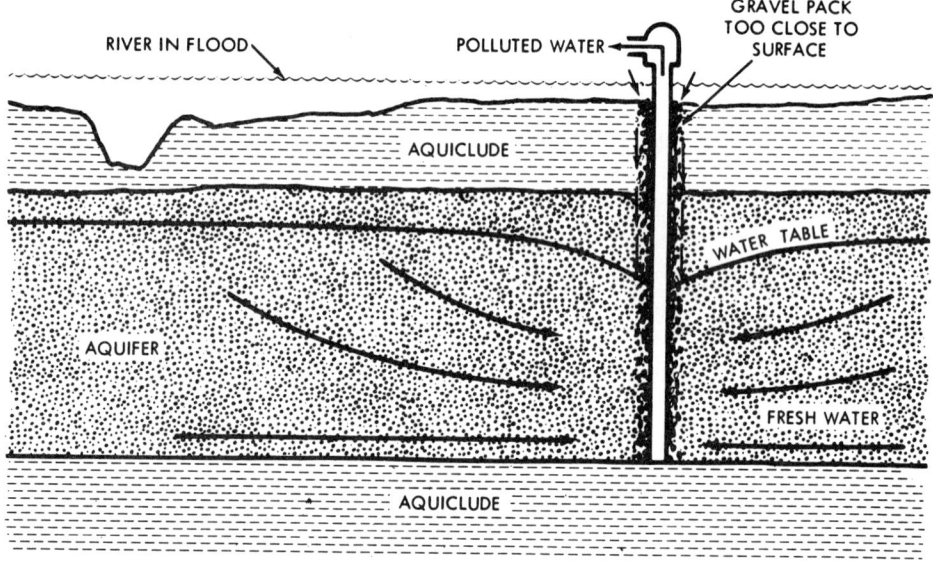

Figure 2-10. Diagram showing floodwater entering a well through an improperly sealed gravel pack (modified after Deutsch, 1963).

Figure 2-11 indicates how the disposal of pollutants into one well can be transported through the aquifer and lead to pollution of a nearby pumping well. Because a pumping well is a convergence point for groundwater over a large area, this collection mechanism increases the opportunity for obtaining polluted water from a pumping well.

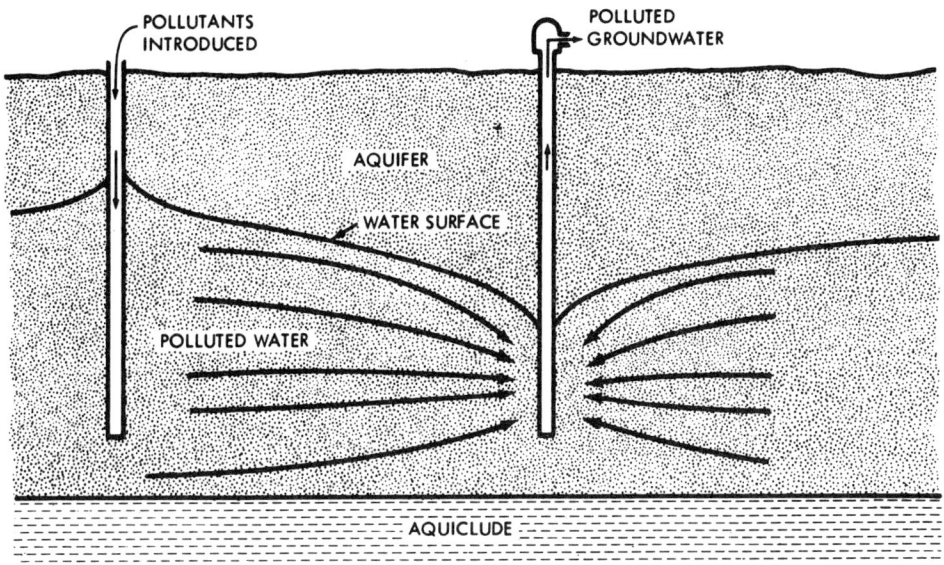

Figure 2-11. Diagram showing movement of pollutants from a recharge well to a nearby pumping well (modified after Deutsch, 1963).

Figure 2-12 outlines the flow patterns of pollutants entering a water table aquifer and a confined aquifer from an injection or recharge well. Note that the pollution tends to spread through the entire thickness of an aquifer and then move radially outward from the well.

Figure 2-13 illustrates the reversal of underground flows due to pumpage from one aquifer and hence the possibility to degrade the groundwater quality by interaquifer flow. Under natural conditions shown in the upper diagram, the water table of Aquifer A is higher than the piezometric surface of Aquifer B; therefore, groundwater tends to move downward through the simipermeable zone separating the two aquifers. In the lower diagram, however, pumping has interchanged the relative positions of the two water levels. As a result, the greater pressure in Aquifer B causes water to migrate upward into Aquifer A. If, as if often the case, the lower aquifer is more saline, this will cause the salt content of the upper aquifer to increase.

Figure 2-14 presents two situations of interaquifer flow through a nonpumping well penetrating more than one aquifer. In Figure 2-14(a) the water table stands above the piezometric surface so that water flows down the well into the deeper aquifer. But in Figure 2-14(b) the situaiton is reversed so that water rises through the well and invades the upper zone.

Figure 2-15 shows plan and profile views of a recharge pond overlying an unconfined aquifer, with a sloping water table and with groundwater flowing from left to right. Under these conditions pollution from the pond extends a short distance upstream and is stabilized. The bulk of the pollutant moves away from the pond in a downgradient direction within clearly defined boundaries. For given aquifer and recharge conditions, the lateral spread of the pollution as it moves downstream can be determined (Todd, 1959). Waste water from a disposal well penetrating an aquifer having the same conditions would move in a similar flow pattern.

Figure 2-16 suggests how underlying saline groundwater can rise due to deepening of a stream channel. This intrusion of saline water occurs because of the reduced head of fresh water.

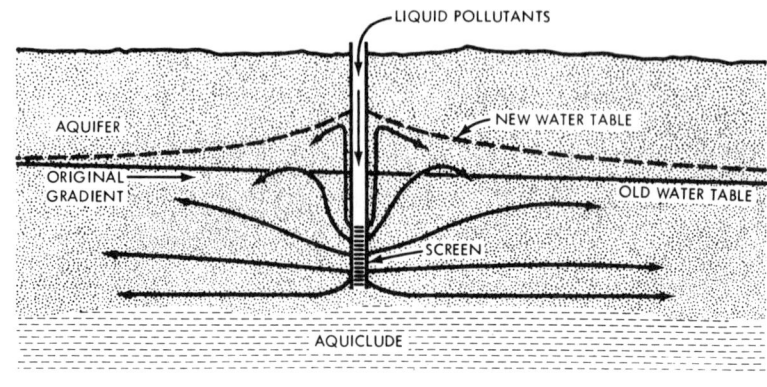
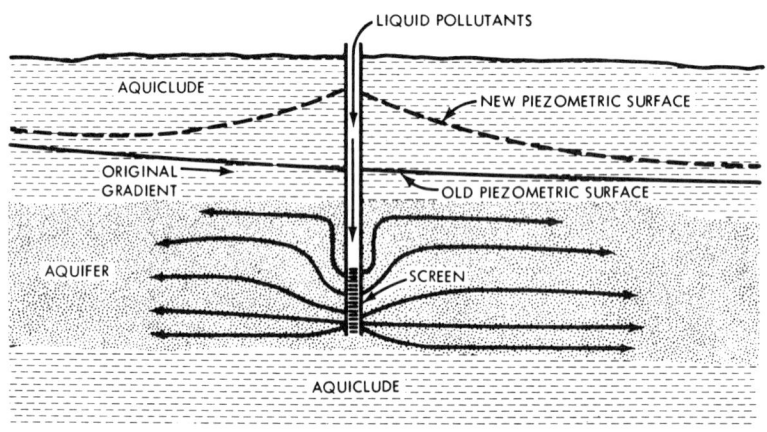

Figure 2-12. Diagrams showing spread of pollutants injected through wells into water table and artesian aquifers (modified after Deutsch, 1963).

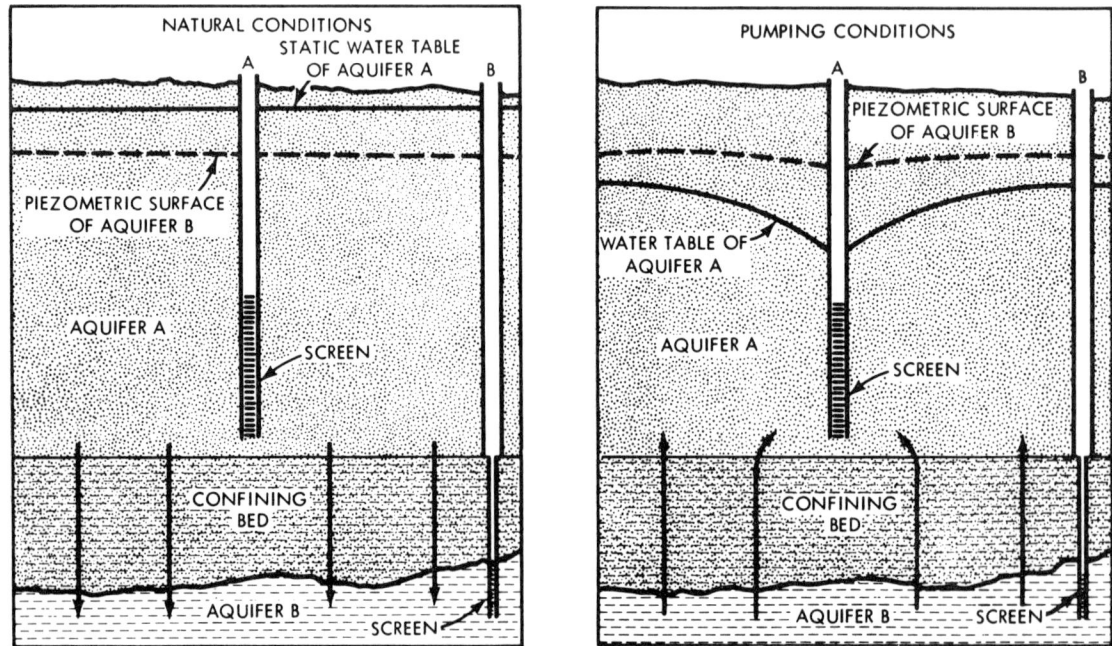

Figure 2-13. Diagrams showing reversal of aquifer leakage by pumping (Deutsch, 1963).

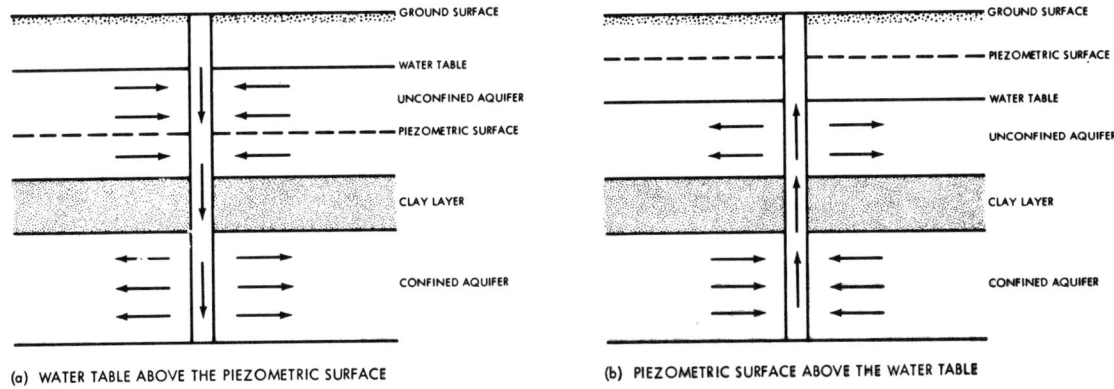

Figure 2-14. Diagrams showing aquifer leakage by vertical movement of water through a nonpumping well.

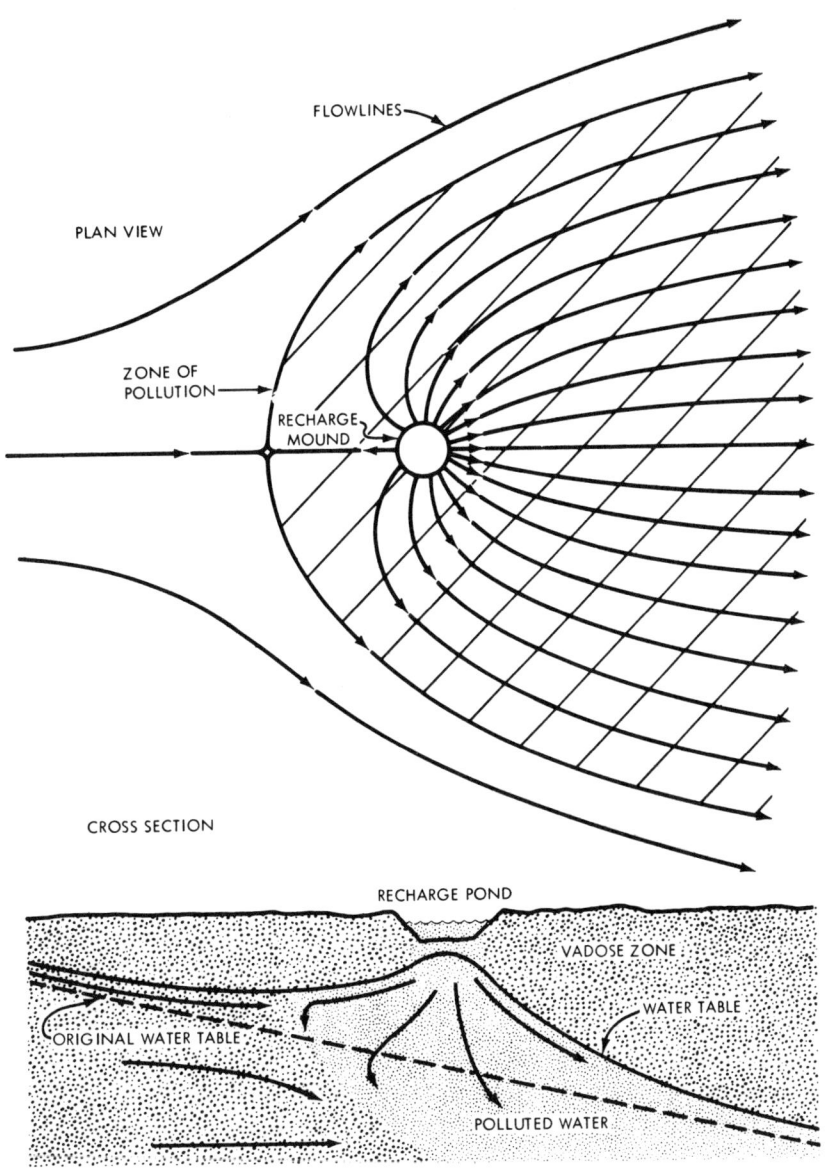

Figure 2-15. Diagrams showing lines of flow of pollutants from a recharge pond above a sloping water table (modified after Deutsch, 1963).

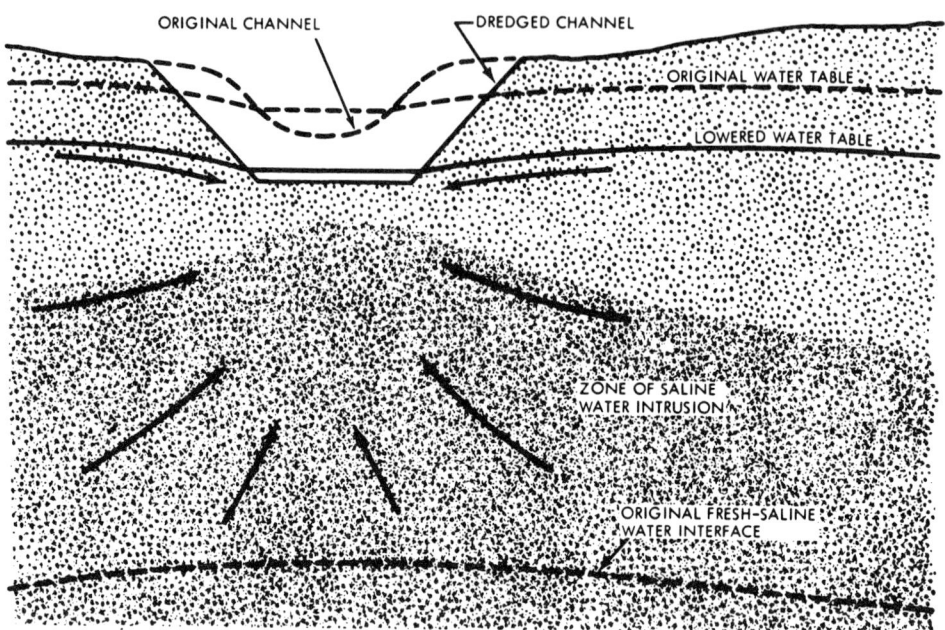

Figure 2-16. Diagram showing migration of saline water caused by lowering of water levels in a gaining stream.

ATTENUATION OF POLLUTION

Pollutants in groundwater tend to be removed or reduced in concentration with time and with distance traveled. Mechanisms involved include decay, chemical processes, and dilution. The rate of pollution attenuation is a function of the type of pollutant and of the local hydrogeologic framework. Predicting the degree to which pollutants will become attenuated is one of the most difficult, but also one of the most important, problems in the design of waste disposal systems utilizing the underground.

- **Decay.** The reduction in the strength or potency of a pollutant with time constitutes a decay mechanism. This may involve oxidation of organic wastes in the zone of aeration, the decrease in concentration of a radioactive waste as a function of its half-life, or death of microorganisms in unsuitable geologic environments.

- **Chemical Processes.** Adsorptive and other physical-chemical interfacial forces combine to remove pollutants from solution and to concentrate them on fine-grained soil and aquifer materials, particularly clays. Soils have a large capacity to adsorb organic materials. Many radioactive wastes, in addition, are strongly adsorbed by alluvial aquifers. Sorption may also include ion-exchange. Furthermore, cations such as potassium and ammonium, anions such as phosphate, and many trace elements tend to be adsorbed by soils and granular porous media.

  Salts can be precipitated from solution under certain conditions of temperature, pH, oxidation potential, and concentration. Thermodynamic relations can be used to determine if precipitation of mineral phases should be occurring under a given set of conditions. Anions most affected by precipitation are carbonate, bicarbonate, sulfate, and phosphate; cations most affected are calcium and magnesium.

- **Dilution.** Pollutants in groundwater flowing through porous media tend to become diluted in concentration due to hydrodynamic dispersion. Microscopic dispersion is mixing caused by the tortuous flow around individual grains as water moves through an aquifer. Macroscopic dispersion is mixing resulting from heterogeneities of geologic formations, such as variable permeability, which cause flow lines to deviate and to converge. The result of these mechanisms is a longitudinal and lateral spreading of a pollutant within the groundwater so that the volume affected increases and the concentration decreases downgradient. An analogous effect would be the dissipation of smoke from a smokestack as it drifts downwind in the atmosphere.

It is important to emphasize the role of aquifers in these attenuation processes. Aquifers composed of fine-grained materials possess very large surface areas which promote sorption processes. These same aquifers also encourage dilution by dispersion because of the large number of small interstices through which the groundwater must flow. On the other hand aquifers with large openings, whether from large-sized materials, cracks, or solution openings, permit a pollutant to advance rapidly underground with little or no reduction in concentration. Thus, in general, adsorption is greater in fine-grained aquifers, while dilution by mixing is greater in coarse-grained aquifers.

Specific statements cannot be made about the distances that pollution will travel because of the wide variability of aquifer conditions and types of pollutants. Yet certain generalizations which are widely applicable can be stated. For fine-grained alluvial aquifers, pollutants such as bacteria, viruses, organic materials, pesticides, and most radioactive materials, are usually removed by adsorption within distances of less than 100 meters. But most common ions in solution move unimpeded through these aquifers, subject only to dilution by mixing and chemical processes.

DISTRIBUTION OF POLLUTION UNDERGROUND

Given the above attenuation processes, pollution from a point source moves outward until a harmless or very low concentration level is reached. Because each constituent of a pollution source may follow a different attenuation rate, the distance to which pollution is effective will vary with each quality component.

A hypothetical example of a waste-disposal site is shown in Figure 2-17. Here, groundwater flows toward a river. Zones A, B, C, D, and E represent essentially stable limits for different contaminants resulting from the steady release of wastes of unchanging composition. Pollutants, once entrained in the saturated groundwater flow tend to form plumes (again analogous to smoke in the atmosphere) of polluted water extending downstream from the pollution source until they attenuate to some minimum quality levels. Only Zone E reaches the river in Figure 2-17 and is subsequently diluted by surface water.

The shape and size of a plume depends upon the local geology, the groundwater flow, the type and concentration of pollutant, the continuity of waste disposal, and any modifications of the groundwater system by man, such as well pumping (LeGrand, 1965b). Examples of various types of plumes are shown in Figure 2-18 and explained in Table 2-4. These are illustrative of the diversity of forms that plumes may assume. Where groundwater is moving relatively rapidly, a plume from a point source will tend to be long and thin; but where the flow rate is low, the pollutant will tend to spread more laterally to form a wider plume. Irregular plumes can be created by local influences such as pumping wells and nonuniformities in permeability.

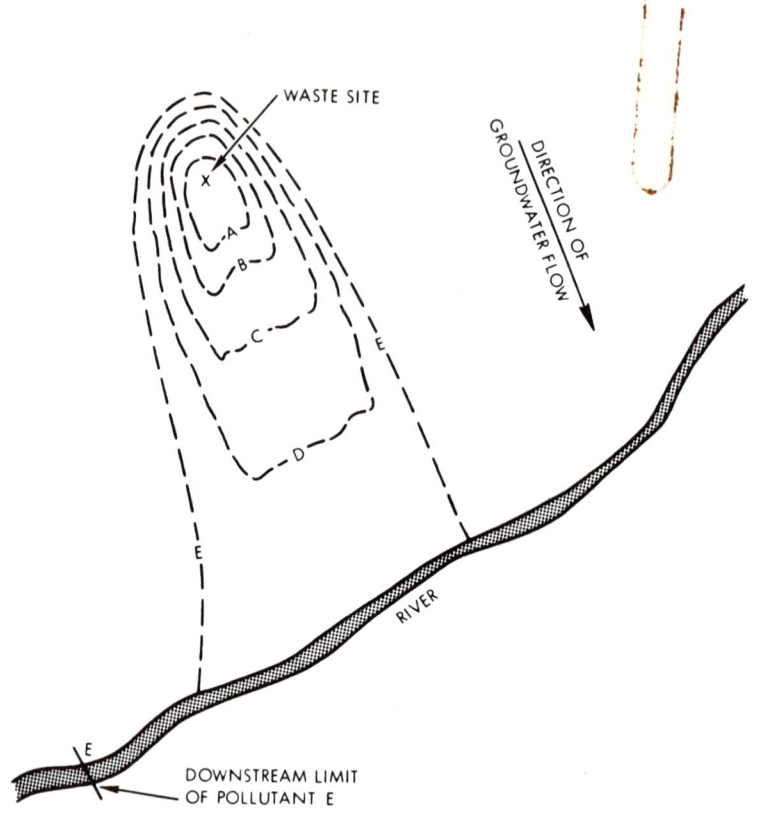

Figure 2-17. Plan view of a water table aquifer showing the hypothetical areal extent to which specific pollutants of mixed wastes at a disposal site disperse and move to insignificant levels (modified after LeGrand, 1965b).

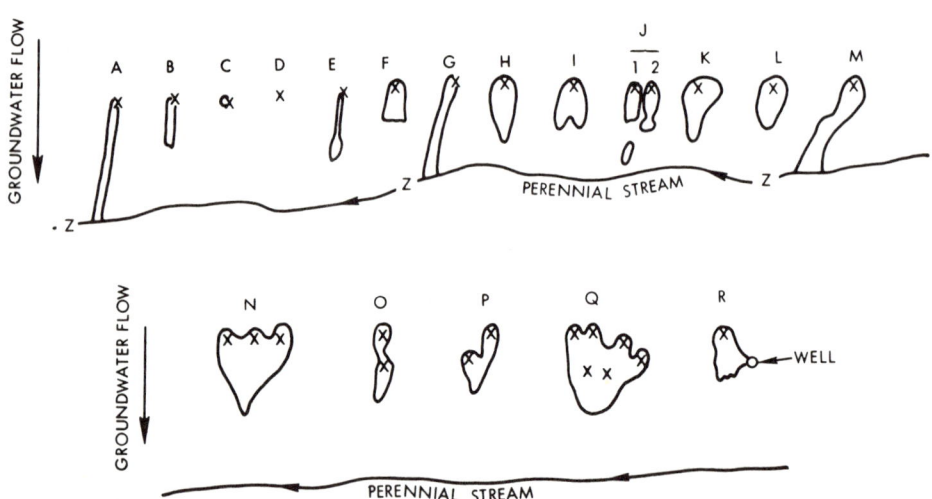

Figure 2-18. Types of pollution plumes in the upper part of the zone of saturation (plan view). An "X" marks the core of contamination beneath a waste site and "Z" the point downstream at which some plumes terminate (LeGrand, 1965b).

Table 2-4

EXPLANATION OF PLUMES SHOWN IN FIGURE 2-18 (modified after LeGrand, 1965b)

Site	Plume Governed By — Dilution	Decay	Sorption	Liquid Waste Recharge Forming Water-Table Mound	Composite Water Sites	Examples of Types of Pollutant	Remarks
A	Not appreciable in ground; some in streams	No	No	No	No	Chlorides, nitrates	Probably small waste release or good attenuation in zone of aeration.
B	Not appreciable	Either decay or sorption or both	Perhaps	No	No	Sewage, radioactive wastes	Pollutant is completely attenuated in zone of aeration and does not reach zone of saturation.
C	Improbable	Perhaps	Perhaps	No	No	Sewage, radioactive wastes	
D	No malenclave formed (See Remarks)	Either decay or sorption or both		No	No	Sewage, radioactive wastes	Lack of dispersion near waste site typical of linear openings in rock; polluted water downgradient disperses into different type of material.
E	Slight near waste site; some at greater distance	Possibly	Possibly	No	No	---	
F	Yes; suggestive of nearly homogeneous porous materials	Improbable	Improbable	No	No	Chlorides, nitrates	Irregularities in permeability cause deviation in plume.
G	Not appreciable in ground; some near and in stream	Not appreciable	Not appreciable	No	No	Chlorides, nitrates	
H	Yes; suggestive of nearly homogeneous porous materials	Probably either decay or sorption or both		No	No	Sewage, radioactive wastes	Downgradient split in plume may be due to dense impermeable rock or great increase in sorptive materials.
I	Yes	Perhaps	Perhaps	No	No	---	
J	Slight	Not appreciable	Probably not appreciable	No	No	Chlorides, nitrates	Downgradient plume is due to shunting of pollutant to land surface at tail of upper plume and reinfiltration of pollutant.
K	Yes; suggestive of nearly homogeneous porous materials	Either decay or sorption or both		Yes	No	Sewage, radioactive wastes	Irregularities in plume caused by changes in permeability and/or sorption.
L	Yes; suggestive of nearly homogeneous porous materials	Either decay or sorption or both		Yes	No	Sewage, radioactive wastes	
M	Some in ground and stream	Not appreciable	Not appreciable	Yes	No	Chlorides, nitrates	Deviation in plume due to impermeable zone.
N	Yes	Either decay or sorption or both		No	No	Sewage, radioactive wastes	Polluted water from three waste sites at right angles to groundwater flow, merging to form a composite plume.
O	Yes	Either decay or sorption or both		No	Yes	Sewage, radioactive wastes	Polluted water from two waste sites parallel to groundwater flow, forming a composite site.
P	Some	Either decay or sorption or both		No	Yes	Sewage, radioactive wastes	Polluted water from two waste sites at an angle with groundwater flow, forming a composite plume.
Q	Some	Either decay or sorption or both		No	Yes	Sewage, radioactive wastes	Large composite plume formed by several waste sites.
R	Yes	Either decay or sorption or both		No	No	Sewage, radioactive wastes	Pumping well draws plume toward it; polluted water is greatly diluted at the well.

Plumes ordinarily become stable areas where there is a constant input of waste into the ground. This occurs for two reasons: the tendency for enlargement as pollutants continue to be added at a point source is counterbalanced by the combined attenuation mechanisms, or the pollutant reaches a location of groundwater discharge, such as a stream, and emerges from the underground. When a waste is first released into groundwater, the plume expands until a quasi-equilibrium stage is reached. If sorption is important, a steady inflow of pollution will cause a slow expansion of the plume as the earth materials within it reach a sorption capability limit.

An approximately stable plume will expand or contract generally in response to changes in the rate of waste discharge. Figure 2-19 shows changes in plumes that can be anticipated from variations in waste inputs.

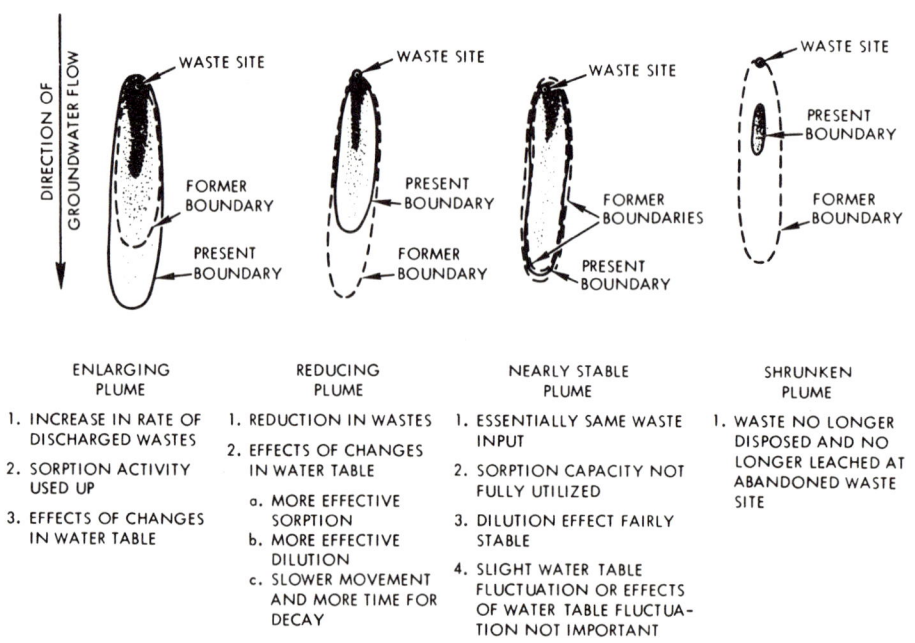

Figure 2-19. Changes in plumes and factors causing the changes (modified after LeGrand, 1965a).

An important aspect of groundwater pollution is the fact that it may persist underground for years, decades, or even centuries. This is in marked contrasts to surface water pollution. The average residence time of groundwater is on the order of 200 years; consequently, a pollutant which is not readily decayed or sorbed underground can remain as a degrading influence on the resource for indefinite periods. The comparable residence time for water in a stream or river is on the order of 10 days; thus, surface water pollution can be rapidly eliminated. Reclaiming polluted groundwater is usually much more difficult and time consuming than reclaiming polluted surface water. Underground pollution control is usually best achieved by regulating the pollution source, and secondarily by physically entrapping and, when feasible, removing the polluted water from the underground.

EVALUATION OF POLLUTION POTENTIAL

To provide guidance in evaluating the potential pollution of a given site, LeGrand (1964) developed an empirical point-count system. The concept is applicable to locations of waste disposal sites and to wells; water table aquifer conditions are presumed.

Local factors influencing pollution include:

- Depth to water table
- Sorption
- Permeability
- Water table gradient
- Distance

The rating chart shown in Figure 2-20 illustrates the evaluation procedure for sites in unconsolidated alluvial materials. A numerical value is read above the line for each of the five factors based upon the corresponding data below the line.

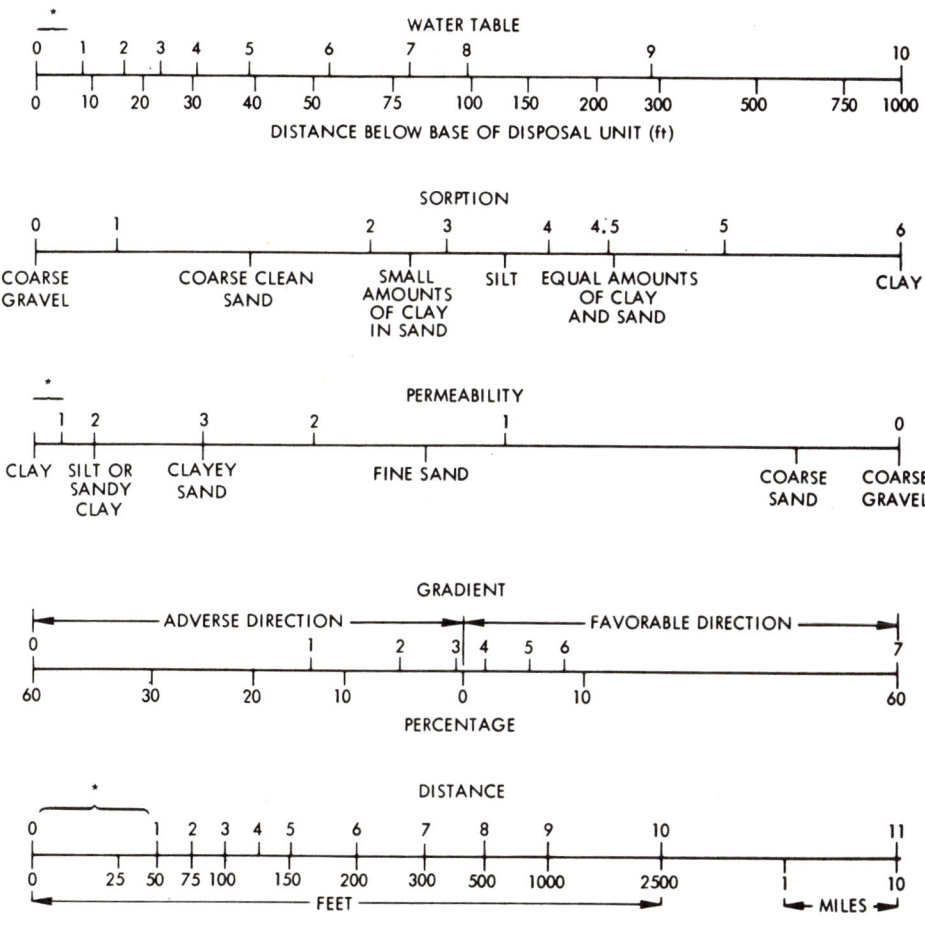

*UNACCEPTABLE RANGES

Figure 2-20. Rating chart for pollution potential in unconfined aquifers of unconsolidated alluvial materials (modified after LeGrand, 1964).

The pollution potential of a given site is the sum of the numerical ratings of the five factors in Figure 2-20. Total point values may be interpreted in terms of possibility of pollution as follows:

33

Total Points	Possibility of Pollution
0-4	Imminent
4-8	Probable or possible
8-12	Possible but not likely
12-25	Very improbable
25-35	Impossible

LeGrand (1964) emphasized that his index system for evaluating pollution was imperfect and only a beginning. Certainly refinements can and should be made in the future, but for the present his approach is useful and sound.

TRENDS IN GROUNDWATER POLLUTION

Because of the many millions of groundwater pollution sources in the United States, comprehensive data defining the extent of the problem are almost nonexistent. Thus, although an overview of the magnitude of the problem is difficult to obtain, basic trends of pollution can be suggested from a demographic approach.

Recognizing that groundwater pollution results from activities of man, one can relate the magnitude of pollution to population and other demographic factors. As an example, Figure 2-21 shows the estimated growth of underground pollution for the United States during the 20th Century. The ordinate is the percent of current (1974) pollution levels. The curve preceding 1974 is based upon the growth of urban population adjusted by a factor of two to account for industrial development, increased per capita water use, development of irrigated agriculture, and introduction of agricultural fertilizers since 1900. The cumulative aspect of pollution, whereby pollution dating from say 1954 may still be underground in 1974, has been neglected.

The curve in Figure 2-21 has been extended from 1974 to 2000 based on estimated urban population growth only, without adjustment. What this analysis suggests is a tenfold increase in groundwater pollution from 1900 to the present, and — assuming no future efforts to control groundwater pollution — a further 60 percent increase by the end of the century. Hopefully, with the implementation of the Federal Water Pollution Control Act of 1972 (PL 92-500) and the Safe Drinking Water Act (PL 93-523) this dire forecast will not become reality.

CONSTITUENTS IN POLLUTED GROUNDWATER

QUALITY CATEGORIES

Water quality is normally expressed in terms of four basic categories: chemical, biological, physical, and radiological. Within each category a large number of individual parameters and constituents can be identified. Table 2-5 lists a wide range of possible pollutants which may be found by analysis of groundwater samples. The possible potential pollutants in the chemical and biological categories are, of course, virtually limitless. On the other hand most analyses of groundwater include only a few key parameters and constituents because they are sufficient to serve as indicators of pollution.

EFFECTS OF WATER USE

The sources and causes of groundwater pollution are by definition related to man's use of water. The diversity of impacts of man's activities on the hydrologic cycle create a complex and interrelated series of modifications to natural water quality. The sources of water and the effects of man on water quality are suggested in the water

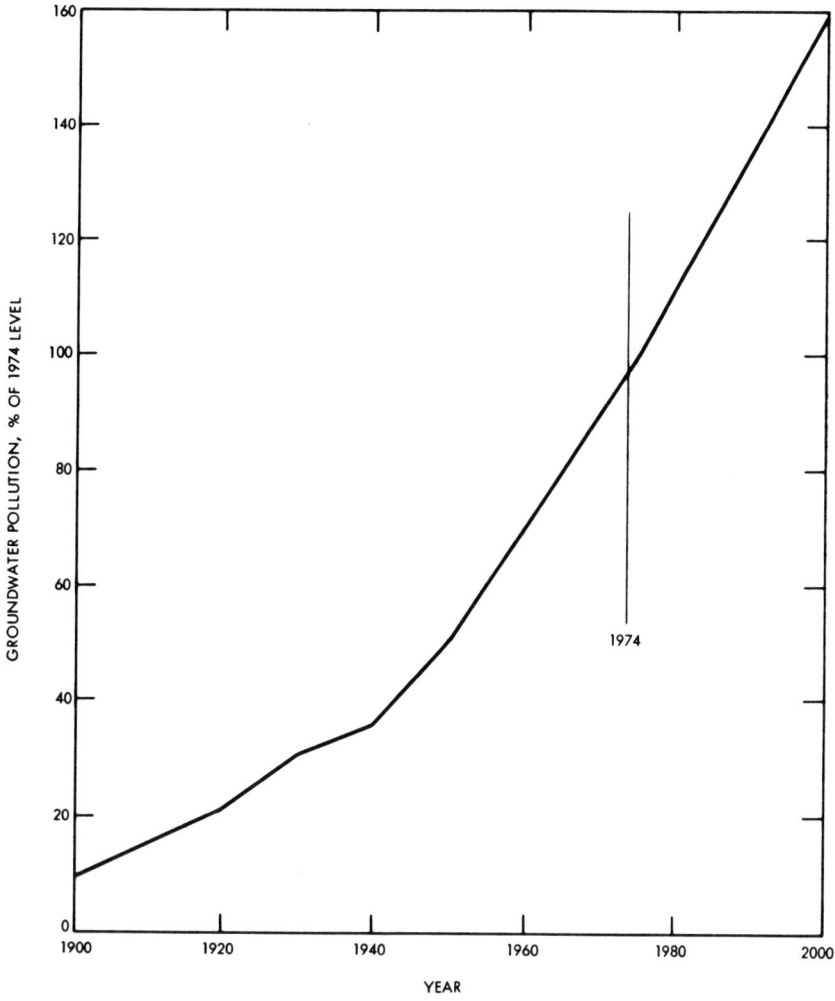

Figure 2-21. Estimated trend of groundwater pollution in the United States during the 20th Century.

quality cycle of Figure 2-22. Note that there are four main uses of water – agricultural, industrial, commercial, and domestic; each of these in turn exerts an influence on groundwater quality.

Agricultural Uses

The various agricultural uses of water and their effects on water quality are shown schematically in Figure 2-23. Chemical constituents from agricultural activities are added at the land surface, as it moves downward through the soil, and as it passes through disposal sites for solid agricultural wastes. In particular, irrigation degrades the quality of water due to evapotranspiration, which concentrates salts in irrigation return water. Addition of fertilizers, soil amendments, pesticides, and other additives applied to the soil also contribute to groundwater pollution.

Solid wastes produced by agricultural activities, particularly dairies and feedlots, are in part deposited in sanitary landfills. Leachates and gases generated by the decomposition of solid wastes may affect the quality. The effect and the areal extent of such leachates vary depending upon the volume of leachate, the rate of mixing with groundwater, the volume of groundwater available, and the rate of groundwater movement.

Table 2-5
PARAMETERS AND CONSTITUENTS WHICH MAY BE INCLUDED IN ANALYSES OF GROUNDWATER QUALITY*

Chemical - Organic	Units	Chemical - Inorganic	Units
Biochemical Oxygen Demand (BOD)	mg/l	Nickel (Ni)	µg/l
Carbon Chloroform Extract (CCE)	µg/l	Nitrite (NO_2)	mg/l
		Nitrate (NO_3)	mg/l
Chemical Oxygen Demand (COD)	mg/l	Nitrogen (N)	mg/l
		Oil and Grease	mg/l
Chlorinated Phenoxy Acid Herbicides	µg/l	Oxygen (O_2)	mg/l
		pH	pH units
Detergents (Surfactants)	mg/l	Phosphate (PO_4)	mg/l
		Potassium (K)	mg/l
Organic Carbon (C)	mg/l	Selenium (Se)	µg/l
Organophosphorus Pesticides	µg/l	Silver (Ag)	µg/l
		Silica (SiO_2)	mg/l
Phenols	mg/l	Sodium (Na)	mg/l
Tannins and Lignins	mg/l	Solids, dissolved	mg/l
		Solids, suspended	mg/l
Chemical - Inorganic		Strontium (Sr)	µg/l
		Sulfate (SO_4)	mg/l
Acidity	mg/l	Sulfide (S)	mg/l
Alkalinity	mg/l	Sulfite (SO_3)	mg/l
Aluminum (Al)	µg/l	Tin (Sn)	µg/l
Ammonia (NH_4)	mg/l	Titanium (Ti)	µg/l
Antimony (Sb)	µg/l	Vanadium (V)	µg/l
Arsenic (As)	µg/l	Zinc (Zn)	µg/l
Barium (Ba)	µg/l		
Beryllium (Be)	µg/l	**Biological**	
Bicarbonate (HCO_3)	mg/l	Coliform Bacteria	Coliforms/100 ml
Boron (B)	µg/l	Fecal Coliform Bacteria	Fecal Coliforms/ 100 ml
Bromide (Br)	µg/l		
Cadmium (Cd)	µg/l	Fecal Streptococci Bacteria	Fecal Streptococci/ 100 ml
Calcium (Ca)	mg/l		
Carbonate (CO_3)	mg/l	**Physical**	
Chloride (Cl)	mg/l		
Chromium (Cr)	µg/l	Color	PCU
Cobalt (Co)	µg/l	Odor	TO
Conductance, specific	umhos/cm at 25°C	Temperature	°C
		Turbidity	TU
Copper (Cu)	µg/l	**Radiological**	
Cyanide (CN)	µg/l		
Fluoride (F)	µg/l	Barium - 140 (^{140}Ba)	pc/l
Hardness	mg/l	Cerium - 141 and 144 (^{141}Ce, ^{144}Ce)	pc/l
Hydroxide (OH)	mg/l		
Iodide (I)	µg/l	Cesium - 134 and 137 (^{134}Cs, ^{137}Cw)	pc/l
Iron (Fe)	ug/l		
Lead (Pb)	µg/l	Gamma Spectrometry	pc/l
Lithium (Li)	µg/l	Gross Alpha	pc/l
Magnesium (Mg)	mg/l	Gross Gamma	nc/l
Manganese (Mn)	ug/l	Iodine - 131(^{131}I)	pc/l
Mercury (Hg)	ug/l	Neptunium - 239(^{239}Np)	pc/l
Molybdenum (Mo)	ug/l	Radium (Ra)	pc/l
		Thorium (Th)	µg/l
		Tritium (^3H)	pc/l
		Uranium (U)	µg/l

*This list is illustrative rather than comprehensive. Various ionic forms of some constituents as well as different analytic expressions for certain constituents have not been included.

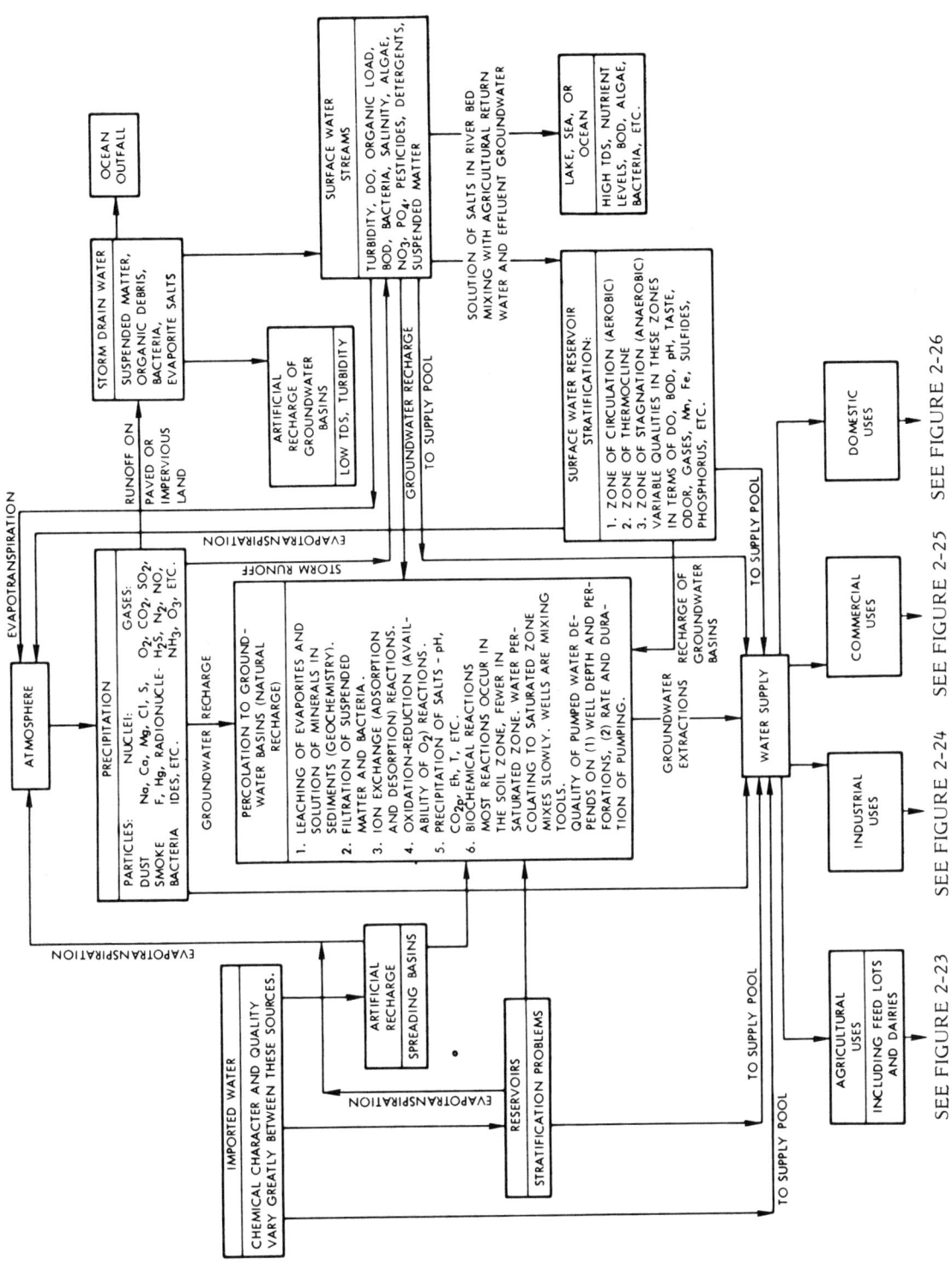

Figure 2-22. Water quality cycle – sources and uses of water and effects on water quality (modified after Hassan, 1974).

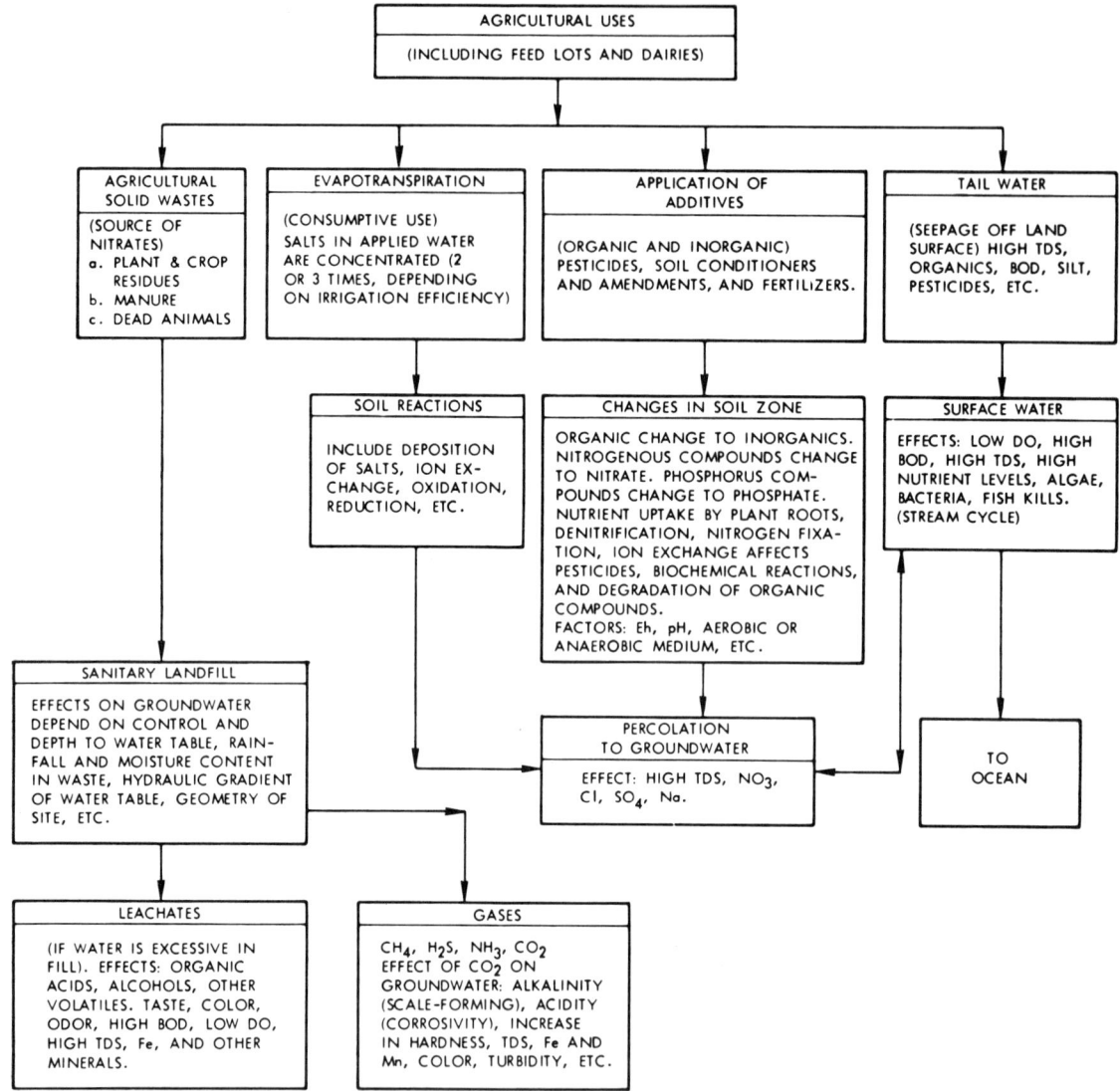

Figure 2-23. Agricultural uses of water and their effects on water quality (modified after Hassan, 1974).

Industrial Uses

The major uses of water in industrial plants are for cooling, manufacturing and processing, and sanitation.

Figure 2-24 shows industrial uses of water and their effects on water quality. As could be expected, the quality of waste water varies with respect to the type of industry and water use.

Water cooling systems vary from "open once-through" to "closed recirculating." The first method requires large volumes of water, and waste water is disposed of after one use. A closed recirculating system uses less water, but generates wastes with higher salinities. Softening of water before use in cooling, to inhibit scale formation, results in generation of brine wastes. Water returned to the underground after use for cooling will possess a higher salt content and a higher temperature.

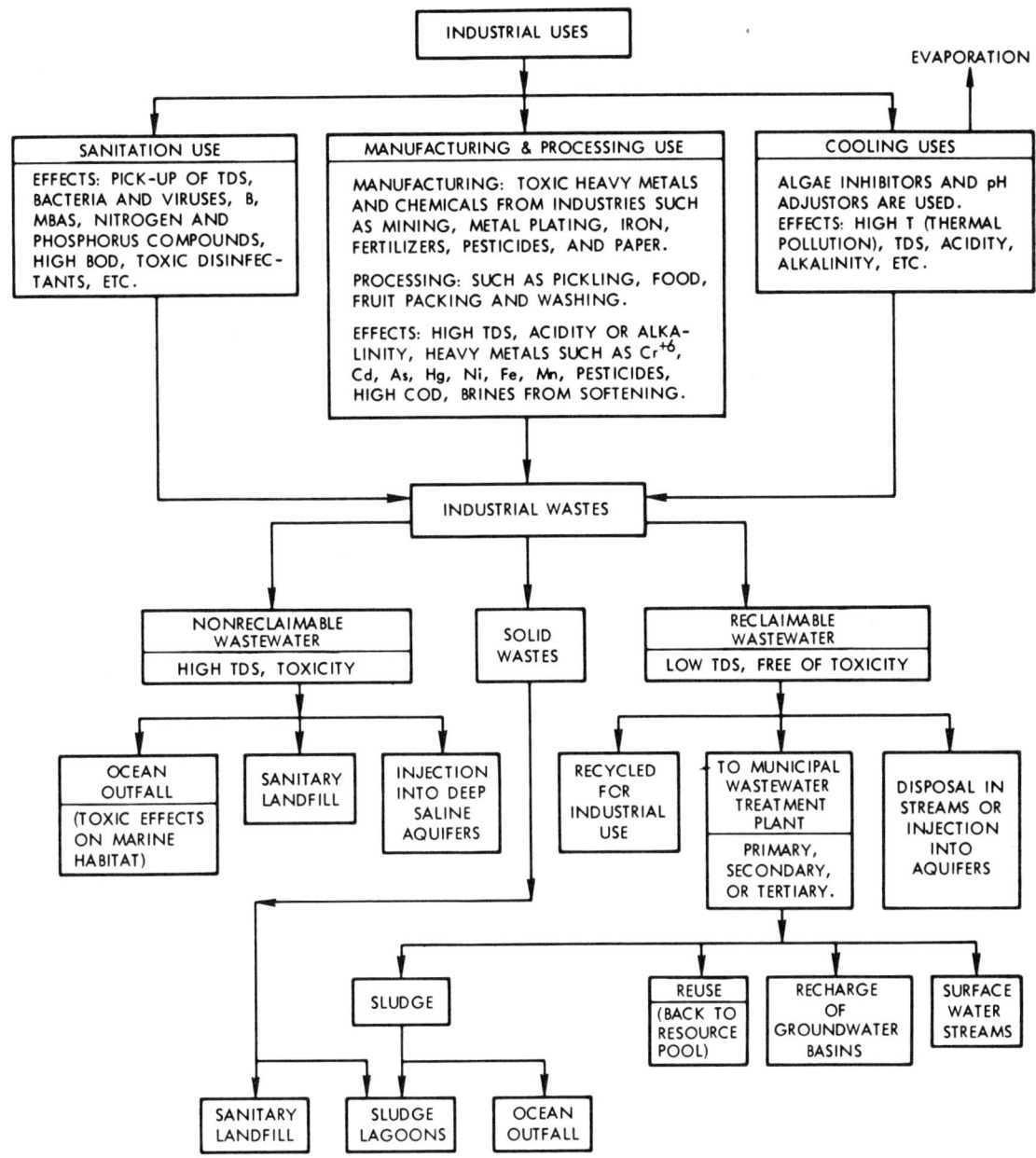

Figure 2-24. Industrial uses of water and their effects on water quality (modified after Hassan, 1974).

Boiler feed systems have similar *open* and *closed* designs and yield waste waters with increased salinities and temperatures. Industrial organic wastes other than sewage generally possess a higher chemical oxygen demand (COD) than domestic wastes. Nonreclaimable industrial waste water may be disposed into deep injection wells, whereas reclaimable waste water may be artificially recharged to groundwater after treatment.

Commercial Uses

Figure 2-25 diagrams commercial uses of water and their effects on water quality. Outside water use includes irrigation of lawns, shrubs, and landscaping as well as activities such as car washes. Irrigation effects are similar to those previously mentioned for irrigated agriculture.

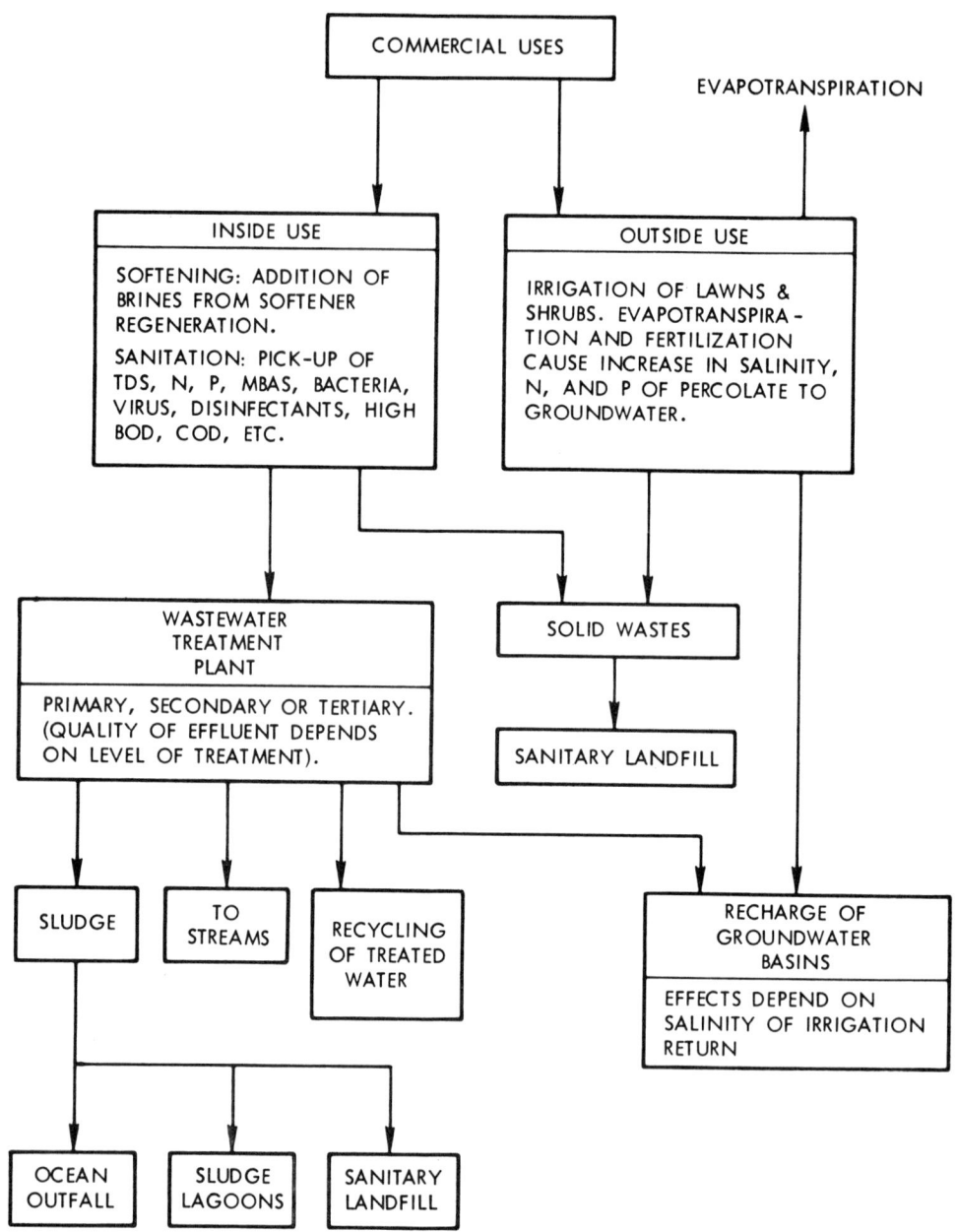

Figure 2-25. Commercial uses of water and their effects on water quality (modified after Hassan, 1974).

Inside uses embrace water softening plants, sanitation uses in schools and public buildings, laundries, and commercial laboratories, among others. Most of the waste water generated from commercial uses is disposed of in sewer systems; from there, after treatment, it may return to groundwater via recharge basins or sanitary landfills.

Domestic Uses

Domestic uses of water include drinking and cooling, cleaning and bathing, laundry, sanitary needs, air conditioning, garden watering, and for swimming pools. The flow of water used domestically and the effects on water quality are outlined in Figure 2-26. Inside domestic water uses are basically nonconsumptive uses. Per capita water uses varies widely between city and suburban areas and with income levels of households. Water softening may also be a quality consideration of domestic use. Effects of outside water uses are similar to those for irrigated agriculture.

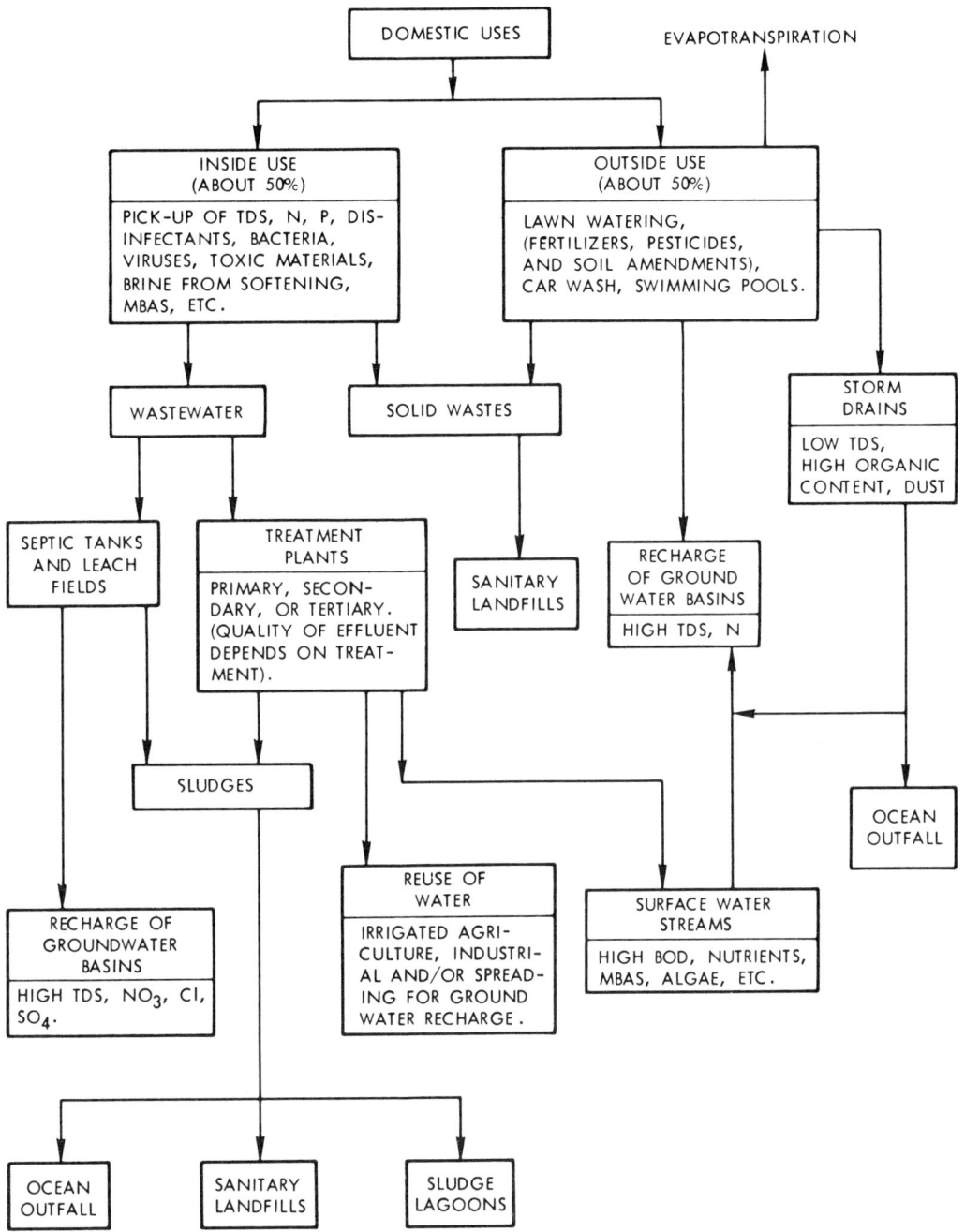

Figure 2-26. Domestic uses of water and their effects on water quality (modified after Hassan, 1974).

SOURCES AND CAUSES OF POLLUTION

Groundwater pollution results from the activities of man, consequently, the various sources and causes of pollution would conceptually form a long and complex list. However, in terms of basic causes and primary influences on groundwater quality, the list can be condensed to a reasonable size. Table 2-6, therefore, contains the principal causes and sources of groundwater pollution grouped into four general categories. Each item listed is discussed in the following sections with respect to its magnitude, location, and effect on groundwater quality.

Table 2-6

PRINCIPAL SOURCES AND CAUSES OF
GROUNDWATER POLLUTION

Agricultural
1. Irrigation return flow
2. Animal wastes
3. Fertilizers
4. Crop residues and dead animals
5. Pesticide residues

Municipal and Industrial
1. Surface disposal of solid wastes
2. Surface disposal of liquid wastes
3. Sewer leakage
4. Tank and pipeline leakage
5. Disposal wells
6. Injection wells
7. Stockpiles
8. Mining activities
9. Oilfield brines

Groundwater Basin Management
1. Saline water intrusion
2. Aquifer interchange through wells

Miscellaneous
1. Spills and surface discharges
2. Septic tanks and cesspools
3. Highway deicing

AGRICULTURAL SOURCES AND CAUSES
Irrigation Return Flow

Approximately one-half to two-thirds of the total water applied during irrigation is used consumptively. The remainder is termed irrigation return flow, which returns to surface streams or groundwater. Irrigation degrades the quality of applied water. The increase in salinity of irrigation return flow results from the addition of salts by dissolution during the irrigation process, from salts added to irrigation water as fertilizers or soil amendments, and from the concentration of salts by evapotranspiration of applied water. The salinity of irrigation return flow due to these processes may range from three to ten times that of the applied water (Jenke, 1974).

The principal cations of return flow are calcium, magnesium, sodium, and potassium. Minor amounts of iron, aluminum, manganese, and other cations may also be present. The dominant anions include carbonate, bicarbonate, sulfate, chloride, and nitrate.

Irrigation return flow is a nonpoint source of groundwater pollution because of its large areal extent. The problem has been identified as the major cause of pollution in the Southwestern United States (Fuhriman and Barton, 1971) and is significant throughout the arid and semiarid portions of the entire Western United States where irrigated agriculture is practiced.

Animal Wastes

In recent years the feedlot has become an important method for beef production. For the 120 to 150 days that a beef animal remains in a feedlot, it will produce over a half-ton of manure on a dry weight basis. With thousands of animals in a single feedlot, the natural assimilative capacity of the receiving soil is heavily overtaxed. Runoff from rainfall that comes in contact with manure may carry high concentrations of pollutants into ponds, streams, and groundwater.

Animal wastes release salts, organic loads, and bacteria into the soil. Investigations have shown that nitrate-nitrogen is the most important persistent pollutant that reaches the water table (Fuhriman and Barton, 1971). Where permeability is moderate and the depth to water table is large, most biological feedlot pollutants are removed in the vadose zone before reaching the groundwater.

Feedlot operations are largest in Texas and California; however, they can be found throughout the western and central portions of the United States.

Fertilizers

The application of fertilizers to agricultural land usually results in a portion of the fertilizer being leached through the soil and into the underlying groundwater. The most important constituents are compounds of nitrate and phosphorus. Most phosphate fertilizers, however, are readily adsorbed on soil particles so that they seldom constitute a groundwater problem. But nitrogen in solution is only partially used by plants and also may be adsorbed to only a limited extent by the soil or lost in gaseous form to the atmosphere; consequently, nitrogen has been found to be the primary fertilizer element related to groundwater pollution (Fuhriman and Barton, 1971).

The use of fertilizers is extensive in the United States and will increase in the future. Therefore, fertilizer usage leading to the potential for nitrogen pollution of groundwater on a nonpoint source basis can be expected to continue. Furthermore, high nitrate concentrations observed in groundwaters of suburban areas suggest the possibility of lawn fertilizers as a source (Miller, 1974).

Crop Residues and Dead Animals

Crop residues are those portions of a plant left in the field or processing shed after harvest. For every pound of food marketed, from 2 to 5 pounds of residues are left in the field or in the packing shed (Fuhriman and Barton, 1971). Although such wastes can create groundwater pollution problems, there is little evidence available to show that they do.

Larger farm animals, when dead, are usually disposed of by rendering plants, and seldom affect groundwater quality. Similarly, dead sheep or wildlife on the range are usually quickly disposed of by other forms of wildlife. Bodies of animals used in laboratory experiments, particularly those subjected to radioactive treatments, must be disposed of with special care.

The disposal of dead poultry is probably the only source which can create a serious groundwater pollution hazard. Poultry producers today often handle flocks exceeding 100,000 fowl. The death rate in such an operation can average 35 fowl per day (Fuhriman and Barton, 1971). Many producers dispose of dead birds by burial in a large trench; consequently, with percolating water in the vicinity of such a mass of decaying organic matter, the groundwater pollution potential is considerable.

Pesticide Residues

The term "pesticide" is here broadly interpreted to embrace any material used to control, destroy, or mitigate pests, including insecticides, herbicides, fungicides, nematocides, rodenticides, bactericides, growth regulators, and defoliants. Whenever these materials occur in groundwater, even in minute concentrations, the consequences are serious in terms of the potability of the water.

The magnitude of the threat of pesticides to groundwater quality depends upon the properties of the pesticide residue; the frequency and rate of rainfall or irrigation; the hydrologic characteristics of the soil; and the volume, the state (liquid or solid), and the persistence of the pesticide applied (Todd and McNulty, 1974). Many pesticides are relatively insoluble in water; many also are readily adsorbed on soil particles.

MUNICIPAL AND INDUSTRIAL SOURCES AND CAUSES
Surface Disposal of Solid Wastes

The land disposal of solid wastes constitutes an important source of groundwater pollution. A landfill may be defined as any land area used for the deposit of urban, or municipal, solid waste. Estimates of the production of solid wastes indicate a total of about 1 ton per capita per year, equal to almost 6 pounds per person per day (Meyer, 1973). Because wastes are generated and disposed of where people are living, the pattern of urban population distribution gives an indication of the location and intensity of landfill practice.

There are two basic types of landfills: one is the sanitary landfill, designed and constructed according to engineering specifications, while the other is simply a refuse dump. Of the more than 100,000 landfills in the United States, probably no more than 10 percent can be classed as sanitary landfills.

Leachate from a landfill can pollute groundwater. For this to occur, however, a source of water moving through the fill material is required. Possible sources include precipitation, moisture content of refuse, surface water infiltrating into the fill, percolating water entering the fill from adjacent land, or groundwater in contact with the fill (Meyer, 1973). Leachate is not produced in a landfill until a significant portion of the material has a moisture content equal to field capacity. Ordinary mixed refuse contains a high paper content and usually has a moisture content far below that of field capacity.

It follows, therefore, from the above reasoning that leachate from a landfill can be prevented if refuse can be compacted and covered without becoming saturated, if rainfall and surface runoff can be diverted from the landfill material, and if the fill material can be isolated from nearby groundwater. These are essentially the goals of a well-designed sanitary landfill. With a properly constructed landfill, any leachate generated can be controlled and prevented from polluting groundwater. Accomplishment of these objectives is possible by proper site selection, by placement of appropriate cover material, and by surface and subsurface leachate collection systems. The problem of pollution from landfills is greatest in the Eastern United States where high rainfalls and shallow water tables occur. Conversely, the problem is minimal in portions of the arid Southwest where deep water tables are found.

Analyses of leachate show that its quality can vary widely. The most important pollutants are chemical oxygen demand (COD), biochemical oxygen demand (BOD), iron, chloride, and nitrate. In addition, hardness, alkalinity, and total dissolved solids are often increased. Gases, including methane, carbon dioxide, ammonia, and hydrogen sulfide, are a further byproduct of landfills; these may cause subsurface chemical reactions which can degrade groundwater quality.

Surface Disposal of Liquid Wastes

There are few, if any, industries which do not make use of water, either directly as part of the product, such as in beverage industries and in steam generation, or indirectly as for cooling water and as a transporting medium (paper industry). Water is often used as a solvent, either as a medium for chemical reactions, or for washing products and containers, or machines, apparatus, factory floors, etc. Much of the water is subsequently discharged as waste water, the composition of which differs according to the usage and nature of the process used.

Wastes from a particular type of industry are usually similar enough from factory to factory to be compared directly. The waste composition and volumes are often correlated with production rates. Industrial wastes may be broadly classified as:

- Nonfermentable inorganics and other inert wastes
- Fermentable, mainly organic wastes
- Toxic wastes

Waste pickling-bath solutions from sheet-metal or galvanizing shops are examples of nonfermentable wastes, however they may also be toxic due to low pH and the presence of trace elements. Many food industries discharge readily fermentable wastes, for example, meat-packing plants and canneries. However, these wastes may also be toxic due to low pH from sulphurous acid, or because of plating shop wastes which are discharged along with packing shop wastes. Industries which commonly discharge toxic wastes include dyehouses and electroplaters (cyanides, sulfides, chromium, or copper) and chemical factories manufacturing chlorinated hydrocarbons, etc. Dyehouse wastes are usually mineralized and may also contain significant amounts of fermentable organic matter. Wastewaters from the use of water for cooling purposes may be relatively unpolluted, except for temperature (Imhoff et al., 1971).

Several major types of industrial waste of importance to groundwater pollution are cooling water, process wastes, boiler blowdown, washdown water, storm runoff, stockpiles, water treatment plant effluent, tank and pipeline leakage, and hydrocarbon storage. These wastes are disposed of by a variety of methods and often in combination.

Cooling Water. Cooling water comprises the greatest industrial use of water. Cooling water is often disposed of in percolation ponds and disposal wells, and is sometimes used for irrigation.

Process Wastes. Process wastes probably present the most serious threat to groundwater quality of all types of industrial wastes. These wastes include spent fluids, catalysts, byproducts, and other wastes, and are commonly disposed of in wells and percolation ponds.

Boiler Blowdown. Boiler blowdown is one of the common types of industrial wastes, and may commonly be combined with other types of wastes before disposal.

Washdown Water. Water for floor washing is often highly polluted from acids, solid wastes, and other substances. This waste is also often combined with other types of waste prior to disposal.

Storm Runoff. Precipitation on uncovered surfaces at industrial sites may pick up significant levels of pollutants. This water may be gathered and disposed of in a percolation pond, dry stream bed, disposal well, or used for irrigation. The quality largely depends on housekeeping procedures of the industry and the type of material that may be spilled or leaked at the land surface.

Stockpiles. The piling of solid raw materials, products, byproducts, and waste materials on the ground without protection from precipitation can cause groundwater pollution.

Water Treatment Plant Effluent. Varying amounts of treatment may be required for various industries depending on the quality of source water. Water treatment commonly involves filtration and clarification, water softening, and demineralization or deionization. Disposal is frequently by percolation ponds or disposal wells. This waste may present a substantial pollutant load in the case of waste brines.

Tank and Pipeline Leakage. One of the sources of groundwater pollution currently receiving considerable attention is small leaks in tanks and pipelines.

Hydrocarbon Storage. Facilities for hydrocarbon storage or disposal are often constructed so as to permit inadvertent groundwater pollution.

The disposal of municipal and industrial liquid wastes on the ground surface involves not only a wide range of potential groundwater pollutants but also several methods of release of effluents. The topic can be discussed under three basic techniques of disposal — into land depressions, by spraying, and into stream beds.

Natural or artificial depressions in the ground surface for waste disposal are usually referred to as lagoons, basins, or pits. These may be lined or unlined, but in either case, leakage to groundwater is a possibility. These sturctures are intended to serve any of a variety of purposes; to name a few, storage, processing, waste treatment, cooling, evaporation, and disposal. It is now widely recognized that many so-called "evaporation" ponds are in reality essentially percolation ponds. The once common practice of discharging oilfield brines into open pits is now largely past as a result of State regulatory practices; however, the effects on groundwater quality remain in many localities.

The type of pollution created clearly depends upon the type of waste material, its volume and concentration, the soil and aquifer conditions, and the location of groundwater. Sewage disposal into a lagoon usually results in the removal of bacteria and suspended organic matter after only a few feet of movement underground; however, dissolved organic matter can persist for longer distances. Removal of pollutants may be minimal for percolation through cavernous or highly fractured rocks. The increase in mineral content of groundwater can be considerable, as shown by Table 2-7. Urban runoff collected in storm sewers can contribute BOD, COD, nitrate, lead, gasoline, oil, and grease, among other pollutants.

Table 2-7

NORMAL RANGE OF MINERAL PICKUP IN
DOMESTIC SEWAGE (Todd, 1970)

Mineral Constituents	Normal Range in mg/l (except as noted)
Total Dissolved Solids	100-300
Boron	0.1-0.4
Percent Sodium	5-15%
Sodium	40-70
Potassium	7-15
Magnesium	15-40
Calcium	15-40
Total Nitrogen	20-40
Phosphate	20-40
Sulfate	15-30
Chloride	20-50
Total Alkalinity	100-150

Effluents percolating from industrial lagoons or basins have a greater potential to degrade groundwater from a constituent standpoint than does domestic sewage effluent, but not volumetrically. These industrial liquids may contain brines, arsenic compounds, heavy metals, acids, petroleum products, phenols, radioactive substances, as well as other miscellaneous chemicals (Todd and McNulty, 1974; Meyer, 1973). And some of these materials are capable of traveling long distances in groundwater.

Spray irrigation has become a common practice for disposal of waste water by fruit and vegetable processors and of treated sewage by municipalities. Studies of the effect of spray effluents on groundwater have indicated that vegetation often takes up a substantial fraction of the nitrogen as compared to percolation beneath a bare soil. Thus, the adverse effect on groundwater quality may be limited to an increase in total dissolved solids. Compared to lagoons, however, spray irrigation differs in that it provides a greater opportunity for waste water to contaminate surface runoff during rainfall periods (Meyer, 1973).

Because spraying is basically a method of wastewater disposal, nutrient removal and water reclamation, it can be expected to increase in the future. The practice should be limited to predominantly organic wastes that can be readily assimilated in the soil zone so that any effect on quality of groundwater is minimized.

Disposal of partly treated sewage and industrial wastes into the beds of intermittent and ephemeral streams is practiced mainly in the arid and semi-arid regions of the Southwestern United States (Meyer, 1973). Storm-water runoff is also similarly discharged in many parts of the country. The benefit of these procedures is the replenishment of groundwater resources.

Pollutants that enter an aquifer beneath a stream bed depend on the character of the wastes, the type of stream-bed material, the depth to the water table, and the type of treatment given to the wastes. For domestic wastes the potential pollutants include chloride, organic compounds, nitrogen compounds, phosphates, boron, synthetic detergents, bacteria, viruses, and perhaps pesticides. For industrial wastes, the spectrum of possible pollutants becomes much broader, in general, those of greatest concern include heavy metals and organics such as phenols and polychlorinated biphenyls.

Sewer Leakage
Conceptually, a sanitary sewer is intended to be watertight, and thus to present no hazard to groundwater quality. In reality, however, leakage, especially from old sewers, is a common occurrence. Leakage from gravity sewers may result from poor workmanship, defective sewer pipe sections, breakage by tree roots or other causes, ruptures from heavy loads, rupture by soil slippage, fractures by seismic activity, loss of foundation support due to washouts, shearing due to differential settlement at manholes, and infiltration causing sewage to flow into abandoned sewer laterals (Meyer, 1973).

Sewer leakage releases raw sewage into the ground, often close to groundwater. This can introduce high concentrations of BOD, COD, chlorides, unstable organics, and bacteria into groundwater. On the other hand suspended solids tend to clog sewer cracks, and the surrounding soils tend to become clogged due to anaerobic conditions; therefore, the actual effect of sewer leakage may be less than the theoretical potential.

Tank and Pipeline Leakage
Underground storage and transmission of a wide variety of fuels and chemicals is a common practice for commercial, industrial, and individual uses. These tanks and pipelines are subject to structural failures from several causes, and the subsequent leakage becomes a source of groundwater pollution.

Petroleum and petroleum products, because of their large usage, are responsible for most of the pollution. Thus, 90 percent of the interstate liquid pipeline accidents reported in 1971 involved crude oil, gasoline, liquefied propane gas, or fuel oil (Meyer, 1973). Major causes of these pipeline leaks were corrosion, equipment rupturing the pipelines, defective pipe seams, or incorrect operations by handling personnel.

Leakage is particularly frequent from small installations such as home fuel oil tanks and gasoline stations, where installation, inspection, and maintenance standards may be low. In Maryland some 60 instances of groundwater pollution were reported in a single year from gasoline stations, and in northern Europe, where most homes are heated by oil stored in subsurface tanks, oil pollution has become the major threat to groundwater quality (Meyer, 1973).

Liquid radioactive wastes are sometimes stored in underground tanks. Leakage has been reported from an installation at Hanford, Washington.

Leakage of an immiscible liquid such as oil into the ground will cause the oil to move downward in relatively permeable soils or, if the leak is from a pipeline in relatively impermeable soil, the oil will tend to remain in the trench and move in the down-slope direction. Oil coats soil particles as it advances, therefore, if the quantity of leaked liquid is sufficiently small, the total flow may become immobilized. Subsequent infiltrating water will tend to transport the pollutant from the soil particles downward to the water table. Once oil reaches a water table it spreads to form a thin layer on top of the water table and then to migrate laterally with the groundwater body.

Disposal Wells

Many thousands of wells throughout the United States are used for disposal of pollutants into freshwater aquifers. Examples include electronic industries disposing of metal-plating wastes in Arizona; domestic sewage disposal from individual homes in Florida and Texas; low-level radioactive wastes at one location in Idaho; and heated water from cooling systems in New York, California, and several Midwestern States (Meyer, 1973).

In recent years considerable attention has been given to the possibility of injecting treated municipal sewage into wells penetrating freshwater aquifers. Several of the proposed schemes not only would solve a sewage-disposal problem but also would help to recharge freshwater aquifers or to establish hydraulic barriers against saltwater encroachment in freshwater aquifers. Advanced pilot plant experiments are being conducted in Long Island and in California (Meyer, 1973). Sewage must be given at least secondary treatment and preferably tertiary treatment to prevent clogging of the disposal wells and to reduce or prevent significant chemical and bacteriological pollution of the aquifer.

Modification of the quality of groundwater caused by subsurface disposal of wastes through wells depends on a variety of factors, including the composition of the native water, the amount and composition of the injected waste fluid, the rate of injection, the permeability of the aquifer, the type of well construction, and the kinds of biological and chemical degradation that may occur. Most wells used for disposal of polluted liquids are, for economic reasons, located in the shallowest available aquifer.

Injection Wells

Deep injection wells are employed for disposal of industrial wastes and oilfield brines. At present there are fewer than 300 industrial disposal wells in the United States (Meyer, 1973), but there are more than 70,000 brine disposal wells.

It has been common practice in the past to use abandoned oil production wells for brine disposal. Because the wells were not designed, cased, or cemented for brine injection, there have been numerous instances of injection wells with undetected ruptures beneath the surface, where indicated brines have seeped into freshwater aquifers for many years before being discovered. Some State regulatory agencies have alleviated this problem by requiring an injection tubing inside a casing. The space between the tubing and casing is filled with a fluid which is monitored to detect ruptures.

Currently, most oilfield brines are returned to subsurface formations either for secondary recovery in an oil-producing formation or just as a disposal method. However, even with properly designed and constructed injection wells, brine disposal presents pollution problems because of the numerous oil, gas, injection, and test holes, which for many years were simply abandoned without proper plugging. Unplugged wells provide vertical pathways for injected brines to rise into overlying freshwater aquifers.

Because of the limited number of industrial waste injection wells and the careful attention given to their siting, design, and operation by State regulatory agencies, there have been few reported cases of groundwater pollution from this source; however, the potential for this to occur at distances from such wells should not be minimized.

Stockpiles

Stockpiles of solid materials are often seen around industrial plants and at construction sites. These may be raw materials awaiting use, or they may be solid wastes placed for temporary or permanent storage. Precipitation falling on unsheltered stockpiles causes leaching to occur which may then transport heavy metals, salt, and other inorganic and organic constituents as pollutants to the groundwater. Only by storing solid wastes in bins or shelters can this source of groundwater pollution be effectively controlled.

Mining Activities

A variety of groundwater pollution problems can be associated with mining activities. Mines fall into two basic categories — surface mines and underground mines. Pollution effects depend on the material being extracted and the milling process: coal mines are a major contributor; metallic ores for production of iron, copper, zinc, and lead are of secondary importance; while stone, sand, and gravel quarries are numerous but chemically much less important.

Both surface and underground mines invariably extend below the water table so that dewatering to further mining activity is common. The water so pumped, either directly from the mine or from nearby wells, may be highly mineralized and is frequently referred to as "acid mine waters." Although there is no typical analysis of mine drainage water, normal characteristics include low pH, high acidity, high ferrous or ferric iron, high aluminum, and high sulfates (Miller, 1974). The magnitudes of the various constituents vary with the extent of oxidation and/or neutralization of the drainage water.

Many economic deposits of coal and other minerals found in bedrock are associated with sulfide minerals, pyrite (FeS_2) being one of the most prominent. Pyrite and most other sulfides are stable under the conditions that exist below the water table. However, if the water table is lowered, oxidation of the sulfides occurs in the dewatered zone. Oxidation of pyrite followed by contact with water produces ferrous sulfate ($FeSO_4$) and sulfuric acid (H_2SO_4) in solution. This situation may arise where a mine is abandoned and dewatering activities cease or by downward percolating rainwater. The net result is that this solution is introduced into the groundwater system, causing a drop in pH, and a rise in sulfate and iron content (Miller, 1974).

In the mining and milling of ores groundwater pollution can result from a variety of processes. With copper ore, for example, reagents, including lime, arsenic, cyanide, kerosene, and organic materials, are added during milling. The crushing of sulfide-related ores can release copper sulfate, molybdenum, potassium, and high total dissolved solids. The leaching of ores in oxide form with acids can contribute iron, manganese, low pH, copper, molybdenum, and high total dissolved solids. The use of explosives in fracturing overburden or ore has potential to introduce ammonium nitrate to groundwater.

Pollution of groundwater can also result from the leaching of old mine tailings and from settling ponds. Uranium mill tailings are radioactive so that this could become a source of localized pollution. Serious pollution problems can be associated with both active and abandoned mines.

Oilfield Brines

The production of oil and gas is usually accompanied with the production of waste water in the form of brines. The ratio of brine to oil or gas varies with location and with the age of a well. A recent study in the Southcentral States (Scalf et al., 1973) reported about three barrels of brine were produced for each barrel of oil. The chief constituents of brine typically include sodium, calcium, ammonia, boron, chloride, sulfate, trace metals, and of course, high total dissolved solids.

Until recently the usual methods of oilfield brine disposal consisted of discharge to streams or to "evaporation" ponds. In both cases brine-polluted aquifers became commonplace in oil production areas as infiltrating water from streams and ponds moved to the groundwater. Thousands of unlined brine pits were in use in the Southcentral States until only a few years ago, when they were prohibited by the oil regulatory agencies of the various States. Despite this ban and the fact that few of the brine-affected areas have been mapped, enough have been located to indicate a serious groundwater pollution problem for many years to come (Todd and McNulty, 1974; Scalf et al., 1973). Because of the slow movement of groundwater, brine-polluted groundwater is only now being discovered in many areas of oil development abandoned 20 or 30 years ago.

With the ban on unlined evaporation pits, oil companies were forced to construct pits lined with impervious material or to inject brines back into the oil-bearing formation or into another formation sufficiently removed from freshwater aquifers to prevent contamination (see previous subsection on "Injection Wells"). Nevertheless, there are numerous reported and suspected violations of these regulations, primarily in the form of bypassing brine pits, by accidental or deliberate rupture of pit liners, by overflowing waste pits, or by leakage from broken lines.

In oil-producing areas brines represent one of the major causes of groundwater pollution. Case studies of the problem have been extensively documented (Todd and McNulty, 1974; Scalf et al., 1973).

GROUNDWATER BASIN MANAGEMENT
Saline Water Intrusion

Saltwater may invade freshwater aquifers to create point or nonpoint sources of pollution. In coastal aquifers seawater is the pollutant; in inland aquifers any of several sources of saline water may be responsible.

Under natural conditions fresh groundwater in coastal aquifers is discharged into the ocean. If, however, localized pumpage becomes sufficiently large, the seaward flow is decreased or even reversed, thereby causing seawater to advance inland within the aquifer. Almost all of the coastal States of the United States have some coastal aquifers polluted by the intrusion of seawater (Meyer, 1973). Florida is the most seriously affected State, followed by California, Texas, New York, and Hawaii.

The usual cause of seawater intrusion in coastal aquifers is overpumping. In flat coastal areas, drainage channels or canals can also cause intrusion. They can contribute to the problem in two ways — only by reducing the water table elevation and its associated freshwater flow, and the other by permitting seawater during periods of high tide to advance long distances inland where it can infiltrate into the ground. On oceanic islands freshwater from rainfall percolates to form a lens overlying seawater. If a well penetrating the lens is pumped at too high a rate, the underlying seawater will rise and pollute the water supply of the well.

At the boundary between freshwater and seawater underground, dispersion and diffusion, together with external influences such as recharge from precipitation, pumping of wells, and tidal action, combine to create a transition zone of brackish water. Figure 2-27 shows a vertical cross-section of a coastal aquifer together with the transition zone and associated circulations.

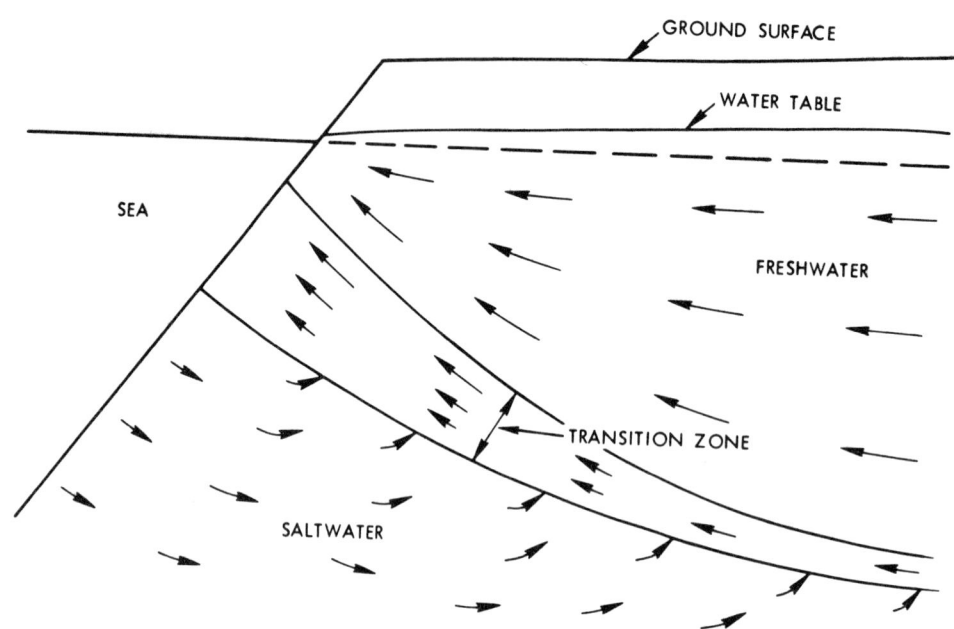

Figure 2-27. Schematic vertical cross section through a coastal aquifer showing freshwater and seawater circulations with a transition zone (Meyer, 1973).

In inland aquifers saline water can be found at increasing depths from seawater which entered aquifers during deposition or during a high stand of the sea in geologic time. Approximately two-thirds of the United States is underlain in part by saline groundwater, often referred to as connate water. In addition, localized saline zones may occur from unique geologic formations or topographic features or from saline wastes contributed by man.

The most common mechanism of intrusion is that caused by overpumping of a well which causes an upconing of underlying saline water toward the well. This upward movement is illustrated in Figure 2-28. Where freshwater and saline aquifers are connected, a lowering of the water table can induce an upward movement of saline water. Thus, dewatering operations, as for quarries, roads, or excavations, or the dredging of a stream channel can contribute to the encroachment of saline water into aquifers.

Aquifers Interchange Through Wells

Because wells form highly permeable vertical connections between aquifers, they can serve as important means for groundwater pollution. This usually happens, however, only where attention is not given to the proper construction, sealing, or abandonment of wells.

Pollution occurs where well screens, perforated casing, or an open borehole interconnects two separate aquifers, or where the surface casing has not been

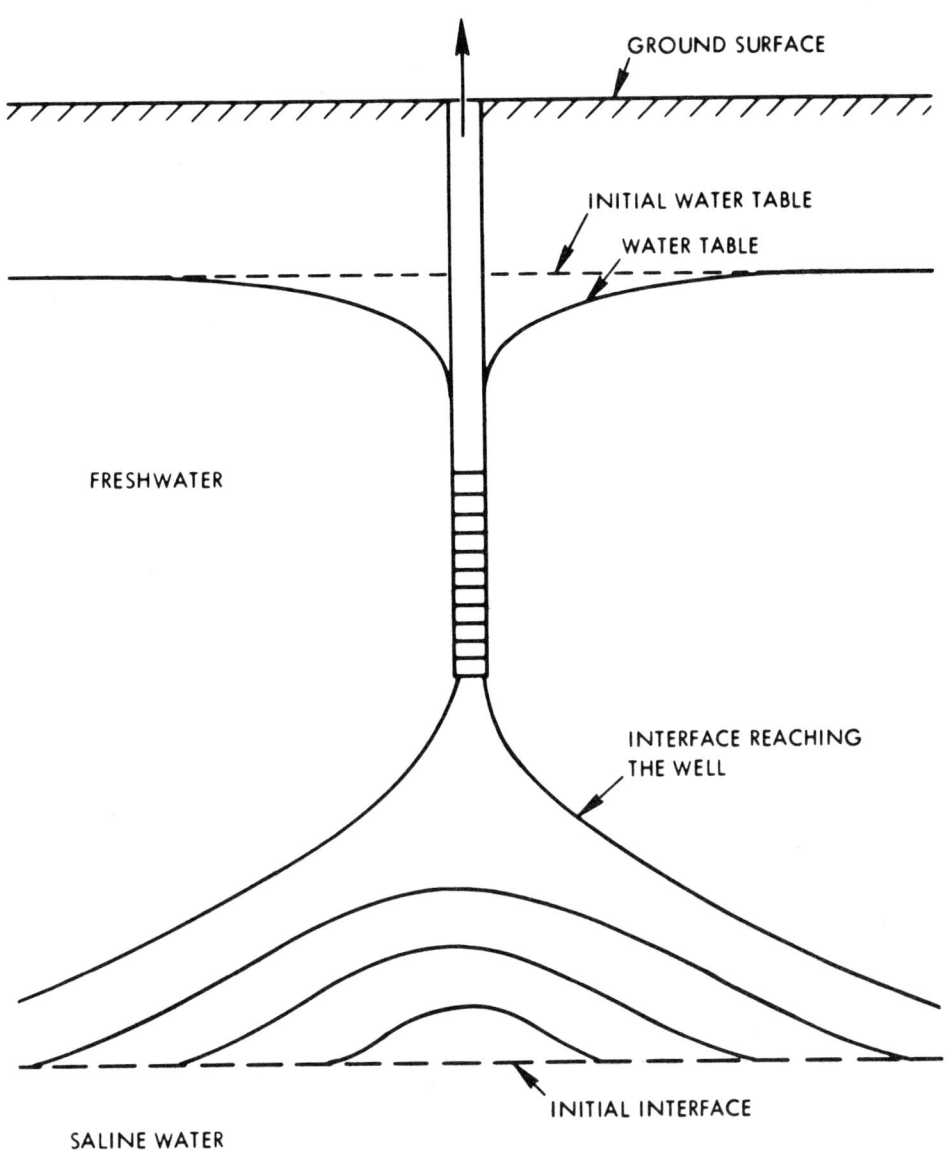

Figure 2-28. Schematic diagram of upcoming and underlying saline water to a pumping well.

adequately sealed. In these instances water wells can serve as mechanisms for transmission of pollutants from one aquifer to another, or from land surface to an aquifer. Interaquifer exchange occurs where there are vertical differences in hydraulic head between aquifers.

A problem of this type is the vertical movement of saline water into a freshwater aquifer. Abandoned and corroded well casings allow the saline water to enter either from an overlying or underlying saline water aquifer, or from an adjacent saline surface water body. The problem is well illustrated by situations in Baltimore, Maryland, (Miller, 1974) and Alameda County, California (Meyer, 1973).

If a well is improperly sealed in the annular space between the casing and the borehole, and connects aquifers of different water quality, polluted water can move along the exterior of the well casing and enter the well. In addition, if provisions are not made to divert surface water away from a well, it can drain downward into a well.

Most public water supply wells are properly sealed, inspected, and maintained, but surveys by health authorities have revealed that private wells serving individual residences often are not protected against pollution from overland runoff containing septic fluids, barnyard wastes, or storm waters.

In most States today, regulations exist requiring plugging of abandoned wells. But there are many thousands of abandoned wells which remain unplugged either because they cannot be located or because changes in property ownership with time make the responsibility for plugging indeterminate. The problem is particularly serious in oil and gas areas where numerous exploration and production wells have been abandoned (Scalf et al., 1973). These holes serve as avenues for brine from injection wells to rise into overlying freshwater aquifers.

MISCELLANEOUS
Spills and Surface Discharges

Groundwater pollution can result from hazardous and nonhazardous liquids that are discharged onto the ground surface in an uncontrolled manner and then seep into the underlying soils. If the volume of the fluid is sufficiently large, the pollutant can migrate down to the water table and degrade the groundwater quality. Any of a variety of activities can lead to spills and surface discharges that may serve as pollution sources.

Poor housekeeping at large industrial plants and airports is a contributory pollution cause. At industrial sites, causal activities may include boilovers and blowoffs, overpumping during transfer of liquids to or from storage and carriers, leaks from faulty pipes and valves in product distribution systems, and poor control over waste discharges and storm-water runoff. At airports the washing of planes with solvents and spills of fuel can form an extensive body of hydrocarbons floating on the water table (Miller, 1974).

Pollution of groundwater also occurs from the intermittent dumping of fluids on the ground, especially at gasoline stations and other types of small commercial establishments. Automotive waste oil is disposed of on the ground by car owners, by commercial garages and gasoline stations, and at construction sites. The total of these many small contributions runs to millions of gallons of oil annually (Miller, 1974). Small industries often dispose of lubricating, hydraulic, and cutting oils by local dumping. It is not uncommon to find small commercial facilities discharging liquid wastes onto undeveloped land, the reasoning being that it is uneconomic to store and to haul the wastes to municipal treatment plants or landfills, and that the liquids may be harmful to local septic tanks or cesspools.

Finally, accidents involving aboveground pipes and tanks, railroad cars, and trucks can cause the release of large quantities of a pollutant at a particular site. The use of water to flush spilled fluids, as from a highway, may actually aid in transporting the pollutant down to the water table.

Septic Tanks and Cesspools

Of all the sources and causes of groundwater pollution, the most numerous and widely distributed is that of septic tanks and cesspools. It is estimated that approximately 40 million persons, or nearly 20 percent of the total population, are served by individual household waste water treatment systems. This means that some 2.5 billion gallons of partially treated sewage is discharged from residences directly into the underground every day. In addition, stores, laundries, small office buildings, hospitals, and industries employ septic tanks in areas where community sewer systems are not available. Furthermore, innumerable summer cabins, Forest Service campgrounds, and organized group camps depend upon subsurface disposal of waste water, primarily during the summer season.

The heaviest concentrations of septic systems are to be found in the suburban subdivisions that developed on the fringes of major cities after World War II. A septic tank is a watertight basin intended to separate floating and settleable solids from the liquid fraction of domestic sewage, and to discharge this liquid, together with its burden of dissolved and particulate solids, into the biologically active zone of the soil mantle through a subsurface percolation system such as a tile field, a seepage bed, or an earth-covered sand filter (Meyer, 1973). A cesspool is a large buried chamber which is walled with a porous material, such as concrete blocks, and designed to receive raw sewage. Although new installations of this latter kind are no longer approved, many thousands of cesspools remain in operation in the United States today wherever soil conditions are favorable.

Domestic sewage adds minerals to groundwater, as indicated in Table 2-7. Bacteria and viruses are normally removed by the soil system. Phosphorus is generally retained by the soil, but significant quantities of nitrogen can, depending upon local soil and vegetation conditions, be added to groundwater. The degree of groundwater pollution has been shown to be related to the density of septic tank installations (Miller, 1974) and to the local hydrogeologic framework.

Highway Deicing
A recent and unique problem that has attracted considerable attention is the pollution of groundwater resulting from application of deicing salts to streets and highways in winter. The region most largely affected includes the Northeastern and Northcentral States. The salt reaches the groundwater from both surface stockpiles and solution of salt that has been spread on roadways.

Salt application quantities by State highway departments of the Northeastern States range from 3 to 20 tons per single-lane mile during a single winter season (Miller, 1974). More than 95 percent of this is sodium chloride, the remainder being calcium chloride. Furthermore, the demand to maintain highways and roads for vehicular use in winter has caused a steady growth in the use of salt for deicing. Data from the State of Massachusetts shows that applied salts increased 700 percent in the 15-year period from 1955 to 1970.

Widespread and long-term degradation of groundwater quality has been the experience with highway deicing salts (Meyer, 1973). In addition, casings and screens of wells have been corroded, necessitating replacement of many wells. The gradual increase in salts has led to some town groundwater supplies exceeding salt limits for persons on low-sodium diets (Todd and McNulty, 1974). Salt concentrations have been found to be highest in areas where salting has been practiced longest, in wells closest to roadways, and during April, the month of greatest snow melt.

QUALITY IN RELATION TO WATER USE

WATER QUALITY STANDARDS
The quality of groundwater is most often evaluated relative to its existing or potential use. Because water quality criteria vary with the type of use, a groundwater may be satisfactory for one use but not for another. Thus, pollution, which is the

manmade degradation of water quality, may or may not restrict a given groundwater for a particular use. On the other hand, increasing pollution leads to increasing impairment for use of water.

The concept of water quality criteria, or standards, can be illustrated by reference to the diagram in Figure 2-29. Here pollution is related to impairment for use. Up to a threshold level of pollution, water is satisfactory for a given use (such as irrigation, drinking, etc.). This level then defines a water quality criteria or standard. With increasing pollution above this level, a relatively narrow cautionary band can be defined. But this terminates at the limiting level of pollution, beyond which increases in pollution constitute a danger for that particular use. Another way of stating this relationship is that pollution up to the threshold level may be regarded as a reasonable impairment, whereas pollution beyond the limiting level becomes an unreasonable impairment. A pollutant may here be defined as any chemical constituent, any biological organism, or any physical characteristic which can adversely affect the use of the water.

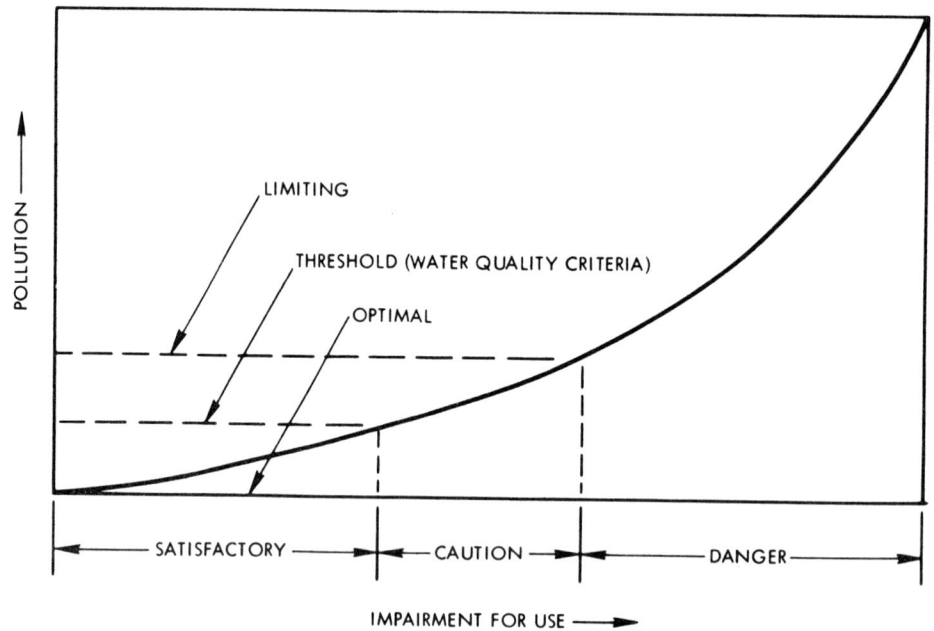

Figure 2-29. Diagram illustrating the relation of water pollution to impairment for a given water use (modified after McGauhey, 1968a).

Standards or guidelines have been established for all types of water use. Those applicable to the major uses of groundwater are briefly described in the following subsections.

Drinking Water

The most important use of groundwater is for drinking water purposes. Drinking water standards have in the past been set by the U.S. Public Health Service in the United States, and most States have adhered closely to these requirements. The standards are summarized in Table 2-8. These criteria are under continual review so that new constituents and revised levels of constituents are to be expected with time.

Table 2-8

DRINKING WATER STANDARDS OF THE U.S. PUBLIC HEALTH SERVICE (McGauhey, 1968b)

PHYSICAL STANDARDS	
	Units
Turbidity	5
Color	15
Threshold Odor Number	3

CHEMICAL STANDARDS		
Substance	Recommended Limits of Concentrations, in mg/l	Mandatory Limits of Concentrations, in mg/l
Alkyl Benzene Sulfonate (ABS)	0.5	---
Arsenic (As)	0.01	0.05
Barium (Ba)	---	1.0
Cadmium (Cd)	---	0.01
Carbon Chloroform Extract (CCE)	0.2	---
Chloride (Cl)	250	---
Chromium (hexavalent) (Cr^{+6})	---	0.05
Copper (Cu)	1.0	---
Cyanide (CN)	0.01	0.2
Fluoride (F)	1.7†	2.2†
Iron (Fe)	0.3	---
Lead (Pb)	---	0.05
Manganese (Mn)	0.05	---
Nitrate (NO_3)*	45	---
Phenols	0.001	---
Selenium (Se)	---	0.01
Silver (Ag)	---	0.05
Sulfate (SO_4)	250	---
Total Dissolved Solids (TDS)	500	---
Zinc (Zn)	5	---

*In areas in which the nitrate content of water is known to be in excess of the listed concentration, the public should be warned of the potential dangers of using the water for infant feeding.
†Varies with average maximum air temperature.

BIOLOGICAL STANDARDS	
Sample Examined	Limits
Standard 10-ml portions	Not more than 10 percent in one month shall show coliforms.†
Standard 100-ml portions	Not more than 60 percent in one month shall show coliforms.†

†Subject to further specified restrictions.

RADIOACTIVITY STANDARDS	
Source	Recommended Limits, Picocuries per Liter
Radium-226	3
Strontium-90	10
Gross Beta Activity	1,000

Irrigation Water

The suitability of groundwater for irrigation depends upon the effects of chemical constituents in the water on the plant and the soil (Todd, 1959). Salts can restrict plant growth physically by modifying osmotic processes and chemically by causing undesirable metabolic reactions. Salts can produce changes in soil structure, permeability, and aeration.

Specific limits of permissible salt concentrations for irrigation water cannot be stated because of variations in salinity tolerance among different plants, as well as in soil type, climatic conditions, and irrigation practices. Quality classifications of water for irrigation usually stress certain ranges for sodium, total dissolved solids, and boron. A general guide for evaluating the quality of water used for irrigation is shown in Table 2-9.

Livestock Water

Poultry and farm animals can live on water of considerably lower quality than human beings. Quality criteria depend on factors such as the type of animal and its age, climate, and feeding regimen. A general guide for evaluating the quality of water used by livestock is shown in Table 2-10.

Table 2-9

GUIDE FOR EVALUATING THE QUALITY OF WATER USED FOR IRRIGATION (Todd, 1970)

[MPN is most probable number. Sodium adsorption ratio is defined by the formula SAR = $Na\sqrt{(Ca + Mg)/2}$, where the concentrations are expressed in milliequivalents per liter. Residual sodium carbonate is the sum of the equivalents of normal carbonate and bicarbonate minus the sum of the equivalents of calcium and magnesium.]

Quality Factor	Threshold Concentrations*	Limiting Concentration†
Coliform Organisms, MPN per 100 ml	1000‡	§
Total Dissolved Solids (TDS), mg/l	500‡	1500‡
Electrical Conductivity, μmhos/cm	750‡	2250‡
Range of pH	7.0-8.5	6.0-9.0
Sodium Adsorption Ratio (SAR)	6.0‡	15
Residual Sodium Carbonate (RSC), meq	1.25‡	2.5
Arsenic, mg/l	1.0	5.0
Boron, mg/l	0.5‡	2.0
Chloride, mg/l	100‡	350
Sulfate, mg/l	200‡	1000
Copper, mg/l	0.1‡	1.0

*Threshold values at which irrigator might become concerned about water quality and might consider using additional water for leaching. Below these values, water should be satisfactory for almost all crops and almost any arable soil.
†Limiting values at which the yield of high-value crops might be reduced drastically, or at which an irrigator might be forced to less valuable crops.
‡Values not to be exceeded more than 20 percent of any 20 consecutive samples, nor in any three consecutive samples. The frequency of sampling should be specified.
§Aside from fruits and vegetables which are likely to be eaten raw, no limits can be specified. For such crops, the threshold concentration would be limiting.

Table 2-10

GUIDE FOR EVALUATING THE QUALITY OF WATER USED BY LIVESTOCK (Todd, 1970)

Quality Factor	Threshold Concentration*	Limiting Concentration†
Total Dissolved Solids (TDS), mg/liter	2500	5000
Cadmium, mg/l	5	
Calcium, mg/l	500	1000
Magnesium, mg/l	250	500‡
Sodium, mg/l	1000	2000‡
Arsenic, mg/l	1	
Bicarbonate, mg/l	500	500
Chloride, mg/l	1500	3000
Fluoride, mg/l	1	6
Nitrate, mg/l	200	400
Nitrite, mg/l	None	None
Sulfate, mg/l	500	1000‡
Range of pH	6.0-8.5	5.6-9.0

*Threshold values represent concentrations at which poultry or sensitive animals might show slight effects from prolonged use of such water. Lower concentrations are of little or no concern.
†Limiting concentrations based on interim criteria, South Africa. Animals in lactation or production might show definite adverse reactions.
‡Total magnesium compounds plus sodium sulfate should not exceed 50 percent of the total dissolved solids.

SECTION 3 – MONITORING METHODOLOGY

CONCEPT OF A MONITORING METHODOLOGY

The basic purpose of a methodology for monitoring groundwater pollution is to provide a framework for the planning and development of a monitoring program. The methodology should serve two roles simultaneously. First, it should assist a local designated monitoring agency (DMA) to design and implement a monitoring program. The second role should be to guide governmental agencies at the State and national levels in establishing realistic monitoring priorities, not only in terms of what should be monitored and where, but also in terms of timing and funding.

The monitoring methodology described in the following sections is predicated upon the technical effort being organized and conducted by or under the direct supervision of personnel with professional training in water resources engineering or groundwater geology. This requirement is essential in view of the fact that it is infeasible to summarize the full background of hydrogeology, involving the occurrence, distribution, and movement of groundwater, as well as its geochemistry, which would be needed to establish a successful monitoring program. Section 1 of this chapter has been prepared specifically for hydrologists and geologists not familiar with groundwater quality problems. It is recommended that these individuals review Section 1 prior to using this section.

Various personnel arrangements can be anticipated for implementing the monitoring methodology. Indicative of the possibilities are the three following situations:

- In an urbanized local area the DMA might possess a sufficiently large technical staff so that the organization and conduct of the monitoring program could be handled entirely by this in-house group. Similarly, for a relatively undeveloped area, where the monitoring program would be small, a single technically competent member of the DMA might personally supervise the program.

- Where the staff of a DMA is inadequate, responsibility for the monitoring program could be subcontracted to a firm of consulting engineers or geologists specializing in groundwater. An arrangement of this sort could function either on a continuing basis or until an adequate in-house staff became available.

- With the passage of time and the full development of a monitoring program, surveillance of existing and potential sources of pollution will assume a standardized routine. Thereafter, attention will tend to be focused primarily on the monitoring of new developments, such as landfills, feedlots, industrial plants, subdivisions, etc. Because plans for each new development will need to be reviewed and approved by the DMA, it follows that inclusion of an adequate monitoring system will become a contingency for approval. Under these circumstances much of the work required to extend the monitoring program for these new developments will be completed by the engineering firms responsible for their design.

Finally, it should be understood that the methodology presented herein must of necessity be somewhat generalized. There is an infinite number of combinations of pollution causes, hydrogeologic situations, and monitoring methods, among other variables, that can govern the implementation of a monitoring program. Therefore, persons involved in a monitoring program will be required to exercise professional judgment in order to interpret and apply this methodology to the specific local situations which they encounter. Examples are given for illustrative purposes and case histories are presented by in Chapter VI.

IMPLEMENTATION OF A MONITORING METHODOLOGY

The following material describes procedures for implementing a groundwater pollution monitoring program. These apply to a specified local area under the jurisdiction of a DMA. The procedures are described as a series of steps arranged in chronological order. In practice, however, activities of different steps will overlap in order to make efficient use of personnel and time.

The steps constitute a series of monitoring objectives for a DMA. Taken together they constitute a monitoring methodology which can assist a DMA at any location in the United States to initiate its groundwater pollution monitoring program.

STEP 1 – SELECT AREA OR BASIN FOR MONITORING

The selection of areas to be monitored will be made within a State by the appropriate State water pollution control agency that, in cooperation with the EPA, carries out the mandates of PL 92-500 and PL 93-523. The basis for selecting areas will be governed, in general, by a combination of administrative, physiographic, and priority considerations. Each of these factors will be reviewed in the following paragraphs.

Administrative Considerations

The initiation of a monitoring program requires that a local DMA be specified. In many situations the requisite agency with the necessary technical staff may be a county, district, or regional water organization. Thus, the area to be monitored can often be made to correspond to the jurisdictional area of the DMA. The size of a particular area may vary from a few square miles to thousands of square miles. Size alone is less important than the ready accessibility of all portions of the area to the DMA as well as hydrogeologic knowledge of the area possessed by the DMA.

It should be recognized that political boundaries frequently create problems in terms of water management. Such a boundary may cross a major groundwater basin so that, for example, pollutants from an adjoining area may be entering from sources not subject to monitoring by the DMA. Clearly, such situations should be minimized as much as possible. Alternatively, cooperation among DMA's sharing common groundwater pollution problems will be essential to the success of their respective monitoring programs.

Physiographic Considerations

The physiographic basis for selecting monitoring areas recognizes that groundwater basins are distinct hydrographic units containing one or more aquifers. Such basins usually, but not always, coincide with surface water drainage basins. By establishing a monitoring area related to a groundwater basin, total hydrologic inflows to and outflows from the basin are fully encompassed. This permits all pollution sources and their consequent effects on groundwater quality to be monitored by a single DMA. Where basins are extensive, monitoring areas become too large to be practical. Boundaries should then be drawn parallel to groundwater flows or where cross-flow components are insignificant. Most groundwater basins in the United States have been mapped, based on hydrogeologic investigations, and information is available from State water agencies and/or the U.S. Geologic Survey.

Priority Considerations

It is recognized that establishment of a national program to assess the impact of man's activities on groundwater quality will develop gradually because of administrative, budgetary, and personnel constraints. Since it is the stated intent of the EPA to rely on the States to select the areas to be monitored and to conduct the appropriate monitoring activities, any national program which evolves will, as a consequence, be built upon the data and information generated by these State monitoring activities.

A first consideration of the State will be to select and prioritize those groundwater aquifers subject to the greatest pollution threat. This first level of prioritization is necessary to provide a starting point for application of the groundwater monitoring methodology. Rarely will sufficient data and information be available from the start to make anything but a gross appraisal of the threat to a State's groundwater resources.

In order for the application of the methodology to be most effective on a spatial basis, areas which have the largest number of identified or potential pollution sources and where there is a high utilization of groundwater should be ranked and sectioned off as areas within which to apply the monitoring methodology. By utilizing the above two criteria in combination with the administrative and physiographic considerations previously set forth, the total area of a State can be divided into areas which may require a monitoring program.

Illustrative of this procedure is an expanded priority scheme for groundwater monitoring developed by the State of California. Here groundwater basins are ranked according to the following five criteria.

- Basin population
- Agricultural use of groundwater
- Estimated usable groundwater capacity
- Availability of an alternate water supply source
- Number of existing quality problems which threaten the groundwater

This list essentially restates the two criteria described above, but attempts to add further insight into water resource management rather than a groundwater pollution monitoring program. This approach developed in California is cited as an example only and is not necessarily recommended.

In summary, the administrative, physiographic, and priority considerations provide a rationale for selecting monitoring areas. To be applied within a given State they must be interpreted as guidelines because various special or local conditions will frequently have to be accommodated. Compromises will be inevitable, but careful initial selection of monitoring areas can simplify and make more effective the subsequent monitoring program undertaken by each DMA.

Example – Basin Boundary Area

Figure 2-30 shows the boundaries selected for defining the groundwater basin in the Santa Clara-Calleguas area in southwestern Ventura County, California. The northern and southern boundaries were placed along mountain ridges where groundwater flow would be either nonexistent or parallel to the boundary. The eastern boundary was drawn at the county line, a political boundary, but here this also coincides closely with the drainage divide of the steam systems. The western boundary was the seacoast. The

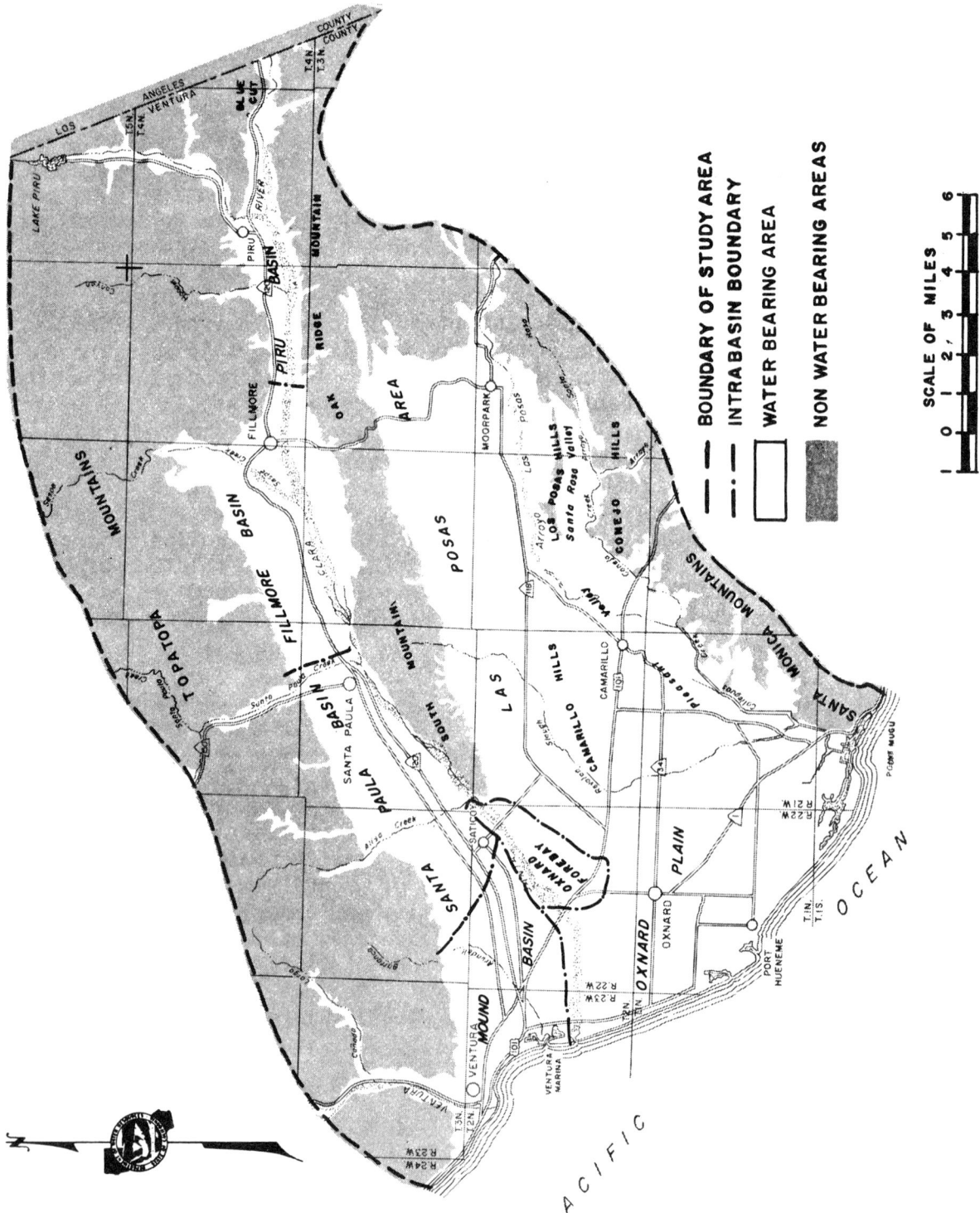

Figure 2-30. Boundaries of the Santa Clara-Calleguas groundwater basin, Ventura County, California (California Department of Water Resources, December 1974).

predominant groundwater flow is in a southwesterly, or down-valley, direction toward the sea, except locally, where major pumping centers cause deviations or reversals of the flow direction.

It should be noted in Figure 2-30 that the water-bearing area occupies some 192,000 acres and that this is only a fraction of the total area, hence monitoring would be restricted to only those portions of the area containing groundwater. Six population centers lie within the area (Ventura, Oxnard, Santa Paula, Fillmore, Piru, and Moorpark), with a total 1970 population of 330,000. The aquifiers are all composed of alluvial materials with unconfined, confined, and multiple types being present. The area of urban land use totals 27,000 acres, while agricultural use covers 106,000 acres. Average annual precipitation is 15.05 inches, which occurs chiefly in winter from North Pacific storms.

STEP 2 – IDENTIFY POLLUTION SOURCES, CAUSES AND METHODS OF WASTE DISPOSAL

The design of a monitoring program requires that the potential sources and causes of groundwater pollution and methods of waste disposal within an area be identified. Groundwater pollution sources can be conveniently placed into six major groups. Municipal, agricultural, and industrial are three major groups. For purposes of monitoring groundwater pollution, oilfield wastes and mining wastes are considered of sufficient importance to also be listed as major groups. The remaining sources are grouped under miscellaneous. Of considerable importance to monitoring efforts is the identification of the type of source, as to whether it is a point, line, or diffuse, source. Several sources have more than one primary disposal method. Table 2-11 summarizes the source and causes of pollution, and common methods of waste disposal, where applicable.

Municipal

Three urban groundwater pollution sources are associated with sewage: sewer leakage, sewage effluent disposal, and sewage sludge disposal. Urban runoff, solid wastes, and lawn fertilizers are other sources of importance. Since septic tanks also occur in rural areas, they are discussed under a separate heading.

- **Sewer Leakage.** One of the more recently recognized sources of groundwater pollution is leakage from sewer lines. Most monitoring programs for detection of sewer leakage are not designed for groundwater pollution evaluations. In the case of leakage from deep sewers, all of the topsoil and a significant part of the vadose zone may be bypassed. Information on the location of sewered areas and major sewers can be obtained from local public works departments, sanitation districts, and regulatory agencies. Pertinent information on the sizes of major sewers, type of pipe and joints, age, pressurization, and leaks should be compiled.

- **Sewage Effluent.** Sewage effluent has been disposed of to groundwater by a number of methods. Crop and forest irrigation, percolation basins, dry stream beds, and disposal wells have been used for disposal of sewage effluent. Information on the location of treatment and disposal facilities, methods of treatment and disposal, and effluent volumes can be obtained from local public works departments, sanitation districts, and regulatory agencies. Percolation basins and dry stream beds used for disposal can be seen on aerial photographs, and crop surveys may indicate lands where sewage effluent is used for irrigation.

Table 2-11

MAJOR SOURCES AND CAUSES OF GROUNDWATER POLLUTION
AND METHODS OF WASTE DISPOSAL

SOURCE	CATEGORY			COMMON METHOD OF DISPOSAL							
	Point	Line	Diffuse	Percolation Pond	Surface Spreading and Irrigation	Seepage Pits and Trenches	Dry Stream Beds	Landfills	Disposal Wells	Injection Wells
Municipal										
Sewer Leakage	X	X		NOT APPLICABLE						
Sewage Effluent	X	X		X	X	X	X	X	X	
Sewage Sludge	X	X			X					
Urban Runoff	X		X	X	X		X	X	X	
Solid Wastes	X				X					
Lawn Fertilizers			X		X					
Agricultural										
Evapotranspiration and Leaching (Return Flow)			X		X					
Fertilizers			X		X					
Soil Amendments			X		X					
Pesticides and Herbicides			X		X					
Animal Wastes (Feedlots and Dairies)	X		X	X	X	X		X		
Stockpiles	X			NOT APPLICABLE						
Industrial										
Cooling Water	X		X	X					X	
Process Waters	X			X	X	X	X		X	X
Storm Runoff	X		X	X	X				X	
Boiler Blowdown	X			X						
Stockpiles	X			NOT APPLICABLE						
Water Treatment Plant Effluent	X			X				X	X	X
Hydrocarbons	X			X					X	
Tanks and Pipeline Leaks	X	X		NOT APPLICABLE						
Oilfield Wastes										
Brines	X	X	X	X	X	X	X	X	X	X
Hydrocarbons	X	X		X					X	X
Mining Wastes	X	X	X	X			X	X	X	X
Miscellaneous										
Polluted Precipitation and Surface Water		X	X	NOT APPLICABLE	X					
Septic Tanks and Cesspools			X	NOT APPLICABLE		X		X		
Highway Deicing		X	X	NOT APPLICABLE						
Seawater Intrusion			X	NOT APPLICABLE						

- **Sewage Sludge.** Sewage sludge is often allowed to dry in open basins from which percolation may occur. Also, the dried sludge is often applied to agricultural lands as fertilizer. Information on the weight of sludge production and method of disposal can be obtained from local public works departments and sanitation districts. Sludge dry beds may be located from aerial photographs.

- **Urban Runoff.** Precipitation falling on paved and other impermeable areas can pick up significant pollutant loads. Much storm water runoff is now being collected, treated, and disposed of separately from sewage. Methods of disposal include percolation ponds, dry stream channels, disposal wells, and irrigation. Information on treatment and disposal methods and volumes of waste water can be obtained from local flood control districts, public works departments, and regulatory agencies.

- **Solid Wastes.** Locations of municipal solid waste disposal sites can be obtained from public works departments, local sanitary districts, and regulatory agencies. Regional solid waste planning documents may also be available. The type of site (sanitary landfill or refuse dump), the type of wastes, the annual volume or weight, and the provisions for control of leachate should be determined.

- **Lawn Fertilizers.** Applications of water and fertilizer in amounts greater than that used by plants in urban areas can produce groundwater pollution. Information on types of fertilizers and amounts is difficult to obtain, because of the diversity of practice from one household to the next. Fertilizer consumption may be estimated from sales records by manufacturers or retailers. Information on irrigation rates would be important but are usually not available.

Agricultural

- **Evapotranspiration and Leaching.** The process of irrigation of agricultural lands results in some water percolating past the root zone. This percolating water is usually degraded with respect to the applied water due to concentration by evapotranspiration, dissolution of mineral matter in the soil, and additives that are applied at the land surface. Fertilizers, soil amendments, and pesticides are considered separately in this section due to their importance in groundwater pollution. Irrigated lands can be identified from aerial photographs or crop surveys, available from the USDA Agricultural Research Service, farm advisors, or irrigation districts. Data on the quality of return flow is rarely available. However, estimates can be made based on the water application rates and consumptive use. The volume of return flow can be calculated from records of precipitation, applied water, and evapotranspiration.

Records of applied water volumes may be obtained from irrigation districts, farmers, power companies, and governmental agencies concerned with water delivery. The calculation of return flow volumes is more fully discussed in Step 7 of this chapter.

- **Fertilizers.** Fertilizers in modern agriculture are applied to almost all crops, whether irrigated or not. Fertilizer application rates usually vary with crop type, soil conditions, and irrigation practice. Local farm advisors can provide information on types and amounts of fertilizers applied. Records of fertilizer sales from manufacturers or retailers can be used to estimate application rates. Regulatory agencies may also have data on fertilizer use.

- **Soil Amendments.** Soil amendments are applied to irrigated lands to alter the physical or chemical properties of the soil. Acid soils may be treated with lime to raise the pH, thus creating more favorable growing conditions for plants. Gypsum or sulfur is widely used to increase infiltration rates on some soils. Irrigation waters in which sodium is the predominant cation may produce dispersion of the soil structure and result in low infiltration rates. The calcium in the applied gypsum replaces sodium in the soil and counteracts this soil clogging. Substantial amounts of these soil amendments may eventually be leached into the groundwater, thereby increasing its salinity. Records of application rates may be obtained from farm advisors and manufacturers.

- **Pesticides and Herbicides.** Pesticides may be significant in agricultural areas as a diffuse source of groundwater pollution. Pesticides, insecticides, herbicides, fungicides, and other chemicals are used to control nuisance organisms. Although many pesticides can be retained in the soil or degraded, some have been found to be readily leached into the groundwater. Farm advisors may have information on the types of pesticides used and application rates. Manufacturers, retailers, and regulatory agencies may have records on amounts sold in certain areas.

- **Animal Wastes.** Livestock wastes constitute a minor groundwater pollution source, except where large numbers of animals are confined within small areas. This situation is usually limited to feedlots and dairies. Dairies present additional problems related to the disposal of wash water. Animal waste disposal by percolation ponds and crop irrigation is common. Feedlots and dairies can be located from land-use maps, zoning maps, and aerial photographs. Information on the number of animals, annual weight of solid waste, and volume of liquid waste can be obtained from farm advisors, dairy and feedlot operators, cattle and dairy associations, and regulatory agencies.

- **Stockpiles.** Unprotected stockpiles in agricultural areas may result in groundwater pollution, particularly where substantial leaching into the soil occurs. These stockpiles include manure; solid chemicals, such as gypsum or sulfur; and miscellaneous waste solids and liquids, such as pesticide containers. Field surveys are often necessary to detect such occurrences.

Industrial

Industrial waste flows are usually related to the amount of raw material processed, to the amount of finished product, or to the number of people employed in the factory. Data on industrial waste flows can be obtained from local industries and regulatory agencies. Stockpile and percolation ponds can be seen on aerial photographs. Data on injection wells have been compiled by the U.S. Geological Survey.

Oilfield Wastes

Brines withdrawn from the ground during oil and natural gas production must subsequently be discarded. Disposal of brines in percolation pounds and injection wells has been widespread. Brines may also be disposed of in dry stream beds, while some brines of low salinity may be used for crop irrigation. Hydrocarbons may be another source of groundwater pollution where wastes are confined in ponds at the land surface. Improperly constructed producing or injection wells can cause significant groundwater pollution. Information on quantity and type of wastes is often available from State oil and gas commissions, regulatory agencies, and local operators.

Mining Wastes

The mining and milling of coal and metallic ores, whether from surface or underground mines, can cause groundwater pollution, therefore, these installations need to be clearly identified if they occur in a monitoring area. Information on mines, their types; their depth; their areal extent; and the amounts and composition of wastes produced, is usually available from a State mining agency, the U.S. Bureau of Mines, mining companies, and regulatory agencies.

Miscellaneous

- **Polluted Precipitation and Surface Water.** Precipitation can be polluted from municipal sources, such as industries and automobile exhaust, as well as agricultural activities, such as the application of ammonia fertilizers. In humid areas, where rainfall is readily leached into the groundwater, pollutants from precipitation may reach the water table. In arid areas, the pollutants in precipitation may accumulate on the land surface and subsequently be leached by irrigation water. Areas where infiltration and percolation to the groundwater may occur from polluted surface water should be identified. Infiltration and percolation rates may be available from hydrogeologic studies by the U.S. Geological Survey, State geological or water agencies, or others.

- **Septic Tanks and Cesspools.** In unsewered areas large numbers of septic tanks and cesspools may be found. Although each is a point source with respect to an individual lot, over large areas septic tanks are diffuse sources of groundwater pollution. Methods of disposal include seepage pits, seepage trenches, and occasionally disposal wells. Location densities, and estimated volume of effluent can be obtained from zoning maps, public works departments, sanitation districts, and regulatory agencies.

- **Highway Deicing.** Sodium chloride and calcium chloride salts are applied to roads to inhibit or to remove ice. Information on the application of salts for deicing of roadways in winter can be obtained from State and local transportation offices. Data should be collected on which roadways are regularly treated with salt and on the estimated application rate, such as tons of salt per lane mile.

The above listing provides an orderly basis for identifying the principal sources and causes of groundwater pollution within a monitoring area. An important footnote is that the listing stresses present-day sources and causes; however, historical sources and causes which are no longer present or active, such as former refuse dumps, brine disposal ponds, and feedlots, among others, may still be responsible for pollution today. This occurs because of the relatively long residence time of pollutants in most groundwater systems, which may be on the order of decades or centuries. It follows, therefore, that while identifying present sources and causes, an effort should also be made to find previous ones that may no longer be visible on the landscape but may be quite evident underground. Examination of old land use or zoning maps and early aerial photographs should prove helpful for this purpose. Furthermore, conversations with long-time residents who are knowledgeable of local activities may reveal information about pollution sources in previous times.

STEP 3 – IDENTIFY POTENTIAL POLLUTANTS

Having identified the pollution sources and the methods of disposal, the next step is to identify potential pollutants for each source. Major determinations in a water analysis are classified into physical, inorganic chemical, organic chemical, bateriological, and radiological (Table 2-12). Table 2-13 summarizes the major sources of groundwater pollution and the types of pollutant present. The trace element portion of the inorganic chemicals has been separated into another category for discussion purposes.

Table 2-12

CLASSIFICATION OF MAJOR POTENTIAL POLLUTANTS
IN GROUNDWATER

Physical	Organic Chemical
Temperature	Carbon
Density	Chlorophylls
Odor	Extractable Organic Matter
Turbidity	Methylene Blue Active Substances
Inorganic Chemical	Nitrogen
Major Constituents	Chemical Oxygen Demand
Other Constituents	Phenolic Material
Trace Elements	Pesticides (Insecticides and Herbicides)
Gases	
Bacteriological	Radiological
Coliform Group	Gross Alpha Activity
Pathogenic Micro-organisms	Gross Beta Activity
Enteric Viruses	Strontium
	Radium
	Tritium

Physical parameters, such as temperature and density, are of most concern in industrial and oilfield wastes. Inorganic chemicals are of primary concern in virtually all wastes. Trace elements are of concern in most categories of sources other than agricultural. Organic chemicals appear to be of concern in wastes from many types of sources. Bacteriological parameters are of primary concern in municipal sources, animal wastes, and in septic tank effluent or cesspool wastes. Radiological parameters are generally only of concern in some industrial and mining wastes. Major attenuation mechanisms can operate on many physical, organic chemical, bacteriological, and radiological constituents in passage through the topsoil and adequate thicknesses of the vadose zone. However, in the case of inorganic chemicals, these attenuation mechanisms are less effective. The attenuation mechanisms are described in more detail in the following steps, however, they have been considered in identifying potential groundwater pollutants.

Table 2-14 represents the inorganic chemical pollutants and gasses common in groundwater or wastes. Pollutants are grouped on the basis of major constituents often found in groundwater, additional constituents of interest in water use, trace elements of interest for drinking water quality, and additional trace elements common in some wastes.

The reports by Todd and McNulty (1974) and Meyer (1973) contain information regarding case histories of groundwater pollution. Data on specific pollutants in the major pollution sources are discussed in the following paragraphs.

Table 2-13

MAJOR SOURCES OF GROUNDWATER POLLUTION
AND TYPES OF POLLUTANTS

SOURCE	TYPE OF POLLUTANT					
	Physical	Inorganic Chemical	Trace Elements	Organic Chemical	Bacteriological	Radiological
Municipal						
Sewer Leakage	Minor	Primary	Secondary	Primary	Primary	Minor
Sewage Effluent	Minor	Primary	Secondary	Primary	Primary	Minor
Sewage Sludge	Minor	Primary	Primary	Primary	Primary	Minor
Urban Runoff	Minor	Secondary	Variable	Primary	Minor	Minor
Solid Wastes	Minor	Primary	Primary	Primary	Secondary	Minor
Lawn Fertilizers	Minor	Primary	Minor	Minor	Minor	Minor
Agricultural						
Evapotranspiration and Leaching	Minor	Primary	Minor	Minor	Minor	Minor
Fertilizers	Minor	Primary	Secondary	Secondary	Minor	Minor
Soil Amendments	Minor	Primary	Minor	Minor	Minor	Minor
Pesticides	Minor	Minor	Minor	Primary	Minor	Minor
Animal Wastes (Feedlots and Dairies)	Minor	Primary	Minor	Secondary	Primary	Minor
Stockpiles	Minor	Primary	Minor	Variable	Variable	Minor
Industrial						
Cooling Water	Primary	Minor	Primary	Minor	Minor	Minor
Process Waters	Variable	Primary	Primary	Variable	Minor	Variable
Storm Runoff	Minor	Secondary	Variable	Primary	Minor	Minor
Boiler Blowdown	Primary	Secondary	Primary	Minor	Minor	Minor
Stockpiles	Minor	Primary	Variable	Variable	Minor	Variable
Water Treatment Plant Effluent	Minor	Primary	Secondary	Minor	Minor	Minor
Hydrocarbons	Secondary	Secondary	Secondary	Primary	Minor	Minor
Tank and Pipeline Leakage	Variable	Variable	Variable	Variable	Minor	Variable
Oilfield Wastes						
Brines	Primary	Primary	Primary	Minor	Minor	Minor
Hydrocarbons	Secondary	Secondary	Secondary	Primary	Minor	Minor
Mining Wastes	Minor	Primary	Primary	Variable	Minor	Variable
Miscellaneous						
Polluted Precipitation and Surface Water	Variable	Variable	Variable	Variable	Variable	Variable
Septic Tanks and Cesspools	Minor	Primary	Minor	Secondary	Primary	Minor
Highway Deicing	Minor	Primary	Minor	Secondary	Minor	Minor
Seawater Intrusion	Primary	Primary	Primary	Minor	Minor	Minor

Table 2-14

INORGANIC CHEMICAL POLLUTANTS

Major	Others	Drinking Water Trace
Calcium	Silica	Iron
Magnesium	Boron	Manganese
Sodium	Fluoride	Arsenic
Potassium	Nitrogen Forms	Barium
Carbonate	Phosphorus Forms	Cadmium
Bicarbonate	Hardness	Hexavalent Chromium
Sulfate		Copper
Chloride		Cyanide
Nitrate		Lead
Total Dissolved Solids		Selenium
pH		Silver
Electrical Conductivity		Zinc
Oxidation Potential		Mercury
Other Trace	**Gases**	
Vanadium	Methane	
Molybdenum	Hydrogen Sulfide	
Bromide	Carbon Dioxide	
Iodide	Dissolved Oxygen	
Nickel	Residual Chlorine	
Aluminum		
Cobalt		
Lithium		
Sulfide		
Beryllium		

Municipal

Five major municipal sources of pollutants can occur.

- **Sewage Effluent.** For most secondary treated sewage effluent, specific pollutants are similar to those found in raw sewage. Thus, this discussion pertains to leaking sewers also. The inorganic composition of sewage effluent usually reflects the inorganic composition of the water supply. Little or none of the inorganic chemicals is removed during secondary treatment. Many heavy metals such as zinc, copper, iron, manganese, chromium, cadmium, lead, mercury, cobalt, and arsenic may enter sewage as a discharge from industry. However, many of the trace metals are removed from the sewage sludge. About 50 milligrams per liter (mg/ℓ) of organic matter is usually present in secondary sewage effluent. More than one-half of the effluent organics is generally soluble. Hunter and Kotalik (1973) and the American Chemical Society (1969) presented data on constituents in sewage effluent. The effects of sewage effluent disposal on groundwater were discussed by Ellis (1973), Hughes et al (1974), Schmidt (1973), Bogan (1961), and Bouwer et al. (1972).

- **Sewage Sludge.** Peterson et al. (1973) discussed the constituents in sewage sludge. Most constituents considered under "sewage effluent" would be applicable, except there is more emphasis on trace elements, as those found in raw sewage are usually concentrated in the sewage sludge. There have been few reported groundwater quality investigations beneath sludge drying beds; however, there is little doubt that this is a major point source of groundwater pollution.

There have been few reported groundwater quality investigations beneath sludge drying beds; however, there is little doubt that this is a major point source of groundwater pollution.

- **Storm Runoff.** The pollutants in storm runoff have been more widely recognized subsequent to the large-scale separation of this waste from sewage. Of concern to groundwater pollution is the presence of nitrogen and phosphorus, trace elements, and organic chemicals. High contents of lead and zinc have been found in storm runoff. Gasoline, oil, grease, and pesticides are also common. (Sartor and Boyd. 1972).

- **Solid Wastes.** Leachate analyses are highly variable; however, the major inorganic chemical constituents are usually present. Trace elements include iron, manganese, barium, chromium, lead, selenium, zinc, and possibly others. Organic chemicals are abundant in leachate. Hughes et al. (1971), Schneider (1970), California Department of Water Resources (1969), Weaver (1964), and Salvato et al. (1971) discussed constituents in solid wastes. There are a large number of evaluations concerning the effects of leachates on groundwater quality. Zanoni (1971), Apgar and Langmuir (1971), Kimmel and Braids (1974), Zenone et al. (1975), Weist and Pettijohn (1975), Coe (1970), Seitz (1972), and Andersen (1972) have reported on these evaluations.

- **Lawn Fertilizers.** Primary concerns in lawn fertilizers are nitrogen, phosphorous, and potassium. Fertilizers are discussed more extensively under the section on Agricultural Sources.

Agricultural

Water percolating past the root zone in irrigated areas is degraded compared to the applied water. This degradation is due to evapotranspiration, leaching of salts from the soil, and additives applied in the irrigation water, such as fertilizers, soil amendments, and pesticides.

- **Evapotranspiration and Leaching.** Of major concern here are the major inorganic chemical constituents. In general the most mobile or soluble constituents will be concentrated the most by evapotranspiration. Of the major chemical constituents, sodium, chloride, nitrate, and boron are the most mobile in usual soil-groundwater systems. Law et al. (1970), Skogerboe and Law (1971), Flack and Howe (1974), and Fuhriman and Barton (1971) discussed return flow from agricultural lands. Mineral dissolution can occur in all areas. Salts in the topsoil and vadose zone in arid areas can also be dissolved by waters percolating from irrigation. Of most concern are sodium, calcium, chloride, bicarbonate, boron, and total dissolved solids. Doneen (1967) and the University of California at Davis (1968) discussed these phenomena for newly-developed lands in California.

- **Fertilizers.** The primary fertilizers are nitrogen, phosphorus, and potassium. Significant amounts of other major chemical constituents can be added in compound form with the fertilizer, such as chloride in potassium chloride. Secondary fertilizers are calcium, magnesium, sulfur, boron, manganese, copper, zinc, molybdenum, chloride, cobalt, vanadium, and sodium. The major concerns in most groundwater systems are nitrate and the accompanying increase in salinity related to the application of nitrogen.

Ammonia also has the ability to replace many other cations on the exchange complex of the soil thus allowing them to be leached into the groundwater. The literature is voluminous, and often contradictory, on the effects of fertilizers on groundwater quality. Mink (1962), Moore (1970), Willrich and Smith (1970a), Tisdale and Nelson (1966), Fitzsimmons et al. (1972), and Shaw (1968) have discussed fertilizers and groundwater pollution.

- **Soil Amendments.** The most common soil amendments of importance to groundwater pollution are gypsum, lime, and manure. Manure, in part, falls under the fertilizer category. Gypsum contains calcium and sulfide, and lime contains calcium or calcium and magnesium carbonate. Sulfur may be applied to calcic solid in the elemental form. The pH and bicarbonate are important parameters, as lime is frequently applied to change the soil pH. When gypsum is applied, it frequently results in the leaching of sodium held in the soil, along with sulfate. Buckman and Brady (1969) discussed soil amendments and their use in agriculture.

- **Pesticides and Herbicides.** Pesticides, herbicides, fungicides, and other chemicals are used for nuisance organism control. Drinking water standards are proposed for chlorinated hydrocarbons and total organophosphorus and carbamate compounds. Thus, for monitoring groundwater pollution, DDT; 2,4-D; lindane; and herbicides are major concerns. Willrich and Smith (1970b), California Department of Water Resources (1968), Johnson et al. (1967), Scalf et al. (1968), Dregne et al. (1969), and Schneider et al. (1970) discussed pesticides and groundwater pollution.

- **Animal Wastes.** Constituents of concern in animal wastes are somewhat similar to those found in human sewage, including many of the major inorganic chemical constituents, particularly nitrate, chloride, sulfate, and total dissolved solids. Organic matter and bacteriological constituents are also present. Willrich and Smith (1970c), Steward et al. (1967, 1968), Loehr (1967), Concannon and Genetilli (1971), Gillham and Webber (1969), and Adriano et al. (1971) have discussed animal wastes and groundwater pollution.

Industrial

Pollution from industry includes the following six types.

- **Cooling Water.** Major degradation of cooling water results from temperature increases and possibly from some trace elements introduced as additives, such as rust inhibitors. If large amounts of evaporation occur brines may result, and concerns for overall salinity and the major inorganic chemical constituents would be increased.

- **Process Water.** Industrial process water is highly variable; however, in general, trace elements will be of major concern. These could be introduced from raw materials and chemicals used in the process. Temperature and density could be important. The major inorganic chemical constituents would usually be of concern, as well as overall salinity. Organic chemicals, such as hydrocarbons, reagents added for processing, etc. would be common constituents. Radioactive components are important in the disposal of nuclear plant wastes and certain other operations. Radium, strontium, and alpha and beta activity are major criteria for drinking water.

- **Storm Runoff.** The quality of storm runoff is highly influenced by housekeeping procedures and drainage systems. If significant amounts of solids and liquids are spilled, blown by the wind, and left to accumulate on the topsoil or impermeable surfaces, they can be subsequently mobilized by storm runoff. The major concerns are organic chemicals, such as hydrocarbons, and possibly trace elements. Major inorganic chemicals are also important.

- **Boiler Blowdown.** Major concerns with boiler blowdown water are temperature and several of the major inorganic chemicals, namely, silica, calcium, magnesium, bicarbonate, and selected trace elements.

- **Stockpiles.** Stockpiles for storage of raw materials, products, byproducts, and wastes at industrial sites would be highly variable, somewhat similar to process water. However, physical parameters would generally be of less importance.

- **Water Treatment Plant Effluent.** Some of the major inorganic chemicals would be of concern, particularly sodium, chloride, calcium, sulfate, bicarbonate, and nitrate, and total dissolved solids. In some cases, selected trace elements may be of concern.

Studies of industrial waste pollution of groundwater are numerous. Examples are Tucker (1971), Smith (1971), Williams and Wilder (1971), Matis (1971), Sweet and Fetrow (1975), Walker (1961), Deutsch (1961), Walton (1961), and Hanby et al. (1973).

Oilfield Wastes

Brines and hydrocarbons are present in oilfield wastes.

- **Brines.** Density and possibly temperature could be important for oilfield brines. The major inorganic chemical constituents are of concern, particularly sodium, chloride, calcium, sulfate, bicarbonate, and total dissolved solids. Boron, ammonium, fluoride, bromide, and iodide are also common in brines. Many trace elements are often present such as lead, iron, strontium, zinc, manganese, nickel, and aluminum.

- **Hydrocarbons.** Materials such as oil can pollute groundwater, and thus hydrocarbons are of major concern.

References by Mattox (1970) and White et al. (1963) detail the composition of saline waters. Many references have documented groundwater pollution from disposal of oilfield wastes, such as Fryberger (1975), Pettyjohn (1972), Bain (1970), Krieger and Hendrickson (1960), Knowles (1965), and McMillion (1965).

Mining Wastes

Waste production from mining activities has been discussed in detail by Williams (1975). In general the major inorganic chemical constituents will be important and often trace elements will be of primary concern. Nitrogen forms may be used in explosives and could appear in the groundwater. Trace elements will depend on constituents of ore processed and reagents added. Organic chemicals may be added as reagents and radiological constituents may be present in the ore.

Miscellaneous
Miscellaneous sources of pollution are the following.

- **Septic Tanks and Cesspools.** The constituents of importance are similar to those in sewage effluent, except trace elements will usually be absent. Also the chances for nitrification in the topsoil or vadose zones are generally better in the case of septic tanks. Polta (1969), Waltz (1971), Schmidt (1972), Quan et al. (1974), and Pitt (1974) have discussed the groundwater pollution due to individual sewage disposal systems.

- **Highway Deicing.** Of primary concern in highway deicing are sodium, chloride, and calcium. Some organic chemicals are periodically used.

- **Saline Water Intrusion.** The same constituents are of interest as for brines under "Oil Field Wastes."

STEP 4 – DEFINE GROUNDWATER USAGE

To evaluate the impact of pollution or potential pollution of groundwater, the usage of the resource becomes a key item. Thus, it is important to define both the quantities of groundwater being extracted, or projected to be extracted, and the locations of major pumping centers within a monitoring area. Pumpage data on a gross basis are usually available from the U.S. Geological Survey, State water agencies, and some local water agencies. Often it will be necessary to refine the data in order to make them useful in developing an effective monitoring program.

Pumpage will vary on a weekly and on a monthly basis; however, in many cases mean annual extractions will be sufficient to evaluate groundwater flows and pollutant movement for monitoring purposes. These data will serve as inputs to the hydrogeologic analysis in a subsequent step.

Determination of groundwater pumpage depends on the type of use. The basic categories used – municipal, industrial, agricultural, and rural – are described below in terms of how their pumpage quantities can be defined.

Municipal Use
Groundwater pumpage for an urban area is often available from records of water purveyors. Furthermore, locations and yields of individual wells and well fields are usually known.

Industrial Use
Many industries purchase water from municipal sources; consequently, under these circumstances groundwater use is incorporated in the municipal category. On the other hand, some industries use self-supplied water, and a portion of these draw upon groundwater through wells. On a national basis the following types of industries are the heaviest users of groundwater:

Rank in Terms of Total Groundwater Use	Type of Industry
1	Oil Refining
2	Paper Manufacturing
3	Metal Working
4	Chemical Manufacturing
5	Building, Air Conditioning and Refrigerating
6	Distilling
7	Ice Manufacturing and Cold Storage
8	Food Processing
9	Rubber Manufacturing
10	Meat Packing
11	Brewing
12	Railroad Yards
13	Gas and Electricity
14	Dairying
15	Electric Equipment Manufacturing

If no local summary of industrial use of self-supplied groundwater is available from appropriate public agencies, pertinent industries will have to be identified. Industrial areas can be located from aerial photographs or zoning maps; industries which do not receive municipal water can then be contacted to obtain their annual groundwater pumpage data. Often these are not directly available and must be estimated from the type of industry, number of workers, units of production, or other factors.

Agricultural Use

In the Western United States groundwater is used mostly in irrigated agriculture; however, data on such usage are also generally difficult to obtain. Accurate records of pumpage are rarely kept by farmers, and, in fact, most irrigation wells are not equipped with flow meters. Pumpage estimates can be made by either of two methods. One is to determine cropping patterns of areas irrigated by wells, using aerial photographs and crop surveys. By knowing the consumptive use of the crops and representative irrigation efficiencies (from local farm advisors) and mean annual growing season rainfall (from U.S. National Weather Service records), the total applied water for irrigation purposes can be computed. In areas where both surface water and groundwater are used, the portion of applied water supplied by groundwater can usually be determined from irrigation district records. A second approach is to obtain records of power consumption for representative pumps and of pump discharges per unit of power use from electric utility companies. These data enable the quantities of water pumped to be calculated. Where such data are available for only a fraction of the irrigated area, extrapolation to the gross area can be made; however, results can be subject to error if significant differences in pumping levels exist.

Rural Use

Data will rarely be available on pumpage from individual domestic water supply wells in rural areas. Estimates of pumpage can be made by multiplying rural population figures by typical per capita water use. These pumpage rates are so scattered over agricultural areas that they do not influence the groundwater flow. However, the existence of the rural use should be recognized, particularly in the event of a localized pollution problem.

In summary, the total of the four above groundwater uses provides the usage data for a monitoring area. Locations of major pumping centers, such as well fields of municipalities and large industries, should be identified. Average pumpage rates over uniform subareas, such as irrigated lands, should be defined from irrigation well pumping rates and well densities. Great precision in determining groundwater pumpage is not as important as a comprehensive coverage of all significant pumpage within an area.

STEP 5 – DEFINE HYDROGEOLOGIC SITUATION

To understand where and how groundwater pollution occurs and moves within a monitoring area, the hydrogeologic framework must be understood. This information will aid in the design of an effective as well as an efficient groundwater quality monitoring system.

Because some subsurface data are available in most areas of groundwater development, initial hydrogeologic work will consist of gathering, organizing, and analyzing existing information. On other occasions, geologic or hydrologic investigations requiring field work will be necessary. Specific materials needed for the monitoring program, as well as how they are obtained, include:

- Aquifer locations, depth, and areal extents – from geologic data
- Transmissivities of aquifers – from well pumping tests and geologic data
- Map of groundwater levels – from observations of well levels
- Map of depths to groundwater – from water levels and topographic data
- Areas and magnitudes of natural groundwater recharge – from precipitation, evapotranspiration, soils, land use, and water level data
- Areas and magnitudes of artificial groundwater recharge – from irrigation and recharge data
- Areas and magnitudes of natural groundwater discharge – from steamflow and water level data
- Directions and velocities of groundwater flows – from water level and transmissivity data

Information sources for preparation of the above items should normally include State geologic and water agencies, local water agencies, the U.S. Geological Survey, and the U.S. National Weather Service.

In preparing the above materials it should be kept in mind that these serve as tools for monitoring. Consequently, the hydrogeologic effort should not become an obstacle to the completion of the monitoring program. All hydrogeologic data are incomplete in a relative sense. What is needed is an overall picture of the hydrogeologic situation in the monitoring area. Initially, refinement is less important than comprehensive coverage, no matter how preliminary or approximate. Categories indicating ranges rather than specific values, such as for transmissivity or dissolved solids, are often sufficient. Also, it should be recognized that, with time and with increasing amounts of groundwater data, knowledge of the hydrogeologic situation will gradually improve.

Example 1 – Pollution Plume Geometry. Figure 2-31 shows examples of the configurations of pollution plumes resulting from a pollution source, such as a landfill, located near a river. Figure 2-31(A) the angle between the groundwater flow direction and river is relatively large. This limits the extent of the plume. Furthermore, where the source is located close to the river, the plume is smaller. In Figure 2-31(b) the angle between the groundwater flow direction and the river is relatively small, creating elongated plumes, particularly where the source is some distance from the steam.

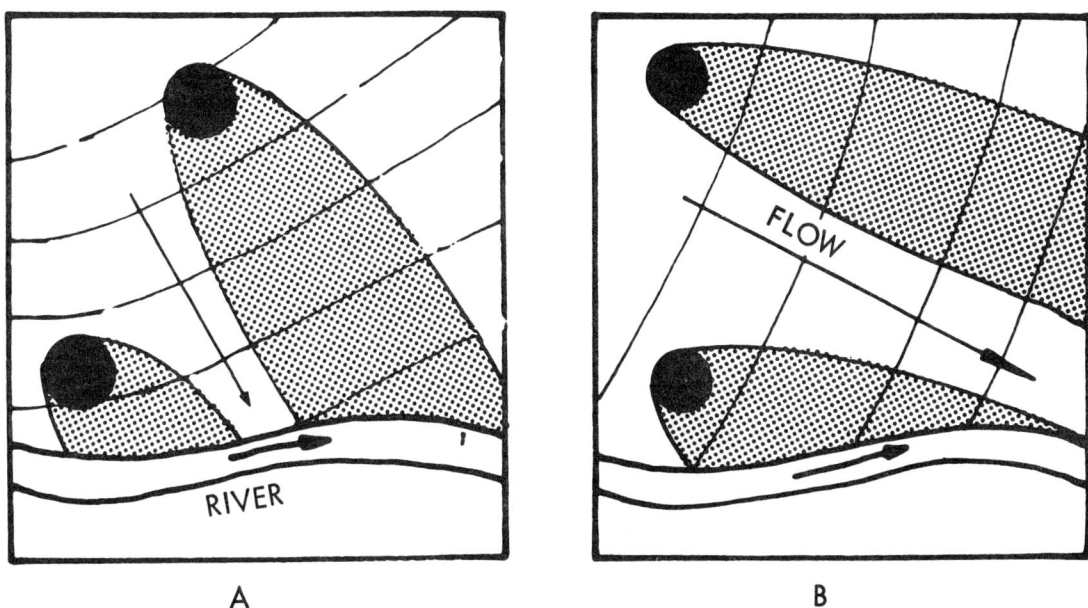

Figure 2-31. Relation of groundwater pollution plum size and orientation for a pollution source near a river with different groundwater flow directions (Palmquist and Sendlien, 1975).

Example 2 – Groundwater Flow System. Figure 2-32 depicts schematically a groundwater flow system under idealized homogeneous aquifer conditions. Groundwater travels along flow paths, which extend from areas of groundwater recharge to areas of groundwater discharge. Horizontal and vertical gradients are indicated by equipotential lines; potential, or total head, is simply the elevation to which water will rise in a cased well from a point source below the water table. Groundwater moves in the direction of decreasing total head. In recharge and discharge areas, movement may have an appreciable vertical component; between these end areas, flow is predominantly horizontal. Any pollution will travel along the flow line where it occurs.

Example 3 – Complex Plume Geometries. Figure 2-33 presents an idealized vertical aquifer cross-section, somewhat similar to Figure 2-32, together with the pollution plumes for various locations of pollution sources. Note that Sources A-1 and A-2 create short plumes because they are close to streams, while B and C are longer because they are more distant. Source F, located in a discharge area, has no effect on groundwater quality. Sources D and E-2 cause two plumes in different directions, while E-1 is responsible for three distinct plumes. Clearly, recognition of the hydrogeological conditions controlling number, shape, and location of pollution plumes is essential in selecting groundwater monitoring sites.

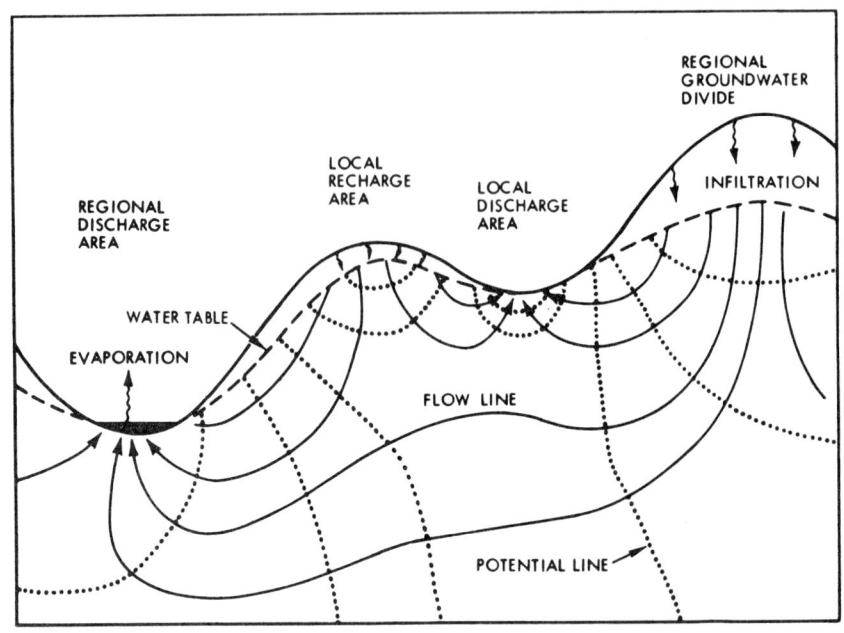

Figure 2-32. Groundwater flow system under idealized homogeneous aquifer conditions (modified after Born and Stephenson, 1969).

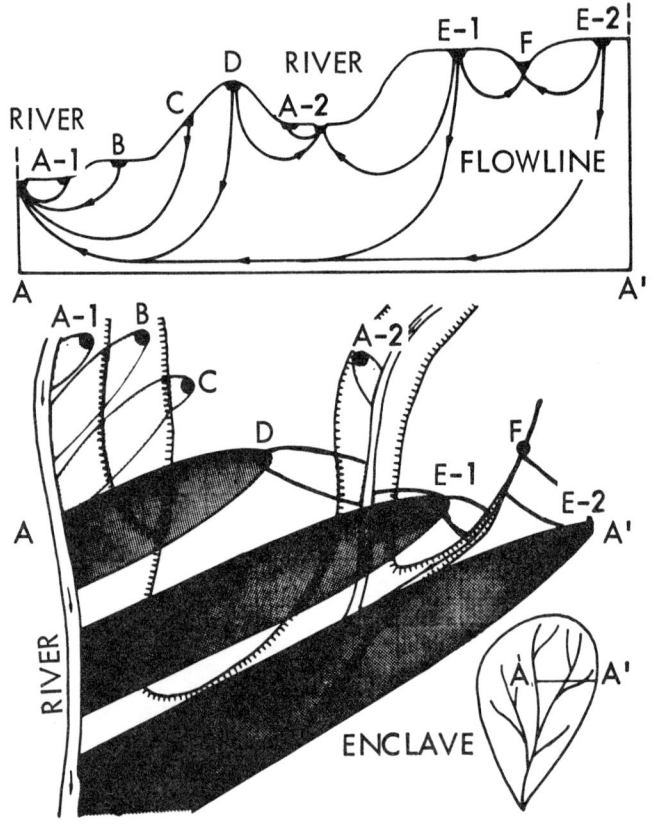

Figure 2-33. Idealized pollution plume configuration for various locations of surface pollution sources (modified after Palmquist and Sendlein, 1975).

Example 4 – Pollution From a Percolation Pond. Figure 2-34 illustrates how hydrogeologic conditions can influence the movement of pollution from a percolation pond. Here the aquifer is recharged by a lake on the north side, while on the southwest a river both recharges and receives discharge from the aquifer. Near the center of the aquifer a well field creates a radially converging flow pattern, and south of it is a recharge pond, which is maintained at a constant water level. Flow lines of groundwater for steady-state conditions are shown on Figure 2-34. In addition, it is assumed that at a certain time, pollution is introduced into the pond. By means of an electronic analog model, the advancing pollution front can be calculated. Positions of the front for 40 and 80 days after the beginning of pollution are shown in Figure 2-34. Note that between these two time intervals pollution has reached both the well field and the river. The dashed lines define the ultimate limits of pollution from the percolation pond.

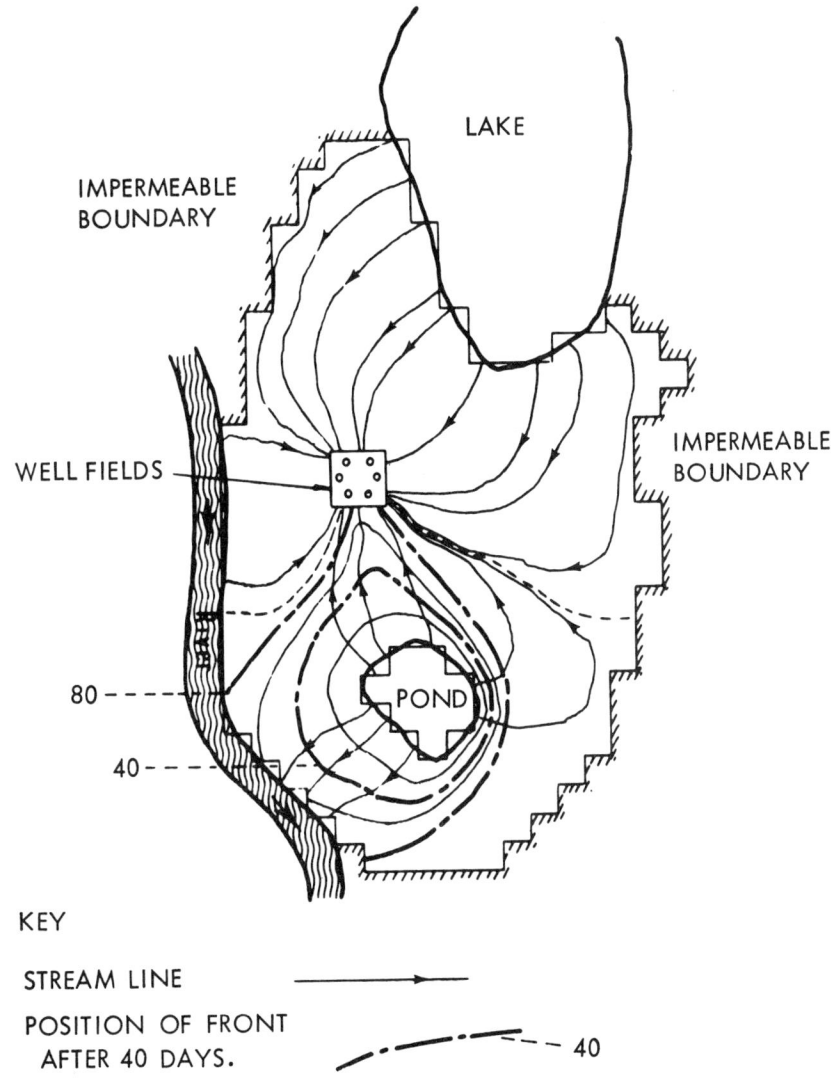

Figure 2-34. Flow lines for steady-state condition in an aquifer and positions of a pollution front advancing from a percolation pond (Cole, 1975).

Example 5 – Effect of Pumping on Groundwater Flow. Figure 2-35 provides dramatic evidence of how heavy pumping in an urban area can influence groundwater flow directions. Solid lines are water table contours, which show that the water table is more than 70 feet below sea level in the center of Stockton, California. It can be seen that any pollutants in the groundwater are transported toward the city from a large area. If such a cone of depression results from pumping of water supply wells, the concentrating effects of the wells moves all pollution into them, which is clearly undesirable.

Example 6 – Density Effect on Pollution Distribution. Figure 2-36 shows the distribution of saline water below a brine disposal pit in an oil field in southwestern Arkansas. Because the brine is denser than native groundwater, it has traveled to the bottom of the sandy confined aquifer, and thereafter has spread laterally. The primary extension of pollution is southward due to the natural groundwater gradient in this direction. If the pollutant had possessed a density more nearly that of the native groundwater, it would be expected to drift horizontally near the top of the aquifer. Thus, the physical character of a pollutant can have a significant effect on how it moves within an aquifer.

STEP 6 – STUDY EXISTING GROUNDWATER QUALITY

In order to define the groundwater quality problems within a monitoring area, an assessment needs to be made of the background quality situation. To do this, recent groundwater quality data need to be collected and reviewed. Attention should be focused first on the natural groundwater quality. A map of indicators of pollution in groundwater, taking into account variations as a function of depth, should be prepared from available well water analyses. This map will typically show a limited range of values within an extensive aquifer. However, if considerable variability or isolated anomalies are evident, these may be indicative of the presence of pollution.

To verify possibilities of pollution, regions or localities displaying unusual quality data must then be examined in conjunction with the inventory of pollution sources and methods of disposal (Step 2) and with hydrogeologic data (Step 5). Where these jointly suggest the physical feasibility of pollution, it is reasonable to assume that pollution exists. Once pollution is tentatively identified, the specific pollutants involved, the areal extent, and the direction and rate of movement can be defined at least approximately.

It is often surprising how many quality data, after investigation, are actually available. A certain amount of ingenuity is required to locate fragmented data, and to interpret them in terms of the subsurface situation. Judgment, however, must frequently be exercised so as to select meaningful data, and to avoid those which are erroneous.

In addition to collecting current data, past groundwater quality records should be reviewed wherever possible. Historical quality records are helpful in establishing the quality of native groundwater, in evaluating quality trends with time, and in relating changes in groundwater quality to source and causes.

Information sources for groundwater quality data should, normally, include State geologic and water agencies, the U.S. Geological Survey, local water and regulatory agencies, and industries with self-supplied groundwater.

It should be recognized that significant portions of a monitoring area may entirely lack groundwater quality data, so that no direct indications of pollution can be

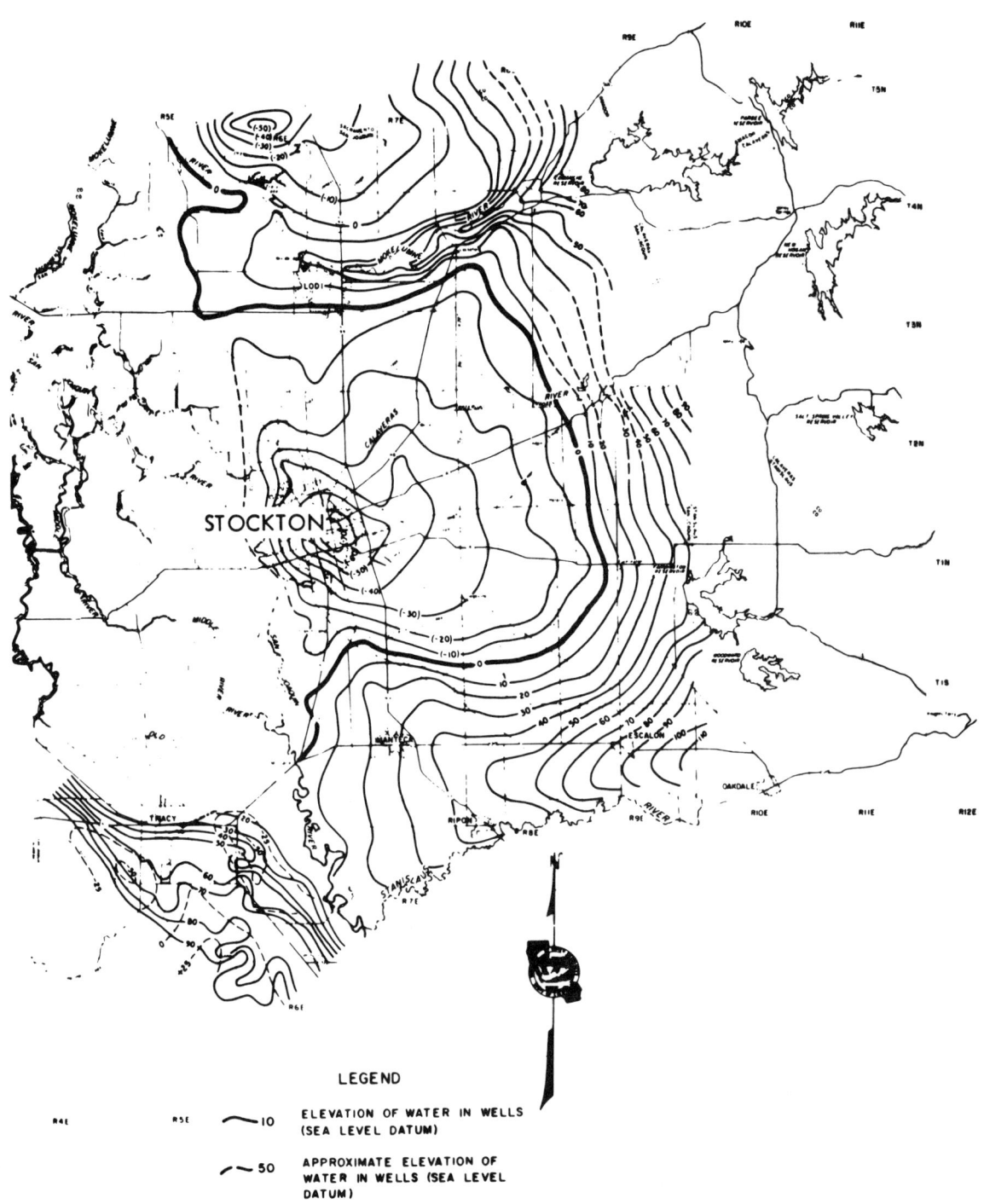

Figure 2-35. Water table contours in the vicinity of Stockton, California, Fall 1964 (California Department of Water Resources, 1967).

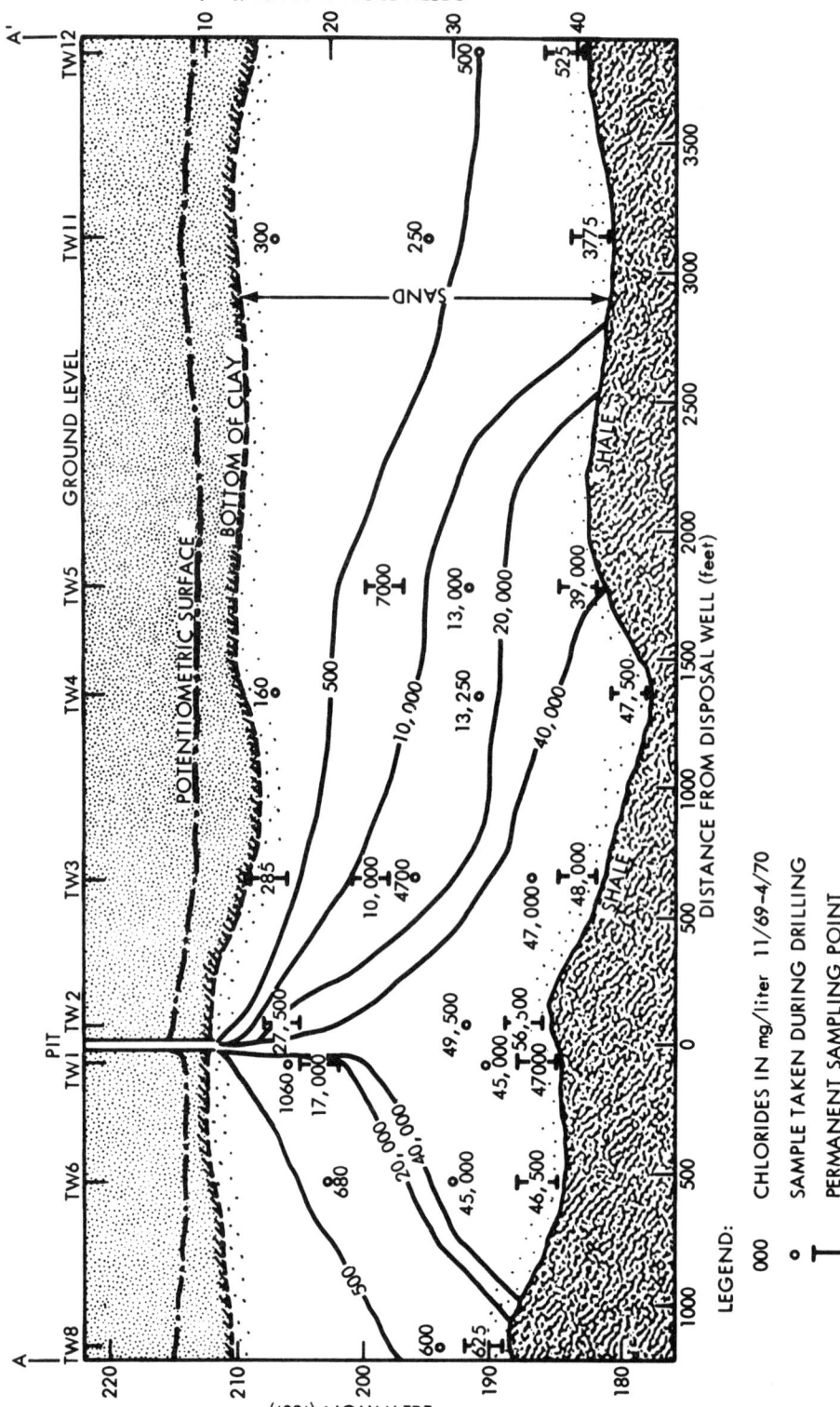

Figure 2-36. Distribution of saline water in a confined aquifer resulting from an oilfield brine disposal pit in southwestern Arkansas (modified after Fryberger, 1975).

obtained. However, sources and causes as well as hydrogeologic evidence may suggest locations of potential pollution. Verification can only come with the implementation of the monitoring program.

STEP 7 – EVALUATE INFILTRATION POTENTIAL OF WASTES AT THE LAND SURFACE

Following the development of an understanding of the hydrogeological framework, a key step in the methodology will be to determine the volume of polluted water which will pass through the vadose zone into the zone of saturation. This volume will vary depending on the method of waste disposal used and the infiltration characteristics of the soil. The monitoring methods applicable to the above determination are discussed in detail in Section 3 of Chapter II.

Considering the common methods of waste disposal presented previously in Step 2, Table 2-1, the total volume of polluted water associated with surface spreading, irrigation, or disposal into dry stream beds, will move directly into the topsoil or stream bed. The volume of polluted water associated with percolation ponds, seepage pits, trenches, or landfills will bypass the topsoil, and enter the underlying portion of the vadose zone or saturated zone, depending on the depth to the water table.

Disposal wells used to dispose of storm water and domestic sewage frequently are terminated in the vadose zone. The volume of polluted water reaching the zone of saturation will be equal to the volume discharged into the well, minus that portion retained by the soil. A properly designed injection well for disposal of industrial wastes or brines will completely bypass the vadose zone and all freshwater aquifers. Monitoring procedures for injection wells are discussed at length in Chapter V.

The water budget method can be used to calculate infiltration from surface spreading or irrigation. In the water budget approach, waste water discharge and precipitation are volumetrically summed as the input, and infiltration is computed as the difference between this input and evapotranspiration. Applied water volumes can be determined from records of waste water discharge/surface water deliveries, and groundwater pumpage. Precipitation volumes can usually be extrapolated from rainfall gaging stations in nearby areas. Monthly rainfall determinations are often suitable; however, in areas with highly variable precipitation, such as in Southwestern States, onsite measurements may be necessary.

Evapotranspiration from areas where surface spreading or irrigation is used can be determined by a number of methods (Cruff and Thompson, 1967; Blaney and Criddle, 1962; Lowry and Johnson, 1942; Penman, 1948; and Thornthwaite, 1948). These methods are generally based on different groupings of climatological parameters. For example, the Blaney-Criddle method depends primarily on temperature and percentage of daylight hours. In general the Penman and Thornthwaite methods are more applicable to humid areas, whereas the Blaney-Criddle method is more applicable to semiarid areas.

Different values of evapotranspiration and consumptive use are usually obtained for different crops and soil conditions. Thus, the cropping pattern must be known, and factors such as double-cropping considered. Sufficient field tests have been conducted in many areas so that evapotranspiration and consumptive use for major crops are well established. In some areas, data may have to be extrapolated from similar areas. Evapotranspiration rates will generally be needed on at least a monthly basis, and sometimes weekly.

In summing up the water budget components for surface spreading and irrigation, the infiltrating component is divided into two components. A portion of this component is diverted into soil moisture retention. This soil mosture component will be gradually depleted by transpiration during periods of zero recharge. When the field capacity requirements have been satisfied the remaining portion will percolate to the zone of saturation.

In many parts of the arid and semiarid West, significant volumes of waste waters are disposed of into normally dry stream channels. Disposal of mining wastes and sewage effluent in Arizona and oilfield wastes in California by this method is common. That portion which does not leave the area as runoff or evaporate infiltrates into the stream bed. Part of the percolate will be held in storage and the remainder will flow on to the zone of saturation. The portion held in storage will be quite small relative to that of a normal soil profile because of the high permeability associated with stream bed materials.

Stream flow records at different gaging stations along the particular reach under investigation, combined with records of precipitation, evaporation, estimates of soil moisture content, and stream flow diversions, can be used to calculate percolation to the water table. In some areas the U.S. Geological Survey, State geological or water agencies, or others may have already made estimates of infiltration.

The water budget method can also be used to calculate seepage from percolation ponds. In this context, "percolation pond" is used to signify any pond which permits significant movement of water into the underground. Waste discharge and precipitation are volumetrically summed as the input, and seepage is calculated as the difference between input and evaporation. Storage changes in the pond must also be taken into account. Evaporation from free-water surfaces can be determined from measurements using land pans or floating pans (Harbeck et al., 1958; Kohler et al., 1955; Follansbee, 1933; and Rohwer, 1933). Monthly values will often suffice; however, in some cases weekly or daily values are necessary.

Factors such as salinity of waste water can affect the evaporation rate. In general, with increasing salinity the vapor pressure of water decreases, resulting in a lower evaporation rate. In considering evaporation from free-water surfaces from ponds of different sizes, consideration should be given to edge effects. That is, evaporation rates depend on the characteristics of the surrounding land, for example, whether it is cultivated or undeveloped.

In some ponds such as mine tailings ponds, water may occur in several states, such as free water, moist tailings, and wet tailings. Where ponds are periodically dried to improve infiltration rates, evaporation from moist areas and wet areas must be considered separately from the free-water evaporation. Methods for determining such evaporation are not well developed. Field measurements using lysimeters can be used as an independent check on evaporation from soil surfaces. It may be sufficient to apply correction factors to the evaporation rate calculated from lysimeters for use in free-water surfaces.

Effluent from septic tanks and other wastes is often disposed of in seepage pits or trenches which extend only a few feet beneath the land surface. As such, these wastes usually are subject to the effects of evapotranspiration. Determination of evapotranspiration is difficult because of varying soil conditions, septic tank disposal practices, vegetation type and distribution, and other factors. Evapotranspiration is the sum of evaporation from bare soil surfaces and transpiration by vegetation growing in the area. Rooting depth would be a key factor in plant uptake of effluent from trenches. If the land surface is irrigated this must also be considered. If leach lines are placed at depths greater than about 5 feet in most soils in the absence of deep rooted plants, seepage is virtually the same as the waste discharge. If trenches are only a few feet beneath the land surface, then a water budget analysis may be made. This analysis would usually be done for groups of septic tanks, such as for a subdivision in an urban area. Precipitation is combined with waste discharge as the input, while evapotranspiration and runoff are losses. The residual seepage is subject to a soil moisture retention loss, depending on the depth to the water table from the bottom of the trench.

Landfills pose a threat to groundwater quality depending on the volume of leachate leaving the fill. This in turn is a function of the leachate control methods in use for a particular landfill, such as clay liners, collection and treatment, impermeable rubber or plastic barriers and caps. May landfills have not been properly designed and, as a consequence are almost certain to produce leachate.

The water budget method may be applied when estimating leachate production from landfills. Precipitation volumes must be obtained and the portion that infiltrates the landfill determined. This portion will first go into meeting the moisture storage requirements of the solid waste. For this reason the moisture content of the solid waste must be estimated. When the solid waste reaches field capacity leachate will result. Whether this leachate will reach the zone of saturation will depend on the leachate control method in use.

The leachate produced from a landfill can be estimated using the following relationship:

$$\text{Leachate} = P - R - ET - S$$

where
P = precipitation

R = runoff

ET = evapotranspiration

and

S = The field capacity of the solid waste minus its existing moisture content

STEP 8 – EVALUATE MOBILITY OF POLLUTANTS FROM THE LAND SURFACE TO WATER TABLE

The topsoil and materials of the vadose zone may have a significant capacity to remove pollutants from downward-percolating waters. The extent of groundwater pollution due to waste percolation from the land surface depends strongly on the rate and volume of recharge water. In a semiarid or arid climate, pollutants may be retained above the water table in a nearly permanent fashion. On the other hand, in humid areas pollutants may be rapidly carried downward from the land surface to the water table. In a homogeneous porous media, percolating water will pass vertically through the vadose zone. However, in a heterogeneous, stratified material, such as most of the alluvial deposits of the western U.S., percolating water may become perched above layers of low permeability. In this situation, lateral movement for substantial distances can occur above the water table.

The capacity for attenuation of many potential pollutants is greatly limited by the amount and characteristics of the geologic materials present in the vadose zone. This is especially true for the sorption capacity of many organic chemicals, trace elements, and radionuclides. This limited capacity for removal of some pollutants is in sharp contrast to the almost unlimited ability of many unconsolidated materials to remove bacteriological pollutants. The existance of many documented case histories of groundwater pollution indicates that the vadose zone may often not provide complete protection. Problems can occur when the zone is bypassed during waste discharge, or when attenuation capacity is exceeded due to high waste loadings.

Pollutant attenuation in the subsurface commonly occurs due to the following processes: dilution, filtration, sorption, buffering, precipitation, oxidation and reduction, volatilization, biological degradation and assimilation, and radioactive decay. Each of these processes must be evaluated with respect to specific pollutants.

Dilution

Dilution above the water table can be substantial in humid areas and almost nonexistent in arid areas. Sources of water for dilution include precipitation, seepage from streams, lakes and canals, and artificial recharge. A water budget analysis can be used to evaluate the extent of dilution. The analysis is done in a similar manner to that discussed in the previous step concerning evaluation of infiltration potential. However, in this case, all items of recharge from the land surface are considered, such as canal seepage, percolation from streams and lakes, and artificial recharge in the area. The quality of water from each source of recharge must be determined. A comparison of the respective quantities of water and constituents concentrations in a waste discharge at the land surface can indicate the extent of subsequent dilution of pollutants.

An example is presented in the following discussion. Assume an agricultural area in the West, where irrigation is practiced in the dry summer months and rainfall occurs in the wet winter months. Return flow over the area averages 18 inches per year, and the salinity of this water is 300 parts per million (ppm). Rainfall is 12 inches per year, and its salinity is 10 ppm. Consideration of the water budget analysis at the land surface indicates that 9 of the 12 inches of rainfall percolates to the water table. The subsequent dilution can be calculated by the equation:

$$AV_A + BV_B = C$$

Where V_A and V_B are the respective percentages of water from return flow and precipitation. A, B, and C are the respective salinities of the return flow, precipitation, and the mixture of the two. In this example:

$$C = 300(18/27) + 10(9/27)$$
$$C = 200 + 3 = 203 \text{ ppm}$$

This dilution reduces the pollutant concentration by about one-third of the original value. This simple concept can be expanded to encompass dilution and a number of sources of pollution.

Filtration

Filtration removes virtually all of the suspended materials that would be of more concern in surface water pollution. However, this process is generally not effective for most of the inorganic chemical species. Exceptions include iron and manganese, which may be present in aerated waters as hydroxides in particulate matter. Similarly, as precipitates form due to chemical reactions, they may be effectively filtered out as water moves through porous media. For wastes with high iron or manganese contents, laboratory tests can be performed on soils or geologic materials of the vadose zone. However, other attenuation factors affecting the pollutants would have to be evaluated.

Sorption

Sorption is probably one of the most effective processes for attenuating groundwater pollution. Clays, metallic oxides and hydroxides, and organic matter can all be suitable materials for sorption of various pollutants. With the exception of chloride, and to a lesser extent nitrate and sulfate, most pollutants can be sorbed and removed to some extent under favorable conditions. Under other circumstances, however, the pollutants can move freely through the porous media. The pH and oxidation potential often govern the extent of sorption for specific constituents. The sorption process depends on the type of pollutant and the physical and chemical properties of both solution and the containing materials.

When a pollutant in ionic form is sorbed, some other change must occur to compensate for loss of the ion from solution. In ion-exchange processes, a different ion is released by the solid to the water. However, this release is not required if the pollutants are sorbed or electrically neutral, such as most organic and neutral complexes of various metals.

The sorptive capacity can be estimated based on the density, clay content, and cation exchange capacity of the soil and geologic materials above the water table. Values for these parameters can be calculated from available data in soils and groundwater reports on the area of interest. In exceptional cases, these parameters can be determined from detailed onsite measurements. For calculation purposes, the thickness of the vadose zone is known or determined from water level data. For simplicity, the vertical path of polluted water from the land surface beneath the waste disposal area to the water table can be assumed to be the distance traveled.

As an example, assume the average density of materials in the vadose zone is 1.6 grams per cubic centimeter, the clay content is 20 percent by weight, and the clay has a cation exchange capacity of 70 milliequivalents per 100 grams. Each gram of clay will have the ability to remove 0.70 milliequivalents of the pollutant of interest. For example, for potassium (equivalent weight of 39), each gram of clay will have the ability to remove 27.3 (0.70x39) milligrams of potassium from the percolated waste water. Each gram of solid material will have the ability to remove 5.5 (0.20x27.3) milligrams of potassium from the percolated waste water. With a density of soil of 1.6 grams per cubic centimeter, one acre-foot (1.2335×10^9 cubic centimeters) of soil would contain 1.97×10^9 grams of solid material. This soil could sorb 23,900 pounds of potassium. For a vadose zone 50 feet thick, one acre of the vadose zone could sorb over one million pounds of potassium.

To determine the actual extent of adsorption, laboratory tests can be performed utilizing soils and geologic materials typical of the waste disposal site. The actual waste discharge can be used or a similar synthetic solution prepared. Hajek (1969) summarizes laboratory procedures for such tests.

It should be noted that percolating fluids may subsequently remobilize species that have been sorbed. The sorptive capacity of soils and geologic materials is finite for most inorganic substances which cannot be biodegraded. However, for substances which are biodegradable, such as many bacteriological constituents and nitrogen, the sorptive capacity may be renewed indefinitely.

Buffering

The pH is a critical factor in many reactions involving pollutants. Buffering is the resistance to a pH change of the soil solution. The basis of buffer capacity lies in the adsorbed cations on the exchange complex of the soil. The higher the exchange capacity, the greater will be the buffer capacity. The portion of the cation exchange capacity occupied by exchangeable bases is termed base saturation. There is a correlation between base saturation and pH, with higher base saturation for higher pH. The degree of buffering is lowest at the extremes of base saturation, and highest at intermediate base saturation values.

The extent of buffering in most cases will be relatively unimportant if the pH of the waste discharge is between 6 and 9. These pH values correspond to those found in natural groundwater. Wastes with a pH in this range will generally be buffered to an extent that the percolating waste water will present no unusual problem. Consideration of buffering is thus of foremost importance in cases of disposal of very acidic and basic wastes. Detailed considerations are presented in Buckman and Brady (1969).

Chemical Precipitation

It is theoretically possible to precipitate almost any dissolved species from solution. However, in soil-groundwater systems, the necessary species often are not present in sufficient quantities to precipitate potential pollutants. Certain constituents are normally present and available for reaction in most groundwater, soil, and geologic materials. Calcium, magnesium, sodium, potassium, bicarbonate, sulfate, chloride, and silica are usually the major species in groundwater. Iron, aluminum, nitrogen, and carbonate, in addition to the previous constituents, may be found in soil and geologic materials.

There are important precipitation reactions for calcium, magnesium, bicarbonate, and sulfate. Calcite, aragonite, gypsum, and magnesium carbonates are major compounds which may precipitate in the soil and vadose zone. In arid areas, virtually all major constituents could be precipitated at or near the land surface in some situations; however, in this case virtually all of the waste has evaporated and percolation is minimal. The following trace constituents have important precipitation potential: arsenic, barium, cadmium, copper, fluoride, cyanide, iron, lead, mercury, molybdenum, zinc, and radium.

No rigorous procedure is given herein to evaluate chemical precipitation. However references such as Hem (1970), Stumm and Morgan (1970), Faust and Hunter (1967), and Gould (1967), detail thermodynamic calculations which may be used to evaluate this phenomenon. In many field situations data are commonly lacking on some parameters of importance, thus judgment is often necessary.

Oxidation and Reduction

The oxidation of organic matter in the topsoil is one of the most important pollutant attenuation mechanisms. Oxidation and reduction reactions often work in conjunction with other mechansisms for pollutant attenuation. Besides those reactions causing precipitation, reducing conditions can also theoretically cause the formation of native elements such as arsenic, copper, mercury, selenium, silver, and lead, which are quite insoluble. Reducing environments can convert dissolved sulfate to sulfide and dissolved nitrate to nitrogen gas. Sulfides can react with certain metals to produce highly insoluble precipitates, such as sulfides of arsenic, cadmium, copper, iron, lead, mercury, molybdenum, nickel, silver, and zinc.

Oxidation and reduction reactions are also susceptible to analysis by thermodynamic considerations. These reactions are also subject to laboratory experimentation and no further detail is presented here.

Volatilization

Volatilization and loss as a gas can be effective for sulfate and nitrate. Mercury in solution can be volatilized in anaerobic environments or by reaction with dissolved humic acids. Several organic compounds of arsenic are volatile, and the escape of arsenic as a gas has been demonstrated for both aerobic and anaerobic soils. Selenium may be subject to volatilization, because of its chemical similarity to sulfur. The microbial reduction of nitrate to gaseous forms of nitrogen is well documented. No quantitative procedure is proposed to evaluate the extent of this phenomenon. It is more important to be aware of the pollutants that may be affected.

Biologic Degradation and Assimilation

These processes are very important in the removal of organic and biologic pollutants. Many organic chemicals can be attenuated or removed by biological activity in the vadose zone. Sulfate, nitrate, arsenic, cyanide, mercury, and selenium are likely candidates for biologic fixation or volatilization. Molybdenum is strongly assimilated and concentrated by plants. Crop uptake can remove many of the nutrients in waste waters, particularly nitrogen, phosphorous, and potassium. However, the crop has to be removed or the pollutants may be introduced into the soil-groundwater system. Previous references, given under the step for identification of pollution sources and causes, contain relevant information as to evaluations of the extent of this phenomenon. Again, a general knowledge of the pollutants likely to be affected is of most use.

Radioactive Decay

This mechanism is of great potential value in the attenuation of radioactive wastes by subsurface storage. Storage may be possible for thousands or tens of thousands of years, during which time the wastes would lose much of their activity through decay. Half-lives of many radionuclides are presented in references such as Davis and DeWiest (1966) and Hem (1970). The half-life represents the time required for a given quantity of the radionuclide to decay to one-half the original quantity. Consideration of this phenomenon is unnecessary for most cases of groundwater pollution monitoring, as radioactive pollutants are relatively uncommon.

In summary, attenuation above the water table is greatest for biological constituents. Many organic chemicals and trace elements and some radionuclides and inorganic chemicals will not generally reach the water table. The most mobile constituents would generally be the major inorganic chemical constituents and tritium. Chloride, nitrate, sulfate, sodium, and boron can be fairly mobile. Other major chemical constituents in the topsoil or vadose zone can be mobilized during waste disposal opertions. Changes in pH, oxidation potential, and ion-exchange can cause this mobility. Potassium and phosphorus are ordinarily immobile in alluvial sediments, but could be mobile in flow-through consolidated rocks. Iron and manganese are often found in groundwater and are thus fairly mobile. Trace elements such as chromium, cadmium, arsenic, molybdenum, and selenium, which form anions in water, appear to be fairly mobile in some cases. Others such as barium, mercury, and cobalt are relatively immobile in most soil-groundwater systems. Some pesticides are mobile under certain conditions. Bacteriological pollutants are generally removed above the water table unless this zone is bypassed.

The evaluation of pollutant mobility above the water table requires considerable judgment on soils physics and chemistry, hydrogeology, and water chemistry. Such an evaluation is essential in order to accurately select what portion of the system should be monitored, and to what degree. This judgment may often require a team of specially trained investigators. Experience gained from case histories may be combined with this judgment to effectively perform this step.

STEP 9 – EVALUATE ATTENUATION OF POLLUTANTS IN THE SATURATED ZONE

Many of the attenuation processes which occur in the vadose zone can also occur below the water table, but in a modified manner. For example, the lower oxygen content below the water table reduces the possibility of oxidation of organic matter. Some pollutants, such as iron, may be more mobile in the reduced state. Reducing conditions are favorable, however, in some cases for pollutant removal from water, particularly for sulfate and nitrate. Another major consideration is that organic matter, common in the topsoil, is virtually absent in many types of geologic materials comprising the aquifer. This would ordinarily decrease the extent of sorption as well as reactions such as denitrification. In addition, certain geologic materials, such as granite or limestone, may lack many of the common substrates for sorption. The dilution process below the water table differs greatly from that operative in the vadose zone.

Processes Other Than Dilution

The attenuation processes other than dilution do not have to be considered in detail if there is an adequate thickness of the vadose zone for treatment. In cases where the water table is shallow or a large part of the vadose zone is bypassed during waste disposal, detailed consideration of these processes may be necessary. Filtration and sorption are discussed in Step 8; however, in saturated flow, pollutant movement is generally horizontal. Thus, instead of utilizing a thickness of vadose zone beneath a waste disposal site, a volume of the aquifer is selected. This will often correspond to the projected volume and location of recharged waste water plume at a specific time. This volume can be estimated by utilizing flow net analysis to determine the vertical and horizontal direction of groundwater movement from beneath the waste disposal site. Specific distances from the waste discharge site, such as 100 feet, 500 feet, and 1000 feet, can be chosen and volumes of material calculated for each.

Buffering can be handled as discussed in Step 8. This is not of great concern unless extremely acidic or alkaline wastes are disposed of directly to the saturated zone, i.e., the vadose zone is bypassed. Chemical precipitation can be handled as described in Step 8. One major difference in this process compared to that occurring in the vadose zone is the general lack of evapotranspiration as a factor in concentrating solutions. In addition, the materials are continuously saturated below the water table and are usually not exposed to drying. Oxidation and reduction can be handled as discussed in Step 8. However, in the saturated case, oxidation is generally less important and reduction is more important than in the vadose zone. Volatilization and radioactive decay can be handled as discussed in Step 8.

Dilution and Related Factors

Once percolating wastes reach the zone of saturation, in most dynamic groundwater systems there will be a physical attenuation of pollutant concentrations with distance from the intersection with the water table. This attenuation can be of much greater magnitude than that due to the previously discussed factors. This attenuation is one of the primary factors that mitigates against widespread groundwater pollution. In one sense, this restricted dilution is analogous to that of a plume of polluted air. The attenuation occurring in most cases is determined by the following factors:

1. The volume of waste water reaching the water table
2. The waste loading, i.e., the weight per unit area of pollutant reaching the water table
3. Areal hydraulic head distribution, as indicated by water-level elevation contour maps
4. Transmissivity of aquifer materials
5. Vertical hydraulic head gradients and vertical permeabilities through confining beds which are present
6. Quality of native groundwater in a three-dimensional sense
7. Quantity of recharge reaching the water table from other sources at the land surface
8. Quality of recharge reaching the water table from other sources
9. Well construction
10. Pumpage volumes and patterns

The first two factors determine the concentration of pollutants reaching the water table. The third and fourth factors, along with porosity, determine the direction and magnitude of horizontal groundwater inflow and outflow from the area. The fifth factor determines the direction and magnitude of vertical groundwater flow in the area. The sixth factor comprises the quality of groundwater with which the recharge waste water will mix. The seventh and eighth factors determine the effect of recharge from other sources on pollutant concentration. Lastly, well construction can allow short-circuiting of aquifer materials, and well pumping can drastically alter the hydraulic head distribution.

A first approximation of dilution can be obtained by assuming that the recharged waste enters a certain part of the aquifer, for example, the upper 10 feet, 50 feet, or 100 feet, over a certain area. Knowledge of the extent of groundwater pollution in historical situations in the area, or a comparable area, can be used to make this evaluation. Secondly, water reaching the water table from other sources of recharge and groundwater inflow from nearby areas usually tends to dilute the recharged wastewaters. The dilution can be calculated if the volume and quality of the various sources of water are known. Conservative constituents, such as chloride, can be used for a first approximation of dilution. In most cases, the pollutant of interest will be less mobile, and thus occupy a smaller plume than a mobile constituent such as chloride. Groundwater outflow tends to carry pollutants away from the waste disposal site.

Water level evaluation maps and flow nets can be used to consider whether the waste discharge is in an area of converging or diverging groundwater flow, which affects dilution. Vertical head gradients indicate whether wastes could move to deeper levels of the aquifer or whether deeper aquifer water could move up and dilute the wastes. Both cases tend to accentuate mixing or dilution. Aquifer transmissivity can be used to calculate groundwater flow rates into and out of an area. The quality of sources other than the waste discharge and native groundwater will obviously affect dilution as the lower concentration waters will exert relatively more dilution. The foregoing factors can be integrated into a mass balance analysis, for both the recharge wastewater and the individual pollutants.

Wells affect dilution in several ways:

- Gravel packs or perforations in certain situations can act to short-circuit confining beds and allow vertical movement of pollutants near the well
- Well pumping can drastically alter flow patterns, both horizontally and vertically
- Well pumping can remove pollutants from groundwater and expose them to subsequent loss at the land surface or in the topsoil, by processes such as volatilization, crop uptake, and precipitation

Most of these factors are significant for all types of pollutants that reach the water table, whether they are inorganic chemical, physical, organic chemical, bacteriological, or radiological. In some cases, certain factors do not attenuate the pollutant concentration, but rather increase it. An example is the development of a large depression cone in an agricultural area, whereby pollutants are drawn into the area from many directions but are effectively prohibited from leaving by the depression. The factors leading to attenuation or concentration of pollutant concentrations in the saturated zone have seldom been analyzed in case histories.

Plumes or zones of polluted groundwater may behave as a slow-moving viscous mass, but they may also be quite dynamic, especially where influenced by recharge and/or well pumping.

Evaluation of pollutant attenuation mechanisms in the saturated zone requires considerable hydrogeologic judgment. Such an evaluation is essential in order to properly determine the location and construction of monitor wells. Hydrogeologic judgment combined with experience gained from case histories may be effectively used to perform this step.

STEP 10 – PRIORITIZE SOURCES AND CAUSES

The monitoring methodology presented thus far consists of a sequence of monitoring objectives or steps which must be performed in order to prioritize the identified sources and causes of groundwater pollution for monitoring within each of the selected monitoring areas. These monitoring objectives are presented not only in a chronologically preferred order for accomplishment, but also in an order of increasing difficulty, both in terms of the monitoring methods available to accomplish these objectives and in terms of capability to interpret the data obtained from using these methods. Concomitantly, the costs increase with implementation of these methods and the analytical procedures used in their evaluation.

In the early stages of a monitoring program it will not be possible to make an accurate prioritization of the pollution sources because of incomplete data and knowledge regarding the fate and transport of pollutants in site-specific situations, unless considerable monitoring is already underway. The prioritization scheme is dynamic in nature, and, with time, will tend toward an optimum as more information is gained from a monitoring program.

Each objective of the methodology has been carefully selected to identify and build up the prioritization scheme which follows. In Step 1 criteria for selecting the monitoring areas are set forth. Step 2 prescribes that each monitoring area should be fully inventoried for known or potential sources and causes of groundwater pollution, and the method of waste disposal associated with each source determined. Completion of Step 3 will result in the identification of the biologic, chemical, physical, and radiological characteristics of the pollutants associated with each source, and, as a result, will allow a trial ranking to be made of the sources and causes in terms of the potential of the various pollutants to violate drinking water standards. Completion of Step 4, which has as its objective identification of groundwater usage in the vicinity of each source, will allow a further refinement of the trial ranking of the sources and causes made in Step 3, based this time on their threat to existing groundwater usage.

The various uses of groundwater, including the applicable water quality standards, are discussed in Section 2 of this report under the heading "Quality in Relation to Water Use." Of the possible uses of groundwater, it is apparent that usage for potable water supplies stands out as paramount. Therefore, sources and causes which pose a health threat to potable water supplies will have priority over nonhealth-related damages to potential users. Of foremost concern then in the prioritization will be pollution sources expected to result in the presence of pollutants exceeding the U.S. Public Health Service mandatory drinking water limits. For other sources of pollution the damage to users can be estimated on a monetary basis and the prioritization carried out based on the cost of anticipated damages to existing or potential users. The cost of groundwater quality degradation has been estimated, in areas such as the Los Angeles Basin Southern California, in terms of the following considerations:

- Quality-related consumer costs, i.e., the cost of various treatments such as water softening or chlorination
- The cost of removing the pollutant to the background level of the groundwater resource
- The cost of removing the pollutant in the groundwater to the limits of drinking water standards

In another instance, a detailed investigation was made of an incident where a freshwater aquifer has been polluted by accepted disposal of oilfield brine through an "evaporation" pit and later a faulty disposal well (Fryberger, 1972). Several rehabilitation methods were evaluated in detail, including controlled pumping to the Red River and deep-well disposal. Although real economic damage, both present and future, resulted from this brine pollution, rehabilitation was determined to not be economically justified.

With completion of Steps 2, 3, and 4 of the methodology a much improved appreciation of the potential for groundwater quality degradation at the land surface will be realized.

One of the more difficult tasks facing a DMA charged with monitoring groundwater pollution sources will be to gain an accurate understanding of the fate and transport of pollutants associated with particular sources in site-specific situations. The presence of a pollutant at the land surface does not necessarily demonstrate a threat to groundwater quality. In most situations important data deficiencies will exist relative to pollutant mobility. The application of generalizations regarding the fate and transport of pollutants subsurface should be used with great care. Sometimes under tight budgetary constraints no other choice will exist. In such cases the application of rule of thumb generalizations should be left only to the specialist.

The objectives of Steps 5 and 6 are to gain a site-specific appreciation of the hydrogeology and existing water quality in relation to the pollution sources. In general, it will not be an objective of Steps 5 and 6 to require that regional appraisals of hydrogeology and groundwater quality be conducted, unless a diffuse source covering many square miles is involved. Regional appraisals of hydrogeology are outside the overall objective of the groundwater pollution monitoring methodology described in this report. This does not preclude that regional studies will be conducted by other agencies in cooperation with the DMA.

Steps 7, 8, and 9 comprise the basis for estimating pollutant mobilities and their ultimate presence at a point of groundwater discharge. Based on the estimations of pollutant mobilities determined as a result of the analyses carried out in these three steps, it will be possible to make a more accurate revision of the prioritization made in Step 4. The final three-phase adjustment in the prioritization will be made by considering first, the potential pollutants at the land surface (Step 7), second, pollutants potentially reaching the water table (Step 8), and last, pollutants potentially reaching water supply wells (Step 9).

The refinement in the prioritization carried out as the result of Step 8 will have particular significance in terms of the nondegradation mandates of PL 92-500 and for the protection of sole source aquifers as required under PL 93-523. The refinement in the prioritization carried out as the result of Step 9 will have particular significance in demonstrating whether or not existing users are being threatened and for estimating potential damages to these users. Solute transport models will play an increasingly important role in the years to come as a means for estimating the damage potential to the groundwater users.

Prioritization of potential pollutants at the land surface (Step 7) will require an estimation of the annual volume of liquid waste or weight of solid waste subject to leaching, pollutant concentration, solid or liquid waste loading per unit area, and pollutant loading per unit area.

Completion of the procedural steps and resulting prioritization scheme suggested above will allow a DMA to direct its monitoring resources in a most effective manner toward the primary goals of PL 92-500 and PL 93-523. The prioritization obtained will need to be reevaluated on at least a yearly basis as pollution controls are implemented and as new sources of pollution are detected.

STEP 11 – EVALUATE EXISTING MONITORING PROGRAMS

Every effort should be made to incorporate these ongoing activities in a new monitoring program. In fact, such inclusion is essential if the program is to be both comprehensive and cost-effective.

Existing monitoring activities in an area may be carried on by the following agencies or organizations:

- U.S. Geological Survey – groundwater and surface water quality data
- State geologic and water agencies – groundwater and surface water quality data
- Local water districts (flood control, irrigation, water conservation, etc.) – groundwater and surface water quality data
- Health departments (city, county, and State) – data on quality of groundwater from water supply wells

- Sanitation districts — data on quality of treated waste water effluent before disposal on land

- Industries — data on quality of self-supplied groundwater and of treated waste water effluent before disposal on land

- Local consulting firms working in the water resources field

It should be noted that many of the existing monitoring programs are not specifically oriented toward monitoring groundwater pollution. Thus, their value is often limited. With the review of existing monitoring programs completed, it will be possible to determine monitoring deficiencies. This is accomplished by noting the availability of existing monitoring activities serving the high priority sources and causes. Clearly, if the surveillance systems for these are inadequate or incomplete, monitoring deficiencies exist.

The final result of this step will be a priority listing of sources and causes having monitoring deficiencies. The sequence of this listing will be identical to that developed in Step 10, except that sources which are adequately monitored will be eliminated from the list.

STEP 12 – ESTABLISH ALTERNATIVE MONITORING APPROACHES

As a result of making a first pass through the first 11 steps of the methodology a DMA will have a good appreciation of those sources and causes of groundwater pollution which hold the highest priority for monitoring. Although in some cases the accuracy of this first-cut prioritization will be questionable, an important objective will have been accomplished – that of identifying data and information deficiencies.

In Step 11 the existing monitoring programs were checked to ascertain if it would be possible to modify these existing monitoring activities to collect the needed data. Whether or not this can be accomplished will usually be determined on legal or institutional grounds. In some instances it may be a least-cost approach to pay to have specific data collected during the course of ongoing monitoring programs.

Based on the data and information deficiencies identified in the prioritization of Step 10, it will be possible to identify the required alternative monitoring approaches. To establish the alternative monitoring approaches it will be necessary to first identify and develop pollutant-specific technical monitoring objectives for the pollutants known to emanate from each source. For example, consider a landfill producing leachate high in dissolved solids. A pollutant-specific technical monitoring objective could be to determine total dissolved solids (TDS). A first step would then be to identify the monitoring zone (surface, vadose, or saturated zone), the monitoring methods that could be used to monitor TDS in each zone, the sample analysis techniques for each monitoring method, and sampling frequencies.

Since it is possible to have several combinations of monitoring zones, monitoring methods, sample analyses, and sample frequencies to meet the same technical objective, more than one monitoring approach may exist for each technical objective. By definition, a monitoring approach consists of a mix of monitoring methods. If only one method is feasible for meeting the technical objective, it follows that this method will be the preferred monitoring approach. When several monitoring approaches exist it will be necessary to make a cost-effective comparison between them in order to select the preferred approach.

Selection of Portion of System to be Monitored

The portion, or portions, of the system to be monitored for each source depends on the method of waste disposal, infiltration potential, travel time of waste waters from the land surface to the water table, and the mobility of pollutants in the topsoil and the vadose zone. Obviously, no monitoring is required in the topsoil or vadose zone if they are bypassed during waste disposal. If pollutants undergo significant attenuation above the water table, little monitoring may be necessary in the saturated zone. For example, in the case of phosphorus, which is readily adsorbed on soil particles, little monitoring in the saturated zone would be necessary for most diffuse sources. However, where overloading may occur, such as at point sources of wastes with a high phosphorus concentration, monitoring in the zone of saturation would often be necessary. In cases where significant storage capacity exists in the vadose zone and travel times of pollutants to the water table are long (decades or centuries), sampling in the vadose zone may be stressed. Likewise, in cases where waste water application and infiltration rates are very small (several millimeters per year), monitoring in the vadose zone may be emphasized.

In cases where there is rapid movement through the vadose zone and high waste water application and infiltration rates (tens or hundreds of feet per year), sampling in the vadose zone is minimal and water well sampling is maximal. Where sources are highly variable in potential pollutant composition or in cases where the composition is poorly known, intense source sampling may be necessary. In cases where the composition is fairly constant or relatively well known, little source sampling may be required. Selection of the portion or portions of the system to be monitored requires judgment on the soils, hydrogeology, and water chemistry.

It is also necessary to determine the area over which to sample. This is self-explanatory for land surface monitoring. In the case of the vadose zone, generally, sampling can be confined to the area beneath the pollution source. This is particularly true in cases where geologic materials allow primarily vertical movement to the water table. In cases where layers of low permeability inhibit vertical flow, significant lateral flow may take place. If this factor is predominant, sampling in parts of the vadose zone only under the source of pollution may be ineffective. In this case sampling of a much larger area may be warranted, and batteries of piezometers and tensiometers may be necessary to determine this area.

In the saturated zone, the area to sample depends largely on pollutant attenuation factors in the aquifer. These factors have been discussed previously in Step 9 of the methodology. Application of the methodology of Step 9 will allow estimation of the horizontal extent of pollutant travel in the aquifer.

Selection of Nonsampling Methods

Nonsampling methods at the land surface almost always will include the waste load inventory. The testing of liners, pipelines, and tanks for leakage is obviously only applicable to cases where the type of waste disposal involves these items. Aerial surveillance is generally useful when sources of pollution and methods of waste disposal are poorly known or when the method of waste disposal is highly variable. Nonsampling methods in the vadose zone largely comprise in-place measurements of moisture characteristics, such as by neutron logging, tensiometers, and moisture blocks. These methods are most useful for point sources of pollution. Experienced soils scientists and hydrologists can best select the appropriate tool. In the saturated zone, most nonsampling methods are of direct use only in special cases, for example, resistivity surveys which are generally applicable only to cases of high salinity waste disposal.

Determination of Required Analyses

The pollutants in the major sources have been previously described. For the determination of required analyses, consideration is given to liquid wastes, solid wastes, soil and geologic materials, and water in the soil-groundwater system.

- **Liquid Wastes.** Often the analyses of liquid wastes can be based on known compositions from the literature. Specific parameters of importance to groundwater pollution can be selected. The more mobile constituents would have priority, as well as those of greatest importance in subsequent reuse of polluted water. There are great advantages to running relatively complete chemical analysis for the major species commonly found in groundwater, namely calcium, magnesium, sodium, potassium, carbonate, bicarbonate, chloride, sulfate, nitrate, and silica. Electrical conductivity, total dissolved solids, and pH are also routinely determined. With the determination of other major ionic species in polluted waters, such as ammonium and fluoride, certain procedures can be used to check the chemical analyses. Determination of these parameters also allows for limited chemical equilibrium calculations. In case of a drinking water supply, fluoride and arsenic would also be important parameters. For agricultural supply, boron would also be important. Some of the more common trace elements in polluted groundwater are arsenic, chromium, cadmium, molybdenum, iron, and manganese. With respect to nitrogen, phosphorus, and sulfur, various forms may have to be determined, such as ammonia, organic nitrogen, nitrite, and nitrate in the case of nitrogen. Oxidation potential is also important in this case.

    Electrical conductivity monitored on a regular basis can be used as an indicator for concentration changes in major constituents. Monitoring for total salinity in salt balance studies may only require determination of electrical conductivity. Another widely used measurement is chloride content. Chloride is common in many types of liquid waste, is extremely mobile in the soil-groundwater system, and is therefore of special value as a tracer. Representative constituents of interest in groundwater pollution are given by Everett et al. (1976). If wastes have certain characteristics and insufficient attenuation occurs, secondary effects of crucial importance to groundwater pollution monitoring could result. For example, acidic wastes, which if not buffered sufficiently above the water table, could mobilize certain trace elements in minerals in the aquifer.

    Bacteriological determinations are not of great importance in waste water, unless specific pollutants have been found in groundwater near a pollution source. The bacteriological composition of most waste waters is generally known from the literature. Radiological determinations are very important in selected cases. The composition of organic chemicals possibly entering groundwater could assume a major role in the future. At present most organic chemicals are known and will rarely have to be sampled in waste liquids. Again if a problem is found in well samples, a specific pollutant may need to be monitored in waste waters.

- **Solid Wastes.** Analyses to be run on solid wastes are highly variable. In general, the major chemical constituents are well known from the literature (for example, gypsum stockpiles). Thus analyses usually focus on trace element content and radioactivity.

- **Soil and Geologic Materials.** Physical analyses on soil are performed in order to monitor moisture characteristics and water movement in the vadose zone. The selection of these parameters is based on soil and hydrogeologic judgment. Chemical analyses will usually be limited to materials absorbed on or precipitated in the materials. Trace elements and radionuclides would be commonly analyzed. Trace elements and radionuclides in geologic materials which might be mobilized by disposal of certain toxic wastes may also be analyzed.

- **Water.** The analyses for water in the vadose zone and aquifer are related to those for waste water at the land surface. However, analyses of wastes at the land surface can considerably narrow down the required analyses for water beneath the land surface. The relative immobility of some pollutants in the topsoil often renders analysis of water for these constituents in the vadose zone unnecessary. For example, many bacteriological constituents and trace elements can be removed in the topsoil under a favorable set of circumstances. Sampling in the vadose zone indicates additional attenuation of pollutants. Thus, in most respects, analyses for aquifer water may require at least determinations, since many potential pollutants will not reach the aquifer. In any case, the major inorganic species commonly found in groundwater should generally be determined. This is primarily for the sake of checking chemical analyses, equilibrium calculations, and providing data for the use of trilinear diagrams and other analytical tools.

As a rule, in the early part of a monitoring program, sampling for detailed determination is made. This decision is based on the knowledge of trace elements, radionuclides, bacteriological constituents, and organic chemicals in the aquifer of the selected area. Nationwide, there is a dearth of data on these substances. The substances of major concern will be elucidated by source monitoring, and secondarily by monitoring in the topsoil and vadose zone.

Determination of Sampling Frequencies

The selection of sampling frequencies often involves a trial and error procedure. The fluctuations with time in composition over short periods (days or weeks) are greatest for waste water at the land surface, and least for groundwater samples. Passage of pollutants through the topsoil and vadose zone will tend to smooth out fluctuations in composition at the land surface. In sampling at the land surface, the foregoing should be kept in mind. Often the extremes in concentration are not desired, but rather an integrated or typical value. For waste water sampling, open pond water subject to mixing processes can often be sampled monthly to determine the major characteristics. Compositing may be advisable, but is usually done on the waste stream. A trial and error procedure is often used, and the main idea is to establish seasonal changes or patterns in quality. Chemical hydrographs can be plotted and used as an aid in selection of optimal sampling frequencies.

Geologic and soil samples are normally analyzed less frequently than water. In many cases geologic sampling may only be done once. Easily retrievable soils samples (upper 5 to 10 feet) may be analyzed several times a year or annually. Water in the vadose zone in areas of rapid percolation should be done monthly or quarterly. The frequency of sampling in this case is limited by the slowness of water movement into sampling devices. In areas of slow percolation to the water table, measurements may be made only every 5 or 10 years.

Water samples in the aquifer can often be collected on a semiannual basis for large-capacity wells once seasonal trends are established. In the early part of sampling programs, analyses on a weekly, daily, hourly, or more frequent schedule may be necessary. In cases of point sources, monthly analyses may be necessary. For diffuse sources, annual sampling of wells pumping long time periods will suffice, when seasonal trends have been established. For low-capacity wells, more frequent sampling of a greater number of wells is necessary, and monthly sampling may be necessary in many cases. Chemical hydrographs can also be used in this case as an aid in selection of optimal sampling frequencies.

Specific examples which illustrate the selection and application of alternative monitoring approaches using the stepwise procedure outlined in this chapter are given in Section II of Chapter VI.

STEP 13 – SELECT AND IMPLEMENT THE MONITORING PROGRAM

In Step 12 a procedure for developing pollutant-specific preferred monitoring approaches based on cost-effectiveness was discussed. When working with a source containing many pollutants it will not be unusual to go through several suboptimizations before a source-specific preferred monitoring approach can be selected. By definition, a source-specific preferred monitoring approach consists of the set of pollutant-specific monitoring approaches. The reason for this suboptimization is that in many cases it will be possible to measure more than one pollutant using the same method. The normal procedure when dealing with several pollutants from a single source will be to examine each in terms of the available methods for sampling and analyses and select the one best suited.

After a source-specific preferred monitoring approach has been selected for the highest priority source, the cost of implementing this monitoring approach will need to be computed and compared against the monitoring budget. If funds remain, the procedure for selecting a source-specific preferred approach and its implementation will be repeated for the next highest priority source and so on to each lower priority source until the budget is depleted.

A preferred monitoring program will consist of the set of selected source-specific preferred monitoring approaches. Monitoring approaches, e.g., remote sensing, which provide information pertaining to more than one source, may assist in achieving cost-effectiveness at the program level. Certain other economies of scale will also exist at the program level when discounts for materials and chemical analyses are available, and the cost of personnel and equipment is spread over many sources.

When implementing a monitoring program for an area, a DMA must see that the necessary monitoring activities are carried out. This will involve making personnel assignments, purchasing monitoring equipment and data handling supplies, and letting contracts for services, such as drilling, performing well tests, and performing chemical analyses. A most important action on the part of a DMA will be to insure that a strong quality assurance program is carried out. A key consideration in this program will be for the DMA to provide valid data to support enforcement actions in situations where environmental harm is projected.

It is important to recognize that additional monitoring may not imply large expenditures. Innovative techniques may produce useful quality data at low cost. For example, consider a large unmonitored sanitary landfill, which potentially could be a major pollution source. If the DMA decides that monitoring of this source is important, the first move should be to search for existing wells near and on the downgradient side of the landfill. Assume that three wells — two domestic and one irrigation — are found which satisfy the locational conditions. The appropriate action would be for the DMA to arrange for regular sampling and analysis of the three well waters. If no wells were available then a more expensive alternative would be to drill one or more monitoring wells.

A data management system applicable to groundwater quality is described in Chapter IV. The selection of a system for a given area requires consideration of several factors. One is the volume of data to be handled by the system. Second is the use and distribution that will be made of the data. Third, the compatibility of a system with others that may be in use in the area must be considered. Fourth, decisions may already have been made at a higher administrative level, such as by a State, which will specify the monitoring system to be employed. Fifth, the cost of the system is an important factor. Where the monitoring program is small, as in an undeveloped rural area, the data management system can be extremely simple — perhaps only xeroxing and filing of a few data sheets. In a large urban area which is heavily pumped, a sophisticated system involving digital computers may be essential to handling the input of monitoring data.

STEP 14 — REVIEW AND INTERPRET MONITORING RESULTS

A key function of a DMA, after the implementation of a monitoring program is underway, will be to collect and to review all current monitoring data in its area. The data should be analyzed and interpreted to define quality trends, new pollution problems, regions of improvement, and effectiveness of pollution control activities. Assessment, such as those portions of the groundwater resource not meeting water quality standards and predictions of future quality under projected population and land use conditions, should be prepared.

The responsibility for analyzing monitoring data to convert it to water quality information will be a continuing activity for a DMA. As new data are received, they should be studied promptly in order to detect changes rapidly, particularly those requiring immediate attention or action.

STEP 15 — SUMMARIZE AND TRANSMIT MONITORING INFORMATION

The final result of a monitoring program organized in an area by a DMA is information on groundwater quality. Thus, the final task of a DMA is to disseminate the information gained in usable forms to the agencies and organizations concerned with such information.

Monitoring should be summarized in appropriate forms for convenient study before distribution outside of the DMA. This may involve preparation of tables showing averages and/or changes in water quality. Similarly, graphs prepared to readily display long-term trends may be helpful. Maps showing, for example, locations of major known sources of pollution, areal distributions of concentrations of key pollutants, and regions having groundwater with qualities not meeting drinking water standards, can also be shown to be both useful and effective.

Monitoring information should be distributed regularly to appropriate public agencies — local, State, and Federal. Major industries in the area should receive the material as well as cooperating agencies and organizations that contribute monitoring data. In addition, the general public should be informed of the results of the monitoring program; therefore, reports should be sent to local newspapers, citizens' groups,

Chambers of Commerce, and conservation and environmental organizations. It can be expected that the public awareness of groundwater quality created by such publicity efforts will indirectly act to encourage individuals and organizations to preserve and protect the underground resource which they, perhaps for the first time, more fully understand.

Finally, a DMA has the responsibility to alert action and enforcement agencies of critical problems or situations discovered within the monitoring program. This may involve, for example, detection of hazardous or toxic pollutants, which could imminently affect a nearby municipal well field. Prompt reporting of such instances is essential as well as following up with specialized monitoring efforts for documenting and controlling emergency situations.

Example 1 – Tabular Presentation of Groundwater Quality. Table 2-15 illustrates a format for presenting groundwater quality data in tabular form. With the exception of temperature, only chemical quality data are included. Note that the ionic constituents are presented both in milligrams per liter and milli-equivalents per liter, and that hardness includes both total and noncarbonate values. The well-numbering system is based on township and range designations so that the well location is known from the table to the nearest quarter-quarter section. In this table, some wells have been sampled semiannually in the spring (for high groundwater level conditions) and in the fall (for low groundwater conditions), while others have been sampled only once a year.

Example 2 – Graphical Presentation of Groundwater Quality. Graphical representations of analyses of the chemical quality of groundwater are useful for display purposes, for comparing analyses, and for emphasizing similarities and differences. Graphs can also aid in detecting the mixing of waters of different composition and identifying chemical processes occurring as groundwater moves through the underground. A variety of graphical techniques is available; some of the more useful ones are described and illustrated in the following paragraphs.

Figure 2-37 illustrates the most widely used bar graph in the United States. Each analysis appears as a vertical bar graph whose total height is proportional to the total concentration of anions or cations, expressed in milliequivalents per liter. The left half of the bar represents cations, and the right half anions. These segments are divided horizontally to show the concentrations of major ions or groups of closely relative ions, which are shown by distinctive patterns. Numbers at the tops of the bars serve to identify the analyses.

Figure 2-38 shows the same bar graph as Figure 2-37 but with addition of a bar for hardness. Values of hardness are expressed in milligrams per liter as $CaCO_3$, which is equivalent to the sum of the calcium and magnesium segments, in milliequivalents per liter, multiplied by 50.

Figure 2-39 also shows the same bar graph as Figure 2-37 but with the addition of a horizontal bar for silica. Here the concentration of silica is given in millimoles per liter because milliequivalents cannot be used for uncharged solutes.

Figure 2-40 illustrates a system of plotting quality by radiating vectors in which the lengths of the six vectors from the center represent ionic concentrations in milliequivalents per liter.

Table 2-15

ANALYSES OF CHEMICAL QUALITY OF GROUNDWATER PRESENTED IN TABULAR FORM
(California Department of Water Resources, December 1974)

Owner and Use Sources; State Well and Other Numbers; Dates Sampled	Temp (°F)	Specific Conductance (micromhos at 25°C)	pH	Mineral Constituents in mg/l											Total Dissolved Solids (ppm)	Sodium (%)	Hardness as $CaCO_3$		Analyzed by
				Calcium (Ca)	Magnesium (Mg)	Sodium (Na)	Potassium (K)	Carbonate (CO_3)	Bicarbonate (HCO_3)	Sulfate (SO_4)	Chloride (Cl)	Nitrate (NO_3)	Fluoride (F)	Boron (B)	Silica (SiO_2)			Total ppm	NC ppm	
TIA JUANA VALLEY BASIN (9-19)																				
Calif. Water and Telephone Co. 18S/2W-32H1 4-11-60	70	9852	7.3	504 25.15	292 24.05	1266 55.07	8 0.20	0 0.00	544 8.93	663 13.81	2870 80.93	5 0.09	1.0 0.05	0.72	32	7220	53	2160		DWR
11-15-60	70	13780	7.3	626 31.25	375 30.75	1870 81.30	2.5 0.06	0 0.00	528 8.65	506 10.54	4290 121.00	0 0.00	0.5 0.03	1.26	19	9630	57	3100	2667	Lein
18S/2W-32P# 4-11-60	69	22860	7.3	608 30.34	707 58.16	3710 161.40	90 2.30	0 0.00	2.96 4.86	1017 21.18	7900 222.80	9 0.14	1.0 0.05	0.83	30	16320	64	4425	4090	DWR
11-14-60	68	23180	7.3	661 30.34	654 53.75	3850 167.45	108 2.77	0 0.00	303 4.96	1104 22.98	7890 222.50	0 0.00	0.4 0.02	1.16	19	15618	65	4338		Lein
19S/2W-4A5 4-12-60	68	1965	7.6					0 0.00	326 5.35		402 11.35							510	242	DWR
12-6-60	66	2007	7.6					0 0.00	303 4.96		408 11.51							491	243	DWR
19S/2W-5Q6 4-12-60	71	22260	7.0	680 33.93	681 56.07	3630 157.90	24 0.61	0 0.00	300 4.96	1075 22.40	7700 217.10	7 0.11	0.7 0.04	0.66	28	16250	64	4500	3896	DWR
11-14-60	70	20780	6.7	714 35.63	573 47.09	3690 160.50	29 0.74	0 0.00	293 4.80	1066 22.20	7700 217.10	26 0.42	1.1 0.06	0.69	19	15692	66	4136		DWR
Henry Schaffner 18S/2W-35L1 4-12-60		3830	7.6					0 0.00	567 9.30	1326 37.40									1245	236	DWR
12-6-60		2829	7.5	131 6.54	52 4.27	430 18.71	5.6 0.14	0 0.00	368 6.04	214 4.46	678 19.12	2.5 0.04	0.58 0.03	0.48	28	1740	63	540		DWR
19S/2W-2E1 4-12-60		2550	7.6	226 11.30	87 7.15	589 25.60	9.4 0.24	0 0.00	424 6.94	455 9.48	909 25.65	57 0.92	1.2 0.06	0.53	16	2650	58	923	575	DWR
Aballo and Wright 19S/2W-3A1 4-12-60		2130	7.7	187 9.35	80 6.55	370 16.10	7.8 0.20	0 0.00	339 5.55	309 6.44	709 20.00	6.2 0.10	1.0 0.05	0.14	22	1424	50	795	517	DWR
Knox Dairy Farm 19S/2W-5G18 4-12-60	71	7072	7.5	328 16.37	190 15.63	955 41.54	5.6 0.14	0 0.00	405 6.64	472 9.83	1980 55.84	14 0.23	0.96 0.05	0.62	24	4967	56	1600		DWR
11-15-60	70	7210	7.0	406 20.25	157 12.85	975 42.40	2.5 0.06	0 0.00	387 6.35	845 17.61	1915 54.00	0 0.00	0.6 0.03	0.66	19	5390	56	1655		Lein

104

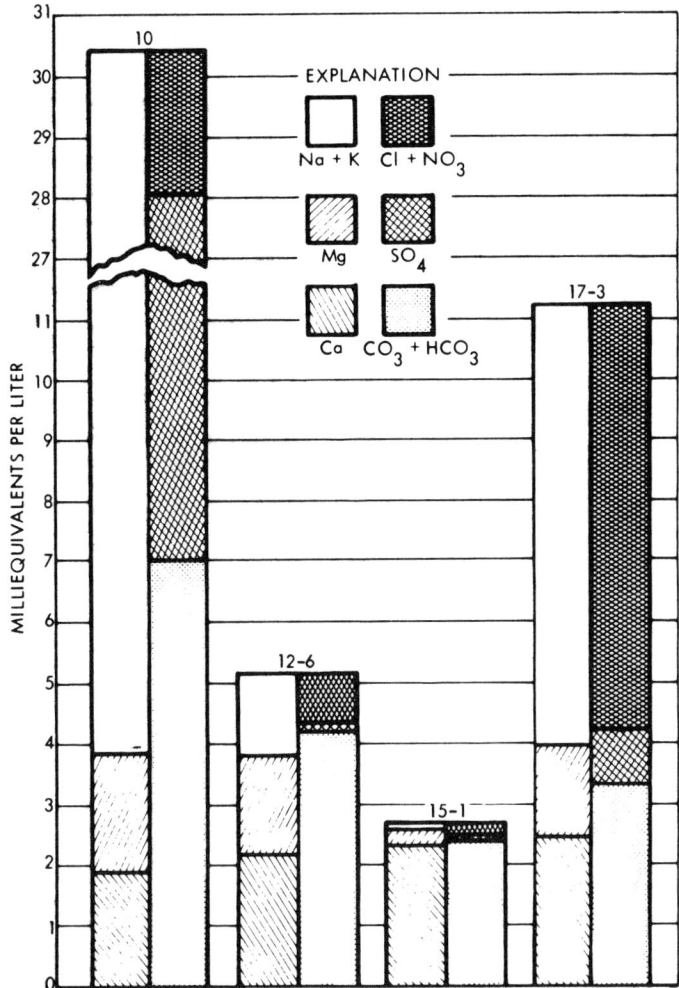

Figure 2-37. Vertical bar graph of chemical quality expressed in miliequivalents per liter (Hem, 1970).

Figure 2-41 shows a method for presenting analyses using four parallel axes extending on each side of a vertical zero axis. Concentrations of cations are plotted to the left and anions to the right, all in milliequivalents per liter. The resulting points are connected to form an irregular polygonal pattern; waters of similar quality define a distinctive shape.

Figure 2-42 indicates still another graphical representation of water quality. Here a circular "pie" diagram is drawn with a scale for the radii which makes the area of the circle represent the total ionic concentration. Subdivisions of the area show the proportions of the different ions as percentages of the total milliequivalents per liter.

Figure 2-43 illustrates a trilinear diagram, which is a useful method for representing and comparing water quality analyses. Cations, expressed in percentage of total cations as milliequivalents per liter, plot as a single point on the left triangle, while anions, similarly expressed as a percentage of total anions, appear as a point in the right triangle. These points are projected into the central diamond-shaped area parallel to the upper edges of the central area. This single point is thus uniquely related to the total ionic quality, and at this

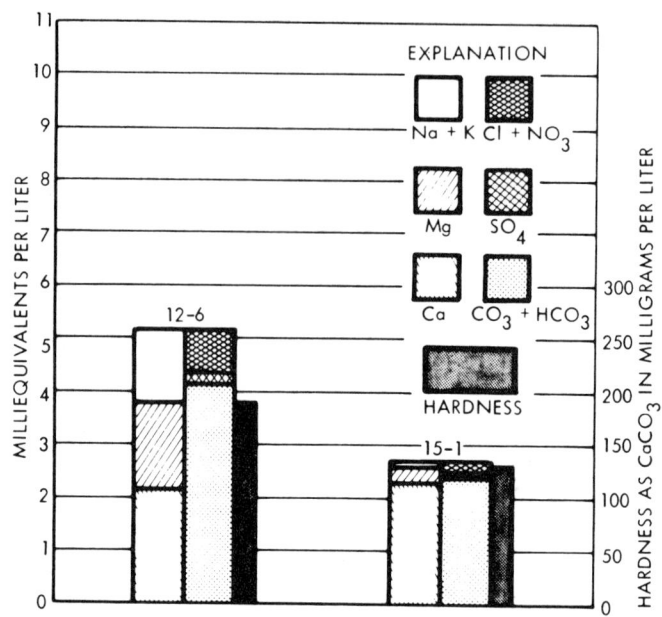

Figure 2-38. Vertical bar graph of chemical quality expressed in milliequivalents per liter which also shows hardness as CaCO$_3$ in milligrams per liter (Hem, 1970).

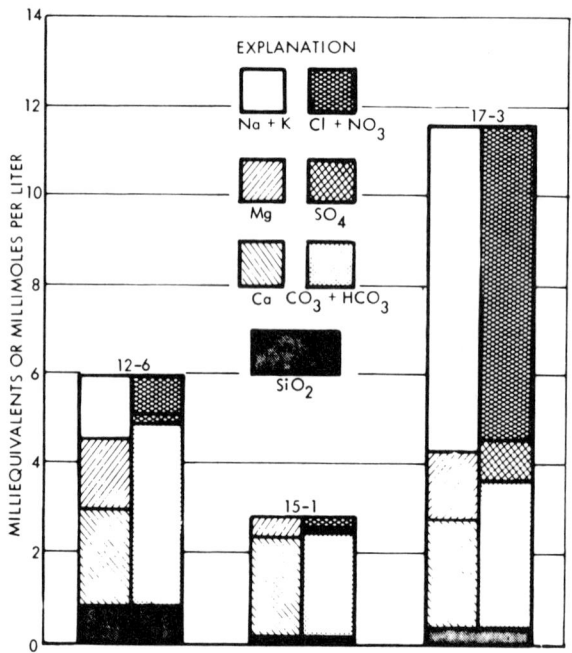

Figure 2-39. Vertical bar graph of chemical quality expressed in milliequivalents per liter which also shows silica in millimoles per liter (Hem, 1970).

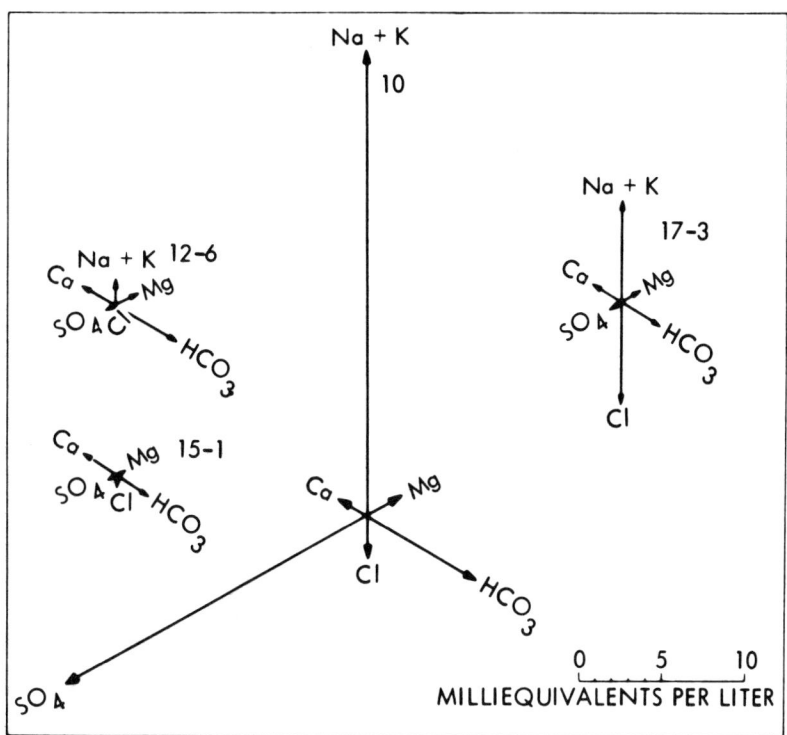

Figure 2-40. Radial vector diagram of chemical quality expressed in milliequivalents per liter (Hem, 1970).

point a circle can be drawn with an area proportional to the total dissolved solids concentration. The trilinear diagram is a convenient way to distinguish similarities and differences among various groundwater samples as waters with similar qualities will tend to plot together as groups. Also, simple mixtures of waters can be identified. For example, an analysis of any mixture of waters A and B will plot on the straight line AB on the diagram, where A and B are the analyses of the two component waters.

Figure 2-44 is a modification of the trilinear diagram, which is convenient where a large number of analyses are to be graphically displayed. The triangle and diamond area are the same as in Figure 2-43, except that they are shifted in position. The difference is that analyses are shown as points rather than circles on the diamond area, and total dissolved solids are plotted to the right on a parallelogram with a logarithmic scale. In addition, hardness as $CaCO_3$ is plotted on a second parallelogram to the right of the cation triangle.

Example 3 – Quality Variation with Time. Figure 2-45 shows graphically the variation in chloride concentration with time for groundwater at Burlington, Massachusetts. The increase in chlorides is associated with the advent of road salting in the area. Salt was first stored uncovered in 1961, approximately 400 feet upgradient from the town's well field. At that time the chloride concentration was less than 15 mg/ℓ. By 1963 the concentration began to increase notably. Remedial measures by Burlington included construction of a salt-storage shelter in 1968, when the chloride reached 170 mg/ℓ, and banning the use of deicing chemicals on city streets in 1970, when the concentration exceeded the 250 mg/ℓ upper limit, recommended by the U.S. Public Health Service. Chlorides decreased after 1970 and were down to 85 mg/ℓ in 1972, perhaps

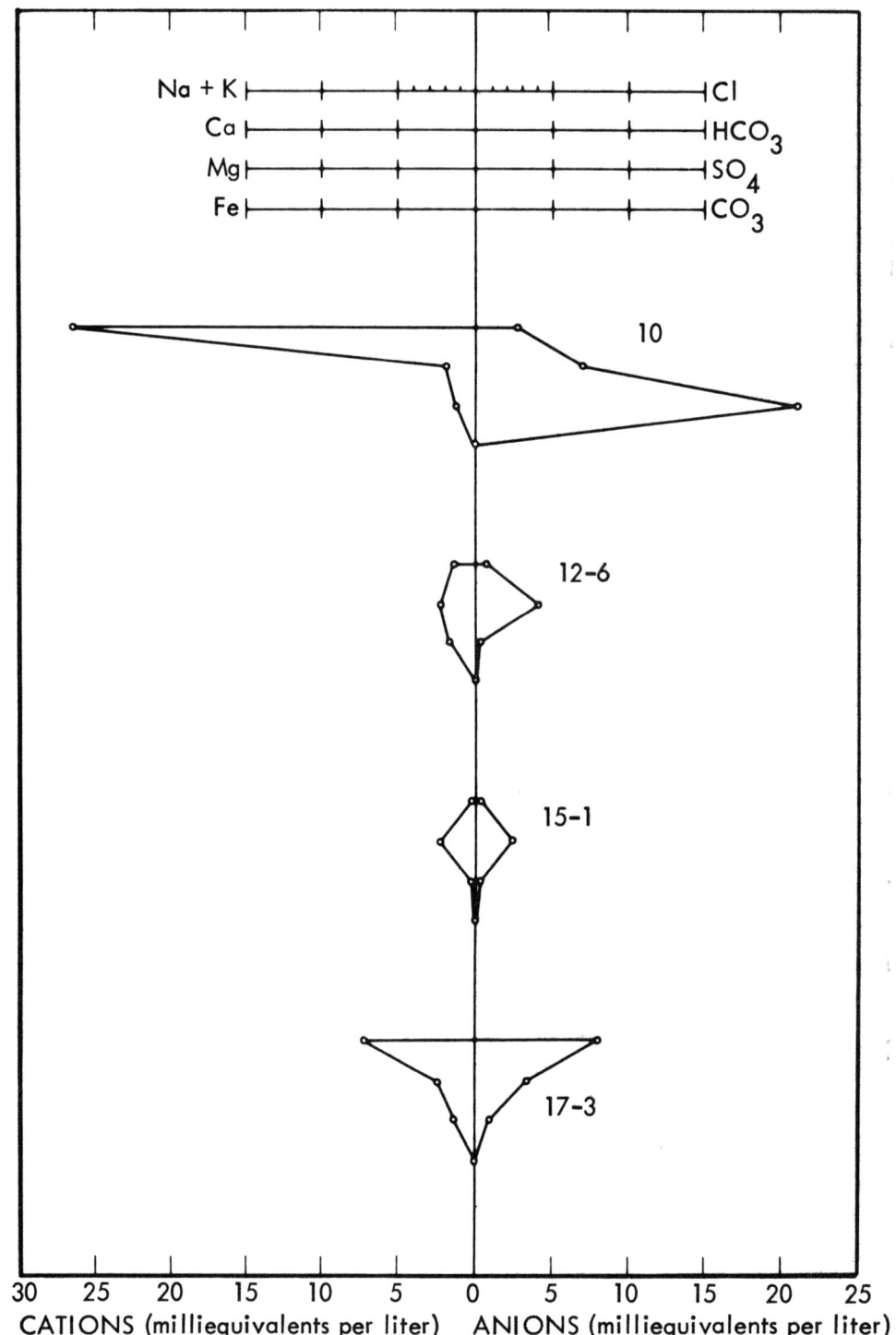

Figure 2-41. Pattern diagram of chemical quality expressed in milliequivalents per liter (Hem, 1970).

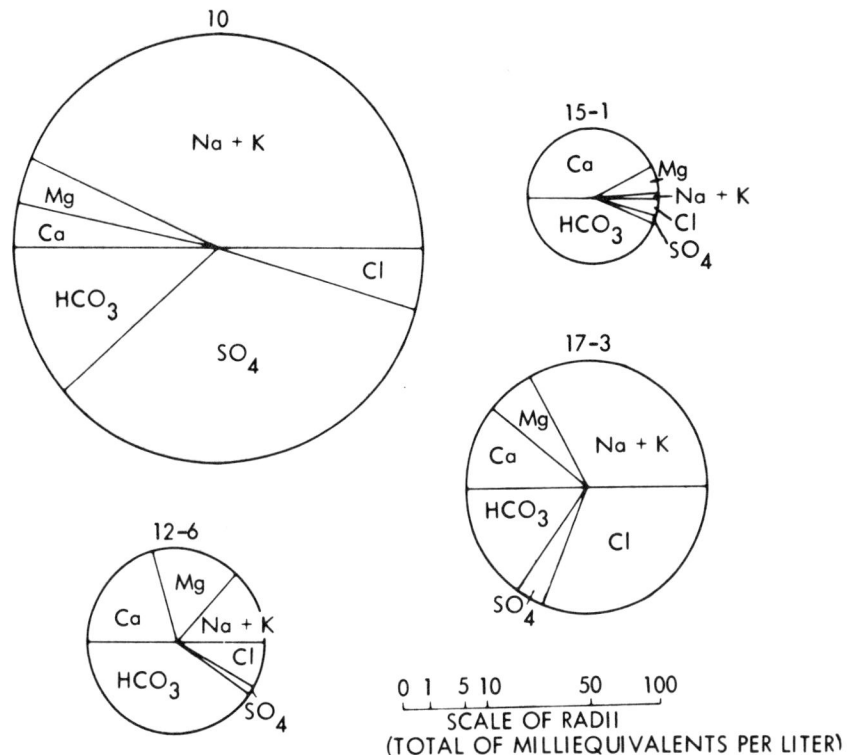

Figure 2-42. Circular diagram of chemical quality with subdivisions showing percentages of total milli-equivalents per liter (Hem, 1970).

assisted by abnormally high rainfall that year. The deicing ban was lifted in December 1972.

Figure 2-46 illustrates another change of groundwater quality with time. Here, seasonal variations in nitrate measured in groundwater near a sewage treatment plant in Fresno, California, are plotted. A significant annual cycle is apparent, with a maximum in the fall and a minimum in the spring. This fluctuation can be attributed to the discharge of high nitrogen content wastes from wineries in the area during the fall of each year. It is clear that frequent sampling is necessary to define these short-term changes.

Example 4 – Quality Variation with Distance. Figure 2-47 presents the variation in electrical conductivity of groundwater along the length of the Santa Ana River Basin in California. Electrical conductivity (EC) is an inexpensive measure of total dissolved solids; normally TDS is about 60 percent of EC. The pronounced increase in conductivity in a downstream direction can be attributed to a combination of urbanization, wastewater disposal, and artificial recharge of poorer quality imported water.

Example 5 – Quality Variation with Depth. Figure 2-48 shows the variation of total dissolved solids with depth for groundwater in one portion of the Santa Clara-Calleguas Area, Ventura County, California. At this location there are three distinct aquifers present. The samples from wells penetrating the aquifers indicate that there is little significant difference in quality in the two lower aquifers; however, the shallow perched water has been appreciably degraded, probably from waste disposal or seawater sources.

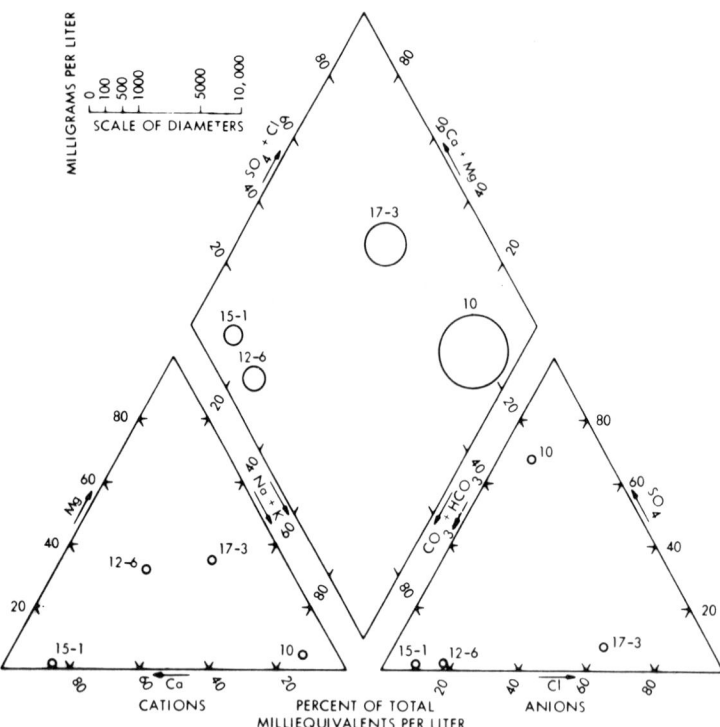

Figure 2-43. Trilinear diagram of chemical quality expressed in percentages of cations and anions as milliequivalents per liter and represented by two points and a circle (Hem, 1970).

Example 6 – Area Variation in Quality. Figure 2-49 is a map of total dissolved solids in groundwater of the Santa Clara-Calleguas area, Ventura County, California. Isosalinity lines cover the water-bearing portions of the area, and are drawn on the basis of quality analyses from well water samples gathered throughout the basin.

Example 7 – Vertical Cross Section of Quality. Figure 2-50 presents a vertical cross section of an alluvial aquifer in Michigan. Pollution resulted from disposal of industrial wastewaters into disposal ponds; chloride is used as an indicator of the pollution. Control of spread of the pollutant underground was accomplished by a combination of recharge ponds for cooling water and purge wells, both shown in Figure 2-50. These provided an effective hydraulic barrier to the spread of the pollution. Monitoring wells throughout the area enabled hydraulic and quality variations of the groundwater to be evaluated.

Example 8 – Maps of Pollution Plumes. Figure 2-51 outlines the plume of pollution in groundwater resulting from an oilfield brine disposal pit in southwestern Arkansas. This map shows that highly saline water, expressed by contours of chloride concentration, has spread generally southward in the direction of groundwater flow. The irregular shape of the pollution zone may be attributed to variations in permeability within the aquifer and to irregularities in the top surface of the shale which forms the bottom of the alluvial aquifer.

Figure 2-52 shows a long narrow plume of groundwater pollution produced by a landfill near Munich, West Germany. Some 70 monitoring wells were measured

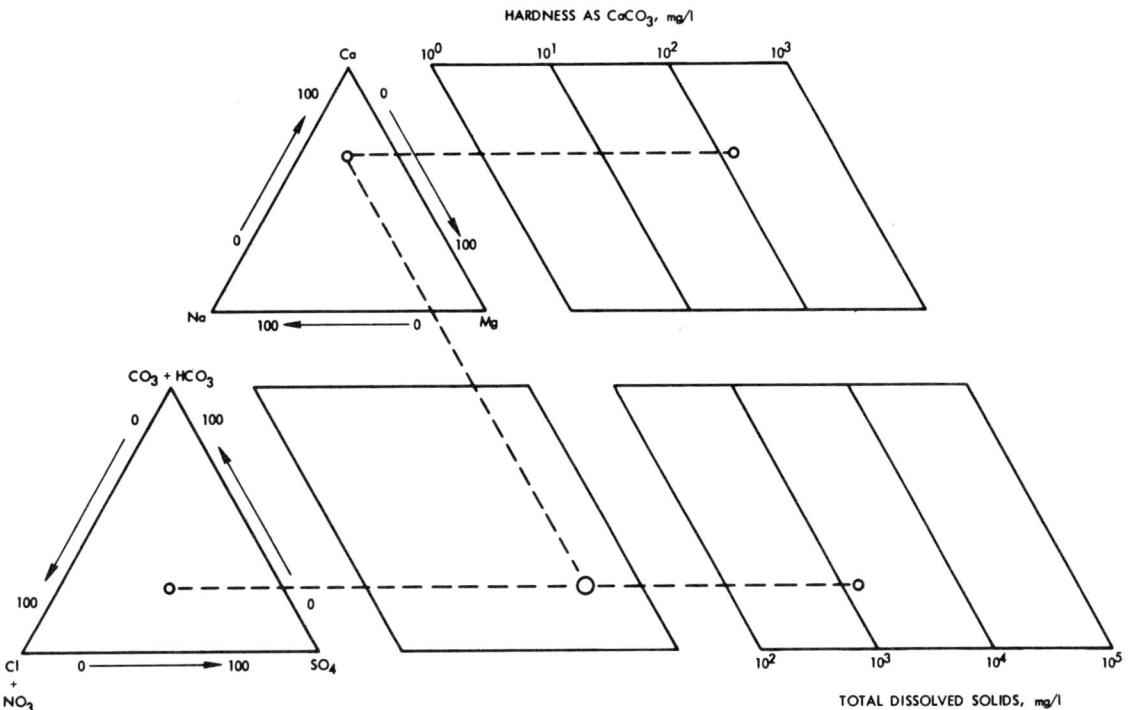

Figure 2-44. Comprehensive trilinear diagram of chemical quality in which five points represent the proportions of cations and anions together with total dissolved solids in milligrams per liter and hardness as CaCO$_3$ in milligrams per liter.

biweekly at this site for chloride, electrical conductivity, and temperature. Only chloride concentrations are shown in Figure 2-52; however, all three indicators of pollution display nearly identical configurations, suggesting that they are conservative and hence useful parameters for detection of subsurface polution.

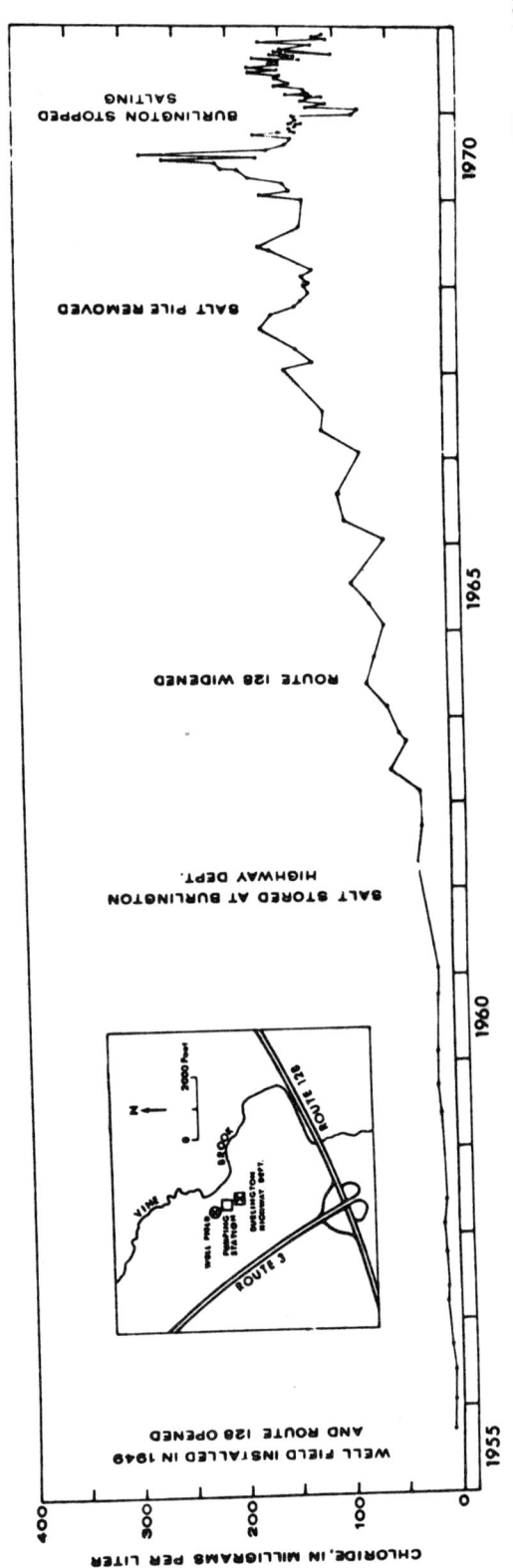

Figure 2-45. Variation in chloride concentration with time for groundwater at Burlington, Massachusetts (Terry, 1974).

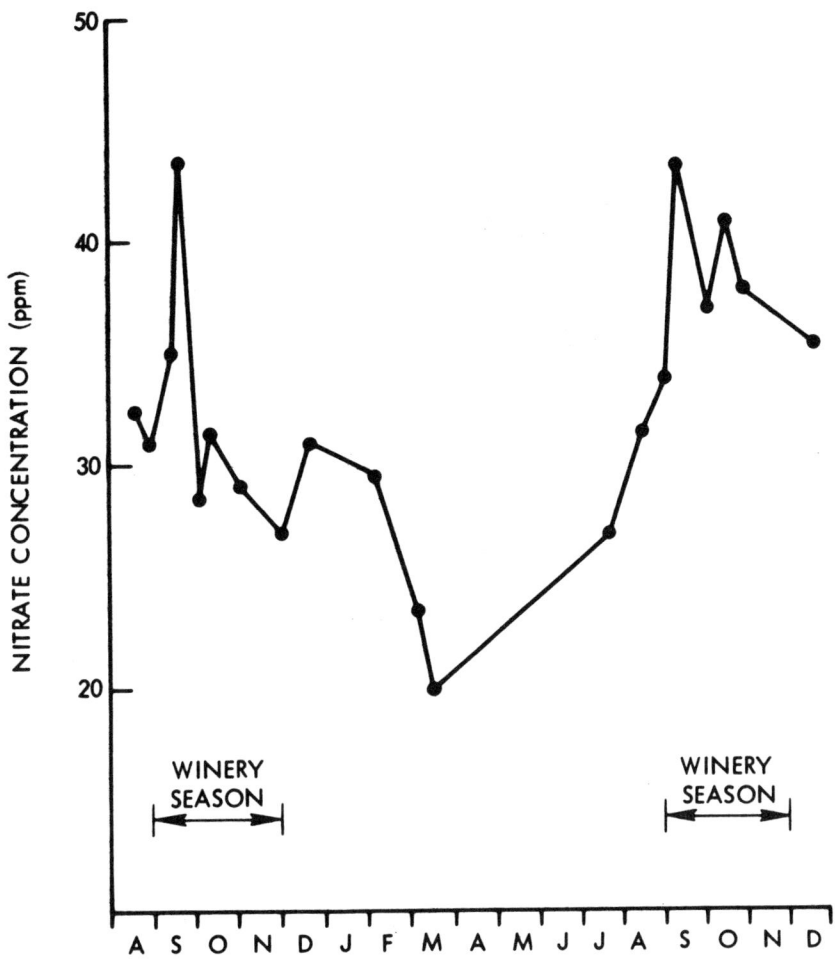

Figure 2-46. Seasonal variation in nitrate concentration for groundwater at Fresno, California (Schmidt, 1972).

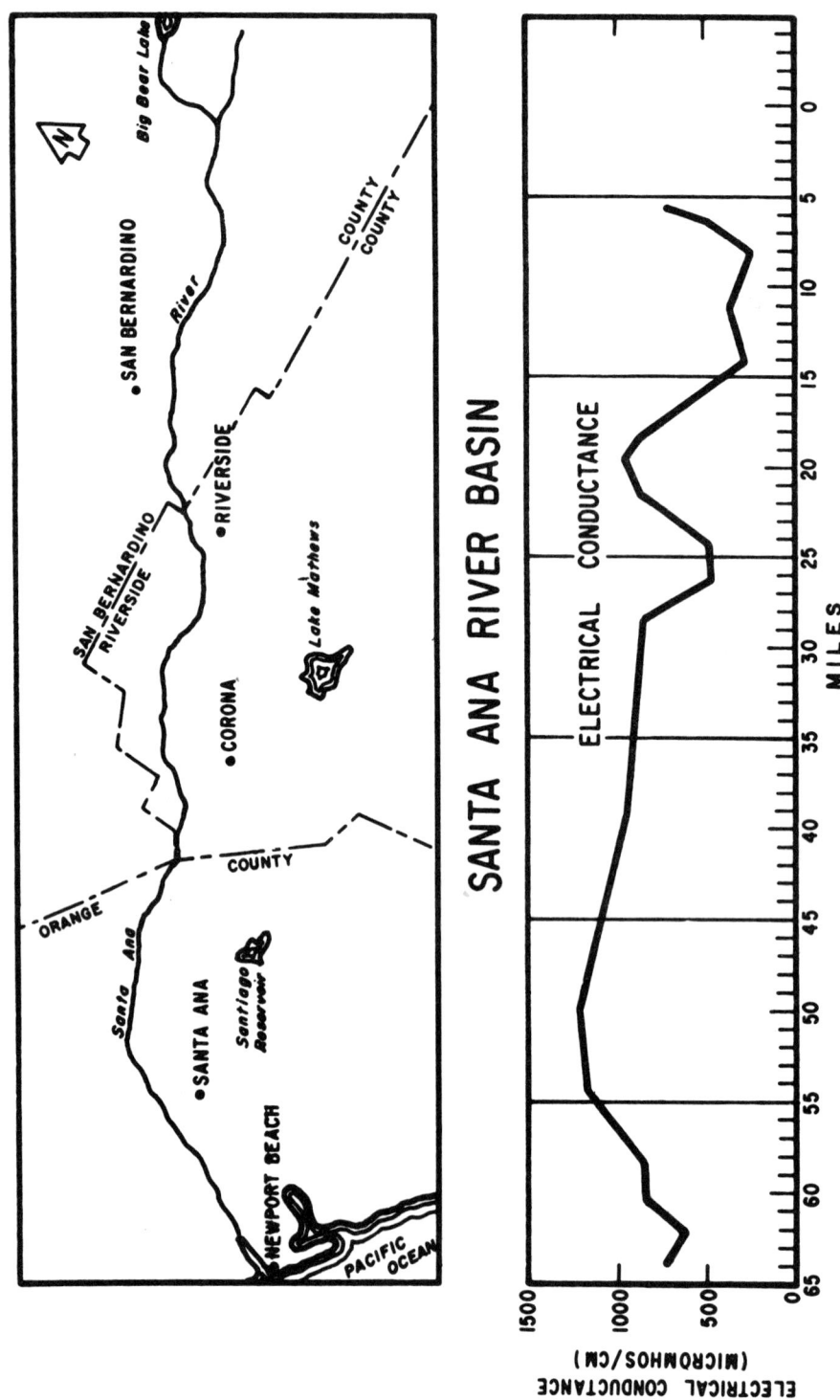

Figure 2-47. Variation in electrical conductivity of groundwater along the length of the Santa Ana River Basin, California

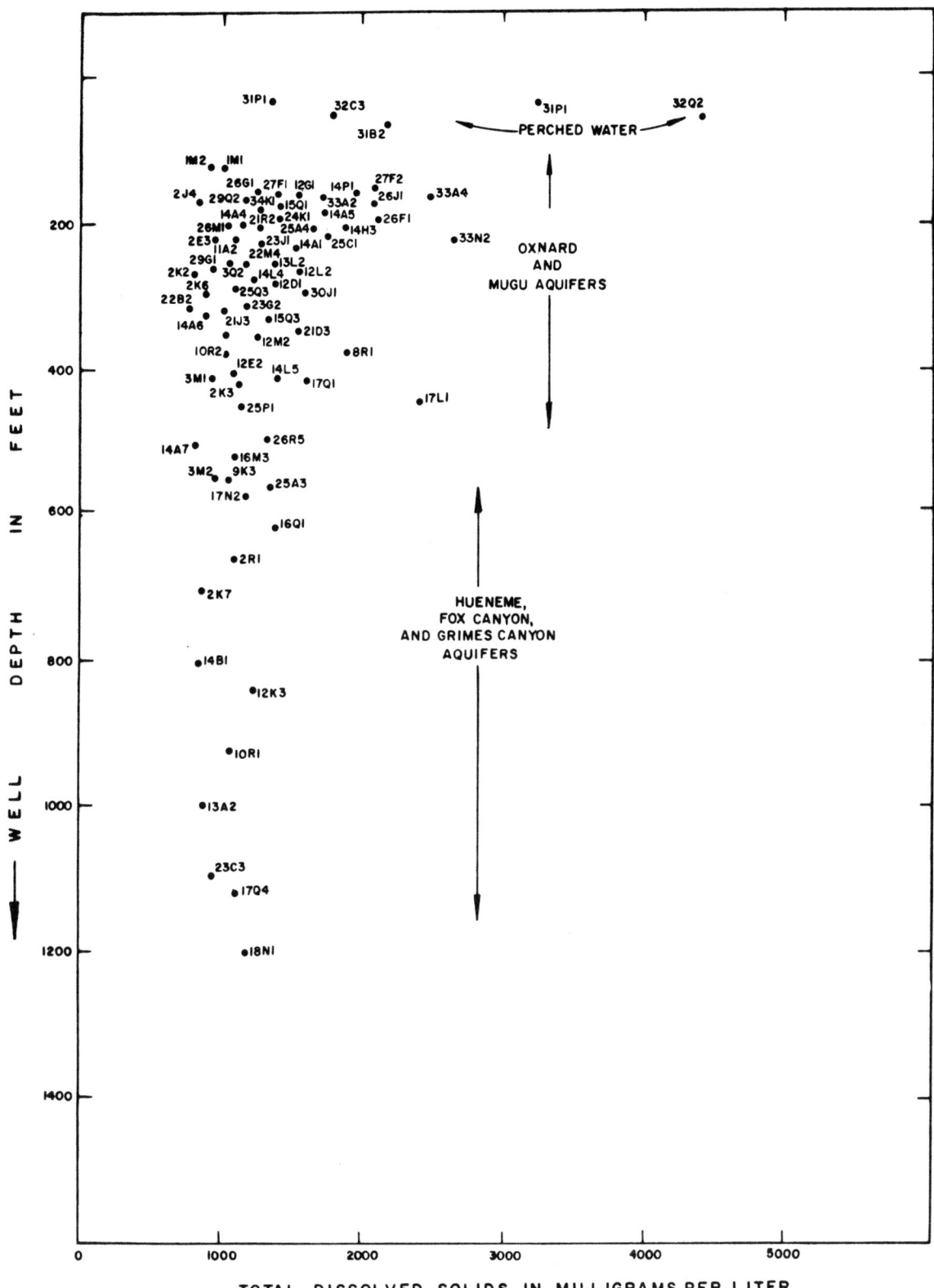

Figure 2-48. Variation of total dissolved solids with well depth in a portion of the Santa Clara-Calleguas groundwater basin, Ventura County, California (California Department

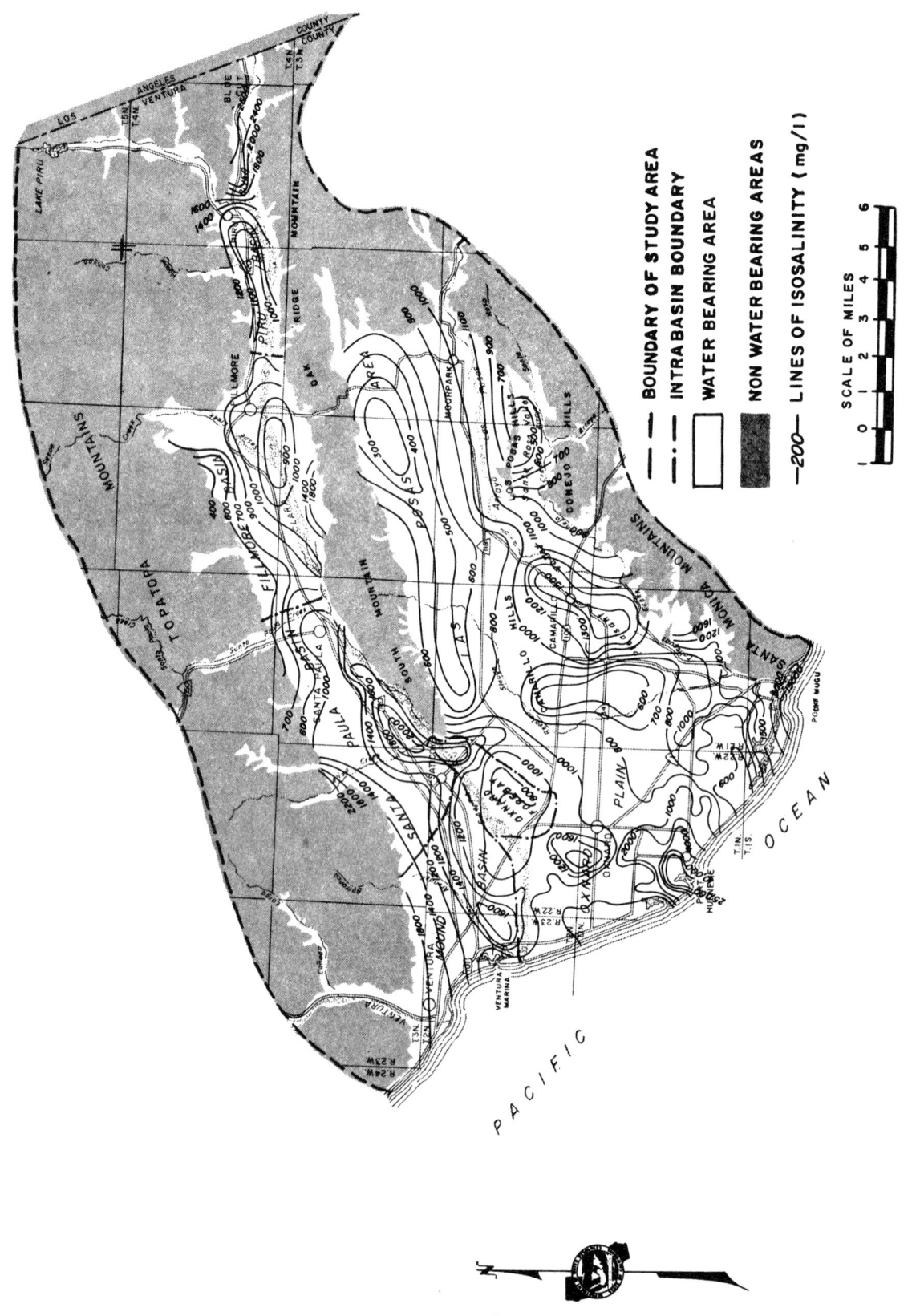

Figure 2-49. Isosalinity map of groundwater in the Santa Clara-Calleguas groundwater basin, Ventura County, California, as of 1966 (California Department of Water Resources, August 1974).

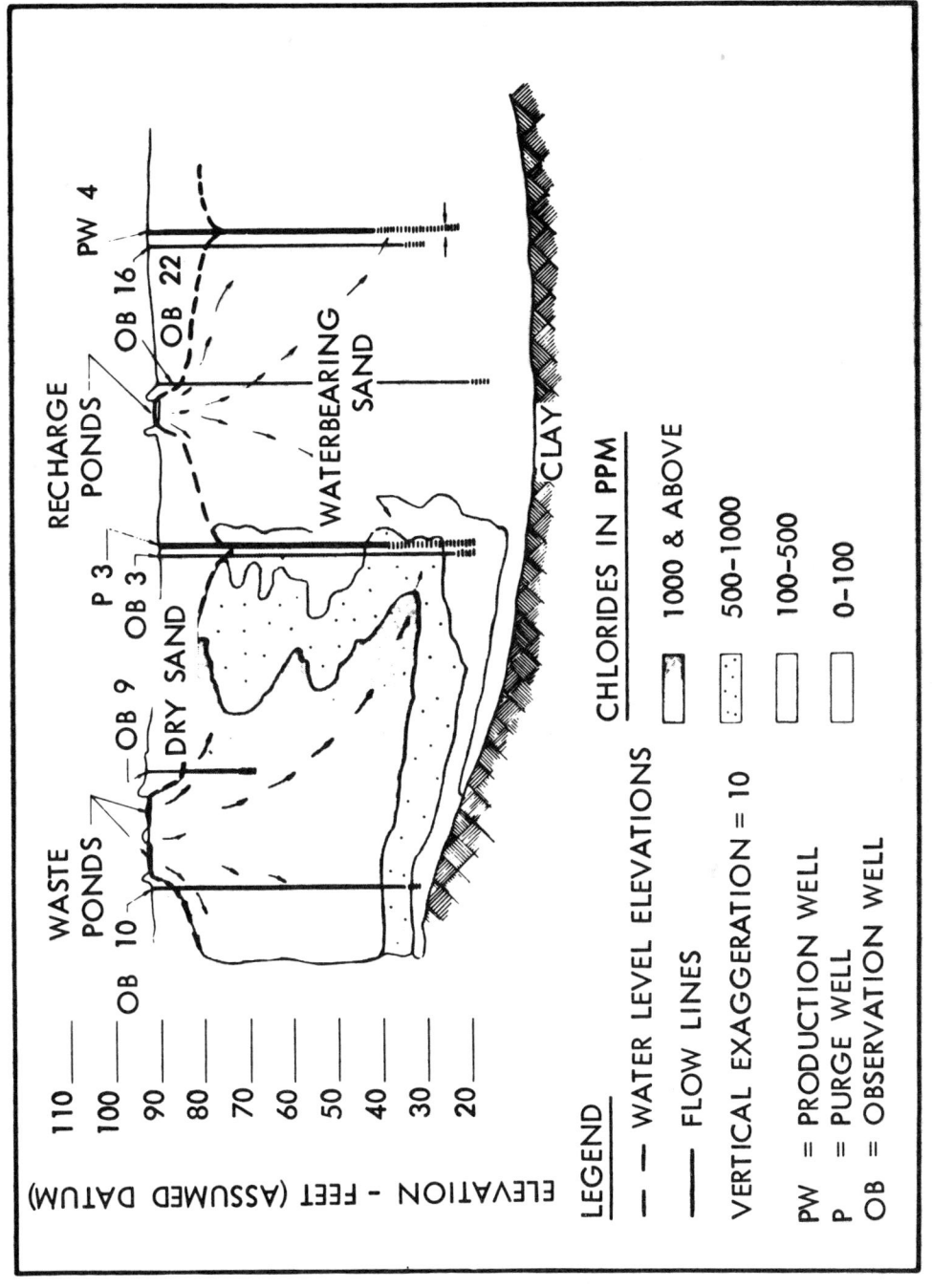

Figure 2-50. Vertical cross section showing groundwater pollution movement from waste disposal ponds and control by cooling water recharge ponds and purge wells (Burt, 1972).

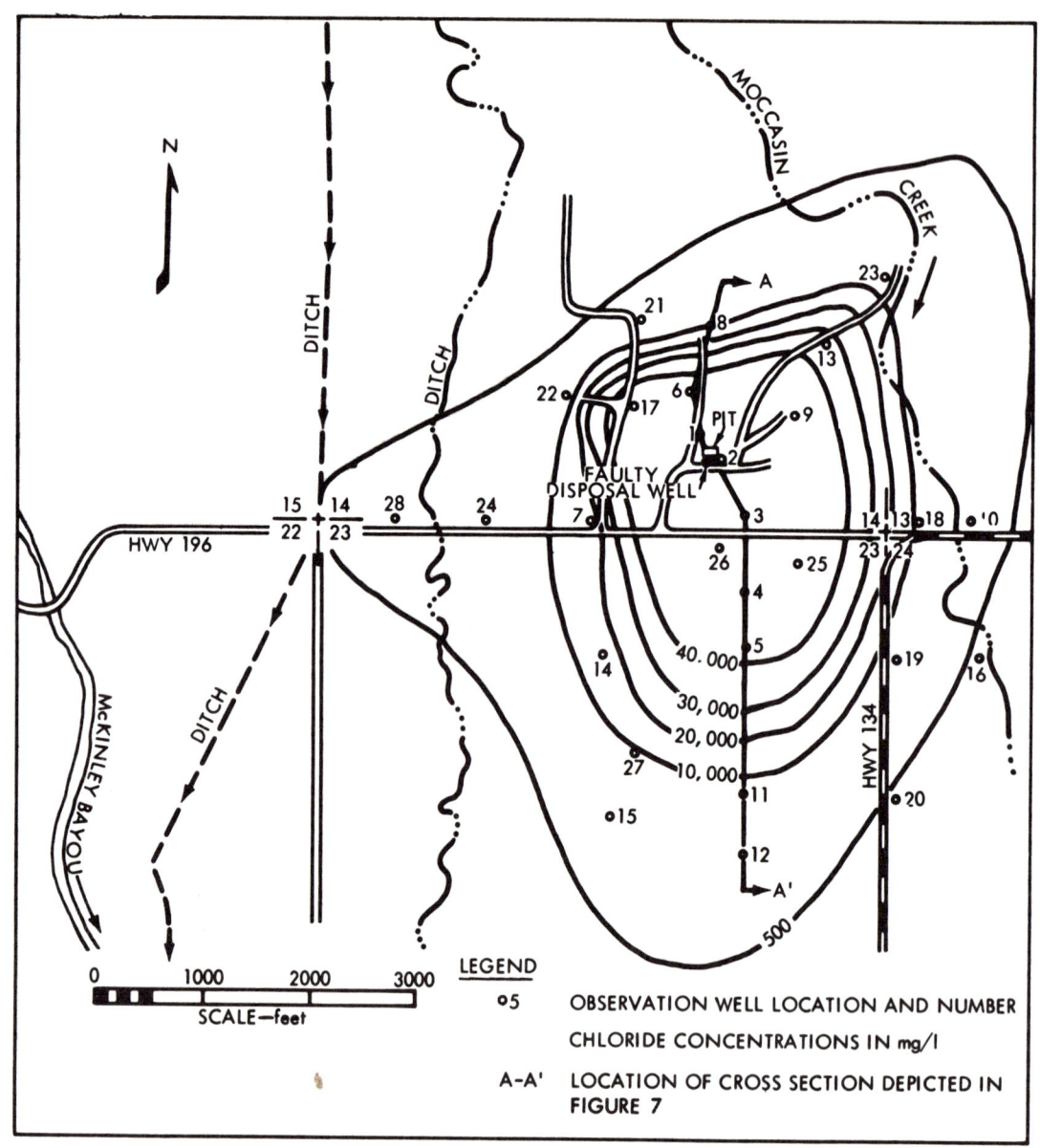

Figure 2-51. Contours of chloride concentration in groundwater surrounding a brine disposal pit in southwestern Arkansas (modified after Fryberger, 1975).

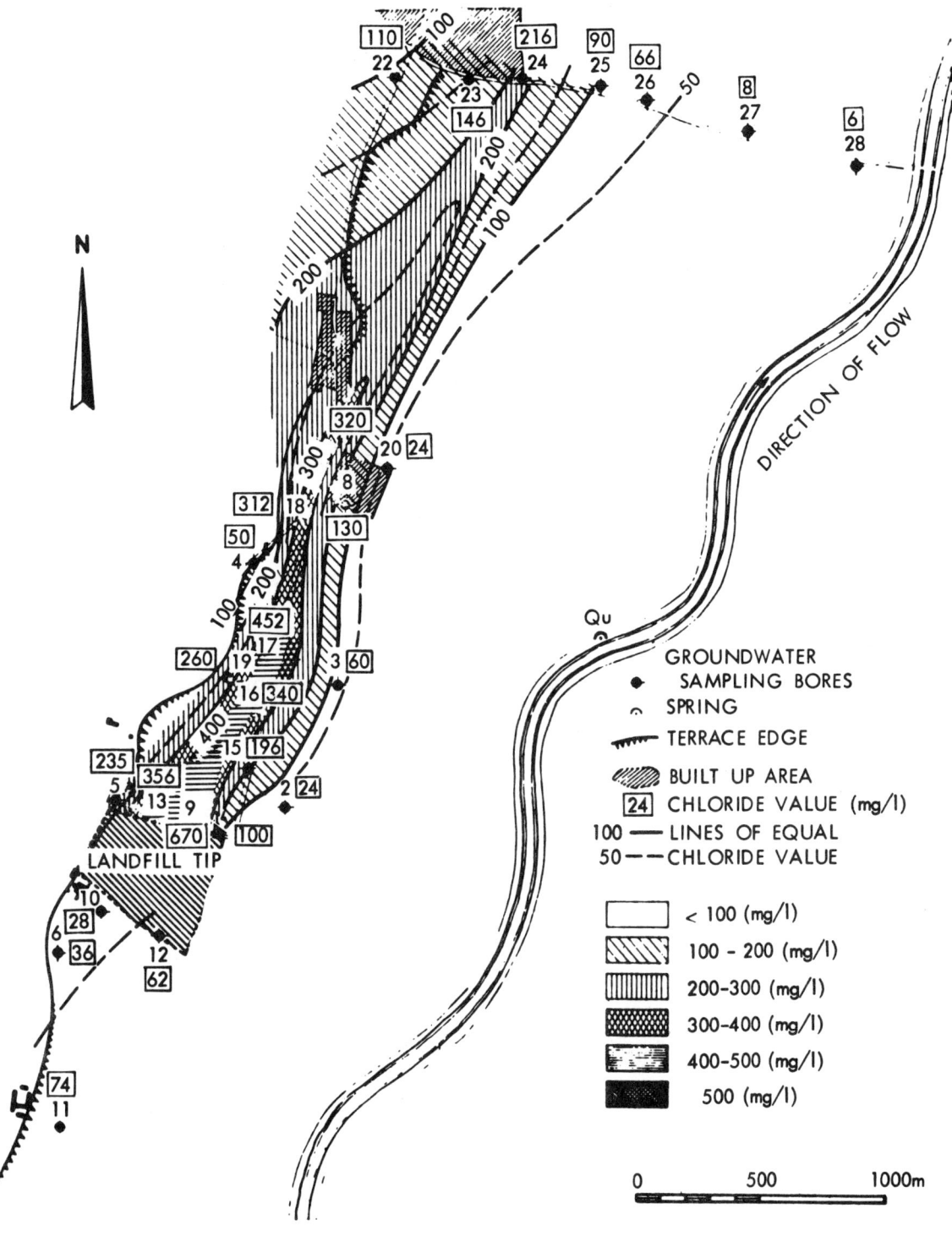

Figure 2-52. Plume of groundwater pollution from a landfill near Munich, West Germany, shown by lines of chloride concentration (Cole, 1975).

SELECTED BIBLIOGRAPHY

D.C. Adriano, et al., "Nitrate and Salt in Soils and Ground Waters from Land Disposal of Dairy Manure," *Soil Science Society of America Proceedings*, Vol. 35, pp 759-762, 1971.

American Chemical Society, *Cleaning Our Environment,* The Chemical Basis for Action, Report of the Committee on Chemistry and Public Affairs, Washington, D.C., 1969.

J.R. Andersen, *Studies of the Influence of Lagoons and Landfills on Groundwater Quality,* Water Resources Institute, South Dakota State University, Brookings, NTIS:PB-214 138, December 1972. (47 pp.)

M.A. Apgar, and D. Langmuir, "Ground-Water Pollution Potential of a Landfill Above the Water Table," *Ground-water* Vol. 9, No. 6, pp 77-94, 1971)

G.L. Bain, *Salty Groundwater in the Pocatalico River Basin,* West Virginia Geological and Economic Survey Circular Series, No. 11, October 1970. (31 pp.)

H.R. Blaney, and W.D. Criddle, *Determining Consumptive Use and Irrigation Water Requirements,* U.S. Department of Agriculture Technical Bulletin 1275, 1962. (59 pp.)

R.H. Bogan, *Problems Arising from Ground Water Contamination by Sewage Lagoons at Tieton,* Washington, U.S. Public Health Service Technical Report W61-5, pp 83-87, 1961.

S.M. Born and D.A. Stephenson, "Hydrogeologic Considerations in Liquid Waste Disposal," *Journal of Soil and Water Conservation*, Vol. 24, No. 2, pp 52-55, 1969.

H. Bouwer, R.C. Rice, and E.D. Escarcega, *Renovating Secondary Sewage by Ground Water Recharge with Infiltration Basins,* U.S. Environmental Protection Agency, Water Pollution Control Research Series, 16060 DRV 03/72, 1972. (102 pp.)

H.O. Buckman, and N.C. Brady, *The Nature and Properties of Soils,* Seventh Edition, Macmillan Publishing Company, New York, 1969. (653 pp.)

E.M. Burt, "The Use, Abuse and Recovery of a Glacial Aquifer," *Ground Water*, Vol. 10, No. 1, pp 65-72, 1972.

California Department of Water Resources, *Water Conditions in California,* Bulletin 120-74, December 1974.

California Department of Water Resources, *Mathematical Modeling of Water Quality for Water Resources Management:* Vol. II—*Development of Historic Data for the Verification of the Ground Water Quality Model of the Santa Clara-Calleguas Area, Ventura County,* August 1974. (114 pp.)

California Department of Water Resources, *Sanitary Landfill Studies:* "Appendix A—Summary of Selected Previous Investigations," Bulletin 147-5, 1969. (115 pp.)

California Department of Water Resource,s *The Fate of Pesticides Applied to Irrigated Agricultural Land,* Bulletin 174-1, 1968.

California Department of Water Resources, *San Joaquin County Ground Water Investigation*, Bulletin No. 146, Sacramento, July 1967. (177 pp.)

J.J. Coe, "Effect of Solid Waste Disposal on Groundwater Quality," *Journal of the American Water Works Association*, Vol. 62, No. 12, pp 776-783, December 1970.

J.A. Cole (ed.), *Groundwater Pollution in Europe*, Water Information Center, Port Washington, New York, 1975. (347 pp.)

T.J. Concannon and E.J. Genetelli, "Groundwater Pollution due to High Organic Manure Loadings," *Livestock Waste Management and Pollution Abatement*, Proceedings of International Symposium on Livestock Wastes, Ohio State University, pp 249-253, April 1971.

C. Corker (ed.), *Groundwater Law, Management and Administration*, National Water Commission, Final Report, NWL-L-72-06, October 1971.

R.L. Crouch, R.D. Eckert, and D.D. Rugg, *Monitoring Groundwater Quality: Economic Framework and Principles*, (in-press) U.S. Environmental Protection Agency, Las Vegas, Nevada, 1976.

R.W. Cruff and T.H. Thompson, *A Comparison of Methods of Estimating Potential Evapotranspiration from Climatological Data in Arid and Subhumid Environments*, U.S. Geological Survey Water Supply Paper 1839-M, 1967. (28 pp.)

S.N. Davis, and R.J.M. DeWiest, *Hydrogeology*, John Wiley & Sons, New York, 1966. (463 pp.)

M. Deutsch, *Groundwater Contamination and Legal Controls in Michigan*, U.S. Geological Survey Water Supply Paper 1691, 1963. (79 pp.)

M. Deutsch, *Incidents of Chromium Contamination of Ground Water in Michigan*, Public Health Service Technical Report W61-5, pp 98-104, 1961.

L.D. Doneen, "Properties of Deep Substrata Materials in the West Side of the San Joaquin Valley, California," in "Quality of Percolating Waters," *Hilgardia*, Vol. 38, No. 9, pp 285-305, 1967.

H.E. Dregne, et al., *Movement of 2, 4-D in Soils*, Western Regional Research Project Progress Report, New Mexico Agricultural Experiment Station, University Park, November 1969. (35 pp.)

B.G. Ellis, "The Soil as a Chemical Filter," *Recycling Treated Municipal Wastewater and Sludge through Forest and Cropland*, W.E. Sopper and L.T. Kardos (eds.), Pennsylvania State University Press, 1973.

L.G. Everett, K.D. Schmidt, R.M. Tinlin, and K.D. Todd, *Monitoring Groundwater Quality: Methods and Costs*, U.S. Environmental Protection Agency, Las Vegas, Nevada, 1976.

S.D. Faust, and J.V. Hunter, *Principles and Applications of Water Chemistry*, John Wiley & Sons, Inc., New York, 1967. (643 pp.)

D.W. Fitzsimmons, et al., "Nitrogen, Phosphorus, and Other Inorganic Materials in Waters in a Gravity-Irrigated Area," *Transactions of the American Society of Agricultural Engineers*, Vol. 15, No. 2, pp 292-295, 1972.

J.E. Flack and C.W. Howe, "Salinity in Water Resources," *Proceedings of the 15th Annual Western Resources Conference*, July 1973, University of Colorado, Merriman Publishing Company, Boulder, Colorado, 1974. (177 pp.)

R. Follansbee, "Evaporation from Reservoir Surfaces," *American Society of Civil Engineers Transactions*, Paper No. 1871, pp 707-710, 1933.

J.S. Fryberger, "Investigation and Rehabilitation of a Brine-Contaminated Aquifer," *Ground Water*, Vol. 13, No. 2, pp 155-160, 1975.

John Fryberger, *Rehabilitation of a Brine-Polluted Aquifer*, Office of Research and Monitoring, U.S. Environmental Protection Agency, Washington, D.C. 20460, Report No. EPA-R2-72-104, December 1972.

D.K. Fuhriman and J.R. Barton, *Groundwater Pollution in Arizona, California, Nevada, and Utah*, U.S. Environmental Protection Agency Water Pollution Control Research Series, 16060 ERU, Washington, D.C., December 1971. (249 pp.)

R.W. Gillham, and L.R. Webber, "Nitrogen Contamination of Groundwater by Barnyard Leachates," *Journal of the Water Pollution Control Federation*, Vol. 41, No. 10, pp 1752-1762, 1969.

R.F. Gould, *Equilibrium Concepts in Natural Water Systems*, Advances in Chemistry Series, American Chemical Society, 1967. (344 pp.)

B.F. Hajek, "Chemical Interactions of Wastewater in a Soil Environment," *Journal of Water Pollution Control Federation*, Vol. 41, No. 10, pp 1775-1786, 1969.

N.F. Hampton, *Monitoring Groundwater Quality: Data Management*, U.S. Environmental Protection Agency, Las Vegas, Nevada, 1976.

K.P. Hanby, R.E. Kidd, and P.A. LaMoreaux, "Subsurface Disposal of Liquid Industrial Wastes in Alabama – A Current Status Report," *Underground Waste Management and Artificial Recharge*, Vol. I, J. Braunstein (ed.), American Association of Petroleum Geologists, pp 72-90, 1973.

G.E. Harbeck, Jr., M.A. Kohler, G.E. Koberg, *Water-Loss Investigations, Lake Mead Studies*, U.S. Geological Survey Professional Paper 298, 1958.

A.A. Hassan, "Water Quality Cycle–Reflection of Activities of Nature and Man," *Ground Water*, Vol. 12, No. 1, pp 16-21, 1974.

J.D. Hem, *Study and Interpretation of the Chemical Characteristics of Natural Water*, 2nd edition, U.S. Geological Survey Water Supply Paper 1473, 1970. (363 pp.)

G.M. Hughes, R.A. Landon, and R.N. Farvolden, *Hydrogeology of Solid Waste Disposal Sites in Northeastern Illinois*, U.S. Environmental Protection Agency, Report SW-12d, 1971. (154 pp.)

J.L. Hughes, L.A. Eccles, and R.L. Malcolm, "Dissolved Organic Carbon (DOC), an Index of Organic Contamination in Ground Water near Barstow, California," *Ground Water*, Vol. 12, No. 5, pp 283-290, 1974.

J.V. Hunter and T.A. Kotalik, "Chemical and Biological Quality of Treated Sewage Effluents," *Recycling Treated Municipal Wastewater and Sludge Through Forest and Cropland*, W.E. Sopper and L.T. Kardos (eds), Pennsylvania State University Press, pp 6-25, 1973.

K. Imhoff, W.J. Muller, and P.K. Thistlethwayte, *Disposal of Sewage and Other Water-Borne Waste*, Ann Arbor Science Publisher, Inc., Second Edition, 1971. (405 pp.)

A.L. Jenke, *Evaluation of Salinity Created by Irrigation Return Flows* U.S. Environmental Protection Agency, Washington, D.C., January 1974. (128 pp.)

W.R. Johnston, F.T. Ittihadieth, K.R. Craig, and A.F. Pillsbury, "Insecticides in Tile Drainage Effluent," *Water Resources Research*, Vol. 3, No. 2, pp 525-537, 1967.

G.E. Kimmel, and O.C. Braids, "Laechate Plumes in a Highly Permeable Aquifer," *Ground Water*, Vol. 12, No. 6, pp 388-393, 1974.

D.B. Knowles, "Hydrologic Aspects of the Disposal of Oil-Field Brines in Alabama," *Ground Water*, Vol. 3, No. 2, pp 22-27, 1965.

M.A. Kohler, T.J. Nordenson, and W.E. Fox, *Evaporation from Pans and Lakes*, U.S. Weather Bureau, Research Paper 38, May 1955.

R.A. Krieger, and G.E. Hendrickson, *Effects of Greensburg Oilfield Brines on the Streams, Wells, and Springs of the Upper Green River Basin, Kentucky*, Kentucky Geological Survey Report on Investigation 2, Ser. X, 1960. (36 pp.)

J.P. Law, Jr., et al., *Degradation of Water Quality in Irrigation Return Flows*, Bulletin B-684, R.S. Kerr Water Research Center, Oklahoma Agricultural Experiment Station, Ada; and Oklahoma State University, Stillwater, Department of Agronomy, October 1970. (26 pp.)

H.E. LeGrand, "Environmental Framework of Groundwater Contamination," *Ground Water*, Vol. 3, No. 2, pp 11-15, 1965a.

H.E. LeGrand, "Patterns of Contaminated Zones of Water in the Ground," *Water Resources Research*, Vol. 1, pp 83-95, 1965b.

H.E. LeGrand, "System for Evaluation of Contamination Potential of Some Waste Disposal Sites," *Journal of the American Water Works Association*, Vol. 56, pp. 959-974, 1964.

R.C. Loehr, "Effluent Quality from Anaerobic Lagoons Treating Feedlot Wastes," *Journal of the Water Pollution Control Federation*, Vol. 39, No. 3, pp 384-391, 1967.

R.L. Lowry, Jr., and A.F. Johnson, "Consumptive Use of Water for Agriculture," Transactions of the American Society of Civil Engineers, Vol. 107, pp 1243-1266, 1942.

J.R. Matis, "Petroleum Contamination of Ground Water in Maryland," U.S. Environmental Protection Agency Water Pollution Control Research Series, 16060 GRB 08/71, pp 57-61, 1971.

R.B. Mattox, *Groundwater Salinity,* Contribution No. 13 of Committee on Desert and Arid Zones Research, Southwestern and Rocky Mountain Division, A.A.A.S., New Mexico Highlands University, Las Vegas, New Mexico, 1970. (150 pp.)

P.H. McGauhey, "Manmade Contamination Hazards," *Ground Water,* Vol. 6, No. 3 pp 10-13, 1968a.

P.H. McGauhey, *Engineering Management of Water Quality,* McGraw-Hill, New York, 1968b. (295 pp.)

L.G. McMillion, "Hydrologic Aspects of Disposal of Oil-Field Brines in Texas," *Ground Water,* Vol. 3, No. 4, pp 36-42, 1965.

C.F. Meyer (ed.), *Polluted Groundwater: Some Causes, Effects, Controls, and Monitoring,* U.S. Environmental Protection Agency, Report No. EPA-600/4-73-001b, Washington, D.C., July 1973 (282 pp.)

D.W. Miller, et al., *Ground Water Contamination in the Northeast States,* U.S. Environmental Protection Agency, Washington, D.C. (325 pp.)

J.F. Mink, "Excessive Irrigation and the Soils and Ground Water of Oahu, Hawaii," *Science,* Vol. 135, No. 3504, pp 672-673, February 23, 1962.

T.M. Moore, *Water Geochemistry, Hog Creek Basin, Central Texas,* Baylor Geological Studies Bulletin No. 18, Department of Geology, Baylor, University, Waco, Spring 1970. (44 pp.)

R. Palmquist, and L.V.A. Sendlein, "The Configuration of Contamination Enclaves from Refuse Disposal Sites on Floodplains," Ground Water, Vol. 13, No. 2, pp 167-181, 1975.

H.L. Penman, "Natural Evaporation from Open Water, Bare Soil and Grass," *Proceedings of the Royal Society of London,* Series A, Vol. 193, pp 120-145, 1948.

J.R. Peterson, C. Lue-Hing, and D.R. Zenz, "Chemical and Biological Quality of Municipal Sludge," *Recycling Treated Municipal Wastewater and Sludge Through Forest and Cropland,* W.E. Sopper and L.T. Kardos (eds), Pennsylvania State University Press, pp 26-36, 1973.

W.A. Pettyjohn, "Water Pollution by Oil-Field Brines and Related Industrial Wastes in Ohio," *Water Quality in a Stressed Environment,* Burgess Publishing Company, Minneapolis, Minnesota, 1972.

W.A.J. Pitt, *Effects of Septic Tank Effluent on Ground-Water Quality, Dade County, Florida,* An Interim Report, U.S. Geological Survey, Open File Report, 1974. (50 pp.)

R.C. Polta, "Septic Tank Effluents," *Water Pollution by Nutrients—Sources, Effects, and Controls,* Water Resources Research Center, University of Minnesota, Minneapolis, WRRC Bulletin 13, pp 53-57, June 1969.

E.L. Quan, H.R. Sweet, and J.R. Illian, "Subsurface Sewage Disposal and Contamination of *Ground Water* in East Portland, Oregon," Ground Water, Vol. 12, No. 6, pp 356-368, 1974.

Carl Rohwer, "Evaporation from Salt Solutions from Oil-Covered Water Surfaces," *Journal of Agricultural Research*, Vol. 46, pp 715-729, 1933.

J.A. Salvato, et al., "Sanitary Landfill—Leaching Prevention and Control," *Journal of the Water Pollution Control Federation*, Vol. 43, No. 10, pp 2084-2100, 1971.

J.D. Sartor and G.B. Boyd, *Water Pollution Aspects of Street Surface Contaminants*, U.S. Environmental Protection Agency, Environmental Protection Technology Series, EPA-R2-72-081, 1972. (236 pp.)

M.R. Scalf, et al., *Fate of DDT and Nitrate in Ground Water*, Federal Water Pollution Control Administration, Robert S. Kerr Water Research Center, Ada, Oklahoma and Agricultural Research Service, Southwestern Great Plains Research Center, Bushland, Texas, 1968. (46 pp.)

M.R. Scalf, J.W. Keeley, and C.J. LaFevers, *Ground Water Pollution in the South Central States*, U.S. Environmental Protection Agency Environmental Protection Technology Series, EPA-R2-73-268, Corvallis, Oregon, June 1973. (181 pp.)

K.D. Schmidt, "Groundwater Quality in the Cortaro Area Northwest of Tucson, Arizona," *Water Resources Bulletin*, Vol. 9, No. 3, pp 598-606, 1973.

K.D. Schmidt, "Nitrate in Ground Water of the Fresno-Clovis Metropolitan Area, California," *Ground Water*, Vol. 10, No. 1, pp 50-64, 1972.

W.J. Schneider, *Hydrologic Implications of Solid-Waste Disposal*, U.S. Geological Survey Circular 601-F, 1970. (10 pp.)

H.R. Seitz, "Investigation of a Landfill in Granite-Loess Terrain," *Ground Water*, Vol. 10.4, pp 35-41, 1972.

E.J. Shaw, *Western Fertilizer Handbook*, Fourth Edition, Second Printing, California Fertilizer Association, Sacramento, California, 1968. (200 pp.)

G.V. Skogerboe, and J.V. Law, *Research Needs for Irrigation Return Flow Quality Control*, U.S. Environmental Protection Agency Water Pollution Control Research Series 13030-11/71, 1971. (98 pp.)

H.F. Smith, *Subsurface Storage and Disposal in Illinois*, U.S. Environmental Protection Agency Water Pollution Control Research Series, 16060 GRB08-71, pp 20-28, 1971.

B.A. Stewart, et al., *Distribution of Nitrates and Other Water Pollutants Under Fields and Corrals in the Middle South Platte Valley of Colorado*, U.S. Department of Agriculture, Agricultural Research Service, Report ARS 41-134, 1967. (206 pp.)

B.A. Stewart, F.G. Viets, and G.L. Hutchison, "Agriculture's Effect on Nitrate Pollution," *Journal of Soil Water Conservation*, Vol. 23, No. 13, pp 13-15, 1968.

W. Stumm, and J.J. Morgan, *Aquatic Chemistry, an Introduction Emphasizing Chemical Equilibria in Natural Waters*, Wiley-Interscience, New York, 1970. (583 pp.)

H.R. Sweet and R.H. Fetrow, "Ground Water Pollution by Wood Waste Disposal," *Ground Water,* Vol. 13, No. 2, pp 227-231, 1975.

R.C. Terry, Jr., *Road Salt, Drinking Water, and Safety,* Ballinger Publishing Co., Cambridge, Massachusetts, 1974. (161 pp.)

C.W. Thornthwaite, "An Approach Toward a Rational Classification of Climate," *Geographical Review,* Vol. 38, pp 55-94, 1948.

R.M. Tinlin (ed.), *Monitoring Groundwater Quality: Illustrative Examples,* U.S. Environmental Protection Agency, Las Vegas, Nevada, 1976.

S.L. Tisdale, and W.L. Nelson, *Soil Fertility and Fertilizers,* 2nd ed. Macmillan Publishing Company, New York, 1966. (694 pp.)

D.K. Todd (ed.), *The Water Encyclopedia,* Water Information Center, Port Washington, New York, 1970. (559 pp.)

D.K. Todd, *Ground Water Hydrology,* John Wiley & Sons, New York, 1959. (336 pp.)

D.K. Todd, and D.E. McNulty, *Polluted Groundwater: A Review of the Significant Literature,* U.S. Environmental Protection Agency, Report No. EPA-680/4-74-001, Washington, D.C., March 1974 (215 pp.)

W.E. Tucker, "Subsurface Disposal of Liquid Industrial Wastes in Alabama—A current Status Report," *Proceedings of the National Ground Water Quality Symposium,* U.S. Environmental Protection Agency, Water Pollution Control Research Series, 16060 GRB 08/71, pp 10-19, 1971.

University of California at Davis, *Agricultural Development of New Lands, West Side of San Joaquin Valley, Land, Crops, and Economics,* Report No. 1, Dean's Committee, College of Agriculture and Environmental Science, 1968. (83 pp.)

U.S. Council on Environmental Quality, *Fourth Annual Report,* 1973.

F. van der Leeden, et al., *Ground Water Pollution Problems in the Northwestern United States,* U.S. Environmental Protection Agency, Washington, D.C., May 1975.

F. van der Leeden, et al., *Ground Water Contamination in the Northwest States,* U.S. Environmental Protection Agency, Office of Research and Monitoring, Technology Series EPA-R2-73-268, Washington, D.C., June 1973.

T.R. Walker, "Ground-Water Contamination in the Rocky Mountain Arsenal Area, Denver, Colorado," *Geological Society of America Bulletin,* Vol. 72, No. 3, pp 489-494, 1961.

G. Walton, *Public Health Aspects of the Contamination of Ground Water in the Vicinity of Derby, Colorado,* U.S. Public Health Service Technical Report W61-5, pp 10)-128, 1961.

J.P. Waltz, "Methods of Geologic Evaluation of Pollution Potential at Mountain Home Sites," *Proceedings of the National Ground Water Quality Symposium,* U.S. Environmental Protection Agency Water Pollution Control Research Series, 16060 GRB 08/71, pp 136-143, 1971.

D.L. Warner, *Monitoring Disposal-Well Systems*, U.S. Environmental Protection Agency, EPA-680/4-75-008, Las Vegas, Nevada, July 1975. (109 pp.)

L. Weaver, "Refuse Disposal—Its Significance," *Ground Water*, Vol. 2, No. 1, pp 26-30, 1964.

W.G., Weist and R.A. Pettijohn, "Investigation Ground-Water Pollution from Indianapolis' Landfills—The Lessons Learned," *Ground Water*, Vol. 13, No. 2, pp 191-196, 1975.

D.E. White, J.D. Hem, and G.A. Waring, "Data of Geochemistry," *Chemical Composition of Subsurface Waters*, 6th ed., Chapter F, U.S. Geological Survey Professional Paper 440-F, 1963. (67 pp.)

D.E., Williams and D.G. Wilder, *Gasoline Pollution of a Ground-Water Reservoir — A Case History*, U.S. Environmental Protection Agency Water Pollution Control Research Series, 16060 GRB 08/71, pp 50-56, 1971.

R.E. Williams, *Waste Production in Mining, Milling, and Metallurgical Industries*, Miller Freeman Publications, Inc., San Francisco, May 1975.

T.L. Willrich and G.E. Smith, *Agricultural Practices and Water Quality*, Iowa State University Press, Ames, Iowa, 1970a. (415 pp.)

T.L. Willrich and G.E. Smith, "Pesticides as Water Pollutants," *Agricultural Practices and Water Quality*, Iowa State University Press, Ames, Iowa, Part 3, pp 167-230, 1970b.

T.L. Willrich and G.E. Smith, "Animal Wastes as Water Pollutants," *Agricultural Practices and Water Quality*, Iowa State University Press, Ames, Iowa, Part 4, pp 231-302, 1970c.

A.E. Zanoni, "Ground-Water Pollution and Sanitary Landfills—A Critical Review," *Proceedings of the National Ground Water Quality Symposium*, U.S. Environmental Protection Agency, Water Pollution Control Research Series 16060 GRB 08/71, pp 97-110, 1971.

C. Zenone, D.E. Donaldson, and J.J. Grunwaldt, "Groundwater Quality Beneath Solid-Waste Disposal Sites at Anchorage, Alaska," *Ground Water*, Vol. 13, No. 2, pp 182-190, 1975.

CHAPTER III

GROUNDWATER MONITORING METHODS AND COSTS

SECTION 1 – INTRODUCTION

Chapter II, which was originally entitled *Monitoring Groundwater Quality: Monitoring Methodology* (Todd et al., 1976), outlines a procedure for creating a monitoring program for groundwater quality under the general supervision of the U.S. Environmental Protection Agency. As an essential supplemental reference to the methodology, Chapter III documents the various methods and techniques available to monitor groundwater quality, and presents detailed cost data for the methods and techniques. Monitoring has been broadly interpreted to include all possible measuring techniques, with the intent to make this chapter as useful as possible.

PURPOSE

Developing a groundwater monitoring program in an area requires not only an extensive understanding of the various monitoring techniques available, but also a knowledge of their costs. Previous economic groundwater studies dealt primarily with the cost of supplying groundwater for municipal and industrial usage. More recent cost studies by D.J. Cederstrom (1970), James P. Gibb (1971), and Robert H. Forste (1973) are directed toward costs of producing wells of varying yields. The subject of groundwater monitoring methods and their associated costs is fragmented throughout the hydrologic literature. The purpose of this chapter, therefore, is to provide a general summary of each of the groundwater monitoring methods so that a strategy can be derived, based upon specific methods and actual cost data.

SCOPE

This chapter describes the various monitoring methods, and provides a generalized cost breakdown of the major economic factors for each monitoring method. The itemization of factors for each method is not exhaustive, but does serve to quantify reasonably well the cost of the monitoring technique. A cost interpretation structure to update the 1974 data presented herein and a scheme to cover the national spatial distribution of costs are provided. It should be emphasized that the costs listed are the best available estimates from sources that are believed to be reliable.

The material in this chapter is presented for reference purposes without recommendation for a least-cost technique, a least-cost mix of monitoring approaches, or an optimal information system. Essentially, the cost data are intended to serve as line price guides for the various monitoring methods.* Furthermore, it should be noted that, with few exceptions, costs presented here are those associated with goods and services only. Any organized monitoring program in a local area will have personnel costs – supervisors, technicians, clerical staff, etc. – and overhead costs such as for offices, supplies, communication, and transportation. Thus, the data provide specific technical costs, but do not attempt to encompass total budgets for monitoring programs.

---

*Considerable assistance on cost data was provided by Federal and State agencies and several private companies.

HYDROGEOLOGICAL FRAMEWORK

The following discussion concerns the most common types of groundwater pollution from sources at or near the land surface. Seawater intrusion, deep well waste disposal, and groundwater overdraft are other types of water quality changes that are of major significance in many local areas. There are diffuse, line, and point sources of waste discharge. Figure 3-1 illustrates the sequence of events that occur when a well is polluted by surface wastes.

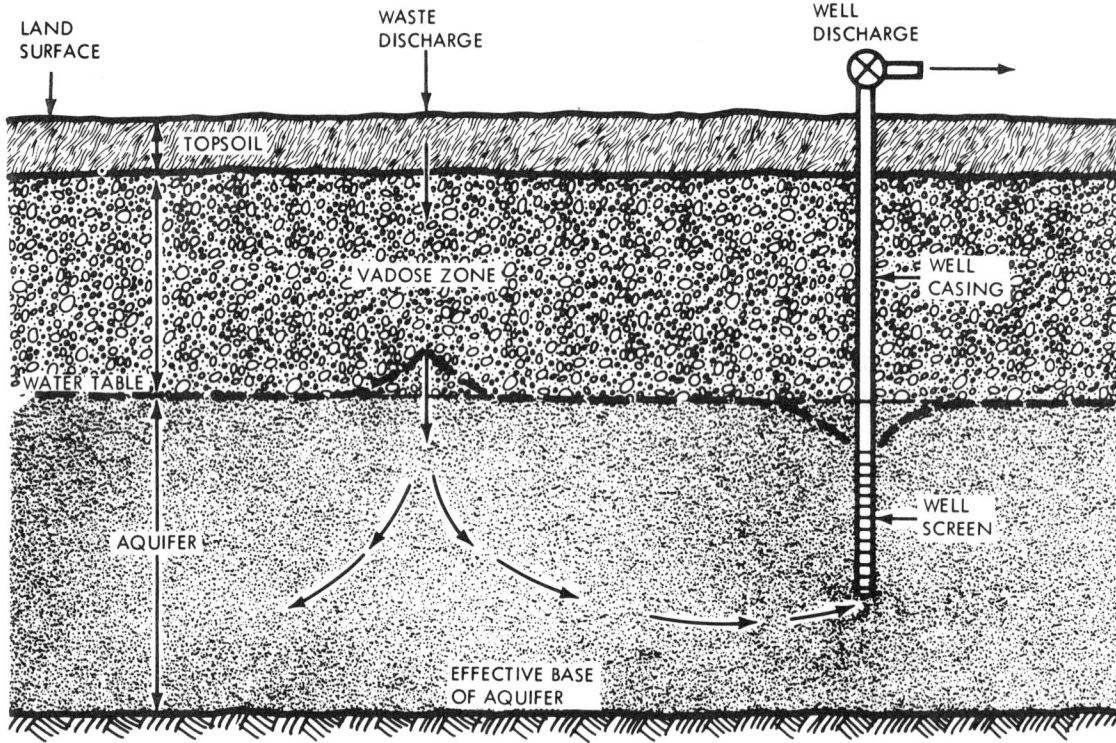

Figure 3-1. Schematic diagram of groundwater pollution from surface sources.

Considerable confusion exists in groundwater quality investigations because of a lack of understanding of the pertinent physical, chemical, biological, soil, and geologic factors. The diagram illustrates the five basic portions of the system: (1) the land surface, (2) topsoil, (3) vadose zone, (4) saturated zone, and (5) well. Sanitary engineers and surface water hydrologists focus attention primarily on water at the land surface. Soil scientists usually study the topsoil and, secondarily, the vadose zone. Geologists are generally concerned with the portion below the water table, while water supply agencies concentrate on well extraction where groundwater is used as a source of supply. The entire system illustrated in the simplified diagram must be studied in order to understand groundwater pollution.

Sampling at the land surface, in the topsoil, in the vadose zone, and from the saturated zone will be discussed in detail. Understanding of quality changes in the topsoil requires a knowledge of soil water, soil physics, soil chemistry, microbiology, and physical chemistry. Historically, soil scientists have been concerned more with growing crops than with the quality of percolating waters. Today, this situation has changed; now many studies concern the quality of percolate. An extensive body of literature is available on water quality changes in the topsoil, such as in *Soil Science Society of America Proceedings and Journal of Environmental Quality*. This information has often been overlooked in groundwater quality studies.

Understanding of groundwater quality below the water table requires a knowledge of hydrogeology, well hydraulics, geochemistry, and physical chemistry. Of principal importance is knowledge of the hydrogeologic framework. The hydraulic and physical features of the groundwater system must be established before groundwater quality can be understood. An extensive body of literature is available on natural groundwater quality.

The vadose zone is commonly approached by the black box method, especially by geologists. Many vadose zones are not composed of soil in the sense of the topsoil known to agricultural workers. The vadose zone often extends to geologic formations beneath the topsoil. These may have different physical characteristics from the topsoil. Soil scientists have made great strides in analyzing the chemical quality changes that occur in the vadose zone (Wilson, 1971; Pratt, 1972; and Stout et al., 1965).

A major difficulty in past groundwater pollution studies has been the prediction of travel times of pollutants from the land surface to the water table. Predictions based on theoretical calculations often yield travel times that are much too slow. Two items argue for a relatively rapid vertical movement in many cases: (1) water balance calculations, i.e., comparison of leaching volumes versus storage capacity of the vadose zone; and (2) hydrologic and water quality evidence in areas where actual field measurements have been made (Wilson, 1971; and Stout et al., 1965). The rate of vertical movement of pollutants in the zone of aeration is of prime importance in developing a groundwater monitoring methodology.

GROUNDWATER MONITORING METHODS

The discussion of monitoring methods is presented in four sections: Section 2, "Monitoring at the Land Surface;" Section 3, "Monitoring in the Topsoil and the Vadose Zone;" Section 4, "Monitoring in the Zone of Saturation;" and Section 5, "Analysis of Samples." This order considers major subdivisions of a monitoring effect; however, it is not intended to convey the impression that the major subdivisions are not closely related. The successful implementation of a groundwater monitoring methodology will largely depend on an appropriate combination of monitoring methods.

Land surface monitoring includes sampling of surface water bodies for pollution, particularly streams and lakes, as well as rainfall sampling. Source monitoring deals with both solids and liquids that may impact on groundwater quality. Land surface monitoring thus includes water sampling, solids sampling, land use surveys, and inventories of amounts of wastes. The tools available, besides sampling, include remote sensing, pipeline and tank testing for leaks, and testing of artificial liners for leakage.

In most cases wastes applied at the land surface travel through significant thicknesses of topsoil and geologic materials before reaching the water table. Pollutants can be significantly retained or attenuated in the topsoil and the vadose zone. The storage capacity of the vadose zone for percolating waters may also be great. Long travel times from the land surface to the water table may necessitate detailed sampling in the vadose zone. Past studies in the vadose zone have largely been accomplished beneath point sources.

The primary site of groundwater quality monitoring lies in the saturated zone, as this is where water is ultimately pumped from wells for use at the land surface. Water sampling from wells is a key item, and along with source monitoring is the primary groundwater pollution monitoring approach. The techniques of well sampling and well drilling (for cases when existing wells do not suffice) are discussed extensively. Past groundwater pollution studies have demonstrated the usefulness of well sampling.

Subsurface sampling in both the vadose zone and in the zone of saturation requires considerable experience and careful judgement due to the complexity of most soil-aquifer systems.

TYPES OF COST

The cost structure is broken into four areas: capital, amortization, maintenance, and operational. The capital costs refer to those items that are fixed by location, such as screens, casings, pumps, or are fixed by their initial investment and subsequent repetitious use, such as pH meters, water samplers, and other equipment. Interest on the initial capital investment may be considerable, especially over long periods of time. The annual write-off or amortization expense of capital plus interest over each 1-year period is dependent upon the interest rate at which the money was loaned and the life of the depreciation fund or the effective life of the capital item. The effective life of water wells is about 20 years and the expected life of water pumps is about 10 years.

Based on the work of Cederstrom (1970), it is estimated that 1 percent of the total capital costs is sufficient to provide for the maintenance of drilled wells. Special consideration, however, is given to small-diameter driven wells as these are more frequently replaced in total. The operational costs of groundwater monitoring are a major item. Capital, amortization, and maintenance costs relate primarily to wells, while the remaining monitoring methods are primarily service functions involving operating costs.

UPDATING COST DATA

Rapid changes in the cost of labor and materials, within the United States, require that published cost data be updated to current costs prior to use. A convenient method to update costs is based upon the *Engineering News-Record* (ENR) indexes. Verification of cost estimates can be made by obtaining bids on specific items.

The *Engineering News-Record* Construction Cost Index was created in 1921 to diagnose the erratic price gyrations that occurred during and immediately following World War I, and to evaluate their effects on construction costs. The index was designed as a general purpose construction cost index to chart basic costs. It is a weighted aggregate index of constant quantities of structural steel, portland cement, lumber, and common labor. This hypothetical block of construction, repriced weekly, was valued at $100 in 1913 prices. The original use of common labor in the Construction Cost Index was based on the idea that it set the trend for all wage rates. In the 1930's, however, wages plus fringe benefits climbed faster for laborers than for the skilled trades.

The *Engineering News-Record* Building Cost Index was introduced in 1938 to weigh the impact of skilled labor on cost trends. For its labor component it uses an average of carpenter, bricklayer, and structural ironworker wages. Its materials component is the same as used in the Construction Cost Index.

Figure 3-2 shows the time variation of well construction costs and the materials component of the ENR index. It can be seen that domestic and farm well costs agree closely with the trend of the index. Similarly, Figure 3-3 shows that large-capacity well costs are associated with the Construction Cost Index.

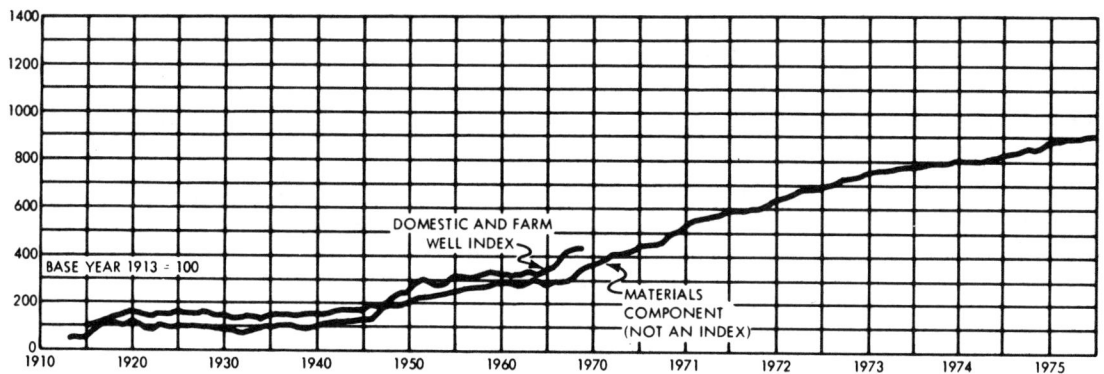

Figure 3-2. Small-diameter well costs index and the ENR materials component (after Gibb, 1971).

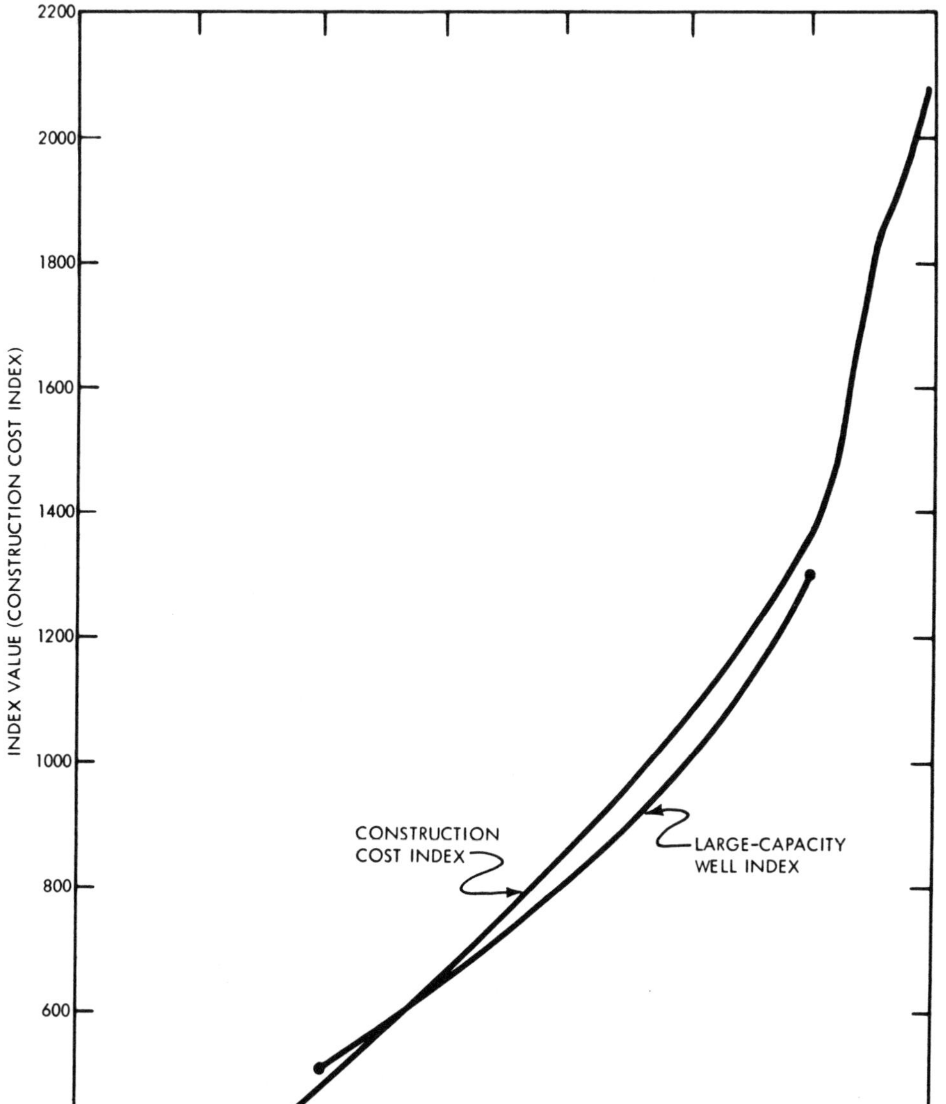

Figure 3-3. Large-capacity well costs index and the ENR construction cost index (modified after Gibb, 1971).

To update costs and to consider regional influences requires the formulation of multiplicative factors based upon the ENR indexes. These multiplicative factors can be formulated as simple ratios from graphical and tabular information. Figure 3-2 enables the determination of the materials component for any prior cost data back to 1913. Table 3-1, taken from the September 12, 1974 issue of ENR, provides indexes on a monthly basis for 20 cities within the United States and the U.S. 20-city average index. At least 1 of the cities listed is found in each of the 10 EPA regions.

Table 3-1

EXAMPLE OF ENR INDEXES FOR SEPTEMBER 12, 1974

City	ENR COST INDEXES IN 22 CITIES (based on 1913 U.S. average = 100)					
	Construction cost			Building cost		
	Sep 74 index	Percent change from last		Sep 74 index	Percent change from last	
		month	year		month	year
Atlanta	1,588.9	+0.2	+5.9	1,080.22	+0.3	+5.5
Baltimore	1,786.94	-0.9	+0.4	1,174.56	-0.7	+0.4
Birmingham	1,562.04	+0.8	+9.6	1,082.28	+1.7	+8.3
Boston	2,174.06	+0.2	+10.4	1,299.15	+0.1	+11.2
Chicago	2,205.91	-1.6	+5.4	1,296.94	-2.8	+6.1
Cincinnati	2,400.44	0	+9.5	1,328.34	-5.0	+9.8
Cleveland	2,375.78	-5.0	+8.5	1,312.39	-5.0	+7.2
Dallas	1,660.59	-0.5	+4.8	1,063.05	-0.8	+4.1
Denver	1,658.39	-0.3	+5.9	1,157.37	-0.4	+6.5
Detroit	2,387.69	-0.1	+7.1	1,341.53	-0.1	+7.4
Kansas City	2,261.63	-1.0	+9.2	1,240.71	-1.7	+7.0
Los Angeles	2,290.23	+6.7	+10.0	1,278.83	+3.1	+8.8
Minneapolis	2,093.94	0	+7.4	1,205.24	-5.0	+7.1
New Orleans	1,610.74	0	+9.4	1,104.52	0	+11.8
New York	2,568.37	+0.1	+6.6	1,461.86	+0.1	+6.8
Philadelphia	2,203.17	-1.0	+12.6	1,341.82	-1.6	+11.1
Pittsburgh	2,067.52	-0.1	+9.3	1,317.06	-0.2	+7.0
St. Louis	2,321.94	+0.2	+9.8	1,179.43	-0.6	+6.9
San Francisco	2,446.14	-0.2	+9.9	1,339.67	-0.4	+8.1
Seattle	2,111.96	+11.5	+14.5	1,149.32	+0.7	+12.2
U.S.—20 Cities' Average	2,088.82	+0.6	+8.3	1,237.71	-0.2	+7.6
Montreal	1,880.84	-5.0	+19.7	1,134.69	0	+21.5
Toronto	1,972.44	+0.4	+9.5	1,109.91	+0.5	+6.7

City	ENR WAGE, MATERIALS AND COST INDEXES IN 20 CITIES (based on each city's 1967 average = 100)						Const cost index	Building cost index
	Common labor		Skilled labor		Materials prices			
	Sep 74 index	% change from Sep 73	Sep 74 index	% change from Sep 73	Sep 74 index	% change from Sep 73		
Atlanta	206.64	+7.7	185.86	+8.6	172.97	+1.7	195.30	180.01
Baltimore	229.39	+2.5	190.69	+4.2	175.14	-4.2	202.58	183.72
Birmingham	198.35	+12.3	193.45	+12.0	167.05	+4.0	187.22	180.28
Boston	205.36	+10.7	189.93	+12.3	187.44	+9.8	200.45	188.90
Chicago	193.79	+5.1	190.98	+5.8	155.14	+6.5	182.92	174.81
Cincinnati	233.59	+9.0	209.29	+8.6	197.42	+11.4	225.90	207.64
Cleveland	196.92	+8.7	193.91	+6.8	172.19	+7.8	169.60	184.62
Dallas	236.65	+7.1	186.50	+8.1	154.75	-0.5	210.14	175.98
Denver	177.88	+6.7	191.47	+8.4	163.37	+3.8	169.71	176.96
Detroit	201.72	+7.0	199.60	+7.3	157.23	+7.8	190.62	181.48
Kansas City	236.98	+9.4	206.37	+5.9	175.84	+8.4	218.67	191.59
Los Angeles	199.55	+7.9	192.14	+7.8	184.86	+10.3	194.24	185.76
Minneapolis	189.51	+7.0	184.24	+5.9	179.23	+8.6	182.61	174.93
New Orleans	199.27	+7.3	183.12	+9.8	181.03	+14.2	192.59	182.13
New York	180.36	+5.3	177.19	+4.2	194.83	+11.5	184.68	183.81
Philadelphia	220.00	+12.2	198.85	+9.4	200.34	+13.5	215.06	199.53
Pittsburgh	191.32	+9.3	183.56	+5.2	195.58	+9.4	185.77	178.83
St. Louis	198.70	+11.2	174.26	+8.4	171.32	+5.0	192.22	172.78
San Francisco	197.22	+10.4	181.24	+7.9	178.41	+8.4	193.25	180.05
Seattle	186.21	+14.9	176.39	+11.6	190.03	+13.1	187.08	181.82

To apply the update method, the ratio of the materials component of the future date to the materials component of the date of the report cost data is required. The materials component associated with the report cost data can be obtained from Figure 3-2 or 3-3, depending upon the well size of interest. The materials component associated with the future date can be determined from the latest issue of ENR. Future costs as a national average can then be computed by multiplying the cost data given in this report by the computed ratio.

For example, assume that in May 1977 the estimated cost of 200 feet of 8-inch diameter polyvinyl chloride (PVC) pipe in Region VII is desired. From Table 3-10 (Section 4) it can be seen that the cost of the pipe was $1120 in Region IX in October 1974. Figure 3-2 shows the materials component for October 1974 as 850. Assume that the materials component of the May 1977 issue of the ENR showed an increase to 1300. The cost of the PVC pipe in 1977 is then determined by multiplying the October 1974 costs ($1120) times the ratio of the 1977 materials component (1300) to the 1974 component (850), or

$$\$1120 \times \frac{1300}{850} = \$1713$$

The approximate cost of the PVC pipe in Region IX in May 1977 is $1713. To obtain the cost of the PVC pipe in Region VII would require a similar procedure using October 1974 and May 1977 ENR cities material prices, an example of which is given in Table 3-1.

Regional variations can be handled in a similar manner. For example, in Table 3-1 the city which is closest in geographical location to the monitor well activity is identified. The index for this city divided by the U.S. 20-cities index gives a ratio. This ratio multiplied by the current costs as a national average provides the updated current costs for a given region.

EFFECTS OF SCALE ON COSTS

Manufacturers generally offer discounts for large purchases or to preferred customers. Water well casing suppliers, for example, do not discount the price per foot of casing for the first 100 feet; however, discounts of 20 percent are common above a 100-foot purchase, and discounts of 30 percent are not uncommon on purchases above 1000 feet. Because pumps and grouting material are not purchased in large quantities for monitor wells, the effect of scale is small. Water analyses are subject to discounting on the basis of the number of routine tests performed on a sample rather than on the number of samples analyzed. As many chemical analyses require similar sample work ups, a routine or batch series of tests usually is considerably cheaper than the total for each of the tests performed individually.

IN-HOUSE VERSUS OUT-OF-HOUSE COSTS

The internalization of costs within the Federal government and universities makes competition with commercial enterprises somewhat unbalanced. Subsidies to many Federal and State laboratories, for example, often are not reflected in water analysis prices. Federal government costs vary with and within agencies. By way of illustration, the Water Resources Section of the U.S. Geological Survey (USGS) has a water resource service geophysical facility that has one price for users within the USGS water resource sector, a second price for other USGS users, and a third price for Federal agencies such as the EPA. On the other hand, many monitoring functions performed within Federal agencies have one set price for all.

ACCURACY AND COSTS

The cost of a monitoring method can vary with the accuracy of the technique. An aquifer model with 1000 nodes per unit area is much more expensive than a model with 10 nodes; the larger the model, the more accurate but also the more costly it is.

The accuracy of water quality analyses has become an increasing problem. Some parameters have accuracies established by the EPA, the Corps of Engineers, and local public health officials. Commercial laboratories performing water quality analyses usually report results in terms of 0.1 mg/ℓ (0.1 ppm). However, if accuracy to 0.01 mg/ℓ or greater is required, an additional charge of approximately one-third is added for each additional decimal point required.

SECTION 2 – MONITORING AT THE LAND SURFACE

INTRODUCTION

Groundwater monitoring programs have concentrated primarily on the saturated zone for indications of pollution. Too often, these programs have not considered the information that is available through analyzing the unsaturated zone and monitoring of the land surface. In addition, surface water monitoring of recharge areas is only beginning to be recognized as a critical variable in pollution detection.

Monitoring of the land surface can be divided into nonsampling and sampling methods. The nonsampling methods can be further divided into wasteload inventory considerations, leaching potential calculations, pipeline and tank tests, artificial liner testing, aerial surveillance, and notification and emergency procedures. The sampling methods are divided into those for surface water bodies, wastewater, and solid wastes.

NONSAMPLING METHODS

WASTE-LOAD INVENTORY

Identification of sources and methods of disposal are key items. An inventory of waste loads comprises data collection and tabulation of the volumes of liquid wastes, and weights of solid wastes and their compositions. For most purposes, monthly data on volumes and weights will be sufficient, and, in some cases, annual data will suffice. Acquisition of these data is necessary as a basis for calculating the amounts of percolate that may occur and concentrations of pollutants in the percolate.

Data on the physical, chemical, bacteriological, and radiological characteristics of the wastes should be collected. For wastewaters, chemical analyses are almost always included. Records of temperature and density measurements are appropriate for cases where the wastewaters have characteristics greatly different from native groundwaters. Turbidity records are important in cases where the wastes bypass the topsoil, such as in disposal or injection wells. Bacteriological analyses are important in the case of disposal of human and/or animal wastes. Radiological analyses are of foremost importance in the cases of nuclear waste disposal and certain mining wastes, among others.

For solid wastes, data on the chemical, bacteriological, and radiological characteristics should be collected. Bacteriological analyses are important in the case of sewage sludge and certain animal wastes.

CALCULATION OF LEACHING POTENTIAL

One of the key portions of a groundwater pollution monitoring program is to calculate or determine the amount of water which percolates through the topsoil and is in transit to the water table. The water budget approach may be used, where the input at the land surface in the form of waste volume and precipitation are compared to the output, or evapotranspiration. For diffuse sources, such as return flow of irrigation water, annual values may suffice. If evapotranspiration exceeds the water input at the land surface, then no percolate may occur. For point sources where input may greatly exceed evapotranspiration, annual values may also be sufficient. However, monthly calculations are necessary where evapotranspiration is greater than the input on an annual basis, but the input is greater than evapotranspiration in some months. This is particularly true where most precipitation or much of the waste disposal occurs during a few months of the year.

Precipitation records are usually available for most areas, or can be extrapolated. Evaporation from free water surfaces can be determined from measurements for land or floating pans (Harbeck et al., 1958; Kohler, 1954; Follansbee, 1933; Rohwer, 1933). Quite often these values are unavailable and measurements of pan evaporation must be made. Evapotranspiration in irrigated areas can be determined by a number of methods (Cruff and Thompson, 1967; Blaney and Criddle, 1962; Lowry and Johnson, 1942; Penman, 1948; Thornthwaite, 1948). The residual value, or percolation, is an important parameter, as it affects the subsequent dilution of wastes that occur in the aquifer. The value also is necessary as an index of potential pollutant load escaping the land surface.

PIPELINE AND TANK TESTS

Previous studies (Osgood, 1974; U.S. EPA, 1974b) indicate that buried pipelines and storage tanks are potential sources of sewage, storm water, and petroleum product contamination. The volume of flow from these sources may be a secondary concern if the pollutant is of a toxic or noxious character. The major consideration is to determine if and where the system is leaking. Small leaks from service station storage tanks can release 15 to 20 thousand gallons of gasoline over time, without the operator being aware of his loss (Osgood, 1974). Since the threat of fire or explosion is of primary concern in oil and gasoline leaks, the National Fire Protection Association (NFPA) has been active in setting up monitoring programs. To date, the NFPA has not approved any pipeline tightness testing equipment, although this equipment is being developed; it has endorsed one kind of tank tightness testing equipment.

Exfiltration and infiltration occurring in sanitary and storm sewers is a recognized engineering phenomenon. Where the system originally is poorly designed and improperly installed or where the pipelines are old and in disrepair, leakage of substantial quantities of poor quality water into the soil system can take place, eventually leading to contamination of groundwater. It is difficult to accurately determine the amount of leakage from these buried sanitary lines. The use of pressure in pipelines which are designed not to leak is a straightforward technique; however, these techniques are difficult to use in sanitary and storm sewers. The cost of performing a pipeline test would approximate the cost for a commercial tank test of the same volume.

In Maryland the county health departments reported 60 cases of pipeline and tank contamination in 1969 – 1970. These were detected from water in surrounding wells and not the sources, however. If it is possible to conduct input-output tests in sewer lines from basic flow data, large leaks may be detected. Large fluctuations in the flow received at the treatment plant which cannot otherwise be explained are another indication of leaks or breaks. However, in most cases the number of leaks in storm and sanitary drains will be large and individual losses will be low.

The testing of underground storage tanks can be done routinely. The Pennsylvania Department of Environmental Resources estimates that 2600 new or replacement subsurface storage tanks are buried in the ground in that State each year. If those replaced have failed, then this source of pollution deserves further monitoring. Most storage tanks are monitored using a calibrated dip stick. This is a rough measurement, better than nothing, but not sufficiently sensitive to detect small leaks. The stick is usually read in the morning and evening, and any difference indicates a possible leak. If a stick is not used, a manometer or float recorder may be used. These monitoring tools generally provide only rough approximations.

A more sensitive car-transportable tank tester, endorsed by NFPA, involves the use of a pressurizing device and a pressure loss recorder. The tank system tightness tester costs about $2875 FOB and is used by various consulting firms throughout the country. The average cost for an 8-hour (1 man-day) tank test is $300. Normally, one man can operate the tester; however, he may need a pipefitter assistant under certain circumstances. The largest tank tested to date has a capacity of about 24,000 gallons. The cost of the tank test varies primarily with the size of the tank in question and the size of the leak. The average cost of on-site testing of various tank sizes is given in Figure 3-4. A tank with a bad leak may require only 15 to 30 minutes for leak detection after the tester has been set up, while a tight tank can require 3-1/2 to 4 hours. The time required to test a series of tanks or one large tank may be more than 8 hours. The rate charged usually does not vary and remains around $35 – $40 per hour for both transportation to the test site and actual testing. The test procedure can also be used to determine if water is seeping into the tank.

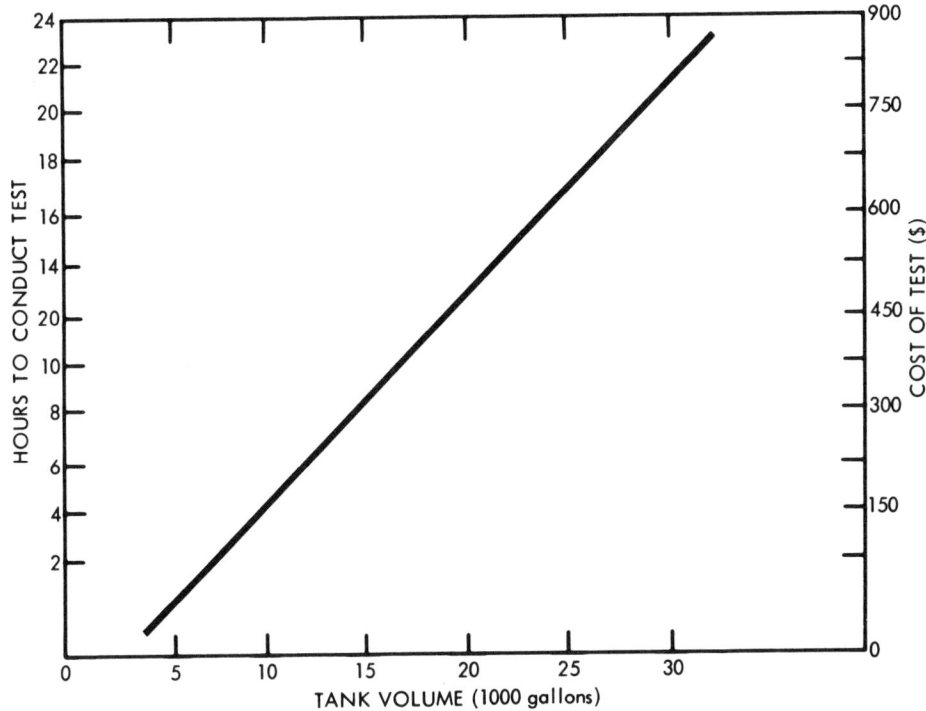

Figure 3-4. Cost of testing submerged tanks of various volumes, November 1974.

Tank testing has not been widely used, possibly through lack of awareness of the existence of test equipment. Owners have little legal responsibility to test their tanks. This situation may be compared to fire extinguishers, which are required by law and/or insurance companies to be routinely tested. Tanks that have been abandoned by gas station operators, industries, etc., are of considerable concern. Many of these abandoned tanks still have hazardous materials in them and, if leaking, are potential pollution sources.

TESTING ARTIFICIAL LINERS FOR LEAKAGE

Large numbers of storage ponds and *evaporation ponds* leak and permit substantial quantities of pollutants to escape the land surface. Artificial liners are coming into wide use to limit such percolation. These liners may be made of compacted clay, plastic, rubber, concrete, bentonite, and other materials.

There are several ways to determine leakage. One method is to fill the pond prior to use with a liquid of like composition to the waste to be contained. Seepage can be calculated after precipitation and evaporation are determined. Another suitable test involves construction of a special collection device beneath the artificial liner. This is usually constructed by placing a relatively impermeable material several feet below the final liner. Drain pipes are then placed above the layer, on a grade toward a central collection point. Coarse-grained permeable materials are then placed in this interval beneath the artificial liner. These collection devices enable rather rapid determination of leakage from artificial liners. Seepage can also be calculated if the permeability of the artificial liner can be determined.

AERIAL SURVEILLANCE

Aerial surveillance can be especially useful in land surface monitoring for groundwater pollution. Stockpiles, disposal and storage basins, and accidental spills can be seen from the air. Aerial surveillance can be used to determine what methods of disposal are being used under routine operation, as well as during accidents or unusual weather conditions. If, for example, sewage effluent is bypassing percolation ponds and flowing into a stream channel or canal, it can be detected from the air.

In disposal operations such as mine tailing ponds, water may be present in the ponds as free water, moist tailings, or wet tailings. Since the calculation of percolation requires estimation of evaporation and the evaporation rate is different in each case, the relative concentrations of pond areas must be periodically determined. As many such ponds cannot be freely traversed on the surface, aerial surveillance is an important aspect. Photographs and visual observations from various directions and heights can provide useful information. Observations and surveys on the ground can be used as a control for the aerial surveillance.

Infrared photography is a proven tool in analysis of shallow groundwater conditions and phreatophyte growth, as well as cropping patterns and irrigation analyses. Chandler, Dowdy, and Hodder (1970) reported on the utility of aerial surveillance methods in water quality monitoring; however, most of their discussion was in reference to surface water.

Remote-sensing technology, usually in the form of aerial and satellite imagery, provides the land-use planner and manager with a new source of data. Remote sensing can be employed in a land-use planning program to: (1) provide an initial source of information, (2) identify the factors responsible for change in the resource base, (3) help define management policy, and (4) monitor the effect of these policies. In particular, septic field pollution could be determined from population estimates which in turn could be obtained from remote sensors. This estimate of population can be based on housing occupancy ratios and counts of houses. Another method involving remote sensing is sampling population densities per unit area of land-use classes, and using the total area of a given land use as the multiplier to calculate its total population.

In-house costs are minimized more in remote sensing than in any other groundwater monitoring technique. For example, the funding required to produce the Earth Resources Technology Satellite (ERTS-1), Skylab, or a U-2 high-altitude sensor airplane are excluded from the purchase costs of raw data gathered by this equipment. In addition, Federal agencies, such as the EPA and the USGS, are active in remote-sensing research. The capital costs incurred by these agencies to develop a remote-sensing capability are difficult to determine; often only operational costs such as man-time, film, airplane fuel, and like expenses are charged for land-use surveys.

A few guiding principles in the use of aerial and remote-sensing surveys are:

- The larger the area, the less the cost per unit area for data acquisition and interpretation. This is true because the cost of mobilization is usually prorated on a per kilometer basis over the actual cost of acquisition, resulting in economies of size.

- In general, the higher the resolution, the larger the scale or the more specific the sensor, the higher the cost of acquisition, processing, and interpretation. This suggests a multilevel approach beginning with large area, low resolution generalized coverage and proceeding to studies of selected smaller areas, with more specific higher resolution techniques. At each level, the available data should be exploited to the maximum degree possible before proceeding to the next step.

- Multipurpose of multiresource programs reduce the cost accorded to each end use. This follows because the costs of data acquisition, processing, and capital investment for interpretation can be shared among users.

- The cost of adding a sensor to an aerial platform is usually relatively small. Most of the cost of data acquisition is the cost of flying.

- Computer and computer-assisted interpretation becomes economical when large areas are covered, a large number of comparisons among several data types are required, and the decisions to be made are relatively simple. The same is generally true of other machine-assisted techniques.

Data acquisition and processing refer primarily to the imagery and tape costs. An example of the standard price list from the Earth Resources Observation Satellite (EROS) Data Center, Sioux Falls, South Dakota is given in Table 3-2. Upon request the Center will provide a standard remote sensing order form. If the interpretation and verification capabilities are available, the cost of the EROS data is minimal. However, if interpretation and verification expertise is not available the costs involved escalate rapidly. For example, the level of detail required may be Level I, II, or III. Examples of the kinds of detail observed at Levels I and II are given in Table 3-3.

A review of work by Sizer,* Simonett,† and Thorley (1973) for land-use mapping studies in the United States resulted in the cost figures for data acquisition, processing interpretation, and map preparation presented in Table 3-4.

NOTIFICATION PROCEDURES

Warning systems to protect against accidental spills have previously been developed by agencies other than those charged with groundwater quality control. The major agency responsible for control of these spills has been the Department of Transportation. Apart from their other effects, spills of very short duration, such as hours or days, can result in groundwater pollution for decades or longer. Thus, prompt action is necessary, and advance warning of potential problems can be given by established notification procedures. However, notification procedures for hazardous substance transportation are poorly documented in most States and nonexistent in others. Notification procedures require that a responsible body be notified when any hazardous materials are to be moved. The intent is to insure that, when they are transported through a certain area, the full potential of the danger involved is realized,

---

*Personal communication, 1973.
†Personal communication, 1973.

Table 3-2

EXAMPLE OF EROS DATA CENTER STANDARD COSTS – SATELLITE PRODUCTS, OCTOBER 1974

AERIAL MAPPING PHOTOGRAPHY		
Image size (inches)	Format	Black & White unit price ($)
9	Film Positive	3.00
9	Film Negative	6.00
9	Paper	2.00
18	Paper	5.00
27	Paper	6.00
36	Paper	12.00
Photo Index	Paper	3.00

NASA RESEARCH AIRCRAFT PHOTOGRAPHY			
Image size (inches)	Format	Black & White unit price ($)	Color unit price ($)
2.2	Film Positive	2.00	5.00
2.2	Film Negative	4.00	NA
4.5	Film Positive	2.00	6.00
4.5	Film Negative	4.00	NA
4.5	Paper	2.00	6.00
9.0	Film Positive	3.00	12.00
9.0	Film Negative	6.00	NA
9.0	Paper	2.00	7.00
9x18	Film Positive	6.00	24.00
9x18	Film Negative	12.00	NA
9x18	Paper	4.00	14.00
18.0	Paper	5.00	15.00
27.0	Paper	6.00	20.00
36.0	Paper	12.00	30.00

MISCELLANEOUS		
	Black & White roll price ($)	Color roll price ($)
Microfilm		
16 mm (100-foot roll)	15.00	35.00
35 mm (100-foot roll)	20.00	40.00
Kelsh Plates		
Contact prints on glass. Specify thickness (0.25 or 0.06 inch) and method of	10.00	

(continued)

Table 3-2 (Continued)

ERTS DATA				
Image size (inches)	Scale	Format	Black & White unit price ($)	Color composite unit price ($)
2.2	1:3,369,000	Film Positive	2.00	NA
2.2	1:3,369,000	Film Negative	2.00	NA
7.3	1:1,000,000	Film Positive	3.00	12.00
7.3	1:1,000,000	Film Negative	3.00	NA
7.3	1:1,000,000	Paper	2.00	7.00
14.6	1:500,000	Paper	5.00	15.00
29.2	1:250,000	Paper	12.00	30.00

COLOR COMPOSITE GENERATION[a]			
(when not already available)			
Image size	Scale	Format	Unit price
7.3 inches	1:1,000,000	Printing master[b]	$50.00

COMPUTER COMPATIBLE TAPES			
Tracks	Density (bits per inch)	Format	Set price ($)
7	800	4-tape set	200.00
9	800	4-tape set	200.00
9	1600	4-tape set	200.00

SKYLAB PHOTOGRAPHY				
S190A image size (inches)	Scale	Format	Black & White unit price ($)	Color unit price ($)
2.2	1:2,850,000	Film Positive	2.00	5.00
2.2	1:2,850,000	Film Negative	4.00	NA
6.4	1:1,000,000	Paper	2.00	7.00
12.8	1:500,000	Paper	5.00	15.00
25.6	1:250,000	Paper	12.00	30.00

Notes:

[a] Color composites are portrayed in false color (infrared) and not true color.

[b] Cost of product from this composite must be added to total cost.

(continued)

Table 3-2 (Continued)

MISCELLANEOUS (continued)		
	Black & White roll price ($)	Color roll price ($)

Kelsh Plates (continued)
 	
   printing (emulsion to emulsion or through film base).
 	
Transformed Prints
 	
   From convergent or transverse low oblique photographs
	7.00	

35-mm Mounted Slide 35-mm mounted duplicate slide where available	.60	

and emergency crews are alerted in case of an accident. Of more importance, however, is the assurance that those doing the transporting will abide by safety regulations governing the move.

The majority of the States do not have a notification procedure tied into their groundwater monitoring program, and thus basic cost data are not available for warning systems and notification procedures. It is estimated that a notification monitor would approximate a safety inspection officer in training and salary, with the latter ranging from $10,000 to $15,000 per year.

Although transportation of hazardous substances can be by boat, rail, highway, air, or pipeline, the regulations for highway transportation are perhaps the most relevant to groundwater problems. Assuring adherence to notification regulations usually falls outside a monitoring agency's responsibility. However, the agency personnel charged with monitoring notification procedures should spot check to see that they are followed.

The best form of notification procedure is a specific permit to transport hazardous substances, which should be based upon applications submitted to a Transportation Board. An application for hazardous material permit should include the following:

- Name and address of applicant
- Business of applicant
- Requested exemptions to regulations
- Reason a specific permit is required
- Points of origin-destination and proposed routes
- Evidence to establish that transportation can be accomplished without undue hazard to public health and safety

- Period during which permit is to be effective
- Additional information as may be required

Table 3-3

LAND-USE CLASSIFICATION SYSTEM FOR USE WITH REMOTE-SENSOR DATA
(Anderson et al., 1972; Poulton, 1972)

Level I		Level II	
01.	Urban and built-up land	01.	Residential
		02.	Commercial and services
		03.	Industrial
		04.	Extractive
		05.	Transportation, communications, and utilities
		06.	Institutional
		07.	Strip and clustered settlement
		08.	Mixed
		09.	Open and other
02.	Agricultural land	01.	Cropland and pasture
		02.	Orchards, groves, bush fruits, vineyards, and horticultural areas
		03.	Feeding operations
		04.	Other
03.	Rangeland	01.	Grass
		02.	Savannas (palmetto prairies)
		03.	Chaparral
		04.	Desert shrub
04.	Forest land	01.	Deciduous
		02.	Evergreen (coniferous and others)
		03.	Mixed
05.	Water	01.	Streams and waterways
		02.	Lakes
		03.	Reservoirs
		04.	Bays and estuaries
		05.	Other
06.	Nonforested wetland	01.	Vegetated
		02.	Bare
07.	Barren land	01.	Salt flats
		02	Beaches
		03.	Sand other than beaches
		04.	Bare exposed rock
		05.	Other
08.	Tundra	01.	Tundra
09.	Permanent snow and icefields	01.	Permanent snow and icefields

Table 3-4

TOTAL COST OF AERIAL SURVEILLANCE FOR LAND-USE MAPPING, JULY 1974

Level of detail	Scale of final product	Method or basis	Data sources	Cost/km^2 ($)
Level I	1:250,000 to 1:500,000	Satellite imagery	1:250,000 ERTS imagery	0.15 - 0.27
		High-altitude aircraft imagery	1:50,000 aerial photographs	2.00 - 3.60
Level II	1:250,000 to 1:63,360	Satellite imagery only	ERTS imagery	0.25 - 0.45
		High-altitude aircraft imagery	1:90,000 aerial photography	2.25 - 4.05
	1:63,360	Multiscaled approach	ERTS imagery, high-altitude aircraft photography and ground truth	0.75 - 1.35
Level III	1:24,000 to 1:63,360	Commercial	Relys on 1:90,000 high-altitude photography	2.75 - 4.95
	1:12,000 to 1:40,000	Commercial	1:24,000 aerial photography	5.75 - 10.35

The Transportation Board should then determine if a permit is required. If a permit is needed it should:

- Describe the situation to which the permit applies
- Authorize such exemptions as may be warranted
- Establish the period of time in which the permit is effective
- Impose any new or special conditions which must be observed

With the establishment of a permit system, a set of rules must be established for both shipper and carrier to cover the safety aspects required by the permit. The rules and regulations for the shipper could specify acceptable containers, marking, labeling, and other precautions, whereas those for the carrier should specify loading, unloading, and vehicle operation procedures, driver qualifications, as well as other safe-handling procedures.

The person charged with monitoring the notification procedures would check both shippers and carriers, according to some sampling schedule to see that the rules and regulations were being followed. The monitoring could involve a review of a few permit applications that were accepted. Follow-up inspections could be done to see that proper containers were used and that the labeling was sufficiently clear. In addition, those permit applications that were turned down could be investigated to determine if potentially dangerous materials were still being transported. Carriers could be monitored for fully operational vehicles. In addition, drivers' logs could be reviewed for compliance with prescribed driving periods.

The transportation of hazardous material by highways is but one example of the kind of notification monitoring that could be implemented. Currently this type of monitoring is relatively rare; however, the reason primarily has been a lack of rules and regulations on the handling of hazardous materials. Many States currently are developing legislation to cover the transportation and monitoring of hazardous materials.

EMERGENCY PROCEDURES

Many States have a disaster office which controls emergency procedures for hazardous material spills. In the event of a hazardous substance spill or serious threat of such a spill, warnings should be given to endangered persons, to local authorities, and to the State Disaster Office (Figure 3-5). Emergency procedures to cover accidents once they have occurred are generally well defined. City and State police officials, as well as Department of Transportation personnel, are informed as to procedures to follow when an accidental spill occurs. The time required to correct the damage is very important because, if the material seeps into the soil, subsurface excavation can be costly. The emphasis, therefore, is on quickly reaching the site and taking action to minimize the infiltration and contamination of the soil matrix. The role of monitoring under these circumstances is to insure that the emergency procedures are operational and responsible personnel and equipment are ready to react to a call. This monitoring responsibility may be compared to that of a fire inspector, who is charged with insuring that fire-fighting equipment is ready should an alarm be sounded.

Regardless of the makeup of the organization or the type of location of a spill, certain basic operations must be carried out. The employment of any, or a combination, of the suggested measures can be undertaken only after technical advice has been sought and safety, feasibility, availability of material and equipment, side effects, and consequences have been considered. Some of the following operations may be conducted a step at a time, but many will of necessity be carried out simultaneously:

- Issue warnings and establish patrols
- Establish operations center
- Gather information
- Secure spill
- Contain, remove, and dispose of material
- Clean up and rehabilitate the area

All of the above operations require logistic support, such as provisions, materials and equipment, transportation, loading, unloading and storage facilities, and security provisions for same; communications; personnel; sampling and analysis; equipment maintenance; medical services; collection and recording of data (including photography) on a day-to-day basis; legal counsel; and administration, record keeping, funding, and accounting.

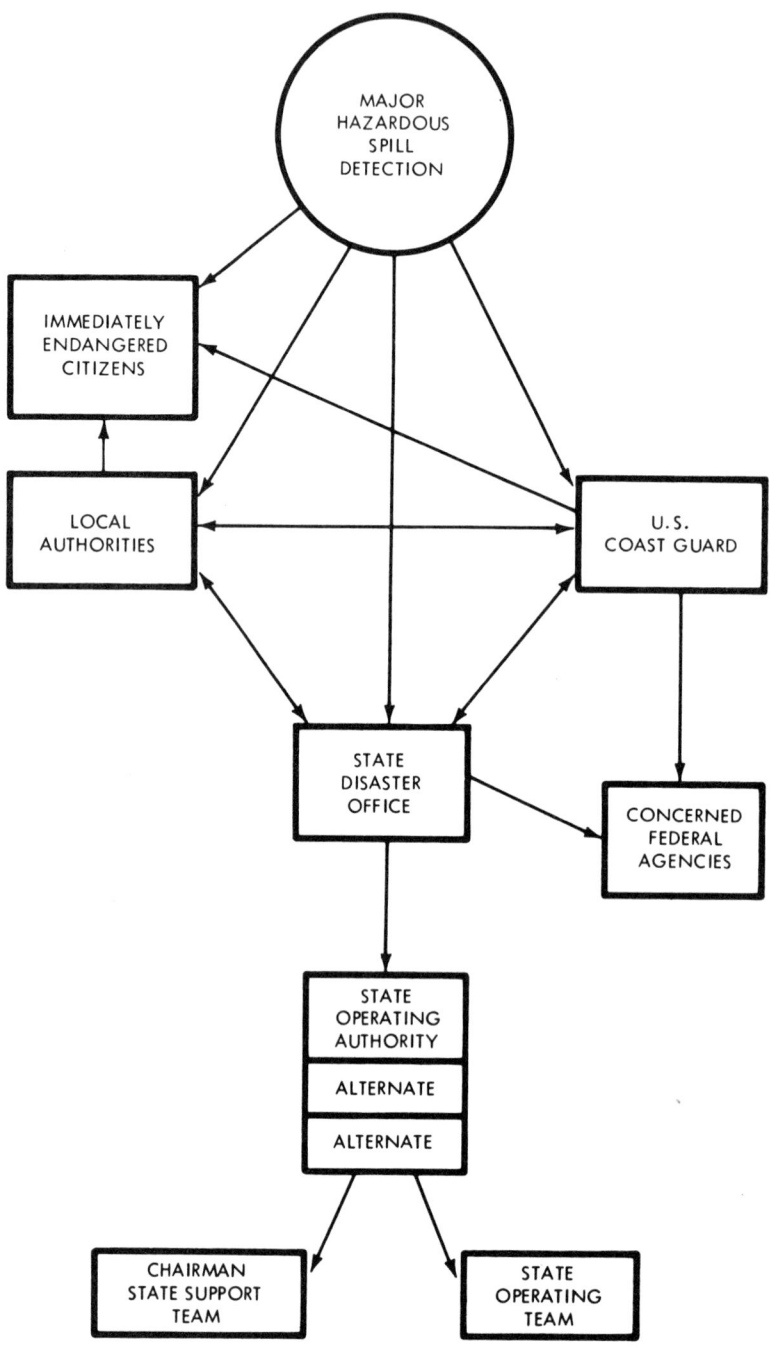

Figure 3-5. Alerting procedure chart for hazardous material spills (after California Emergency Plan, 1970).

The cost of monitoring the emergency procedures consists primarily of the salary and travel expenses of an inspector. The inspector can check to see that each person involved is aware of his responsibilities and all equipment ready for mobilization. Through a sequence of phone calls, personal visits, and actual equipment testing, the inspector can determine if the emergency procedures are workable. An inspector's training would often be derived from industrial safety programs rather than from a county or State program.

SAMPLING METHODS

SURFACE WATER BODIES

Surface water sampling includes rivers, lakes, estuaries, and canals that are sources of groundwater recharge. Of primary concern are polluted surface waters. Precipitation should also be sampled in areas where it directly recharges the groundwater. Surface water sampling will be conducted through federally mandated monitoring programs (U.S. EPA, 1974b), but additional monitoring will be necessary in some areas to evaluate polluted recharge. Sampling of surface water has been discussed by Hem (1970), and Brown, Skougstad, and Fishman (1970).

To adequately determine the composition of a flowing stream, each sample, or set of simultaneous samples, must be representative of the entire flow at the sampling site at that instant. Furthermore, the sampling process must be repeated frequently enough to define changes that occur with time in the water passing the sampling point. For most streams, one sample cannot be safely assumed to represent the water composition closely for more than a day or two, and for some streams, not for more than a few hours. If the average composition of the whole flow of the stream, or its changes in composition over a period of time, are factors of principal significance, sampling locations, where mixing is incomplete, should be avoided.

Daily sampling is generally believed to provide a reasonably complete record for most large rivers. Composited daily samples can be used to calculate mean annual chemical composition of streamflow and total loads of dissolved solids carried with reasonable accuracy. For many kinds of water-quality studies, however, compositing of samples is not desirable. As automated or continuous-recording equipment for conductivity and other variables comes into wider use, properly designed supplementary sampling and chemical analyses will provide many details on water-quality regimes, which are not attainable by daily sampling alone. Discharge or flow measurements should be available for sampling sites.

Water stored in lakes and reservoirs is often poorly mixed. Single samples from lakes or reservoirs can be assumed to represent only a single location within the water body.

The EPA (U.S. EPA, 1974b) has summarized rules for intensive monitoring of navigable waters. Station locations, parameter coverage, and sampling frequencies are considered. Monitoring stations should be located so as to measure inputs, transformations, movements, and outputs of pollutants within a given survey area.

The physical, chemical, biological, microbiological, hydraulic, hydrologic, climatic, and geometric parameters to be measured during monitoring surveys will depend upon the survey purpose and local conditions, and should be tailored to the specific pollution problems of the area. Sampling frequencies must be determined on the basis of variability of each of the parameters associated with the pollution problem, and must be adequate to define the pollution problem within statistically determined confidence intervals. The sampling frequencies must be adequate to determine mass balances of pollutants, where necessary, to define fluctuations of water quality and related parameters in receiving waters and pollutant sources.

Portions of surface water bodies can be selected for detailed sampling in areas where groundwater recharge is significant. This determination can be made on the basis of surface water budget analysis, groundwater level contour maps, and interpretation of soils and hydrogeological factors. It may be desirable to locate surface water sampling sites near wells or other groundwater monitoring facilities, in some cases.

WASTEWATER

The American Public Health Association (1971) reported on the collection of samples of domestic and industrial wastewaters for physical and chemical analyses. Only representative samples should be used for analysis, but the great variety of conditions under which collections must be made precludes prescription of a fixed procedure. In general, the sampling procedure should take account both of analyses to be made and the purpose for which samples are taken.

When testing for average concentrations, a 24-hour composite sample is considered standard for most determinations. When the purpose is to show peak concentrations, the duration of peak loads, or the occurrence of variations, grab samples collected at suitable intervals, and analyzed separately, are more appropriate. The sampling interval should be chosen on the basis of the frequency with which changes may be expected, and may vary from as little as 5 minutes to as long as an hour or more. Under other circumstances, a composite sample representing one work shift, or less, or a complete cycle of a periodic operation may be required. Evaluation of the effects of special, variable, or irregular discharges and operations may require composite samples representing the periods during which such wastes are present.

Composite samples cannot be used for analyses of constituents which are subject to significant and unavoidable changes when stored. Such determinations should be performed on individual samples as soon as possible after collection, preferably at the sampling point. Analyses for all dissolved gases, residual chlorine, soluble sulfides, temperature, and pH are examples of determinations of this type. Additional details are given by the American Public Health Association (1971), part 200, the American Society for Testing Materials (1969), and EPA (U.S. EPA, 1972, 1973, 1974a). Current articles in the *Journal of the Water Pollution Control Federation* discuss wastewater sampling.

Two basic categories of wastewater sampling relate to groundwater pollution. The first category is sampling of the discharge stream at the point of wastewater discharge, which, in some cases, contains solids as well as liquids. This type of sampling has been common, especially in cases where treatment plant performance is being evaluated. The second category is sampling in percolation ponds, and other types of ponds, which may leak. For monitoring groundwater pollution, the sampling of pond waters is often direct, and often requires either construction of special walkways, or the use of boats for sample retrieval. In others, where water is recycled, collection devices allow relatively easy sampling.

Of extreme importance is an understanding of how the wastewater is generated. For example, in sampling mine tailings pond wastewater a knowledge of what reagents are added at the mill, the type of ore being processed, and any unusual occurrences is necessary to correctly interpret the results. In sampling open waters, climatic factors such as rainfall and temperature are very important. Consideration should be given to groundwater recharge in the selection of sampling sites. Thus, wastewaters overlying more permeable soils and sediments deserve primary attention.

Generally, a 3-liter Kemmerer or Van Dorn style water sampler costing from $120 to $200 is used to sample surface water bodies. The cost of sampling varies with the time required to obtain the samples. Travel time and mileage to the site must be added to the time spent sampling the water body. Provided a small boat and motor are available, one man can sample up to about 50 locations per day in an average mine tailing pond. The sampling time at each location depends primarily upon the depth from which the sample must be retrieved. A 3-liter sample hand-drawn from about 20 meters requires about 5 minutes. In a river or a long reservoir, the distance between sample locations becomes important. Assuming that the boat must return to its original position, samples from 15 to 25 locations over a 25-mile stretch can be taken in 1 day. If wastewater or solid waste samples are to be taken from point sources, the travel time and mileage between the locations may be the principal governing cost factor.

SOLID WASTES

The chemical composition of many solid wastes can be categorized on the basis of past experience and available data in the literature. For point source solid wastes, sampling at the land surface is confined to occasional samples of the waste for analysis. The primary monitoring in this case is an inventory of the weight or volume of waste. For diffuse sources of solids, the main monitoring involves a complete inventory of amounts being applied on an annual basis. Monthly or weekly data may be necessary in some cases, such as the application of nitrogen fertilizers, where the amount leached will vary seasonally depending on crop uptake of nitrogen, temperature, irrigation applications, and like factors.

SECTION 3 – MONITORING IN THE VADOSE ZONE

INTRODUCTION

TOPSOIL

The topsoil is the region that manifests the effects of weathering of geological materials, together with the processes of eluviation and illuviation of colloidal materials, to form more or less well developed profiles (Simonson, 1957). Water movement in the topsoil usually occurs in the unsaturated state, where soil water exists under less-than-atmospheric pressures. The physics of unsaturated, soil-water movement has been intensively studied by soil physicists, agricultural engineers, and others. Copious literature is available on the subject in periodicals (*Soil Science Society of America Proceedings, Soil Science, and Journal of Environmental Quality*) and textbooks (Childs, 1969; Hillel, 1971; Kirkham and Powers, 1972). Saturated zones may develop over horizons of low permeability. A number of books are available on the theory of flow in perched water tables (Luthin, 1957; van Schilfgaarde, 1974). Soil chemists and soil microbiologists have also attempted to quantify the chemical-microbiological transformations during soil-water movement (Rhoades and Bernstein, 1971; Dunlap and McNabb, 1973; Bohn et al., 1979).

VADOSE ZONE

The entire region overlying the water table, including the topsoil zone, is defined as the vadose zone. Weathered materials of the topsoil gradually merge with underlying materials, such as igneous and metamorphic rocks, alluvium, lake deposits, eolian deposits, or lacustrine beds. Water in the vadose zone may exist primarily in the unsaturated state. However, saturated regions may develop. For example, perched water tables may develop above interfaces between layers having greatly different textures. Saturated conditions may also develop beneath recharge sites, as a result of prolonged infiltration. In contrast to the large number of studies on water movement in the topsoil, parallel studies in the vadose zone have been few. Meinzer (1942) coined the term *no-man's land of hydrology* to describe the limited knowledge of this zone.

MONITORING TECHNIQUES

A number of techniques have been developed for monitoring water movement and water quality in the topsoil. Detailed descriptions, specifications, and methods of many of these techniques were compiled in American Society of Agronomy Monograph No. 9, "Methods of Soil Analysis" (Black, 1969). Monitoring in the vadose zone requires an extension of topsoil monitoring technology. Examples are available where this approach has been used; Apgar and Langmuir (1971), for example, successfully used suction cups, developed for sampling soil solutions, to sample at depths up to 50 feet below a sanitary landfill. Of primary interest are flow rates to the water table and the storage capacity of the materials above the water table. A report of the EPA (1977) includes a detailed discussion of methods in monitoring in the vadose zone at solid waste disposal sites. L.G. Wilson (1980) has written a state-of-the-art document on monitoring in the vadose zone.

WATER CONTENT

Soil and underlying geologic materials consist of a solid and a porous matrix. Porosity is a measure of the amount of water which could be stored by the materials under saturated conditions. Field capacity of a soil is represented by its water content (on a dry weight basis) at a certain time after the initiation of drainage following an irrigation cycle (Peters, 1965). The definition of specific retention for geologic materials is essentially the same, except that the water content is expressed on a volumetric basis. Specific yield, or effective porosity, of geologic materials is the difference between porosity and specific retention. At a given time, the volumetric water content of a material is a reflection of various forces acting on the soil-water system, including gravity, capillary, and osmotic forces. Soil-water pressure is a measure of capillary forces and is expressed in terms of negative pressure; the relation between water content and pressure may be determined in the laboratory, and when plotted is called a soil-water characteristic curve. This relation is essential to calculate water flow rates in the vadose zone.

WATER FLOW

Infiltration is the movement of water across the soil surface from an applied water source. Factors affecting infiltration capacity, or the maximum rate of infiltration, include soil texture, initial water content, and soil stratification. Determination of the rate of water flow out of a given soil depth may be determined from sequential drainage profiles (curves showing the relation between soil depth and water content for various times after the cessation of infiltration). The salinity and composition of the solution may have a pronounced effect on water movement.

For saturated materials the pressure is positive and measured by piezometers. For unsaturated materials the pressure is negative and measured by tensiometers.

CHEMICAL CHANGES

The interactions of infiltrating water and soil have been reviewed by Ellis (1973), Rhoades and Bernstein (1971), Murrmann and Koutz (1972), McNeil (1974), Fuller (1977), and Runnells (1976). Among the main reactions are: precipitation, dissolution, ion exchange, adsorption, and oxidation-reduction.

Changes in the concentration of a soil solution, following surface application, include water removal by evaporation and transpiration. Precipitation of slightly soluble salts, primarily calcium carbonate or calcium sulfate, may occur as the soil solution is concentrated during evapotranspiration. Dissolution of soil minerals contributes to the salinity of the soil solution. Ion exchange represents an important soil-water interaction for altering the composition of the soil solution. Ion exchange is a function of the cation exchange capacity of the soil, which, in turn, is related to colloidal clay minerals, soil organic matter, iron and aluminum sesquioxides, and hydrous oxides.

Adsorption of metals onto the surface of soils is the most important process for removing some chemicals from wastewater (Murrmann and Koutz, 1972). In contrast to ion exchange, in which ions retain their mobility, in adsorption reactions ions are held so tightly that they become essentially immobile. Methods to quantify the extent of adsorption of ions onto solids have been developed. Oxidation and reduction reactions are of special importance in the case of sulfur and nitrogen compounds in water, as well as for many trace elements. These reactions affect the mobility of some elements, and may result in the production of gases, which can be lost from the soil-groundwater system.

SOIL SAMPLING AND WELL DRILLING

A test drilling program may be necessary to supplement existing data on the vadose zone at a site. Soil sampling and well drilling are generally necessary for all types of monitoring in the vadose zone. Test wells can also function as observation wells, piezometers, or access wells.

SHALLOW WELLS

In general, samples from the soil zone are obtained to (1) characterize the average soil texture, water content, or chemistry in depth increments (e.g., 15 centimeters), (2) observe the precise depth distribution of soil texture, or (3) determine the bulk density, or water-release curves, of soil increments. For the first purpose, hand samplers such as post-hole augers, screw or sleeve-type augers, or power-driven augers are useful. For the second purpose, cores are obtained by driving small-diameter (e.g., 2.5 centimeter) tubes into the soil to the desired depth. For the third purpose, larger diameter core samplers are used, which may be hand-driven or power-driven. Cores of a specific volume are obtained by each method. More information on coring is presented by Blake (1965).

The cost of a bucket-type hand auger is about $35. These augers are available to cut core sizes from 2 to 4 inches in diameter. Four-foot extensions for deeper sampling cost about $6 each. For deeper samples, a hand-driven soil sampler system can be purchased. These systems include soil sampling tubes, a drop hammer, and a tube puller jack to withdraw the tubes. The soil sampling tubes vary from 4 to 16 feet in length and cost from $65 to $115, respectively. The cost of the hammer and puller jack is about $200.

DEEPER WELLS

To sample throughout the vadose zone it may be necessary to drill deep wells using standard techniques. Such techniques include jetting, rotary, cable tool, augering, and air drilling. Of these methods, perhaps augering, using continuous flights, and air drilling provide the most usable samples; problems develop in characterizing the distribution of indigenous salt and water content, with cable tool and rotary methods because of water additions during the drilling process. The cost of power-augering multiple depth and density holes is given in Figure 3-6. These augering costs are developed for 8-inch diameter holes.

One air-drilling technique (Becker Drills, Inc., Denver, Colo.) involves driving a double-wall tube by a pile hammer, while concurrently forcing air under pressure down the annulus of the pipe. Air and entrained material cut by the bit return to the surface through the inside pipe. The sample, available continuously, is diverted into a cyclone sampler, where it is bagged for laboratory analyses. Changes in formation can be determined within a few centimeters; furthermore, water seams can be determined immediately. This feature is advantageous in locating the depth and thickness of perching layers. Whichever technique is used, samples should be taken in specified increments throughout the vadose zone.

DETERMINATION OF WATER CONTENT

Although a major concern of a monitoring program may be the movement of water in the vadose zone, it is also important to account for the storage capabilities of this zone. In western valleys, with deep alluvium, the vadose zone may constitute a vast reservoir for such in-transit storage.

Examining this water content requires the extension of techniques employed in the soil zone downward to the water table. Fortunately, many of the methods may also be used simultaneously to monitor water movement.

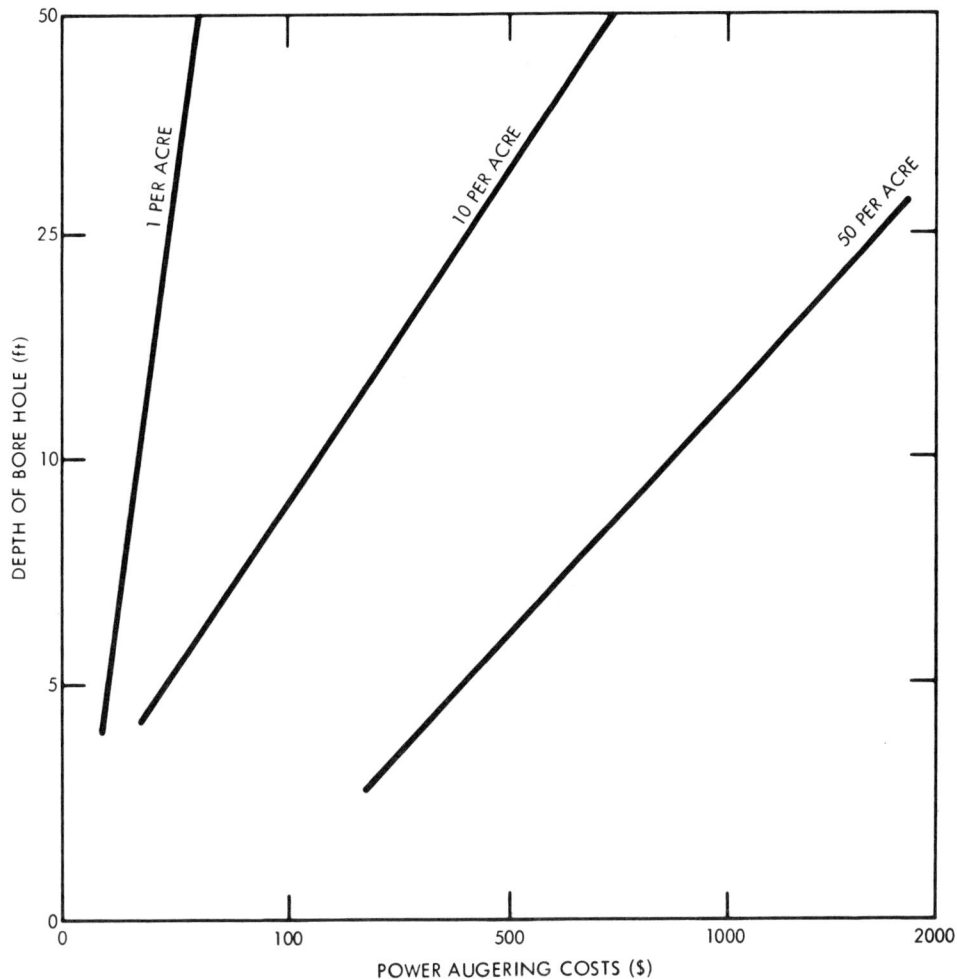

Figure 3-6. Cost of power-augering 8-inch diameter sample holes to various depths and hole densities, November 1974.

NEUTRON MODERATION OR MOISTURE LOGGING

For many years agriculturists have employed the principle of neutron moderation, or thermalization, to measure the volumetric water content of soil in situ. Recently, the technique has been used to monitor water storage in the vadose zone, particularly to delineate perching layers and mounds, and also to estimate flow rates.

The method of water content evaluation by neutron moderation depends on two properties relating to the interaction of neutrons with matter: scattering and capture (Gardner, 1965). High-energy neutrons, emitted from a radioactive source are slowed down, or thermalized, by collisions with atomic nuclei. Hydrogen has a greater thermalizing effect on fast neutrons than occurs with many elements commonly found in soils. This forms the basis for detecting the concentration of water in a soil (Van Bavel, 1963). The second property of interest in the neutron moderation method is capture of slow neutrons, with the release of energy. The property of energy release during capture serves as a means of detecting the concentration of slow neutrons.

When a source of fast neutrons is lowered into a soil through a suitable well bore or casing, a cloud of thermalized neutrons is established. If a suitable calibration is made to isolate the moderating effects of soil nuclei other than hydrogen, changes in the volume of the thermalized cloud will reflect changes in water content. Finally, a detector which relies on capture of thermalized neutrons is used, in conjunction with suitable electronic circuitry, to measure the water content (on a volume basis).

Instrumentation used to measure water content by neutron thermalization requires three principal components: (1) a source of fast neutrons, (2) a detector of slow neutrons, and (3) an instrument to determine the count rate from the detection equipment, as shown in Figure 3-7.

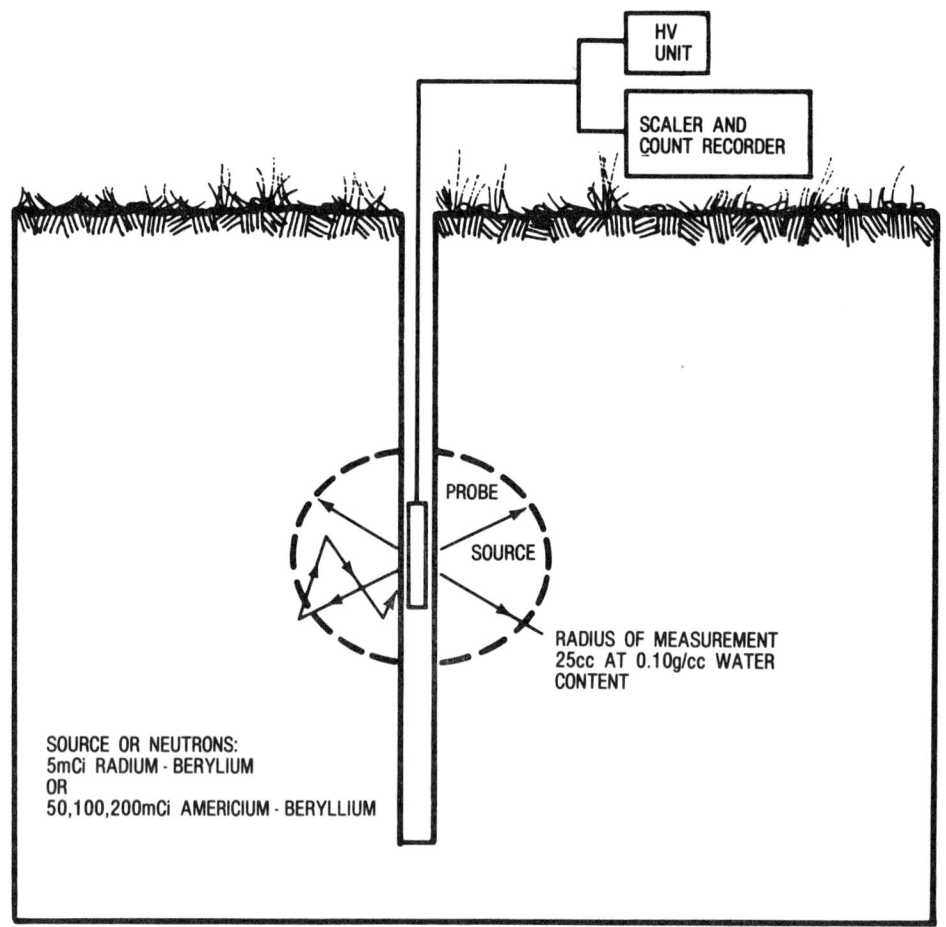

Figure 3-7. Sketch of neutron moisture logger and accessories (after Holmes et al., 1967).

Americium-beryllium is commonly used as a source of fast neutrons. Detectors of slow neutrons rely on high capture cross sections of detecting materials (e.g., boron). A charged particle is emitted during capture, which can be detected by a solid or gaseous counting device. The predominant detector is lithium-enriched BF_3. Both the source and detector are located in a cylindrical probe, or tool, which is lowered via a cable into access tubing, either by hand or motor-drive. The depth of measurement is determined by graduations on the cable or, alternatively, by some type of counter (Van Bavel, 1963).

The pulse emitted during the nuclear reaction resulting from capture is amplified within the down-hole tool, and transmitted through the cable to an above-ground meter for counting. Count rate is converted to water content via suitable calibration curves. The type of equipment that has been available for soils work is hand operated, and requires placing the down-hole tool at a discrete depth, taking a count, or number of counts, and then moving the tool by hand to another discrete depth. This process is not troublesome for shallow soils studies. However, for deeper access wells, in the vadose zone, an inordinate amount of time would be involved. A motorized unit is commercially available, which permits lowering the tool in the well at a constant rate, via a motor drive. Internal electronic components concurrently translate pulse rate into water content, which is automatically recorded.

The individual neutron moisture logger should be calibrated against samples of known water content. Such calibration may be done in the field by comparing a count at a given soil depth, with the water content of soil cores from the same depth. Access tubes for neutron moisture logging are usually constructed of seamless steel or aluminum. Polyvinylchloride-cased wells may cause difficulty in that hydrogen or chlorine atoms in the tubing may moderate the thermal neutrons, interfering with soil moisture evaluation. Aluminum-cased wells may deteriorate in saline groundwaters. The inside diameter of wells should be as close as possible to the outside diameter (OD) of the probe. Work by Ralston (1967) showed that for a 100-millicurie (mCi) Am-Be source in a 3.8-cm OD tool, the water content could not be accurately evaluated in well casings with an inside diameter (ID) greater than 10 cm. For larger casing a source in the multicurie range should be used.

Wells drilled for shallow water content monitoring can be easily installed by successively augering and driving the tube. Myhre et al. (1969) reported on a simple power-driven auger for installing wells to 150-cm depths. For deeper wells standard drilling techniques are necessary. During installation by drilling techniques, it is essential to establish a tight fit between the well shaft and casing, to minimize the amount of vertical leakage of water. Drilling mud, although facilitating the drilling operation, is not recommended since it interferes with water content observations. In situations requiring a tight fit, drilling techniques which do not require a drilling mud (jetting, augering) are used.

The principal advantage of neutron moisture logging is that water content profiles are obtained in situ with minimal soil disturbance. A history of profiles can be established during a monitoring program. Moisture logs clearly show the presence of perched water tables, together with their growth and dissipation (see Figure 3-8). Such logs may also be used to estimate unsaturated flow rates. Water content exchanges at a given depth, in a succession of wells, may provide clues on lateral flow velocities (Wilson, 1971).

Certain limitations should be noted. First of all, the presence of excessive concentrations of other fast neutron moderators (e.g., boron, chlorine) may cause erroneous water content determinations. Neutron moisture logs indicate only the water content of soils, and may not always manifest water movement. However, water content values may be translated to equivalent soil-water suction values via a soil-water characteristic curve.

The cost of a neutron moisture logger, including the neutron source, slow neutron detector, and surface counter is about $3000. Figure 3-9 gives the cost of neutron moisture logging, with various numbers of well loggings performed at depth intervals of 2 feet. The costs in Figure 3-9 do not include the capital cost of the instrument, but only the time required to conduct the log.

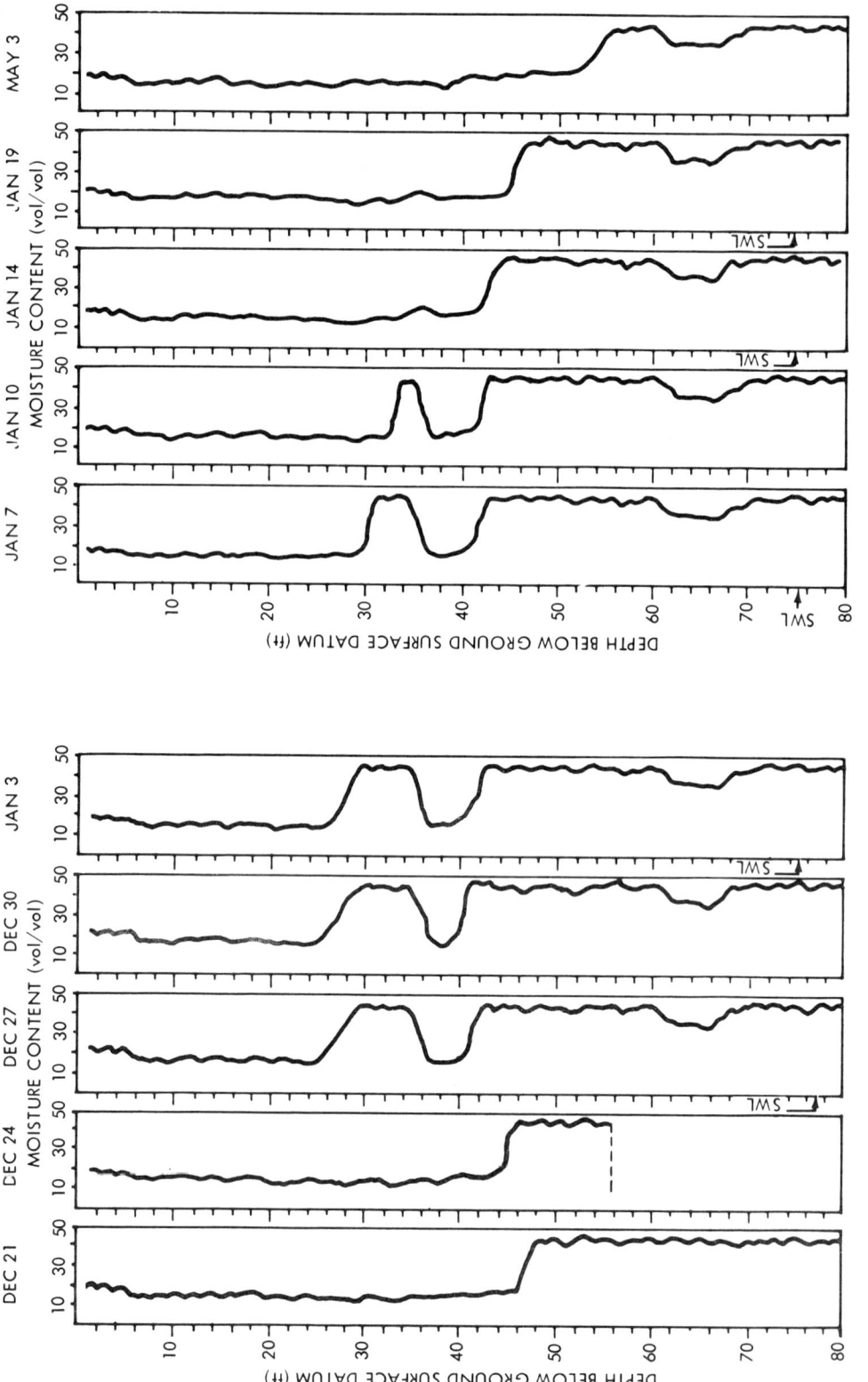

Figure 3-8. Moisture logs showing growth and dissipation of mounds in the vadose zone (after Wilson and DeCook, 1968).

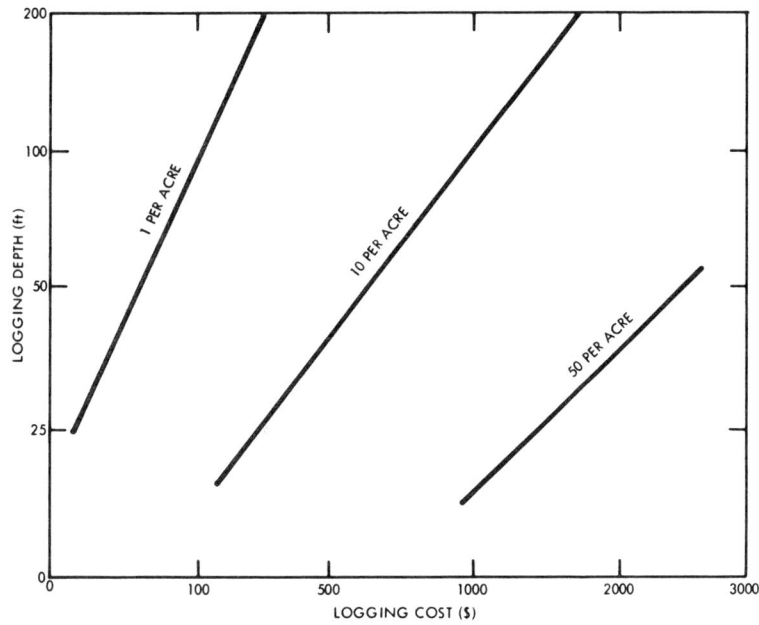

Figure 3-9. Cost of neutron moisture logging for various depths and sample space densities, December 1974.

TENSIOMETERS

Tensiometers are used to measure soil-water pressures during unsaturated flow. If caution is exercised, they may also provide estimates of soil-water content via suitable soil-water characteristic curves. Basically, a tensiometer consists of a porous ceramic cup cemented to a rigid plastic tube, small-diameter tubing leading to a manometer and terminating in a reservoir of mercury, and a filler plug in the rigid plastic tube (see Figure 3-10). Except for the reservoir and the portion of the small-diameter tubing filled with mercury, the internal volume of the system is completely filled with water. When properly emplaced in the soil the pores in the cup form a continuum with the pores in the soil. Water moves either into or out of the tensiometer system, until an equilibrium is attained across the ceramic cup. The mercury level in the manometer tubing adjusts correspondingly.

Holmes et al. (1967) have noted the precautions to be taken during installation of tensiometers. Prior to field installation, the system should be purged of as much entrained air as possible, and its response checked. The tensiometer unit should be placed into soil cavities larger than the outside diameter of the unit. The cup should be forced into the soil at the base of the hole to ensure good contact. The hole is then backfilled with soil.

The principal limitation of tensiometers is that they are useful in measuring soil-water pressures only up to about 1.0 atmosphere. Furthermore, the determination of water content via tensiometer values is subject to hysteresis: i.e., different water content versus pressure curves will be followed depending on whether the soil is wetting or drying.

The cost of a tensiometer including a 6-foot plastic tube is about $30. Figure 3-11 gives the capital costs of tensiometers placed at various depths and at various sampling densities. The costs presented in Figure 3-11 do not include installation.

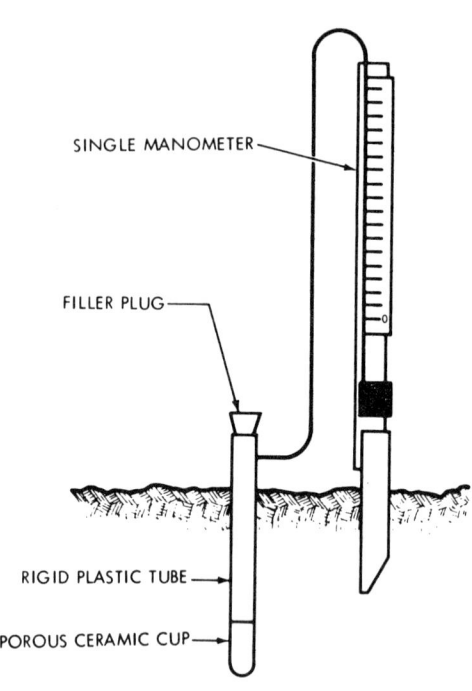

Figure 3-10. A single manometer tensiometer.

ELECTRIC RESISTANCE BLOCKS

Electrical resistance blocks, used to measure either soil-water content or soil-water pressure, consists of electrodes embedded in a suitable porous material, such as plaster of paris, fiberglass, or nylon cloth. The principal of operation of these blocks (Holmes et al., 1967) is that water content (or negative pressure) within the blocks responds to the water content (or suction) of the soil, with which the blocks are in intimate contact, and the electrical resistance properties of the blocks change correspondingly. Moisture blocks are calibrated in soil from the site at which they are to be installed. Such calibration involves evaluating resistance readings against a range of soil-water contents, or negative pressures.

Holmes et al. discuss the advantages of resistance blocks, indicating that (1) they appear to be best suited for general use in the study of soil-water relations, (2) they are inexpensive, and (3) they can be calibrated for either suction or water content. Generally, for agricultural applications the blocks are used for negative soil-water pressure greater than 0.8 atmosphere.

The cost of a soil moisture block is about $4 and the leads to the soil moisture blocks are very nominal in cost. A soil moisture meter costs about $120. The capital cost, excluding emplacement, of electrical resistance blocks, leads, and a soil moisture meter, for various depths and sampling densities is given in Figure 3-12.

CALCULATION OF WATER CONTENT

In situ measurement of water content at some locations, such as beneath waste disposal sites with large areas and relatively deep water tables, may be too cumbersome and expensive to perform. One alternative is to evaluate maximum storage capacity. Porosities for common geologic materials are given in Figure 3-13. The storage capacity in the vadose zone will be less than that indicated by the porosity because all of the materials are not saturated.

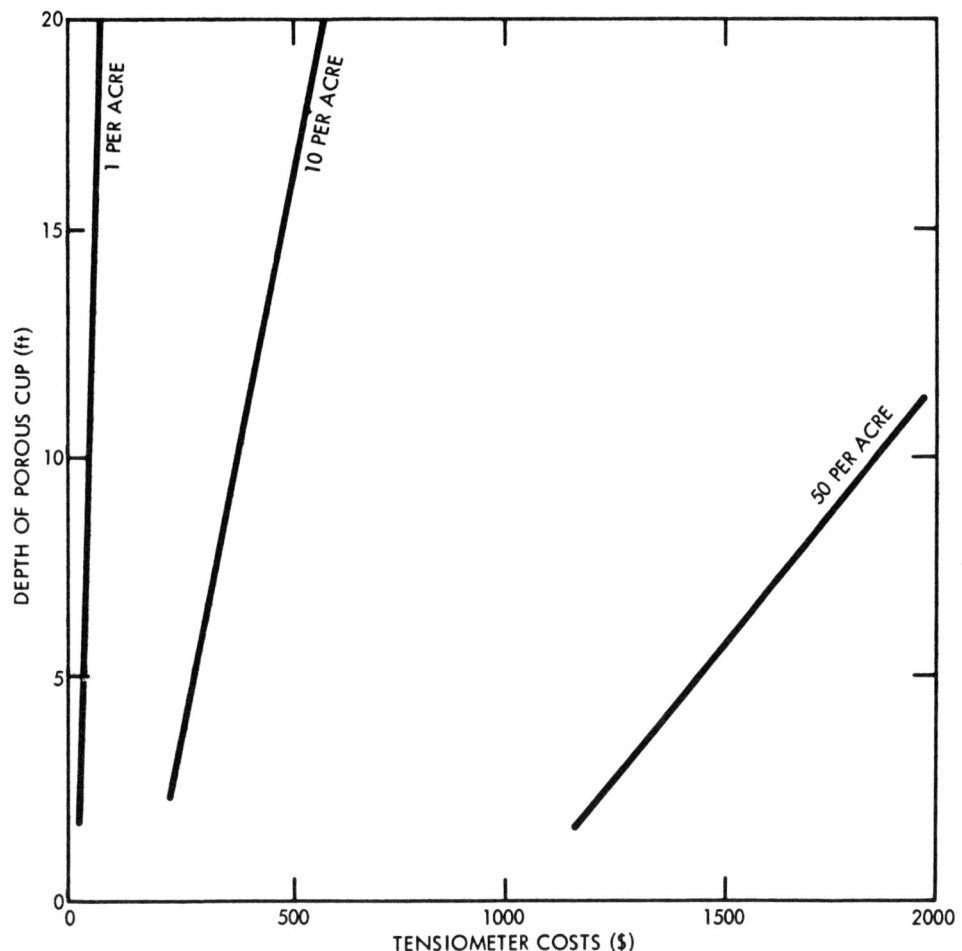

Figure 3-11. Cost of tensiometers for various depths and sample space densities, December 1974.

Many areas can be divided into two types of vadose zones: those where the specific retention of geologic materials has already been satisfied, and those where it has not. The first case occurs primarily in humid areas, where water levels are relatively shallow and an abundant source of recharge has been present for many years. This case also occurs beneath many alluvial basins of the arid west, where water tables are within 100 to 200 feet of the land surface, and water has been artificially applied at the land surface for decades. The second case is common in arid areas, with relatively deep (300 or more feet) water tables. Low natural recharge at the land surface for many years, combined with the presence of soil or subsurface restricting layers, may prohibit natural percolation to the water table. In these cases, the geologic materials may be virtually devoid of water.

The storage capacity for percolated water in cases where the specific retention is satisfied is defined by the specific yield. Representative specific yields for geologic materials are given in Table 3-5. Where the specific retention has not been satisfied, significant amounts of percolated water will be stored in the vadose zone to make up this deficiency. The total storage capability will then be some value less than the porosity.

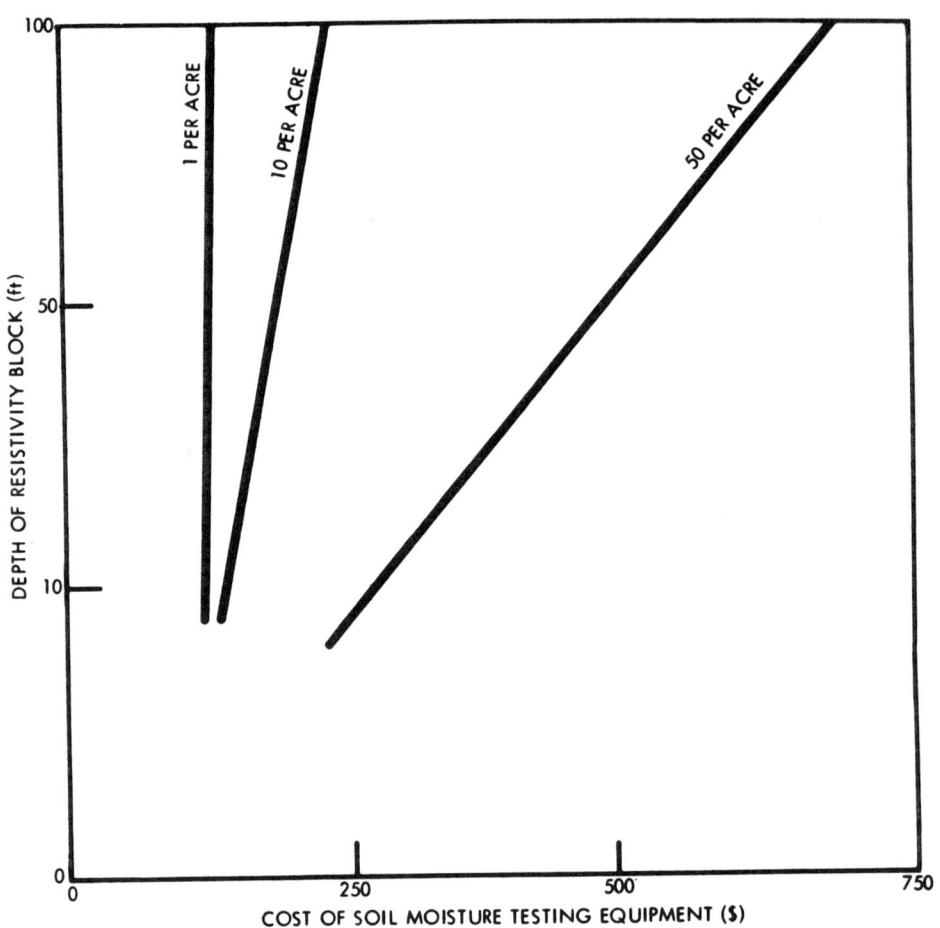

Figure 3-12. Cost of multiple electrical resistance blocks and soil moisture meter for various depths and sampling densities, December 1974.

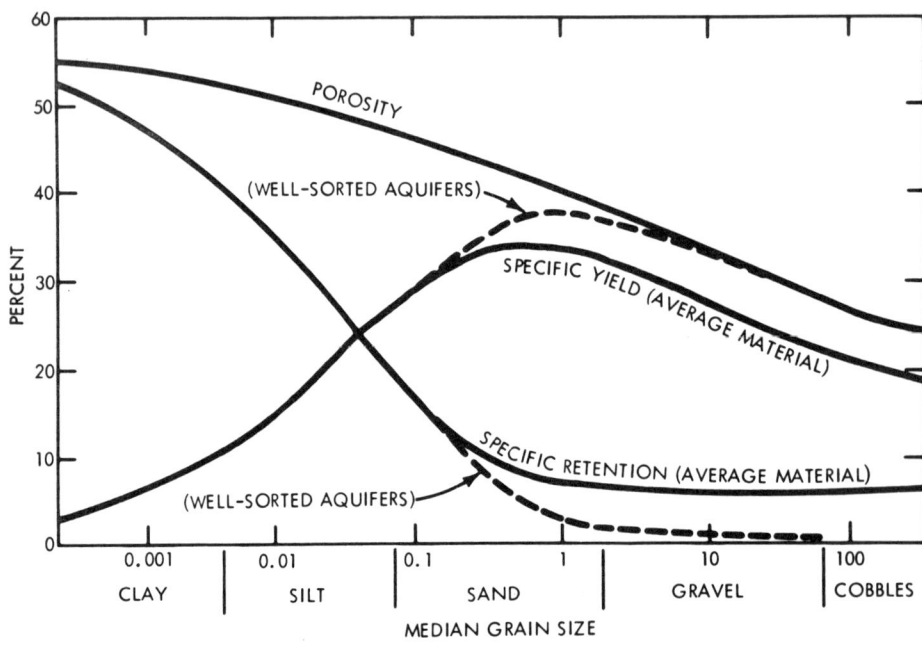

Figure 3-13. Relationship between median grain size and water-storage properties of alluvium from large valleys (Davis and DeWiest, 1966).

Table 3-5

SPECIFIC YIELD VALUES ASSUMED FOR THE SOUTHERN PART
OF THE CENTRAL VALLEY, CALIFORNIA (Davis and DeWiest, 1966)

Driller's Description	Assumed Specific Yield %
Gravel, sand and gravel, and similar materials	25
Fine sand, tight sand, tight gravel, and similar materials	10
Clay and gravel, sandy clay, and similar materials	5
Clay and related materials	3
Crystalline rock (fresh)	0

DETERMINATION OF WATER MOVEMENT

An evaluation of water movement above the saturated zone should account for (1) infiltration across the soil surface and movement through the topsoil, (2) movement in perched water tables, and (3) downward movement through the vadose zone. Flux (volume per unit area per unit time) is the principal flow characteristic to be defined in studying water movement in the vadose zone. An important factor in calculating flux is the hydraulic conductivity.

INFILTRATION ACROSS THE SOIL SURFACE

Flux across the soil surface is termed infiltration, the maximum value of which is the infiltration capacity. This characteristic is important in land disposal operations because it indicates the rate at which effluent may percolate to groundwater. Bouwer (1973) categorized land disposal systems as high-rate or low-rate, reflecting the infiltration rates used to effect effluent treatment.

For existing land disposal operations, effluent can be metered onto each treatment area via suitable flumes, weirs, or flowmeters. Differences in flow rate represent the amount of water which infiltrates on the area. The flux can be calculated (it may be necessary to account for evaporation losses) if the surface area contacted and time that a given volume infiltrated are known.

MOVEMENT IN PERCHED WATER TABLES

Results of neutron logging observations during infiltration tests or examination of data on stratigraphy may indicate the locations of perched water tables in the vadose zone. The principal techniques for monitoring movement in these saturated zones are the use of piezometers or observation wells.

Piezometers

Piezometers consist of small-diameter pipes, driven, augered, or jetted into a known, or expected, saturated soil zone. The level of water measured at the base of the piezometer indicates the water level in the saturated zone.

Reeve (1965) discusses in detail the techniques for installing and cleaning new piezometers. It is important to maintain tight contact between the outer wall of the piezometer and the soil. For shallow units, piezometers may be installed by augering and driving with a sledge hammer. Deeper units will require jetting, or use of standard drilling equipment. It may be necessary to fill the cavity between the well and borehole with grout to ensure tightness of fit. As with regular wells, piezometers should be developed by pumping, bailing, or other applicable methods to open up material at the base of the unit. In some situations, it may be necessary to install piezometers with screened well points to prevent upward movement of saturated material into the unit.

Depth to water in piezometer units is measured by chalked tape, electric sounders, or air lines. (The use and cost of these tools is considered in detail under "Monitoring in the Saturated Zone.") By referencing these measurements to a common datum, groundwater levels can be determined for the perched zones. The slope of these elevations indicates the direction of lateral water movement.

Many field soils exhibit anisotropy. That is, the hydraulic conductivity in the horizontal direction may be much greater than in the vertical direction. For such soils it may be advisable to install more than one array, or battery, of piezometers in a lateral direction away from a recharge source. The bottom openings of piezometers in sequential arrays should terminate at the same elevation. Such an arrangement permits observing lateral head differences resulting from horizontal flow.

The costs of 2- and 4-inch piezometers for various depths and sampling densities are given in Figure 3-14. The piezometer costs presented do not include installation.

Observation Wells
An observation well consists of an uncased borehole, or perforated pipe, extending from ground surface into a perched water table. Diameters and depths of observation wells are generally greater than piezometers.

WATER MOVEMENT IN THE UNSATURATED STATE

Methods for monitoring water movement in saturated regions cannot be used to monitor unsaturated flow because water will not freely enter a soil cavity, unless the soil-water pressure is greater than atmospheric. Consequently, special methods must be used in unsaturated soils. Of the available methods the three most commonly used to infer unsaturated flow employ tensiometers, psychrometers, and neutron moisture logging. Several newer techniques, although not extensively field tested, may also have applicability.

TENSIOMETERS
The principles of operation, methods of installation, and limitations of tensiometers are discussed above. In general, tensiometers appear to be the most effective technique for monitoring flow in the soil-water pressure range down to about 0.8 atmosphere (atm). Nielsen et al. (1973) observe that tensiometers are still routinely used and nothing as reliable, accurate, and inexpensive has replaced them in the past 35 years. The application of tensiometers to depths greater than the soil zone is now possible with the advent of pressure transducer units developed by Watson (1967) and others. In characterizing the direction of water movement, a battery of tensiometers must be installed, with units terminating at successive depths throughout the region of interest.

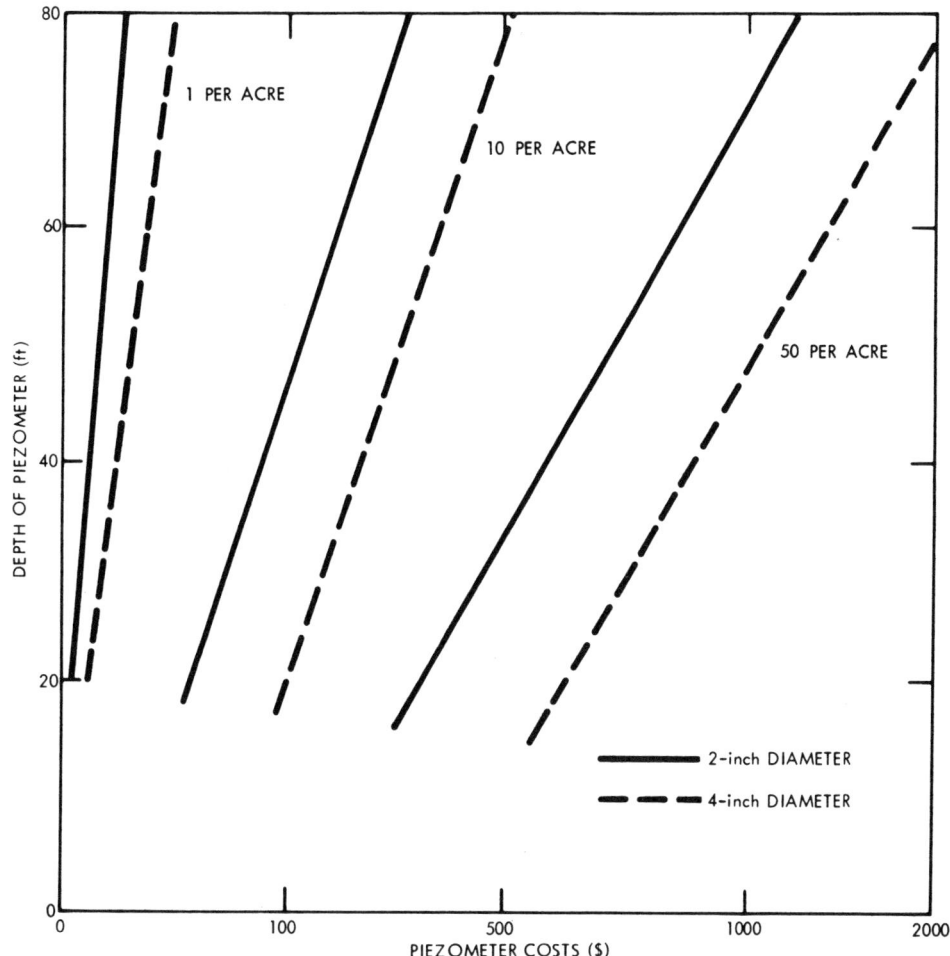

Figure 3-14. Cost of piezometers for various depths and sampling densities, November 1974.

As with piezometers, it is necessary to install more than one array of tensiometers to detect horizontal flow. Thus, if individual arrays terminate at varying depths, differences in total hydraulic heads between corresponding units in successive arrays may suggest lateral movement in the unsaturated state. Because of heterogeneity in soil properties, however, conclusions on lateral flow may not be definitive.

PSYCHROMETERS

Tensiometers cannot be used to measure soil-water pressure below about -1 atm because of air-entry problems. To measure lower negative pressures, therefore, other instrumentation is required. In recent years, progress has been made in developing the thermocouple psychrometer for this purpose. According to Watson (1974) in situ pressure measurements down to -30 atm are possible with these units. The principle of psychrometric measurement of soil-water potential, discussed by Rawlens and Dalton (1967), is that a relationship exists between soil-water potential and the relative humidity of soil water.

Psychrometers consist of a porous bulb comprising a chamber to sample relative humidity of a soil, a sensitive thermocouple, heat sink, reference electrode, and associated electronic circuitry. Two types of thermocouple psychrometers are available. One type, which is installed in access tubes, consists of mounting psychrometers in porous cups at the base of tubing. This unit may be withdrawn for recalibration. The second type, called *sealed-cup* psychrometers by Merrill and Rawlens (1972), contains the thermocouple unit permanently sealed into a porous enclosure.

Each unit must be calibrated prior to field installation; techniques for such calibration are given by Merrill and Rawlens. Also, in field installations, it is important to record and correct for diurnal changes in temperature. Installation of a sensitive thermistor is recommended for diurnal studies. Enfield et al. (1973) have used thermocouple psychrometers to estimate the direction and rate of water flux in material above a water table at a depth of 94 meters in a desert environment in Washington. Individual psychrometric units were placed 3 meters apart within a specially designed well casing.

NEUTRON MOISTURE LOGGING

Although the neutron thermalization technique is used mainly to determine changes in volumetric water content of materials in the vadose zone, the method may also be used to infer water movement. In particular, if a soil-water characteristic curve is available for incremental depths throughout the vadose zone, it may be possible to relate water content values to water pressure. Hydraulic head gradients and, therefore, flow directions can then be inferred. Unfortunately, the accuracy of the method may not be great enough to detect slight water content changes, particularly in the dry range. Wilson and DeCook (1968) and Wilson (1971) have used moisture logs from a network of 30-meter deep access wells to infer the rates of lateral movement of recharge waves in the vadose zone during river recharge and artificial recharge. The arrival of such waves was inferred by the change in water content in a perched mound at about 10 meters.

MOISTURE BLOCKS

Moisture blocks of the type described earlier may be used to measure soil-water pressure, if suitable soil-water characteristic curves are available. A unit relying on the sensing of heat dissipation within the block (Phene et al., 1971) is also available commercially.

QUANTIFICATION OF FLUX IN UNSATURATED MEDIA

Two of the physical characteristics of interest in the vadose zone are soil-water flux and hydraulic conductivity. Under continuous flooding at waste disposal facilities, the steady-state value of infiltration approaches the saturated hydraulic conductivity. Presumably, the steady-state value of flux below the soil surface also approaches this value; therefore, an approximation of the flux value in the wetted soil profile can be obtained during a wetting cycle.

Of equal interest is the flux during drying periods. Three major approaches to evaluating flux in the vadose zone during a drying cycle are: (1) measuring water content changes in profiles as a function of time; (2) using appropriate mathematical expressions and empirically determined relationships between soil-water pressure, soil-water content, and hydraulic conductivity; and (3) direct measurements using flow-meters.

Relating Water Content Changes to Flux

The difference in water content stored in a wetted profile between successive time measurements, above a certain depth, is a measure of the amount of water which has flowed out of that depth. The water content versus depth curves may have been obtained from tensiometer data, moisture logs, or other means. This relationship assumes that the water in storage from the surface to the first tensiometer equals the value at the first tensiometer times the thickness of materials above this point.

Expressions for Flux and Hydraulic Conductivity

Field plots instrumented with tensiometers are required to determine flux and hydraulic conductivity (Nielsen et al., 1973). Infiltration is allowed to progress in each plot to the point that steady-state values are reached. After recession of the water surface, the soil surface is covered to prevent evaporation. When drainage has occurred to the point that the hydraulic gradient becomes unity, tensiometer readings are taken in incremental time steps from successive tensiometer units. At the completion of the study, intact cores of soil are taken in increments throughout the profile. Such samples are used for the laboratory determination of soil-water characteristic curves. Subsequently, the soil-water pressure data taken by tensiometers during the field study are translated to equivalent water content values. Spatial variations occur in the soil properties; consequently, a sufficient number of field plots are needed to provide a measure of this variability. Furthermore, even though the assumption of unit hydraulic gradient is fulfilled, problems may arise if subsurface heterogeneities affect flow. Problems of anisotropy and stratification will be accentuated with depth in the vadose zone. Realistically, it may be expected that flux expressions are most useful in the soil zone.

Direct Measurement of Flux

Attempts have been made in recent years to develop equipment to measure soil-water flux in the unsaturated state which do not require information on the hydraulic conductivity.

Two types of flowmeters reported by Cary (1973) involve (1) direct flow measurement, and (2) the displacement of a thermal field by water in motion. The direct flow unit measures the flow of soil water intercepted by a porous tube containing a sensitive flow transducer. The second unit measures, very accurately, the transfer of a heat pulse in water moving within a porous cup buried in the soil. Because of the intimate contact between the soil and porous cup, the water moving in the cup forms a continuum with soil water. Laboratory calibration curves are prepared to relate the output of a sensitive millivolt recorder, during imposition of heat pulses, to empirically measured flow rates.

For field installation, porous discs containing either flow transducers or heat sources are mounted in cylinders, which are buried in the soil. A limitation, therefore, is that flow is measured in disturbed soils. To date, no data are available on the use of such flowmeters in deep vadose zones.

CALCULATION OF APPROXIMATE FLOW RATES

Vertical flow in the vadose zone is related to the infiltration rate of waters applied at the land surface, storage capacity of the vadose zone, and depth to the water table. If the water table remains at a relatively constant depth, and a specific amount of water percolates from the topsoil, then the flow rate of water in the vadose zone is partly controlled by the available storage space therein. With point or line sources, significant lateral movement can occur in layered deposits above the water table; however, for diffuse sources of large areal extent, such as agricultural return flow, these lateral movements may be neglected.

For example, assume that 25 feet of water is applied per acre per year at the land surface and 20 feet percolate from the topsoil. Assume the depth to water is

55 feet and the vadose zone beneath the topsoil is 50 feet thick. Assuming that the porosity is 30 percent and the specific yield is 15 percent, and that the specific retention has already been satisfied, then the storage capacity in the zone is a maximum of 50 times 0.15, or 7.5 feet. It is clear that the vadose zone could hold this water for only a few months at the most.

As a second example, assume that 5 feet of water is applied per acre per year at the land surface and 2 feet percolate from the topsoil. Assume the depth to the water table is 310 feet and the lower vadose zone is 300 feet thick. Assume that the porosity is 40 percent and the specific retention is 15 percent, but that the moisture content is only 5 percent. Finally, assume piston-like flow of water. For these conditions, the depth of lower vadose zone material wetted each year to the maximum storage capacity would be 2 feet times 100/35, or 5.7 feet. But the saturated profile would drain to a water content corresponding to the specific retention. Thus, the annual depth of penetration at the wetting front would actually be 2 feet times 100/10, or 20 feet. As a result, it would require 15 years for the percolating water front to reach the water table.

USE OF TRACERS

Tracers, such as salts, dyes, and radioisotopes, can be employed to measure flow rates in the vadose zone. Care must be taken to insure that the tracer moves at the same velocity as the water, which, however, is not necessarily the velocity of a specific groundwater pollutant. The practical problem of detecting or sampling a tracer at increasing depths under unsaturated conditions should not be minimized. Smith (1974) reports on using tritium as a tracer in the unsaturated zone.

WATER SAMPLING IN THE VADOSE ZONE

In a program to monitor quality in the vadose zone, samples may be needed from both unsaturated and saturated (perched water table) regions.

SAMPLING IN SATURATED REGIONS
Wells and Piezometers

Preliminary hydrogeological surveys will often delineate existing or potential perched water table regions. Many of these regions may be instrumented with wells or piezometers, which can be modified or used directly for sampling water quality. In general, observation wells are unperforated throughout the vadose zone to prevent cascading water from perched layers affecting water level measurements and water quality at the water table. It may be possible, however, to modify such wells for sampling without causing cascading. For example, the results of neutron logging in a well may reflect the presence of perched layers. It is possible to perforate the casing of abandoned wells at these depth intervals and install packer assemblies to isolate these regions for sampling, and prevent cascading. Alternatively, if information on the location of perched layers is known prior to the construction of a new well, the well can be designed to permit sampling of these regions.

An array of piezometers can be used for water sampling in perched zones. If the locations of such zones are known, individual units can be designed with screened well points. If units in the individual arrays terminate at varying depths, samples from such an array will indicate vertical quality changes from movement of effluent. Similarly, if a number of arrays are distributed in a horizontal transect, samples from individual units terminating at the same elevation will manifest quality changes from lateral spread of effluent.

For programs to monitor the spread of heavy metals in effluent water it is necessary to construct wells or piezometers from materials which will not contribute metals to the sample. Wells cased with PVC pipe will avoid this problem, particularly if screens are also of plastic, but require special construction techniques. A controversy exists regarding the use of PVC wells when organic pollutants are present.

To obtain water samples from wells it is necessary to employ some type of sampling device. A simple bailer may often be used with success. This unit consists of a tube with an open top and a ball check-valve at the bottom to admit water samples. Another commonly used sampler consists of a sampling tube or body, with an air inlet line and a sample outflow line extending to the surface. The lower end of the body contains a spring-loaded flap valve. The sampling line extends to the base of the sampler, and the air inlet line terminates a slight distance below the top. In operation, the sampler is lowered into a well to the sampling depth below the water table. After permitting the sampler to fill, air pressure is applied to the air line and the sample is forced to the surface through the sampler line. Advantages of this unit are that it need not be removed from the well between samples. Also, samples are obtainable at various depths below the water table. A disadvantage is that air is introduced into the sample, possibly interfering with certain chemical determinations. This possibility can be avoided by using compressed nitrogen gas.

Allison (1971) discusses a simple device for sampling water in auger holes, which also could be used in cased wells. Basically, the unit consists of plastic tubing of sufficient length to extend into the water table. The upper end of the tube terminates in a collector flask. A second line from the flask is connected to a hand-held vacuum pump. When the pump evacuates the flask and tubing, water is forced up into the flask. According to Allison, such a system is capable of sampling water to a depth of 7.4 meters.

In addition to these samplers, several other types are available, including positive displacement pumps. Some operations may warrant the purchase of a small-diameter submersible pump mounted on a flexible line.

With any sampling method it is important to ensure that an uncontaminated sample is obtained. The well should be bailed or pumped extensively at the beginning of each collection period to ensure the movement of fresh groundwater into the well. Similarly, the bailer or sampling lines should be flushed out several times before a sample is collected.

Sampling Tile Drain Outflow

If a tile drainage system has been installed to promote rapid drainage of a perched water table, samples can be collected from the tile outfall. In some cases it may be desirable to install commercially available composite or discrete sampling devices.

Willardson et al. (1973) discuss a "flow-path ground water sampler," which enables collection of water in different flow paths around a tile.

Fiberglass Probes

Hansen and Harris (1974) describe a sampler capable of discrete simultaneous sampling at several depths below a water table. The unit consists of a series of isolated fiberglass probes in a wellpoint, with individual lines from each probe running to the surface. Each isolated area is filled with a sand matrix. During collection, all probes are sampled simultaneously, via a vacuum pump, to minimize interference with natural flow within the sampling region.

Multi-Level Sampler

Pickens et al. (1978) presented details of a multi-level sampling device, suitable for sampling in cohesionless deposits with shallow ground water. Basically, the unit consists of PVC pipe, with openings drilled at desired incremental depths. Polypropene tubing is connected to each opening, which is covered with a screen. The polypropene tubing extends to the surface. Units may be designed in the field by locating the openings from stratigraphic data.

Sampling Chambers

In the context of this discussion, a sampling chamber consists of a trench or large-diameter culvert sunk vertically into a waste disposal area. Units are installed to permit sampling of the soil solution throughout a vertical profile, usually several meters in thickness. Culvert type chambers were used by McMichael and McKee (1966) in their studies on wastewater reclamation at Whittier Narrows. Parizek and Lane (1970) used trenches to collect samples from forest soil during effluent irrigation in Pennsylvania.

The chambers developed by McMichael and McKee (1966) employed pan samplers for the collection of water during periods of saturation in a profile to a depth of about 3 meters. These samplers were conical in shape, 61 centimeters in diameter and 23 centimeters deep. The samplers were installed at various depths at a radial distance of about 3 meters from the central well. Individual sampling pans were filled with gravel to prevent clogging, and each was connected to the central culvert via tubing. Excavations used for installation of the samplers were carefully backfilled. During sampling, percolate intercepted by the pans flowed by gravity to the central well and into collection flasks.

The trench described by Parizek and Lane (1970) also used pan samplers. The largest trench was 1.2-meters wide, 4-meters long, and excavated to a depth of about 6 meters. The sides of the trench were braced and lined with wood. Sampling pans were installed at 30-centimeter vertical intervals to the 6-meter depth of the trench. The pans, constructed from galvanized metal, were 30 centimeters by 45 centimeters, with copper tubing to permit sample drainage. Each unit was sunk into the soil a short distance from the side walls. During application of effluent, samples of percolate intercepted by the pans were collected in flasks within the trench.

A basic problem with collection pans, pointed out by Parizek and Lane (1970) is that samples are collected only when the soil-water pressure is greater than atmospheric. Consequently, after cessation of infiltration the soil-water system may shift rapidly to the unsaturated state, prohibiting sample collection. The same problem exists for piezometers, auger holes, or other sampling methods requiring saturation.

SAMPLING IN UNSATURATED REGIONS
Soil Sampling

One method for determining the quality of the soil solution in unsaturated soils is to obtain field samples from which saturated extracts can be obtained. An extensive program of soil sampling throughout the duration of a waste disposal operation could be prohibitively expensive, however, particularly if samples were required from the entire vadose zone.

Suction Cups

An alternative to soil sampling for characterization of the water quality of unsaturated soils is to utilize ceramic cups like those used in tensiometers. These samplers are frequently called lysimeters. When placed in the soil, the pores in these cups become an extension of the pore space of the soil so that the soil-water content in the soil and cup become equilibrated at the existing soil-water pressure. Applying a slight vacuum to the interior of the cup causes the soil solution to flow into the cup. The quality of the soil solution can be determined by bringing this sample to the surface. Although ceramic cups have limitations, at the present time they appear to be the best tool available for sampling unsaturated media.

The basic design of the sampling unit is as follows: a ceramic cup is sealed onto a rigid-plastic body tube that is equal in length to the desired sampling depth. Small-diameter tubing is inserted through a rubber stopper, which seals the top of the body tube, down to the inside base of the cup. The other end of the tubing is attached to a collection flask through a two-hole rubber stopper. Vacuum is applied to a second line in the flask causing the air pressure inside the system to become less than the ambient pressure at the cup's location. Soil solution is drawn into the cup and sucked through the vacuum line to the collection flask. The technique used to emplace suction cups is the same as that for tensiometers. In particular, good contact should be obtained between the cup and the soil.

Recently, alternative designs have been reported in the literature, most notably by Parizek and Lane (1970) and by Wood (1973), that allow raising samples from greater depths than is possible by vacuum withdrawal. The design used by Parizek and Lane is shown schematically in Figure 3-15. The body tube of the sampling unit is 2 feet long and holds about a liter of sample. Two copper lines are forced through a two-hole rubber stopper, which seals the body tube. One line, the sample discharge line, extends to the base of the ceramic cup, as shown, and the other, a pressure-vacuum line, terminates a short distance below the rubber stopper. The discharge line connects to a suitable vacuum-pressure pump. In operation, a vacuum is applied to the system with the discharge tube clamped shut. When sufficient time has elapsed for the unit to fill with soil solution, the vacuum is released, the clamp on the outlet line is opened, and air pressure is applied to the system to force the sample into the collection flasks.

Figure 3-15 also illustrates another feature of the design of Parizek and Lane. The installation method shown does not require an intimate contact between the cup and the soil. Instead, a cavity larger than the cup is drilled to permit backfilling with silica sand around the cup and with tamped backfill for the remainder of the length. This design permits the installation of more than one unit in a hole, although three is about the maximum. Each unit is sealed from adjoining units with bentonite. According to the authors, the silica sand provides a clean medium for soil moisture moving toward the cup, ensures contact between soil and cup, eliminates unevenness in the void space, and precludes clogging of the cup by collodial material. Samples were recovered from a depth of 15 meters.

Wood (1973) reports on a modified version of the design of Parizek and Lane that overcomes a basic problem with their system, i.e., solution being forced back into the soil during the release of vacuum and application of pressure. A sketch of Wood's design is shown in Figure 3-16; details are presented in his paper. The important feature, however, is that a check-valve is included in the pressure-delivery tube assembly within the cup. In operation, vacuum is applied to the system, inducing a flow of water into the cup and tube assembly. Nitrogen gas pressure is applied to one tube and the sample is forced to the surface. The check-valve prohibits pressurization of the porous cup and, therefore, a sample within the cup does not flow back into the soil.

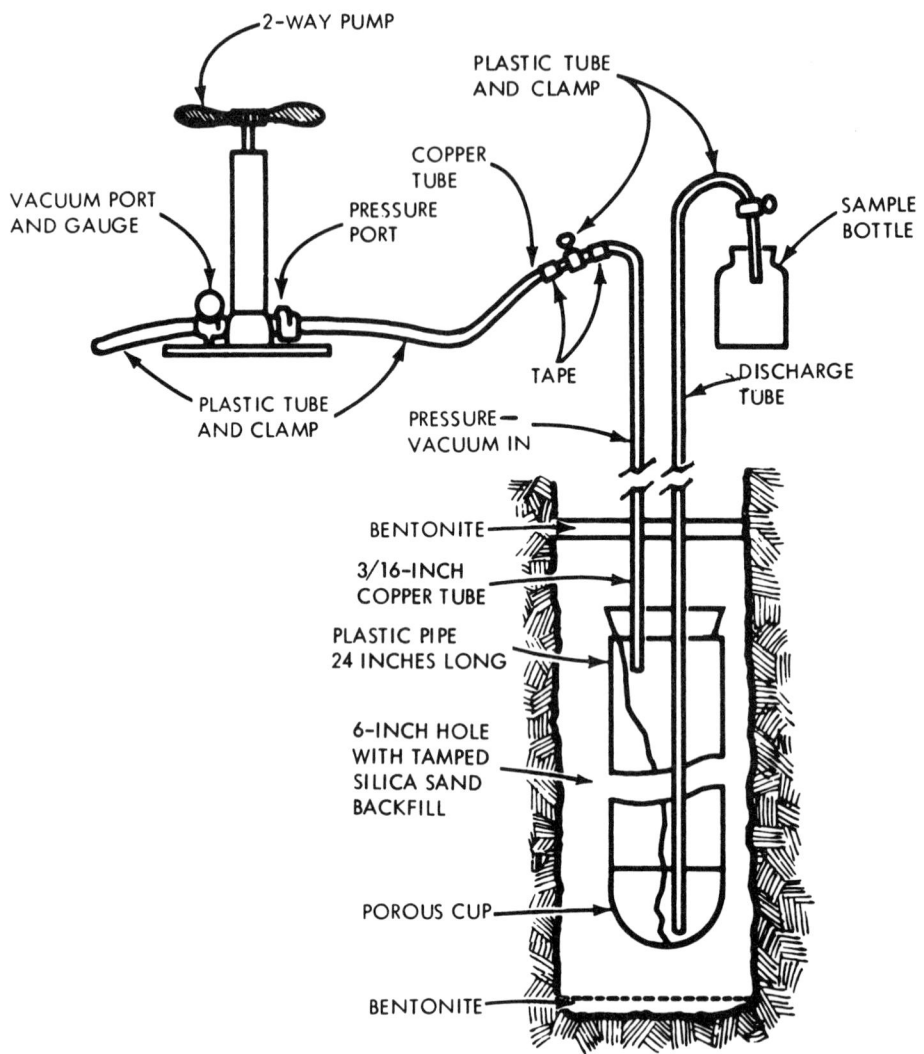

Figure 3-15. Cross section of suction cup assembly and backfilling material (after Parizek and Lane, 1970).

Wood's installation procedure consisted of drilling a 10 centimeter hole with a continuous flight auger to 45 centimeters from the desired depth. The auger was then installed in the lowermost auger flight, and the ensemble pressed back into the hole to the final depth. The assembly was again withdrawn and a suction cup unit placed into the hole. Only one unit was installed per hole; samples were collected from depths ranging to 36 meters.

Although suction cup assemblies are the best available techniques for obtaining samples of the soil solution in situ, certain problems accompany their usage. One is that samples cannot be obtained over the entire range of soil-water pressures: suction cups are capable of sampling only at pressures greater than about -1.0 atmosphere. Because of the very small pores in ceramic cups, inflow of suspended solids is inhibited, making tests for biochemical oxygen demand (BOD) on collected samples inaccurate. Similarly, bacteria may be filtered out. During application of vacuum to the cups, it is important that the negative pressure closely match the pressure of the soil-water system. If the vacuum differential is excessive, the additional suction can influence the movement of soil water in the vicinity of the cup. Finally, there is evidence that

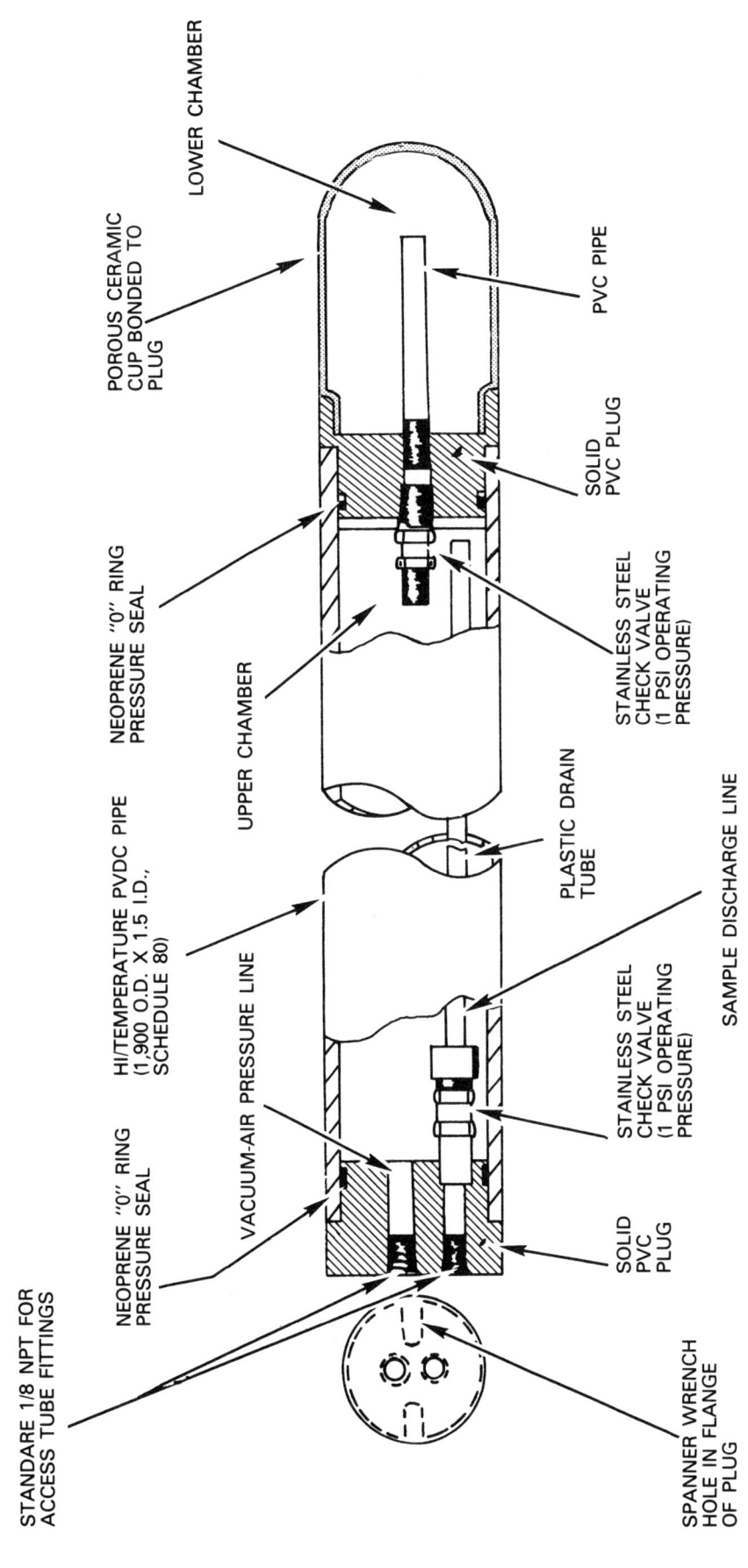

Figure 3-16. Cross section of porous cup hi/pressure-vacuum soil water sampler.

sorption of some ions may occur within suction cups. However, the problem will be important only when the exchange capacity of the cups greatly exceeds that of the surrounding soil.

Salinity Sensors

A method for in situ evaluation of soil salinity is the so-called salinity sensor. The basis of these devices is the relationship between specific electrical conductance (EC) of soil solution and the total concentration of salts in solution.

The salinity sensor described by Richards (1966) uses electrodes embedded in porous ceramic to form a hydraulic continuum with soil water and measure the specific conductance of the soil solution directly. EC values can then be directly related to the total salt content by suitable calibration relations. The unit described by Richards comprises a plate about 1 millimeter thick, with platinum electrodes fixed in place on opposing faces. An interesting feature of the sensor is that the unit is spring loaded to ensure good contact with soil. Because of the strong dependency of EC on temperature, it is important to measure accurately the temperature of the soil solution. Richards used a thermistor to provide temperature compensation.

Oster and Willardson (1971) have reviewed problems arising from calibration and field use of salinity sensors. Of particular importance is their observation that sensors should not be used at soil-water pressures less than -2 atmospheres. Also, they indicate that when sensors are placed in trenches at field sites, the permeability of the materials in the back-filled trench tends to be greater than in indigenous soil. During leaching trials, therefore, the salinity measured with the sensors tended to be lower than in adjacent soil. Differences were attributed to greater leaching in the trench, probably because pore sizes in the trench soil were of a different range than indigeneous soil.

While salinity sensors may be useful to gage changes in the total salinity of a soil solution, in waste disposal operations, concentrations of individual constituents may be of more importance. Furthermore, the water pressure of materials in the lower reaches of the vadose zone may be less than -2 atmospheres, thereby rendering the unit inoperative.

The cost of a salinity bridge is about $620, each salinity sensor costs about $35, and sensor cable costs about $0.50 per foot. Figure 3-17 gives the cost for a salinity bridge and various depths and sampling densities for the sensors.

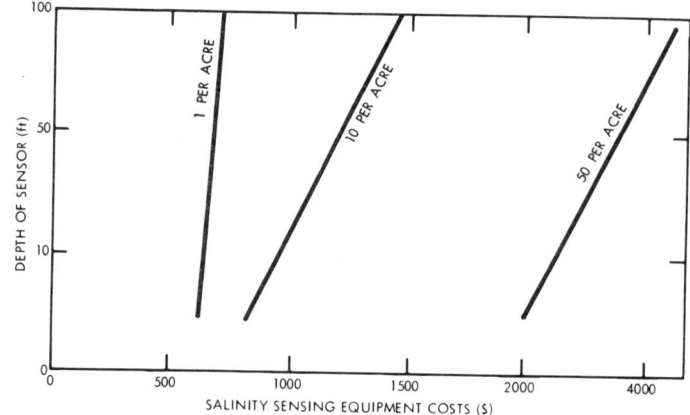

Figure 3-17. Cost of soil salinity sensors and a salinity bridge for various depths and sampling densities, November 1974.

SECTION 4 – MONITORING IN THE ZONE OF SATURATION

INTRODUCTION

Differences in the mineral composition of rocks within the groundwater reservoir, in the composition of soils, and in the chemical composition of recharge waters, cause differences in groundwater quality from place to place. Selection of a sampling site and collection of groundwater quality data is more difficult than for surface water due to the complexity and accessibility of most aquifer systems and the multitude of factors which influence groundwater quality (Brown et al., 1970). The superposition of pollution zones on the natural system imparts an additional degree of complexity.

Knowledge of the hydrogeologic framework is important from two standpoints: (1) prediction of groundwater movement, and (2) geochemical considerations which affect natural groundwater quality. The three-dimensional distribution of hydraulic head and geologic conditions largely determines groundwater flow patterns. The pattern of movement of groundwater has a major effect on the distribution of groundwater pollutants. Both regional and local conditions are significant. Sources of pollution must be identified, and volumes and quality of wastewater or percolate determined. The three-dimensional distribution of groundwater quality must be delineated.

Methods employed in general groundwater studies are summarized in this section, followed by details of specific methods of monitoring pollution in the saturated zone. These methods illustrate some of the data required for groundwater pollution monitoring.

GENERAL MONITORING PROCEDURES

Development of the hydrogeologic framework entails surface phenomena (such as precipitation and streamflow), interactions in the topsoil and unsaturated zone, and considerations of the saturated zone and well hydraulics. In this section the saturated zone is of prime consideration. Brown et al. (1972) have summarized international data on groundwater studies. The U.S. Geological Survey has issued periodic reports under *Techniques of Water Resources Investigations*. Textbooks on groundwater contain abundant information on the subject, including Todd (1959) and Davis and DeWiest (1966). Walton (1970) presented case histories of groundwater studies, while issues of *Ground Water* and other periodicals contain numerous reports on groundwater.

WELL INVENTORY AND WELL DATA COLLECTION

Most groundwater studies require a complete field well inventory and office compilation of existing well data. Data such as driller's logs, electric logs, water-level measurements, pump tests, and chemical analyses are gathered and organized. Field observations and interviews are used to verify the correspondence between data and wells. Considerable confusion can result when this process is not carefully done. It also promotes the proper interpretation of results of a monitoring program. The cooperation and assistance of local water agencies is necessary for successful accomplishment of this initial work.

GEOLOGICAL FRAMEWORK

The geological framework for groundwater studies includes lithology, texture, structure, mineralogy, and the distribution of the subsurface materials through which groundwater flows. The hydraulic properties of earth materials depend on their origin and lithology, as well as the subsequent stresses to which the rocks have been subjected. Weathering, tectonic movement, and fracturing are examples of these stresses.

Surface geology is commonly mapped in the field, if sufficiently detailed reports or maps are not available. Photogeology is a useful tool in analysis of surface or near-surface geologic conditions. Ray (1960) and Seker (1966) discuss the use of aerial photographs in geologic interpretation and mapping and applications to groundwater. Geologic maps are compiled to indicate the distribution of rock types and structural features such as folds or faults. Geophysical methods can provide information about the major features of the underground materials. Zohdy et al. (1974) summarize the application of surface geophysics to groundwater investigations. Geophysics can also provide a good indication of where test drilling should be undertaken to ascertain more information on the configuration of the subsurface materials.

Records of geologic materials penetrated during well drilling are extremely important. Driller's and geologist's logs can often be correlated with borehole geophysics. Keys and MacCary (1971) summarize the application of borehole geophysics to water-resources investigations. The vertical distribution and occurrence of subsurface materials are thus documented at the point where a well is drilled. Electrical resistivity and spontaneous potential logs (electric logs) provide information on the subsurface lithology. A gamma-ray log indicates clay content in sedimentary rocks. This method is used in cased holes, and can be quite effective in alluvial materials (alternating layers of sand and clay).

Once data on subsurface geology are compiled, cross sections are developed on the basis of geological correlation. The three-dimensional variations in the subsurface lithology are thus illustrated. Considerations of importance in water quality studies include chemical composition on the underground materials and contained fluids.

WATER LEVELS AND GROUNDWATER MOVEMENT

Groundwater movement depends on the permeability and the hydraulic gradient within an aquifer. Permeability is related to the nature, size, and degree of interconnection of pores, fissures, joints, and other openings. Water-level measurements are used to draw contours of depth to water and groundwater level elevation over the area of interest. Similarly, water-level measurements at different depths in the aquifer can provide an indication of the vertical flow components. Schematic flow nets are made to illustrate the movement of groundwater in the aquifer (Casagrande, 1937; Davis and DeWiest, 1966). Flow nets provide information on recharge to and discharge from the aquifer, relative direction of groundwater movement, and aquifer geometry and permeability.

Flow lines are perpendicular to water-level elevation contours only in the case of homogeneous and isotropic media. The subsurface lithology is used to evaluate locations where the direction of movement of groundwater may not be perpendicular to these contours, such as in buried stream channels in alluvium. Water-level contours are generally drawn for static conditions, and pumping wells may alter the dynamic situation.

Groundwater velocities can be estimated on the basis of permeability or transmissivity, hydraulic gradient, and porosity. Permeability and transmissivity are determined from analyses of pumping tests. The hydraulic gradient is determined from the water-level contour map, and porosities are known or can be determined for most materials. In the case of confined aquifers and relatively impermeable confining beds, vertical permeability and hydraulic gradients can be used to calculate upward or downward leakage. Similarly, amounts of groundwater flowing through an area can be calculated based upon permeability or transmissivity and the hydraulic gradient. Tracers such as dyes (UNESCO, 1975) or radionuclides (Havely and Nir, 1962; Hanshaw et al., 1965; Kaufman and Orlob, 1956) can also be used as an indication of groundwater flow velocity.

PUMP TESTS AND AQUIFER ANALYSIS

The hydraulic properties of aquifers are determined using pumping or recharge tests, and water-level changes reflecting long-term conditions at the aquifer boundaries. The first method is widely used because it may permit evaluation of the aquifer parameters in a relatively short period of time. Theis (1935), Jacob (1947), Ferris et al. (1962), Johnson Division UOP (1966), Todd (1959), Davis and DeWiest (1966), and Walton (1970) summarize methods of pump testing and aquifer analysis.

The primary aquifer parameters determined from most pump tests are transmissivity and storage coefficient. They are used to define groundwater movement and the reactions of aquifers to pumping or recharge. Permeability can also be determined, which helps indicate flow velocities. Calculation of groundwater flow through certain sections of the aquifer are valuable in order to assess spread and dilution of pollutants from the land surface.

Much of the theory and equations developed for aquifer analysis apply to confined aquifers in porous, granular, unconsolidated materials. Special equations have been developed for unconfined aquifers, leaky aquifers, and aquifers of very irregular geometry. Jointed, fissured, and cavernous rocks are characterized by a high degree of heterogeneity. In rocks with a secondary porosity and permeability, developed by jointing or solution, single pump tests of short duration seldom provide any information, except localized aquifer characteristics.

By considering a large area, variations and heterogeneity of local areas tend to average out, and usable figures on the hydraulic characteristics may be obtained. If continuing records of recharge to and discharge from sizable parts of the aquifer are available, along with records of seasonal fluctuations of the water table, then reliable estimates of the regional hydrologic characteristics can be made.

WATER BUDGET AND MODELING

The law of conservation of matter defines the groundwater basin water budget. All water entering a specific area must go into storage within its boundaries, be consumed therein, be exported therefrom, or flow out as surface or groundwater.

Water budgets can be used to indicate regional conditions in an aquifer. The reaction of the aquifer to pumping or recharge can be determined by comparing the water budget with water-level hydrographs. Water budget analysis can also serve as an independent check on the accuracy of individual items in the budget. Mathematical and analog modeling, along with computer methods, are important tools in analyses of regional groundwater systems (Skibitzke, 1960; Stallman, 1963; Walton and Prickett, 1963). Projections of future changes in aquifer regime can be made based on these model studies.

The groundwater regime is greatly influenced by climatic, hydrologic, biologic, and soil factors. Groundwater is closely interrelated with surface water and soil moisture. The groundwater budget requires quantification of all these factors affecting inflow to, and outflow from, a groundwater reservoir, as well as storage changes. Few of the factors are directly measurable, but some can be determined by residuals of measured variables, and others can be reasonably estimated. Of foremost importance is the proper selection of groundwater basin boundaries based on hydrogeologic factors.

WATER TEMPERATURE AND CHEMICAL QUALITY

Temperature of underground materials is influenced by the heat input and output, as well as by the heat characteristics of the materials. Since the rock and water below the land surface are poor conductors, the temperature distribution is influenced by groundwater movement. Groundwater temperature is an important field of study. Well discharge and down-hole water temperature measurements can indicate directions of water movement and relative permeabilities of aquifer materials. Groundwater temperature affects the chemical quality of groundwater. Groundwater temperatures can be used as an independent check on concepts based on hydraulic and geologic reasoning alone. Bredehoeft and Papadopulos (1965), Cartwright (1968), Sammel (1968), and Bentall (1963) discuss the significance of temperature measurements in groundwater.

Groundwater quality studies also provide an independent check on concepts based on physical reasoning alone. A groundwater quality study is based on a knowledge of the geochemistry of aquifer materials and the chemistry of recharge to the aquifer. Chemical equilibrium and kinetics evaluations may indicate whether certain minerals are dissolving, or precipitating, as well as rates of reactions. Hem (1970) has reviewed pertinent data on natural water quality. Changes in groundwater quality during flow through the aquifer can be assessed. The relative mobility of certain chemical species and biological and radiological parameters is of great concern. Some constituents may move through the aquifer at the same rate as the water, many move at slower rates, and some are virtually immobile.

Contours for constituent concentrations on an areal and vertical basis are often prepared. These can be compared with geologic maps, flow nets, and predominant sources at the land surface. Quite often several constituents such as sodium and chloride, or calcium and sulfate, are interrelated because of the subsurface presence of solid compounds. Subsurface lithology and temperature may correlate well with groundwater quality. Numerous graphical procedures are available for interpretative work. The trilinear diagram is a useful tool to compare water types. The major cations (Ca, Mg, Na + K) are plotted on one triangle and the major anions (Cl, SO_4, $HCO_3 + CO_3$) are plotted on another. This tool is especially helpful in analyzing mixtures of two or more types of groundwater.

SPECIFIC MONITORING METHODS

SURFACE GEOPHYSICS

Often, a major part of an investigation of groundwater pollution is to locate and define its extent. In special situations a resistivity survey can be effectively used. These situations generally occur where:

1. There are either no wells or only a few in the polluted zone

2. The polluted water is of relatively high salinity compared to native groundwater in the area

3. The subsurface geology is relatively well understood

4. The water table is fairly shallow

Griffiths and King (1969) summarize the resistivity method and common applications. Zohdy et al. (1974) summarize the application of surface geophysics to groundwater investigations.

Earth materials containing groundwater of high electrical conductivity will have lower resistivity values than surrounding materials containing natural groundwater. In some cases, the resistivity method can be used to determine quickly the boundaries of the zone of polluted groundwater. Hackbarth (1971) describes the use of resistivity measurements to analyze spent sulfite liquor recharge movement. Merkel (1972) illustrates how resistivity surveys could be used to analyze acid mine drainage in groundwater. Stollar and Roux (1975) summarize the use of resistivity surveys to define groundwater pollution.

Initially, multidepth electrical soundings are made of the area in an attempt to see if the top and bottom of the polluted zone can be delineated. If this can be done, a series of single-depth resistivity measurements are then made at selected points to define the horizontal extent of the plume. By varying the depth of the measurements within the vertical extent of the plume, the three-dimensional extent of the polluted zone can be determined.

The resistivity survey has certain limitations that restrict wide application. These include poor site access, interference of conductors such as pipelines, insufficient contrast between the conductivity of the polluted and the natural groundwater, depth to the top of the polluted zone, thickness of the zone, and the complexity of the surficial geology (Stollar and Roux, 1975). An inherent problem in any resistivity survey which incorporates an electrode configuration consisting of all surface electrodes is that of resolution, when the lithology has high resistivity contrasts. A highly resistive zone inhibits current flow in it or reaching regions below it. Similarly, a highly conducting zone traps the current and again allows little current to flow below it. For examinations at considerable depths, a technique has been developed which incorporates the use of a buried current source in a drill hole.

The primary advantage of the resistivity method in groundwater pollution investigations is to provide information on the extent of the polluted zone at a reduced cost and within a short period of time. Hackbarth (1971) states that resistivity measurements can be useful in supplementing data obtained from piezometers, but cannot replace them.

In conducting a resistivity survey the variables of concern are the depth to the top of the plume, the thickness of the polluted zone, and the areal extent of the polluted zone. If no data are available, a feasibility study can be done for about $1500 to determine whether the technique will work and to estimate the plume size. If the survey appears feasible, the cost that will be incurred by an experienced geophysical team can be estimated from Table 3-6.

The first step in the survey is to determine the electrode spread, which will depend upon the depth of the plume and its thickness. Both of these variables may not be known, but they can be estimated and used in Table 3-6(a) to determine the electrode spacing. The areal extent of the plume must be estimated in acres and applied in Table 3-6(b) to determine the number of surveys to be conducted. From Table 3-6(c) the time required to conduct each study for a particular plume depth, thickness, and areal coverage can be obtained. The cost for a geophysical team including equipment and interpretation is about $75 to $85 per hour. To obtain total costs, the time from Table 3-6(c) is multiplied by $80. This figure does not include travel costs, per diem, or additional time required in difficult terrain.

Table 3-6

EARTH RESISTIVITY SURVEY COSTS, DECEMBER 1974

Depth to top of plume (ft)	Thickness of plume (ft)		
	20	50	100
20	60	100	150
50	90	130	200
100	150	200	250
200	260	300	350
500	560	600	700

(a) Determination of electrode spread to cover depth of plume.

Electrode spread	Areal extent of plume (acres)			
	20	50	100	1000
60	4	8	12	20
90	4	8	12	20
100	3	6	10	15
130	3	6	9	15
150	3	5	8	13
200	2	4	7	10
250	2	4	7	10
260	2	4	7	10
300	2	4	5	8
350	2	4	5	8
560	2	3	4	6
600	2	3	4	6
700	2	3	4	6

(b) Determination of number of surveys to cover area of plume.

Electrode spread from (a)	Time (hrs)	Number of survey from (b) x time (hours)			
		20 acres	50 acres	100 acres	1000 acres
60	2	8	16	24	40
90	2	8	16	24	40
100	3	9	18	30	45
130	3	9	18	27	45
150	3	9	15	24	39
200	4	8	16	28	40
250	4	8	16	28	40
260	4	8	16	28	40
300	6	12	24	30	48
350	6	12	24	30	48
560	8	16	24	32	38
600	8	16	24	32	38
700	10	20	30	40	60

[a] Survey cost = $80 per hour x survey time.

(c) Determination of time (hours) and costs ($) to conduct surveys.[a]

WELL CONSTRUCTION

Most methods for sampling the saturated zone require wells. Where no wells or other devices exist for sampling the saturated zone, well construction must be considered. Wells can be constructed by the following principal methods: (1) dug, (2) augered, (3) driven, (4) jetted, and (5) drilled. The first four methods are usually employed for shallow wells (150 feet deep or less), while drilling techniques are used for both shallow and deeper wells. These five methods and the physical characteristics of wells are described in detail by Johnson (1971), Anderson (1967), Bowman (1911), Gibson and Singer (1971), the U.S. Departments of the Army and Air Force (1957), the California Department of Water Resources (1968), and Campbell and Lehr (1973).

TYPES OF SHALLOW WELLS

Dug wells are excavated by hand or power shovel to depths usually ranging from about 10 to 40 feet. These wells usually are about 5 to 8 feet in diameter, and extend only a few feet below the water table. The shaft may be lined with concrete rings, stone, brick, or wood. Dug wells are used mostly for rural water supplies or for drainage water disposal, and can be polluted relatively easily because large surface openings are generally present. Such large diameter wells are not ordinarily necessary for monitoring groundwater quality.

Shallow wells may be augered by hand or power equipment. Hand augers can be used to penetrate clay, silt, and sands where cave-in will not occur in an open hole. These holes are generally not economical when depths greater than about 10 to 20 feet are encountered. In power augering by the rotary bucket method, the bucket is filled with cuttings and raised to the surface for dumping. In continuous-flight augering, the cuttings rise to the surface on rotating spiral flights. The wall of an uncased auger hole usually collapses below the water table after the auger is removed. Therefore, a drive pipe and attached well point are commonly set in the open hole, and driven below the water table to the required depth immediately after the hole is drilled. Where hollow stem augers are used, a 2-inch pipe and well point can be set inside the hollow stem to the required depth. (Vertical movement of water can occur in the annular space between the wall of the drilled hole and the well casing.) Monitoring wells to depths of 150 to 200 feet can be sunk quickly in unconsolidated deposits by use of a power auger. The size of an auger hole may range from 2 to 32 inches in diameter. Auger holes may be of great value in monitoring relatively shallow polluted zones in unconsolidated deposits.

Driven wells (sometimes called sandpoint wells) generally consist of a casing 1-1/4 to 2 inches in diameter and an attached drive point. The drive point consists of perforated pipe with a steel point at its lower end to break through pebbles or thin layers of hard material. These are installed to depths generally up to 30 feet, and usually cannot be pumped at rates greater than a few tens of gallons per minute. Larger wells (up to 4 inches in diameter) can be driven, however, and pumps installed inside the casing. The chief disadvantages of driven wells are that construction is slow and difficult, when tightly compacted materials are encountered, driving is destructive to the well itself, and yields are small. However, they are useful for monitoring shallow aquifers in unconsolidated rock.

The jetting method employs a high-velocity stream of water to drill a hole in the earth. The water stream loosens earth materials and washes the finer particles upward and out of the hole. The jetting method is most successful in sandy soils with shallow water tables. The technique is ineffective against hard rock and boulders; tight clay and hardpan also present problems. Water is commonly used in jetting wells, but a jetting fluid of greater viscosity and weight can be made by mixing clay or bentonite with water. The casing is usually sunk as the jetting proceeds, and if too much resistance is encountered, the casing may be driven. A screen may be attached to the casing during drilling or installed after the casing is driven to the final well depth. Jetted wells are generally of small diameter and less than 150 feet deep.

COSTS OF SHALLOW WELLS

Monitor wells which are 2 to 4 inches in diameter and less than 200 feet deep can be installed at relatively nominal cost as shown by Figure 3-18. The cost of the first well is heavily weighted by the cost of transportation and setting up the equipment. The cost of a second well, assuming it is located close by, is considerably less. The costs represented in Figure 3-18 represent the average of auger and jet methods and include 2-inch diameter polyvinyl chloride (PVC).

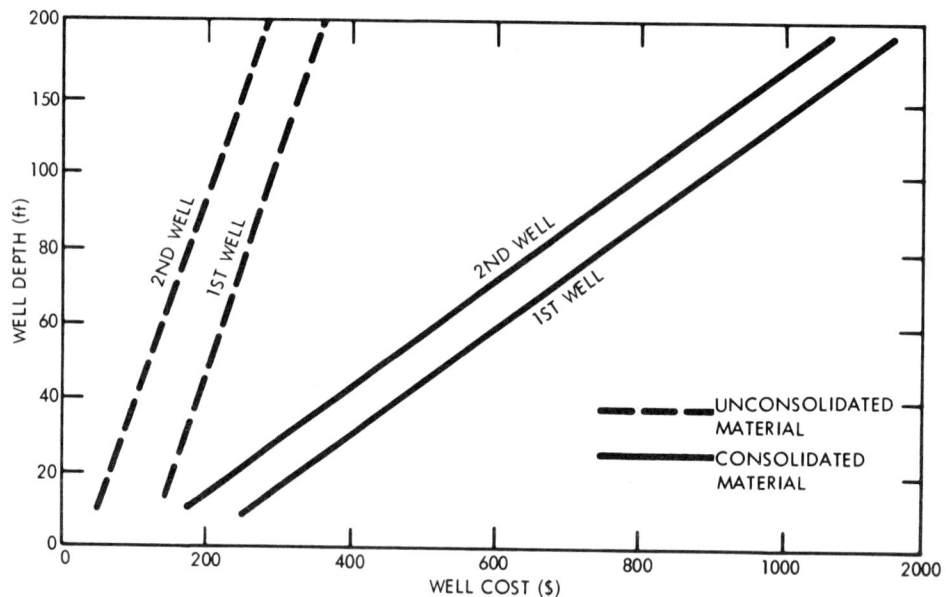

Figure 3-18. Cost of shallow, small-diameter monitor wells, October 1974.

TYPES OF DRILLED WELLS

Drilled wells are constructed by the cable tool (percussion) method or by rotary methods. The relative merits of the various methods of drilling vary, and no one method is superior under all conditions.

In the cable tool method, the hole is formed by the percussion and cutting action of a drilling bit that is alternatively raised and dropped. The drill cuttings are removed at intervals by a bailer or sand pump. An open hole can be drilled in consolidated rock, but in unconsolidated formations, casing is driven down the hole during drilling. Above the water table, water is added to the hole so that the cuttings can be readily bailed. The bottom of the casing is fitted with a heavy-walled, hardened steel drive shoe. As the casing is driven in unconsolidated formations, frictional forces increase until the casing cannot be sunk further without the use of special tools.

Small cable tool rigs can usually drill a 5-inch diameter hole to a depth of about 300 feet. The largest cable tool rigs can drill a 5-inch diameter hole to a depth in excess of 7000 feet. With medium and high capacity rigs, 18- to 24-inch diameter holes are commonly drilled to depths of several hundred feet. The major disadvantage of cable tool drilling is its slow drilling time and depth limitation; however, the latter is well below the monitoring depths for most types of groundwater pollution.

Sampling and logging of geologic formations are simpler and more accurate for the cable tool method than for most other drilling methods. The samples are not contaminated by drilling mud, and clay, shale, and silt, and fractions are not likely to be lost by dispersion in a drilling fluid. When a potential aquifer is encountered, it may be tested by bailing or pumping. The mobility of cable tool rigs give them an advantage in rugged terrain and in isolated areas. Cable tool rigs can usually drill through fractured, fissured, broken, or cavernous rocks. The method requires much less water for drilling than most others, which is an important consideration in arid areas.

A major advantage of this method for water quality monitoring is that water samples can be obtained by bailing as each stratum is opened to the bottom of the casing and upper strata are cased off. The water-bearing characteristics and static heads of various strata can also be determined as the casing is being driven. Special methods have been developed for assuring a tight seal around the casing.

Rotary drilling, used for oil wells, has virtually no depth limit for most fresh groundwater systems. A variant method, reverse-circulation rotary drilling, is also used. The rotary method is distinguished from the reverse-circulation method by pumping a drilling fluid through the inside of the drill pipe and out through the openings in the bit. The drilling fluid is a suspension of solids in liquids and commonly is termed *mud*. The drill stem is rotated as a downward force is applied, and the mud flows back to the surface through the annulus between the drill pipe and the hole wall. The mud cools the bit, transports cuttings to the surface, and maintains hole stability.

The drilling fluid contains solids, the make-up water, and additives, all of which may introduce contaminants into the system. Mud additives are used to make the mud thicker, thinner, or heavier, or to otherwise adapt it to special conditions. Bentonite is one of the most common additives.

New drilling mud additives have been developed in the water-well industry. One such additive is a biodegradable material, which does not contaminate water-bearing sands with clay particles, and can be totally removed from the well during development.

In unconsolidated deposits, the casing (and screen, if used) are installed in a rotary hole after the final drilled depth is reached. The well is then gravel-packed and developed. Gravel packing is done to minimize problems of sanding and to improve the rate of inflow of water. Well developing is done by various methods, with the primary purpose being to clean out drilling mud that has entered the formation during drilling.

Direct-circulation (conventional) rotary drilling may be less desirable than cable-tool for test-hole drilling. Drill cuttings tend to be mixed from different depths and contaminated by drilling mud. The characteristics of drilling mud may be time-variable so that cuttings brought to the surface can vary with mud characteristics, rather than with where the material was penetrated. Sample time lag in deeper holes can also become troublesome in obtaining a reliable geologic log. Measuring static water levels, taking representative water samples, and performing pump tests of individual aquifers during drilling are often not practical with rotary drilling.

The principal advantages of the direct-rotary method are speed, great depth capability, and the ability to run electric logs in an open hole prior to casing installation.

In the reverse-circulation rotary method, the drilling fluid is muddy water instead of drilling mud, and is introduced down the hole outside of the drill pipe. The drill cuttings are pumped up through the drill pipe as drilling proceeds, and are separated from the drilling fluid at the surface. From 20 to 500 gallons per minute (gpm) of make-up water, depending on the hole diameter, may be needed at times when drilling through highly permeable sediments.

The air rotary method is similar to the other rotary methods, except that compressed air is used instead of drilling mud to bring the cuttings to the surface. Mud or other chemical sealants may be required where the materials are loose and tend to cave. The conventional drill bit can be replaced with a pneumatic drill bit, which speeds up the drilling by pulverizing the consolidated rock.

COSTS OF WELL DRILLING

Well-drilling costs vary depending on such factors as site accessibility, labor and material costs, type of well design, use of the well, degree of development, yield, and local geologic conditions. Bids submitted in connection with well-construction contracts may be itemized or may simply quote a lump-sum charge. Itemized bids generally show unit prices for (1) moving in and setting up the rig, (2) drilling, (3) casing, (4) screen or perforated casing, (5) grouting the annular space, (6) coring, (7) developing time, (8) pumping test, and (9) geophysical logging. Providing and installing pumps, pump houses, and controls are additional items whose combined cost, especially in the case of public-supply wells, may exceed the cost of constructing the well. Many of these items, however, can be neglected for small observation wells.

In a well-construction program, the cost of drilling can be separated from well development, and can be generally approximated. The cost of drilling will vary primarily with hole diameter, type of geologic formation, and depth of well. Drilling costs for 40- to 1000-foot depths in unconsolidated formations for various well diameters are given in Table 3-7. Similarly, drilling costs for consolidated formations are given in Table 3-8. Although these drilling costs are very general, they do provide a first approximation. It is common for drillers to drill hole sizes from 4 to 27 inches. The well-drilling costs provided in Tables 3-7 and 3-8 are averages for EPA Regions III and IV, but drilling costs do not vary as much across the country as casing and pump costs.

WELL CASING AND COSTS

Campbell and Lehr (1973) summarized five types of pipe used in the water well and plumbing industries: (1) standard pipe, (2) line pipe, (3) reamed and drifted pipe, (4) drive pipe, and (5) water-well casing. Water-well casing is a thin-walled, fine-threaded casing differing in dimensions and threads from all other types of pipe. Thread dimensions vary among different types of pipe. An improper joint and possible casing failure can result if threading specifications are not adhered to. Specifications for water-well casing often designate ASTM A-120 and ASTM A-53. American Petroleum Institute (API) casing is designated by the outside diameter and the wall thickness. The size and weight of the casing must be designed to assure that drilling tools, well screens, pumps, and other necessary equipment can be inserted. Water-well casing is often selected on the basis of its resistance to corrosion.

Black pipe is used for small-diameter wells since water-well casing is not made below a 4-inch diameter. Four- to 6-inch diameter wells are about the smallest size that will handle a submersible pump. These small-diameter black pipe casing costs are given in Table 3-9. It can be seen that the cost of black pipe in shallow wells compares closely with the drilling costs in unconsolidated formations. On the other hand,

Table 3-7

COSTS IN DOLLARS FOR WELL DRILLING IN UNCONSOLIDATED FORMATIONS, EPA REGIONS III AND IV, OCTOBER 1974

Well hole diameter (in)	Depth of well (ft)																		
	40	60	80	100	140	180	200	240	280	300	350	400	500	600	700	800	900	1000
16	800	1200	1600	2000	2800	3600	4000	4800	5600	6000	7000	8000	10,000	12,000	14,000	16,000	18,000	20,000
14	700	1050	1400	1750	2450	3150	3500	4200	4900	5250	6125	7000	8750	10,500	12,250	14,000	15,750	17,500
12	600	900	1200	1500	2100	2700	3000	3600	4200	4500	5250	6000	7500	9000	10,500	12,000	13,500	15,000
10	500	750	1000	1250	1750	2250	2500	3000	3500	3750	4375	5000	6250	7500	8750	10,000	11,250	12,500
8	400	600	800	1000	1400	1800	2000	2400	2800	3000	3500	4000	5000	6000	7000	8000	9000	10,000
6	300	450	600	750	1050	1350	1500	1800	2100	2250	2625	3000	3750	4500	5250	6000	6750	7500
5	250	375	500	625	875	1125	1250	1500	1750	1875	2188	2500	3125					
4	200	300	400	500	700	900	1000	1200	1400	1500	1750	2000	2500					
3	150	225	300	375	525	675	750	900	1050	1125								
2	100	150	200	250	350	450	500	600	700	750								
1	50	75	100	125	175	225	250	300	350	375								

Table 3-8

COSTS IN DOLLARS FOR WELL DRILLING IN CONSOLIDATED FORMATIONS, EPA REGIONS III AND IV, OCTOBER 1974

Well hole diameter (in)	Depth of well (ft)																		
	40	60	80	100	140	180	200	240	280	300	350	400	500	600	700	800	900	1000
16	640	960	1280	1600	2240	2830	3200	3840	4480	4800	5600	6400	8000	9600	11,200	12,800	14,400	16,000
14	560	840	1120	1400	1960	2520	2800	3360	3920	4200	4900	5600	7000	8400	9800	11,200	12,600	14,000
12	480	720	960	1200	1680	2160	2400	2880	3360	3600	4200	4800	6000	7200	8400	9600	10,800	12,000
10	400	600	800	1000	1400	1800	2000	2400	2800	3000	3500	4000	5000	6000	7000	8000	9000	10,000
8	320	480	640	800	1120	1440	1600	1920	2240	2400	2800	3200	4000	4800	5600	6400	7200	8000
7	280	420	560	700	980	1260	1400	1680	1960	2100	2450	2800	3500	4200	4900	5600	6300	7000
6	240	360	480	600	890	1080	1200	1440	1680	1800	2100	2400	3000	3600	4200	4800	5400	6000
5	200	300	400	500	700	900	1000	1200	1400	1500	1750	2000	2500					
4	160	240	320	400	560	720	800	960	1120	1200	1400	1600	2000					
3	120	180	240	300	420	540	600	720	840	900								
2	80	120	160	200	280	360	400	480	560	600								
1	40	60	80	100	140	180	200	240	280	300								

polyvinyl chloride (PVC) casing is becoming more popular and is available at less cost (Table 3-10). While the cost of small-diameter casing is more competitive than large-diameter casing, these costs can vary with time and regional location.

Some wells are essentially open holes drilled into consolidated rocks and may have only a short length of casing near the land surface. In wells in unconsolidated deposits, the hole must be cased to the bottom. Standard-weight pipe is usually adequate for depths where casings of 6-inch diameter or less are set.

Special types of casing may be needed in areas of certain groundwater qualities and geologic characteristics. Cupronickel alloys are most suited for sea-water production wells. Carbon steel has demonstrated a high resistance to soil corrosion. Stainless steel offers considerable durability and high reliability.

Plastic casing has increased in use recently. Maximum installation depths for PVC pipe are normally less than 200 feet. Rubber-modified polystyrene has also been used as well casing, but is not recommended for depths greater than 300 feet. Plastic well casings are usually not larger than 6 inches in diameter for structural reasons. However, fiberglass-reinforced, epoxy pipe has more desirable characteristics than pure plastics; 8- and 10-inch sizes have been used extensively to depths of about 300 feet for water-well casing. Epoxy plastic pipe has a high resistance to corrosion and reduced incrustation problems.

Wells in highly corrosive environments are sometimes cased with ceramic tile, concrete, asbestos cement, or even wooden pipe, but most of these are heavy, structurally weak, and not readily available in many areas. Concrete and asbestos-cement pipe are subject to rapid deterioration in high-sulfate soils and water.

WELL SCREENS AND PERFORATED CASING

When completing most wells in unconsolidated materials, openings in the casing must be provided to permit entrance of water. In some areas no perforations are used and the pipe is left open, whereby water is drawn through the end of the casing. In areas of relatively thick water-producing strata, the casing is commonly perforated over intervals up to several hundreds of feet. Casings can be perforated in the field by torch or can be factory perforated. Down-hole perforators can be used for blank casing already in a well. Slotted casing has an open area of about 1 percent for 0.030-inch slots to about 12 percent for 0.250-inch slots. Factory perforated casing usually has an open area ranging from 4 to 18 percent.

A specialized piece of equipment, known as a well screen, has been developed for maximizing the open area percentage. For cage type, wire-wound screens, the open areas range from about 2 percent for 0.0006-inch slots to over 60 percent for 0.150-inch slots. Well screens may be made of iron, fiberglass, brass, stainless steel, or PVC. Well screens are especially applicable to sampling relatively thin (less than 50 feet thick) sections of the aquifer.

The sizes of perforations, slots, or screen openings are chosen with respect to particle size distribution of the water-bearing zones. Entrance velocity into the screen or perforated casing is usually recommended to be less than 0.1 to 0.2 feet per second. The amount of open area desired can be calculated based on well discharge.

GRAVEL PACKING

In a naturally developed well, the development process removes finer material from the vicinity of the perforations or screen, leaving a zone of coarser graded

Table 3-9

BLACK PIPE CASING COSTS FOR WELLS IN DOLLARS, EPA REGION IX, OCTOBER 1974

Standard well-black pipe diameter (in)	Depth of casing (ft)																	
	40	60	80	100	140	180	200	240	280	300	350	400	500	600	700	800	900	1000
6	335	502	670	837	1172	1507	1674	2009	2344	2511	2930	3348	4185	5022	5859	6696	7533	8370
5	264	397	529	661	925	1190	1322	1586	1851	1983	2314	2644	3305					
4	190	285	380	472	665	855	950	1140	1330	1425	1662	1900	2375					
3	126	188	251	314	440	565	628	754	899	942								
2	61	92	122	153	214	273	306	367	428	459								
1	29	43	58	72	101	130	144	173	202	216								

Table 3-10

PVC PIPE COSTS FOR WELLS IN DOLLARS, EPA REGION IX, OCTOBER 1974

Well casing diameter (in)	Depth of casing (ft)																	
	40	60	80	100	140	180	200	240	280	300	350	400	500	600	700	800	900	1000
10	347	520	694	867	1214	1561	1734	2081	2428	2601	3035	3468	4335	5202	6069	6936	7803	8670
9	286	428	571	714	999	1285	1427	1713	1998	2141	2498	2854	3568	4281	4995	5708	6422	7135
8	224	336	448	560	784	1008	1120	1344	1568	1680	1960	2240	2800	3360	3920	4480	5040	5600
7	178	267	356	445	623	801	890	1068	1246	1335	1558	1780	2225	2670	3115	3560	4005	4450
6	132	198	264	330	462	594	660	792	924	990	1155	1320	1650	1980	2310	2640	2970	3300
5	93	140	186	233	326	419	466	559	652	699	816	932	1165					
4	61	91	122	152	213	274	304	365	426	456	532	608	760					
3	37	55	74	92	129	179	184	221	258	276								
2	17	25	34	42	59	76	84	101	118	126								

material around the well. This cannot be achieved in a formation consisting of a fine uniform sand, due to the absence of any coarser material. The object of gravel packing a well is to artificially provide the graded gravel or coarser sand that is missing from the natural formation.

Two conditions in unconsolidated materials tend to favor gravel packing: (1) fine uniform sand, and (2) extensively layered deposits. The latter condition is common in alluvial fan deposits of the Southwest. The selection of the grading of the gravel pack is usually based on the finest material in the aquifer.

Gravel packing is also used for formation stabilization. Since drilling by the rotary method through unconsolidated materials, of necessity, results in a hole somewhat larger than the outside diameter of the casing, the annular space around the well screen is gravel packed to prevent silt and clay above the water table from caving or slumping into the well producing zone.

Screen openings or perforations are chosen so as to retain 90 percent or more of the gravel-pack material. The gravel pack should consist of clean, well-rounded, smooth grains. Quartz and other silica-based materials are preferable. Gravel-pack envelopes are usually 3 to 8 inches thick.

WELL SEALING

Grouting well casing involves filling the space around the casing with a suitable slurry of cement of clay and sand. Isolation by grouting is desirable to protect the producing zone from contamination by less desirable fluids from other levels or from the surface. Cementing materials include portland cement, bentonite, pozzolana, perlite, diatomaceous earth, Gilsonite, and mixtures of these.

To assure that grouting provides a satisfactory seal, the slurry must be added continuously to prevent cold joints. The grout should be introduced at the base of the grouting interval to minimize contamination or dilution of the slurry and bridging of the mixture with upper formation material. The grout is usually pumped into the space to be filled; however, placement by gravity is satisfactory in some cases.

An increasing number of States are establishing requirements on sealing wells to protect them against contamination by surface waters. In this case, a near-surface, upper annular-space seal of an aggregate of fine sand and cement is commonly used. This type of seal generally extends from 10 to 50 feet or more in depth.

WELL DEVELOPMENT

The well is developed after it has been drilled, cased and packed. The object of well development is to remove silt, fine sand, and other such material from a zone immediately around the well. This creates larger water-flow passages in the formation around the well. Well development corrects clogging or compaction of the formation which occurs during drilling. Drilling mud used in the rotary method effectively seals the face of the borehole. All drilling methods do some damage to the formation, and well development is used to correct this damage. Well development also grades the material in the aquifer immediately around the casing in such a way that a stable condition is achieved in which the well yields sand-free water at maximum capacity.

To be effective, the development operation must cause reversals of flow, or surging, through the perforation or screen openings. This is necessary to avoid the bridging of openings by groups of particles that can occur when the flow is continuously in one direction. The reversals of flow are caused by forcing water out of the well through the screen, or perforations, and into the aquifer, then removing the force to allow flow to occur from the aquifer back into the well.

Mechanical surging is the name given to a process of operating a plunger up and down in the casing like a piston in a cylinder. The tool used is called a surge plunger or surge block. Before surging, the well should be washed with a jet of water and bailed or pumped to remove the drilling mud cake on the face of the borehole. Dispersing agents, mainly polyphosphates, are added to the water in the well to counteract the tendency of the mud to stick to sand grains. These agents are often applied at the rate of one-half pound to every 100 gallons of water in the well. Materials brought into the well by surging can be periodically removed by bailing. Surging can also be achieved by alternately turning the test pump on and off.

High-velocity jetting, or backwashing of an aquifer with high-velocity jets of water directed horizontally through the screen openings, is generally the most effective method of well development. The energy of the jets can be concentrated over small areas at any particular time, and all of the screen or perforated area can be selectively treated. Thus, uniform and complete development is achieved throughout the length of the perforated interval or well screen.

TEST PUMPING

Test pumping is commenced upon completion of well development operations. Often the well is pumped at low rates during early phases of the test and high rates at later phases. Periodic surging may be necessary if the well does not appear to be developed. Often the well is pumped at a much greater rate than that intended for use. This further acts as a well development procedure. By increasing well discharge in a step pattern, a step-drawdown test can be run. This can be followed by a constant discharge test. In alluvial aquifers, pump tests of 24 to 72 hours may be sufficient. For aquifer analyses in consolidated rocks, pump tests from 1 week to 1 month or longer may be advisable. Pump testing for aquifer analysis requires consideration of aquifer geometries as well as inhomogeneities and anisotropies in the aquifer.

TOTAL WELL CONSTRUCTION COSTS

While general data have been provided for drilling and casing costs, additional costs, such as for screens; gravel packing; grouting; and well development, are highly dependent upon local conditions. It is therefore more appropriate to provide separate cost data for completed wells in sand and gravel and wells in consolidated formations. Detailed well cost analyses have been done by Cederstrom (1970) and Gibb (1971).

Gibb provides a total cost analysis for small-diameter domestic and farm wells. Since these small bore wells approximate the size of most monitor wells, these cost data are presented here. Cost information for 345 wells of various types constructed in sedimentary rocks throughout Illinois during 1967, 1968, and 1969 was collected by Gibb. All data were adjusted to a common 1969 economic level by using a domestic and farm well index developed as a part of the study. This index, shown earlier in Figure 3-2, indicates the increase in the costs of farm and domestic wells in Illinois from 1913 through 1969. Figure 3-2 also illustrates the increase in costs of large-capacity wells. It can be seen that the small-diameter domestic and farm well index closely follows the materials component of the *Engineering News-Record*.

Recent applications of the materials component to Gibb's cost data closely approximate the current cost of small-diameter wells. For this reason his data have been updated to March 1975 using the materials component of the *Engineering News-Record*. Well-cost data were divided into two categories by Gibb, according to the aquifer tapped and the type of well construction. Thus, commercially screened, drilled wells completed in water-bearing sand and gravel deposits are shown in Figure 3-19, while drilled wells completed in water-bearing sandstone, limestone, and dolomite of consolidated bedrock formations are shown in Figure 3-20.

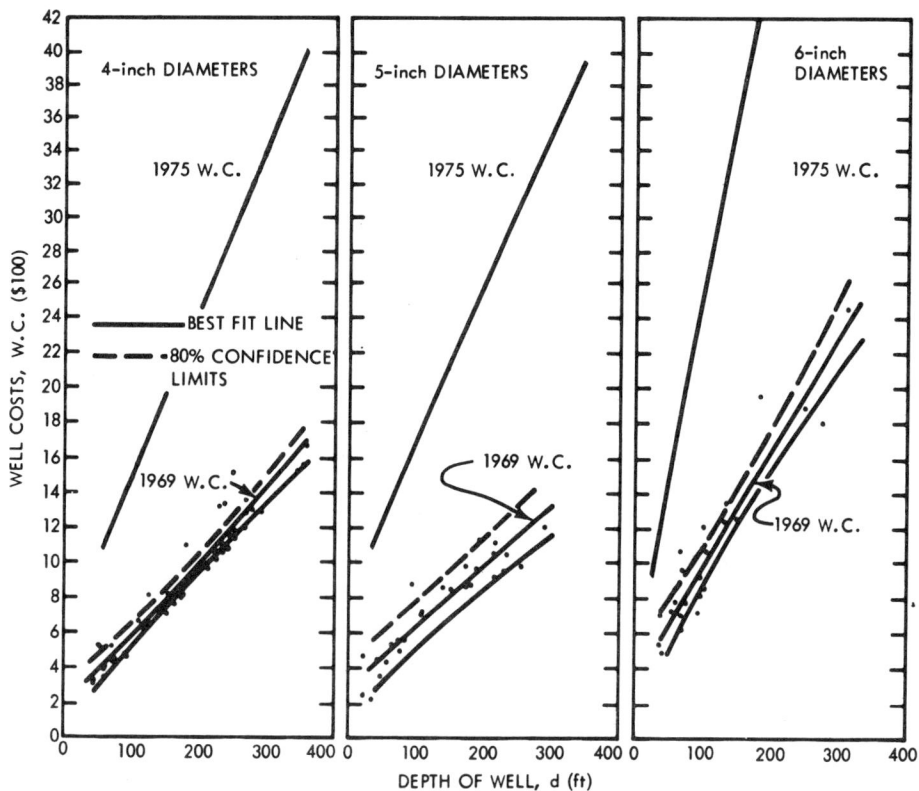

Figure 3-19. Updated (1975) cost of 4-, 5-, and 6-inch wells in sand and gravel (after Gibb, 1971).

The final sets of data may include materials and/or labor cost for the following:

1. Setting up and removing the drilling equipment
2. Drilling the well
3. Installing casings and liners
4. Grouting and sealing the annular spaces between casings and boreholes
5. Installing well screens and fittings
6. Developing the well

TEST DRILLING

Test holes can be drilled to permit water sampling of polluted groundwater. For diffuse sources of pollution, existing wells can often be used. Near point or line sources of pollution, test holes can also be used for delineation of subsurface lithology, water level information, determination of aquifer parameters, and sampling of geologic materials. Test holes and subsequent water sampling and analysis is the most direct method of monitoring groundwater pollution, where existing wells are insufficient in number. The location, method of construction, well specification, and density of monitor wells for a given polluted zone depend upon local conditions.

GEOLOGICAL SAMPLING

Campbell and Lehr (1973) have provided a summary of formation identification and evaluation. The success of a well depends largely upon the degree of care used in obtaining geologic samples. Sampling materials in the saturated zone is used where significant pollutants have been retained in this zone. This will generally occur in cases of shallow water tables (thin vadose zone), or where the vadose zone is bypassed, such as in deep well disposal. The primary strata of interest will usually be the finer grained deposits, such as silt or clay, because of their ability to retain certain pollutants.

Cable-tool rig samples are usually representative of the interval drilled between bailing operations. Some contamination of the cuttings occurs in unconsolidated materials due to the cable rubbing against the upper, uncased part of the hole. When fine-grained saturated sands and silts are encountered, care must be used in bailing to avoid heaving materials into the hole.

Rotary rig samples are usually collected at regular intervals. These samples contain some cavings and fragments circulated in the drilling mud. Differential settling occurs between light and heavy fragments in the drilling mud, and this leads to complications in logging. The collection of samples at the surface lags behind the actual cutting of a given stratum at depth. This lag may amount to more than 20 feet in a 300-foot hole, and depends upon the pump size and speed, the hole size, the drill pipe ID, and the viscosity of the drilling fluid. An electric log can be used in conjunction with a geologist's log to provide information on the time lag. Rotary rig samples from unconsolidated materials may be poor, particularly when relatively thin layers of sand, gravel, and clay are being drilled. Soft clays become dispersed in the drilling mud and are difficult to ascertain. Coring is recommended in such strata.

Samples obtained from compressed-air rotary drilling are generally superior to those from other methods. Casing is usually necessary in order to minimize contamination if soft, fine-grained material is higher in the hole.

Excellent samples can be collected during drilling with reverse-circulation rotary rigs. The highly turbulent fluid velocities up the drill stem result in little or no separation of the fines, with a minimum time lag. The bit tends to loosen materials rather than grind them up, and the cuttings are drawn into the bit and delivered at the surface.

Coring is a sampling method as well as a drilling method. It is done with rotary equipment and individual cores are often less than 4 inches in diameter and 10 feet in length. Numerous rock types can be successfully cored with any core bit, but soft and friable formations are difficult to core.

Unconsolidated deposits of interbedded soft clays, sand, and gravel are difficult to core. However, a split-core, drive sampler will often provide excellent samples of these materials. A number of coring devices are available, ranging from 1-1/2 to 6 inches in diameter. These devices may be forced into the deposits by the weight and rotation from the rotary rig or by *drive coring* with the cable-tool rig. Barton (1974) reports on borehole sampling of saturated unconsolidated sands and gravels. A great variety of samplers have been specially designed for these materials. In extreme situations, it may be necessary to stabilize the strata prior to sampling. Sidewall sampling is gaining wider application for unconsolidated materials.

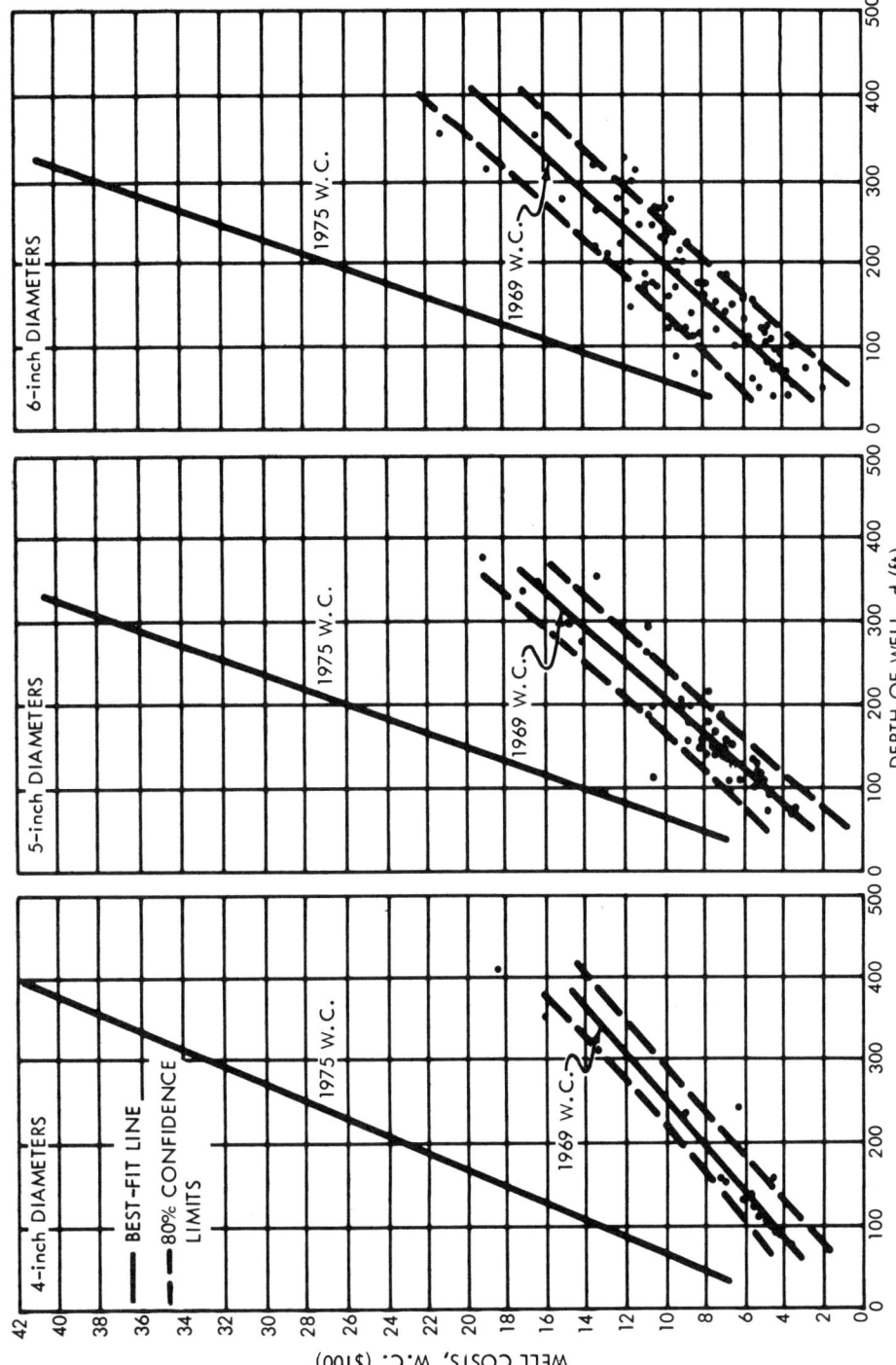

Figure 3-20. Updated (1975) cost of 4-, 5-, and 6-inch wells in consolidated rock (after Gibb, 1971).

Drilling mud should be avoided if possible when sampling formations for groundwater pollution studies. Air rotary and reverse-circulation rotary drilling methods are preferable since neither method requires the use of drilling *additions.* Cable-tool drilling is also useful, but relatively slow. In-place samples are extremely important in aquifer evaluation of permeability and other considerations. Intact, in-place samples are less important in chemical quality and pollution studies; that is, geologic samples can be physically disturbed to some degree.

The cost of obtaining geologic samples other than by coring is primarily a function of the time spent by the geologist who does the logging. Registered geologists charge about $150 to $250 per day for their services. Coring services vary with the kinds of rigs used, the core diameter, and the sampling frequency. A generalized cost diagram is given in Figure 3-21. The cost figures are generalized to cover small-diameter coring (2 to 3 inches), for shallower depths, and larger coring sizes (3 to 5 inches), for deeper samples, since larger bore diameters are required when drilling to depths greater than 100 feet. The first core taken includes a 1-hour setup and travel time cost, while the second includes just the drilling cost (Figure 3-21). It is apparent that in unconsolidated formations the travel and setup constitute a relatively larger part of sampling costs than for consolidated formation sampling. At greater depths, the time involved in taking the samples overrides the drilling time.

BOREHOLE GEOPHYSICS
Geophysical well logging, or borehole geophysics, includes all techniques of lowering sensing devices in a borehole and recording some physical parameter that may be interpreted in terms of the characteristics of the rocks and the fluids in the rocks. Geophysical logs can be interpreted to determine many groundwater parameters. The single most important factor that has limited the use of geophysical logging in groundwater studies in the past is cost. Logging generally is more feasible in deeper, more expensive wells.

The electrical conductivity or resistivity of fluid in a well bore, or the surrounding aquifer, is one of the most useful parameters that can be derived from geophysical logs. If the chemical nature of the water in an aquifer is known from chemical analyses, and the ion ratios are consistent, resistivity or conductivity from logs can be used to determine the approximate quantities of those ions present. The relationship of water in the hole to water in the surrounding materials must be understood in order to interpret a number of geophysical logs. The chemical quality of fluid in the hole is not necessarily similar to that of fluid in the surrounding materials. After a hole is completed, it may be as much as several months before chemical and thermal equilibrium is reached.

The electric log is very useful in determining subsurface lithology, and is the most common geophysical log run in water wells. An electric log can be run only in an uncased hole that is filled with a conducting fluid. The electric log usually includes the spontaneous potential curve, in the left-hand column, and one or more resistivity curves, in the right-hand column. Spontaneous potential logs are records of natural potentials developed between the borehole fluid and the surrounding materials. The use of spontaneous potential for determination of groundwater quality in fresh-water aquifers is not advisable (Keys and MacCary, 1971).

Resistivity logging devices measure the electrical resistivity of a known, or assumed, volume of earth materials under an electric current. Normal logs measure the apparent resistivity of a volume of earth materials surrounding the electrodes. The short normal log records the apparent resistivity of the invaded zone (extent of drilling

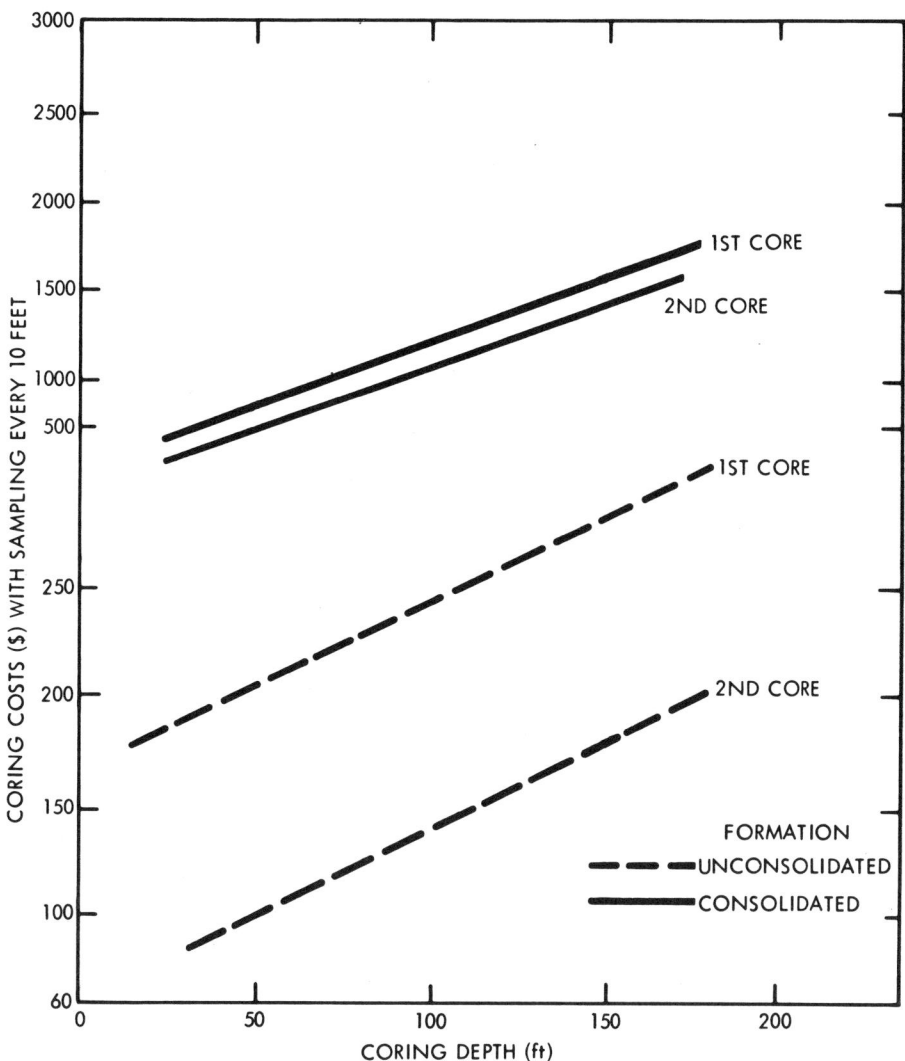

Figure 3-21. Generalized cost of coring in unconsolidated and consolidated formations, October 1974.

mud penetration). The long normal log records the apparent resistivity beyond the invaded zone, and is useful in determining aquifer water quality. The normal logs give poor results in high resistivity rocks.

One method of estimating water quality from electric logs makes use of mathematical expressions which relate the following parameters (Turcan, 1966):

1. Field-formation resistivity factor and fluid resistivity

2. Fluid resistivity and specific conductance

3. Specific conductance and dissolved solids

The field-formation factor is established from preexisting electric logs and water analyses. When a new well is drilled and logged, the resistivity curve is used to determine fluid resistivity in the aquifer by the use of the formation factor. This value is converted to specific conductance at standard temperature from tables or formulas. The specific conductance is then converted to dissolved solids by empirical relationships previously developed.

Electric logs can thus provide an indication of the vertical distribution of groundwater salinity at or near a specific well. However, knowledge of the chemical type of waters at different depths must be available. Interpretation of electric logs can supplement measurements of electrical conductivity made on drilling mud during drilling. For groundwater pollution studies, the tool would be applicable to polluted zones of high electrical conductivity compared to native groundwater.

A portable electric logging system can be purchased for about $6000 to $7000, with an additional cost of $3500 to $4000 for a gamma-ray option. The cost of conducting an electrical conductivity survey including interpretation is given in Figure 3-22. Travel costs are not included, and may be an important item in remote areas.

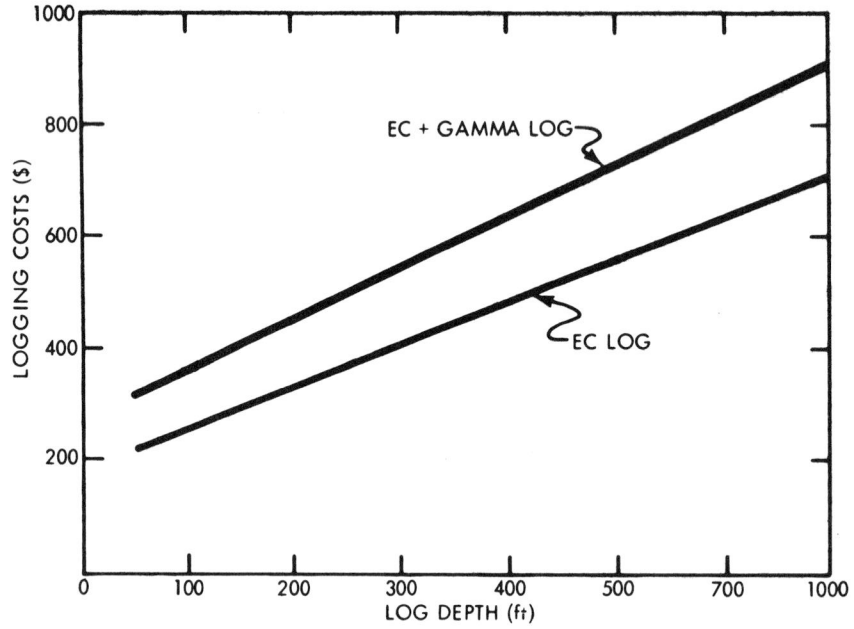

Figure 3-22. Cost of electrical conductivity (EC) and gamma logging, September 1974.

WATER-LEVEL MEASUREMENTS

Water-level measurements for groundwater pollution studies can define a mound beneath a point source, or a ridge beneath a line source. In areas with a paucity of water-level measurements, supplementary measurements may be necessary to define the depth to water and the direction of groundwater movement. Water-level measurements on a regional basis contribute to the hydrogeologic framework; the discussion here will be limited to measurements in localized areas. The most common tools for measurement of water levels are the steel tape, electric sounder, airline, and mechanical recorder.

Steel Tape

The surveyor's steel tape has been used for many years by the U.S. Geological Survey as a water-level measuring device, and is available in lengths of 100, 200, 500, and 1000 feet. Coefficients of stretch and temperature expansion are provided by the manufacturer for each tape. For most groundwater pollution studies, correction for stretch and temperature changes will not be necessary, due to the relatively shallow depths of interest. The tape is usually lowered by hand; however, in deep wells a motor-driven tape reel may be used.

The water level in a well is measured by suspending a known length of tape below a measuring point so that at least the lower few feet of tape are below the water level. The lower portion of the tape is coated with blue chalk or some other substance which exhibits a marked color change when wetted. The water-level measurement is obtained by substracting the length of the wetted portion from the total length suspended below the measuring point.

One disadvantage of using the steel tape is that if the approximate depth to water is unknown, too short a length of tape may be lowered into the well, thereby necessitating a number of attempts. Also, water inside the casing, or cascading water, may wet the tape above the true water table and result in errors in measurement.

The number of wells measured per day depends primarily on well density, access to well sites, depth to water, well ownership, access for the measuring device, and well obstructions. For dense municipal wells under a common manager, up to 40 to 50 water-level measurements per day can be made where the depth to water is less than 100 feet. For irrigation wells spaced at about 1/2-mile intervals in an agricultural area, about 20 to 30 water-level measurements per day can be made where the depth to water is less than 100 feet. For widely scattered stock wells in an undeveloped area, fewer than five water-level measurements may be possible in 1 day.

Measurement time is decreased considerably, when the person performing the measurements is familiar with the area and the wells. For initial measurements in new programs, a much longer period of time is necessary. Prior to initiating well measurements, a comprehensive well data collection and organization program may be necessary. This provides pertinent hydrogeologic data for the wells to be measured.

Costs for steel tapes normally employed to measure water-well levels are summarized in Figure 3-23.

Electric Sounder
The electric sounder consists of a reel with a meter, one or two electric wires, and a sounding tip. The reel and meter remain at the land surface, while the wire and tip are lowered down the well. A battery-powered electrode assembly is usually used; when in contact with water, it causes a sharp needle deflection on a large, sensitive current meter. The wire is mounted on a small reel that is used for lowering and withdrawing the wire and sounding tip. Depth control is aided by metal tags on the wire.

Electric sounders are often used in pump tests where a number of measurements must be taken at small time intervals, and are of great value in cases where cascading water is present. A measurement can be made in less than 5 minutes in most wells less than 100 feet deep.

The average cost of electric sounders is given in Figure 3-24.

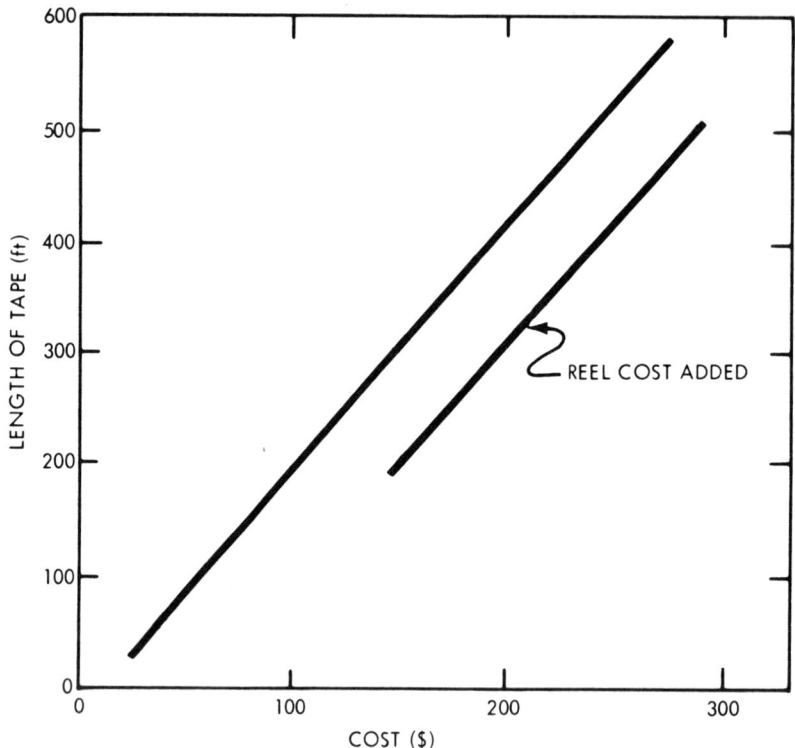

Figure 3-23. Cost of steel tapes, September 1974.

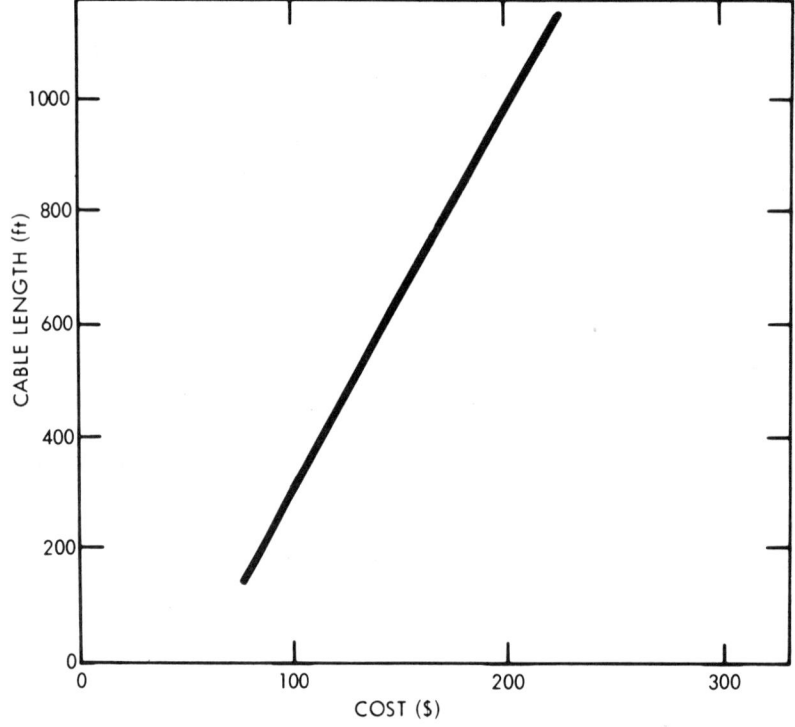

Figure 3-24. Cost of electric sounder and cable, September 1974.

Airline

The airline can be a very practical device in certain situations, such as where pumping produces turbulence inside the well casing. An airline is used by pump testers or well drillers as a rapid means to obtain water-level measurements in deep wells.

The airline consists of a small-diameter copper or iron tube, which is installed in the space between the pump column and the casing. The tube is usually strapped to the pump column, and the two installed simultaneously. The airline should be free of leaks, open at the bottom, and extend several feet below the lowest pumping water level. The airline should preferably be straight and plumb, and the depth of the lower opening must be known. The top of the airline is connected to a source of compressed gas. An air-pressure gauge is used to measure the pressure in the airline.

Pressure is applied until all the water has been expelled from the airline, and the gauge can then be read. The gauge reading indicates the length of the expelled water column (the height of water above the airline opening). To obtain depth to water, the length of this column is subtracted from the length of the airline. This measurement can usually be made in several minutes.

The tubing, which is normally left in the well, may cost as little as $10. The tubing can also be installed for access by the electric sounder. The air compressor and storage tank can cost as much as $250, but a hand pump is frequently satisfactory.

Mechanical Recorder

A simple recording device often used consists of a cylindrical recording chart actuated mechanically by a float that moves with fluctuations in the water level. A small-diameter stranded cable, or tape, is attached to the float, and a counterbalance is suspended over a pulley on the recorder. The pulley is geared to the chart so that the chart rotation is proportional to float movement. A clock drive slowly moves a recording pen across the chart. The mechanical recorder is used when many measurements are necessary over short time periods, and is especially useful when relatively rapid water-level fluctuations occur, such as in observation wells during pump tests. These recorders cost about $375. If a digital system, rather than a chart, is desired, the unit can cost about $675. Most recorders come with about 20 feet of line, and an extra charge of $0.25 per foot is made for additional line.

Garber and Koopman (1968) summarized methods of measuring water levels in deep wells. Accurate measurement of water levels deeper than 1000 feet in wells requires special equipment. Correction for stretch and thermal expansion of measuring tapes must be considered.

WATER SAMPLING

Most of the physical factors which promote mixing in surface waters are absent, or much less effective in groundwater systems. A well can be considered as a sampling point in a large body of slowly moving water, which differs in chemical composition vertically as well as areally. Unfortunately, most techniques for well sampling at discrete depths are usable only in unfinished or nonoperating wells. A more reliable means of evaluating the quality of water tapped by a well is an analysis of a pumped sample.

The volume of water collected for chemical analysis from individual widely scattered monitoring wells is relatively insignificant volumetrically. Therefore, either many more monitoring wells must be sunk, or a smaller number of existing wells must be pumped at higher rates or more frequently, to obtain representative values. Most wells are integrators of complex systems of surface water, soil water, and groundwater. In some cases the integrated sample is of most importance because it represents the water used, or likely to be used in the future. In other cases, an integrated sample is not of interest, but rather a sample from a specific depth zone. This requires either careful selection from existing wells or construciton of special monitoring wells.

WELL HYDRAULICS

The composition of water obtained from a well is likely to be influenced by well construction, well development, and pump operation, and these should be considered in the sampling program. Well construction can determine the depth of zones from which the water is coming and influence the local groundwater flow pattern. Perforations above the water table can permit cascading water, or direct movement of groundwater from shallow perched layers above the water table, to flow down the inside of the well to the water table. Drilling mud remaining from well construction can contain contaminants, as can well casing, seals, and pump parts.

Well hydraulics can cause chemical changes in the composition of a pumped sample. For example, pressure changes associated with velocity changes at the well inlet can cause changes in oxidation state, pH, and temperature which, in turn, affect certain chemical constituents. Thus a sample from the well discharge may differ significantly in quality from groundwater in the aquifer. Also, in cases where chemicals are added at the discharge pipe, care must be taken to ensure the chemical composition of the sample is not altered.

TIME CHANGES IN QUALITY

Quality changes in groundwater are usually much slower than those in surface water. The composition of well samples from a large, relatively homogeneous aquifer usually will not change much over long periods of time. Therefore, changes in groundwater quality can be described satisfactorily by monthly, seasonal, or annual sampling schedules. However, exceptions occur in cases of groundwater pollution.

Just as water levels in a well change drastically soon after a pump is turned on, so can the quality of pumped water. This is particularly true where there is vertical stratification of groundwater quality. These short-term changes, in some cases, may be linear with pumping time, whereas, in other cases, they may be exponential. In the case of high-yielding wells (greater than 500 gpm), several days or weeks of pumping may be necessary to purge the well of atypical water due to local conditions. These local conditions could include poor quality, shallow water cascading down the well casing and accumulating in or near the well during nonpumping periods, or vertical leakage down a gravel pack to a deeper producing zone. Often, field measurements of water temperature and electrical conductivity at the well discharge, as the well is pumped, can yield useful information on short-term changes in water quality. For short-term testing, consideration should be given to a logarithmic frequency of sample collection, analogous to the procedure for water-level measurements during a pump test. Short-term trends should be evaluated before adequate data are collected to determine long-term time trends in groundwater quality.

In the early days of water-level measurement programs, weekly or monthly measurements were made to discover seasonal fluctuations; elsewhere, semiannual measurements, related to high and low water levels for a year, were instituted. In some areas significant seasonal fluctuations occur in the quality of groundwater. In these areas seasonal trends must be established in the first stages of the development of monitoring programs. Specific locations where seasonal changes are likely to occur include:

1. Near large-volume point or line sources of recharge or pollution

2. Near disposal sites for highly concentrated wastes

3. In areas of permeable soils and geologic materials above the water table

4. Where wells tap shallow portions of aquifers

SAMPLING AT THE WELL DISCHARGE

Water samples can be collected directly from pumping wells or flowing artesian wells. Nonpumping wells equipped with a pump require pumping for a substantial period after start-up to obtain a representative sample from the aquifer. In wells without fixed pumps, special portable pumps may be installed to retrieve water samples.

Water-well pumps can be classified into two groups:

1. Constant displacement pumps, which deliver substantially the same quantity of water against any head within their operating capacity

2. Variable displacement pumps, which deliver water in quantity varying inversely with the head against which they operate

Major types of constant displacement pumps include: (1) piston or reciprocating, (2) rotary, and (3) screw or squeeze displacement pumps. Major types of variable displacement pumps are: (1) centrifugal, (2) jet, and (3) suction-lift pumps. A submersible pump is a centrifugal pump closely coupled with a submersible electric motor.

A pump installed above a well is called a suction-lift pump. A pump installed in the well at some depth below the water table is called a positive submergence pump. Suction lift by a pump is accomplished by developing negative pressure head at the pump intake. The maximum suction lift is limited at atmospheric pressure, vapor pressure, head losses due to friction, and the required inlet head of the pump. At sea level the best designed pumps usually achieve a suction lift of about 25 feet, while the suction lift of an average pump varies from 15 to 18 feet. The maximum suction lift decreases with an increase in altitude.

Where the depth to water is greater than the practical suction limit, a hand-operated pitcher pump, a centrifugal pump, a submersible pump, or a shallow-well jet pump is used. Submersible and jet pumps require a minimum casing diameter of 4 and 2 inches, respectively. In large-diameter wells, where the depth to water exceeds the suction limit, a submersible, deep-well jet, or a deep-well turbine pump is used. Before a pump can be intelligently selected for any installation, information on required capacity, location, operating conditions, and total head is necessary.

The time involved in sampling at the well discharge is highly dependent on whether a pump is installed and is operating, travel time between sites, and time required to gain permission to sample the wells. If the pump must be started, additional time is required.

With closely spaced, actively pumped wells in an urban area that are under the managership of only one or a few agencies, as many as 30 to 50 wells can be sampled in one day (without field analyses or treatment). In remote areas with widely spaced nonpumping wells and different owners, fewer than a half dozen samples per day may be possible.

The cost to sample each well, therefore, is a function of the travel time and man-time at each well. The salary of water sampling technicians is from $8000 to $10,000 per year. The cost to sample a number of wells under three categories of well spacings is given in Figure 3-25. It is assumed that a technician can sample 30 to 50 wells daily in a municipal well system, 10 to 15 irrigaiton wells in a rural area, and 4 to 7 in an isolated area.

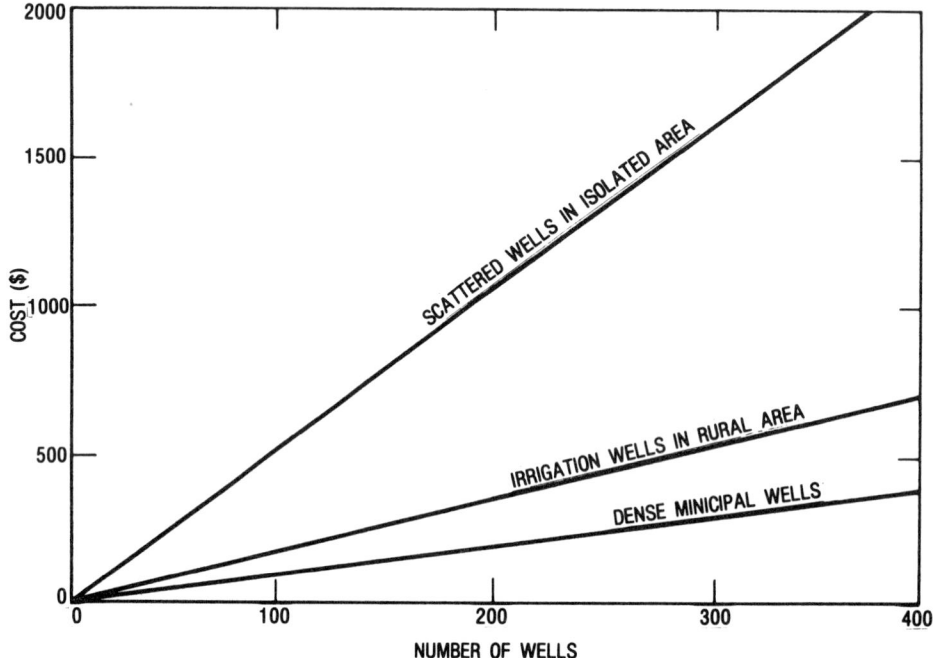

Figure 3-25. Approximate cost of sampling wells (one round), December 1974.

The cost of common field equipment for water quality tests is not significant over the useful life of the instruments. A portable electrical conductivity meter costs about $250 to $400. A portable pH meter can cost from $100 to $300. Specific ion meters can be purchased for about $200 to $250, and additional probes cost about $100 to $150 each. Elaborate continuous recording, digital readout, and multiparameter water quality instruments can be quite expensive ($2000 to $3000), but are not necessary for routine water well sampling.

TYPES OF WELL SAMPLING

Well sampling can be grouped into two broad types: water supply and hydrogeologic. Water supply sampling is concerned primarily with the use of the water. For drinking purposes, a sample might be taken from the household tap rather than from the well discharge. For irrigation purposes, a water sample might be taken from a ditch at the entrance to a field rather than from a well. Hydrogeologic sampling is concerned with natural groundwater quality or pollution. For hydrogeologic sampling, the water sample would, ideally, be taken directly from the aquifer, not at the well discharge. However, for practical purposes, a sample from the well discharge can be used in most cases. Ideally, high-yielding wells (greater than 500 gpm) would be used to evaluate large-scale or regional aquifer conditions; low yielding wells would be used to evaluate local conditions.

A low-yielding well (less than 50 gpm) normally reflects water quality only in the immediate vicinity of the well. In many rural areas, for example, a domestic well would be greatly influenced by local conditions, such as lawn irrigation and septic tank waste disposal. A high-yielding irrigation well in a rural area, pumped continuously, would produce a water sample whose quality would reflect regional conditions, such as fertilizer application, irrigation return flow, and canal recharge.

A 5 gpm domestic well pumped daily for 1 hour produces 300 gallons in 1 day. A 3000-gpm irrigation well pumped continuously for 1 day pumps almost 15,000 times this volume of water. Erroneous interpretations have been made in past groundwater quality studies due to a lack of consideration of this factor.

High-yielding wells also can induce great vertical head differences during pumping, and this should be considered for sampling purposes. In some cases, practical considerations will dictate that wells cannot be placed immediately adjacent to waste disposal sites. A sample from a high-yielding well would be more likely to show the effects of recharged wastewater because it would be more likely to induce flow toward the well. This is especially true in areas of very slow groundwater movement.

SAMPLING IN OPEN WELLS

In wells where a fixed pump is not available, a portable submersible pump can be used. This pump is feasible where a number of monitor wells occur in a small area and only periodic pumping is required. Samples from abandoned or inactive wells, of unknown construction, are often of limited value. The cost of these pumps is a function of the pump size, cable load, flex hose diameter, and sampling depth. The cost of submersible pumps for various pumping lifts and well discharges can be obtained from Table 3-11. In addition to the pump cost, the cost of the support cable and flex hose (Figure 3-26) must be added. The unit is usually placed on a power-driven reel.

The power costs to run submersible pumps are given in Figure 3-27. The power costs vary with the discharge rate, total head, pump efficiency, and electrical costs. Figure 3-27 shows that the power costs for 1 hour operation of a submersible pump are minimal. These costs include a motor efficiency of 75 to 80 percent, a pump efficiency of 70 to 75 percent, and scaled electrical costs of about 3 cents per kilowatt hour.

Water can be pumped from a well by releasing compressed air into a discharge pipe lowered into the well. Air bubbles mix with the water, reducing the specific gravity of the column of water sufficiently to lift it to the surface. For best results, the submergence ratio of the air line (percentage of the total length below water while pumping) should be about 60 percent. Air-lift pumping can be useful in monitoring wells that need to be pumped only at periodic intervals. Johnson Division UOP (1966) presents data on the relation between total lift, submergence of air line, and water discharged per cubic foot of free air delivered by an air compressor.

Water samples may also be taken by bailing. During cable tool well drilling, periodic water samples are retrieved in the bailer, and may provide adequate information on the vertical distribution of groundwater quality. Where pumping equipment is not available or where a pump cannot be installed because the diameter of a well is too small, a container such as a weighted bottle or a short section of pipe capped at the bottom can be lowered into the well to collect a sample of water. This sample will give an indication of the chemical quality of water, but should not be used for bacteriological or detailed chemical analysis because of the likelihood of contamination. This method is not recommended, except where other methods of sampling are unavailable.

Table 3-11

AVERAGE COST OF SUBMERSIBLE WATER PUMPS, FOUND IN EPA REGION IX, OCTOBER 1974

Head in feet (pumping depth) / Horsepower	60	80	100	120	140	160	180	200	240	280	300	350	400	500	600	700	800	Well Diameter (inches)	Cost ($)
								gallons per hour											
1/3	490	460	430	400	360	340	290	230										6	200
1/2			470	450	480	410	380	340	260	200	110							6	250
3/4					480	470	470	480	380	350	320	270	190					6	300
3/4			750		710	690	640	590	470	400	300							6	300
3/4	1450	1350	1230	1100	940	850	520											6	300
1								470	440	420	400	360	310	210				6	250
1			1380	1310	1230	1180	760	730	660	630	580	500	400	400				6	350
1			1590	900			1070	920	510	360								6	300
1	2550	2100																6	300
1-1/2								780	730	700	480	460	430	370	300	230	150	6	550
1-1/2			1450	1400	1340	1320	1250	1170	1000	900	680	620	570	430				6	500
1-1/2			2380	2100	1740	1560	960				780							6	450
1-1/2	2880	2640																6	450
2								1410	1290	1230	1170	770	730	660	570	460	300	6	600
2			2800	2610	2400	2290	2010	1620		1410	1380	1040	880	390				6	500
2		3000							1440			1310	1230	1060	860	620		6	500
3			3000	2880	2760	2700	2520	2350	1930	1650	1320							6	600
3																		6	550

204

Table 3-11 (Continued)

Head in ft (pumping depth) / Horsepower	60	80	100	120	140	160	180	200	240	280	300	350	400	500	600	700	800	900	1000	1100	1200	1400	1600	Well Diameter (in)	Cost ($)
						gallons per minute																			
5	200																							6	1400
5		180																						6	1400
5		160																						6	1300
5		140	120																					6	1300
5					90	80	80	70																6	1300
5						50	50	50	40	40	40	30	20											6	1800
5							70	60	60	50	40	40												6	1300
5									50	40	40													6	1200
5														20	20	20	20	10						6	1300
5															10	10	10	10	10	10	10			6	1400
7-1/2	300	270	230	180																				8	1600
7-1/2			190	160																				8	1700
7-1/2			170	150	120																			6	1500
7-1/2			140	130	120	110	90	80																6	1500
7-1/2						90	90	70	70	70														6	1500
7-1/2									60	60	50													6	1500
7-1/2									50	50	50	40												6	1600
7-1/2											40	40	30											6	1600
10			190	170	140																			6	1700
10					180	160	150																	6	1700

(continued)

Table 3-11 (Continued)

Head in ft (pumping depth) / Horsepower	60	80	100	120	140	160	180	200	240	280	300	350	400	500	600	700	800	900	1000	1100	1200	1400	1600	Well Diameter (in)	Cost ($)
10			350																					6	1700
10				300	280		140	130	120	90	80	80	70	40										6	1700
10					230	270	250	230	190	90	90	70	60	50										6	1800
10							200	190	160		80		50	40										6	1900
15									170	150	130	110	90	80										8	2000
15									140	130	130	90	80	70	60									8	2100
15														60	50	50	40	40						6	2100
15															60	50	40	40						6	2100
15															50	40								6	2200
15																		40	40	30				6	2300
15																			40					6	2300
15																								6	2400
20						340	300	260	240	210	190	150	150											8	2600
20						290	280	270	200	190	180	170	130	110										8	2600
20											190	140		90	80	70	70							6	2800
20															70	70	60	50	40					6	2800
20																	50	50	40	40				6	2700
20																				40	40			6	2700
20																								6	3000
20																								6	3100

Table 3-11 (Continued)

Horse-power	Head in ft (pumping depth) / gallons per minute																							Well Diameter (in)	Cost ($)
	60	80	100	120	140	160	180	200	240	280	300	350	400	500	600	700	800	900	1000	1100	1200	1400	1600		
20																								6	3300
30			570																					8	3300
30				510	420																			8	3400
30						550	500	440																6	3600
30										320	300	240												6	3700
30										260	250	210	180											6	3700
30													190	150										6	3700
30														170	130									6	3600
30														140	130	120	100	50	40	40	40	30		6	4000
40																	90	80	80	70	70	50		6	4100
40											290							70	70	70	60	50	40	8	4400
40													280	250	210				60	50	50	50		8	4700
40													250	200	150	180	140							6	4800
40																160	140	120	110					6	4800
40																140	130	90					70	6	4700
40																				90	80	80		6	4900
40																					80	70	60	6	5300

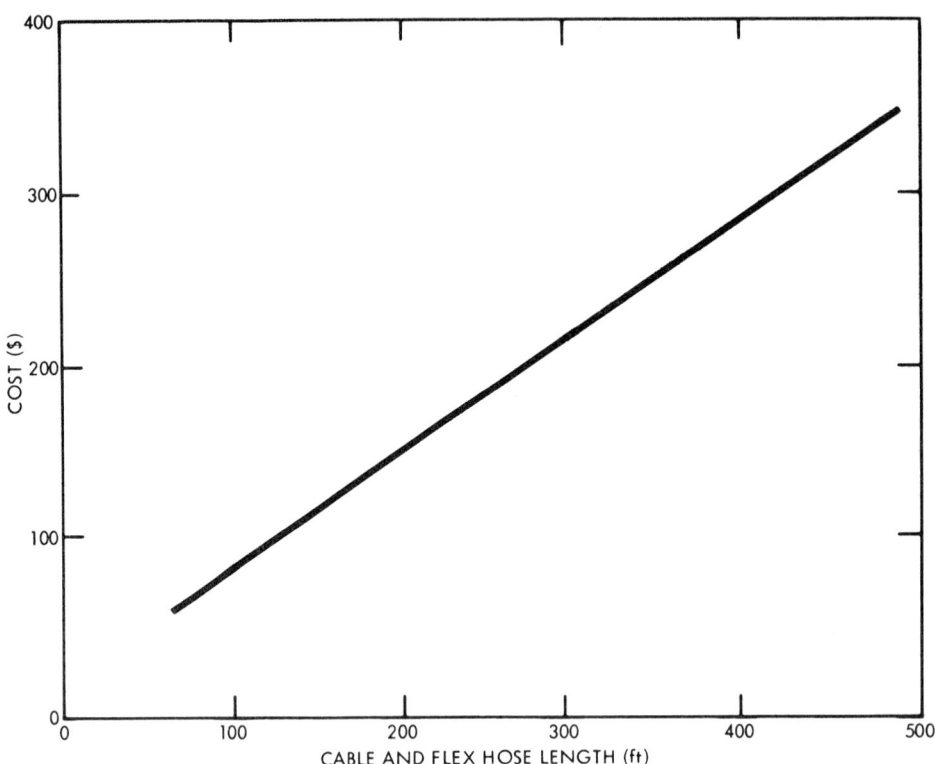

Figure 3-26. Cost of cable and flex hose for use with portable submersible pumps, October 1974.

Where it is desired to collect water from a specific depth inside a well, a special cylinder known as a *thief sampler* can be lowered by cable into the well and closed at a predetermined depth. Designs differ, primarily in their closing mechanism. Most types consist of a cylinder that can be less than 2 inches in diameter. A bar with a plunger or cork on each end runs through the cylinder. When the sampler is lowered below the water table to the desired depth and a continuous upward motion is imparted, the two ends close. The most common type of thief sampler is spring loaded. With this type a messenger is lowered along the cable to release the two end cups, which in turn trap the water sample. Many thief samplers are custom made for a specific use. The cost of these samplers is about $100 to $200.

Another method of collecting depth-controlled samples from uncased holes is by installation of mechanical or inflatable packers, which temporarily isolate selected water-bearing zones for pumping. This method is of limited value in gravel-packed wells due to vertical movement of groundwater in the gravel pack.

Under special circumstances, continuous monitoring of wells may be necessary. Among the devices used for this purpose are probes designed to measure conductivity, temperature, and selected constituents. These probes may be operated down the well or at the discharge outlet.

Yare (1975) describes a procedure for water sampling during rotary drilling. This technique consists of:

1. Drilling a borehole to the base of a sampling horizon

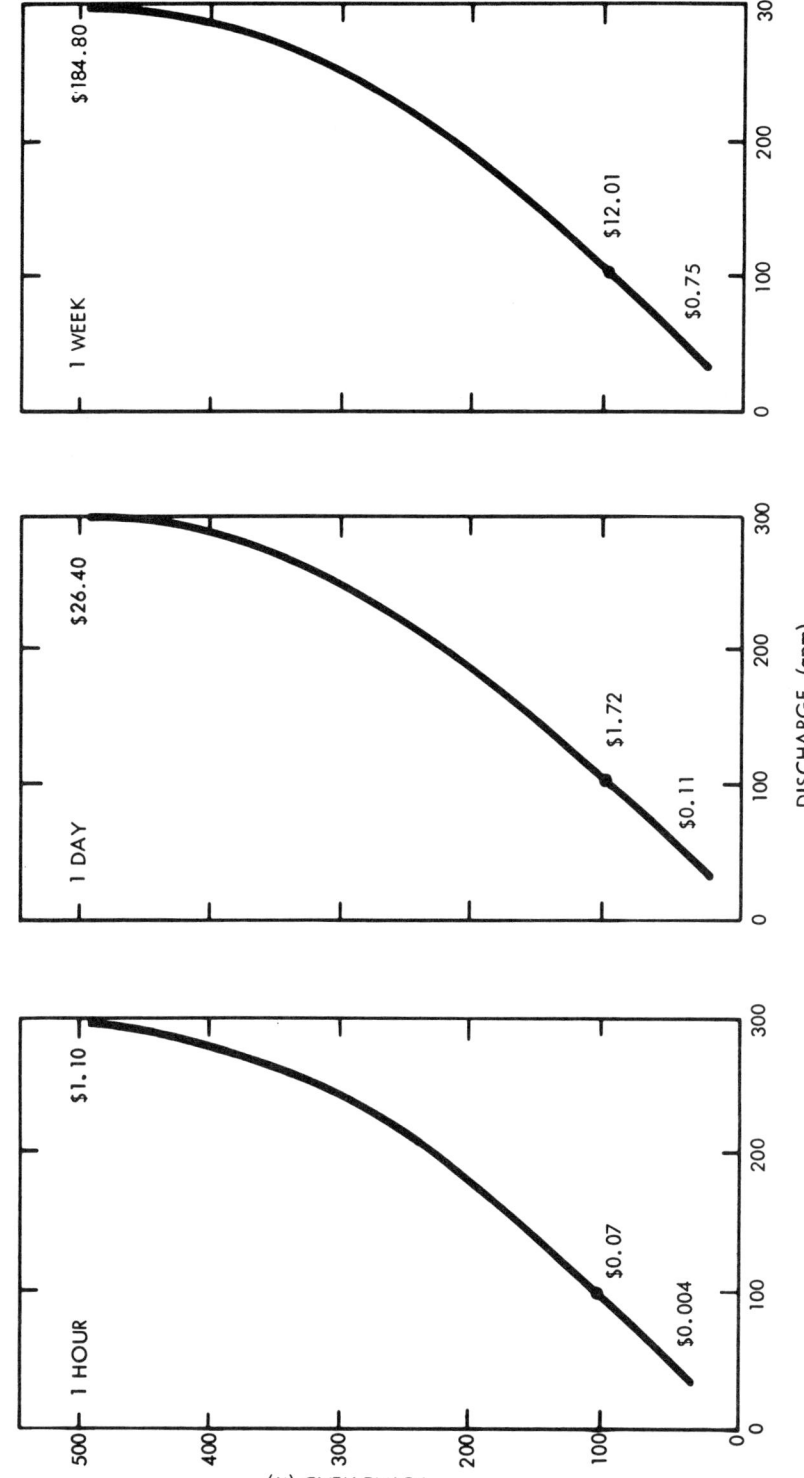

Figure 3-27. Power costs for continuous pumping of 1 hour, 1 day, 1 week for specific pumping heads and specific discharge rates, December 1974.

2. Lowering a wire-wound well screen and riser pipe to the bottom of the borehole, and gravel packing the screen

3. Pumping the borehole well until the discharge is clear of drilling fluid

4. Pumping at least 100 gallons of formation water before collecting the sample

The March-April 1975 issue of *Ground Water* contains at least six articles dealing with monitor well construction and water sample retrieval. These articles indicate how local conditions can influence the type of monitoring which should be undertaken.

OTHER METHODS OF SAMPLING GROUNDWATER

Tile or ditch drains can provide samples of shallow groundwater. Thomas and Barfield (1974) discuss the limitations of tile effluent for monitoring percolate from soils. The primary limitation for monitoring below the water table is that these samples may represent only the shallowest portion of the aquifer. In some cases, this provides very useful information. Springs represent groundwater discharge and can be sampled at the land surface. Interpretation of spring water qualities depends on knowing details of the hydrogeologic framework and groundwater flow patterns. Water sampling of streams under base flow conditions can represent an integration of large groundwater systems, and thus be very useful. Proper selection of sampling sites and times requires careful hydrologic judgment.

SECTION 5 – ANALYSIS OF SAMPLES

INTRODUCTION

Samples, including soil and water, can be taken from the land surface, the vadose zone, or the zone of saturation. In spite of the location of the sample site, many of the sample analysis techniques are similar. A water sample taken at the surface, in the vadose zone, or from the zone of saturation is analyzed in the laboratory using much the same analytical techniques for each parameter. The sample preparation, however, is often quite different.

Soil tests can be divided into physical and chemical analyses. The physical tests are not routinely handled by many chemical analysis laboratories. Agricultural laboratories often provide these services. The physical tests include water content, bulk density or porosity, particle size distribution, soil-moisture characteristic curve, and hydraulic conductivity. The chemical analyses of soil samples include soluble salts, soluble ions, cation exchange capacity and exchangeable ions, and specific surface.

The water tests can be divided into physical, chemical, bacteriological, and radiological analyses. The chemical analyses are further subdivided into inorganic and organic tests. In the discussion of water analysis, consideration is given to sample containers, sample preservation and treatment, and quality control.

CUSTODY CONTROL

The EPA's Office of Water and Hazardous Materials has prepared a procedure (U.S. EPA, 1975) for a recommended *Chain-of-Custody* that will minimize legal complications in obtaining and analyzing water samples. The following comments are abstracted from this document.

Quality assurance should be stressed in all monitoring and examination of self-monitoring programs, no matter what the impetus for the spot check or inspection. The successful implementation of a monitoring program depends to a large degree on the capability to produce valid data, and demonstrate such validity. No other area of environmental monitoring requires more rigorous adherence to the use of validated methodology and quality control measures.

It is imperative that laboratories and field operations involved in the collection of primary evidence prepare written procedures to be followed, whenever evidence samples are collected, transferred, stored, analyzed, or destroyed. A primary objective of these procedures is to create an accurate written record, which can be used to trace the possession of the sample from the moment of its collection through its introduction into evidence. The procedures described here have been successfully employed, and are presented as suggested procedures insofar as they fulfill the legal requirements of the appropriate State legal authority.

PREPARATION

The evidence-gathering portion of a survey is characterized by the absolute minimum number of samples required to give a fair representation of the effluent or water body sampled. The quantity of samples and sample locations are determined prior to the survey.

Chain-of-custody record tags are prepared prior to the actual survey fieldwork, and contain as much information as possible to minimize clerical work by field personnel. The source of each sample is also written on the container itself prior to any field survey work.

Field logsheets used for documenting field procedures and chain-of-custody, and to identify samples should be prefilled to the extent practicable to minimize repetitive clerical field entries. Custody during sampling is maintained by the sampler or project leader through the use of the logbook. Any information from previous studies should be copied (or removed) and filed before the book is returned to the field.

Explicit chain-of-custody procedures are followed to maintain the documentation necessary to trace sample possession from the time taken until the evidence is introduced into court. A sample is in your *custody* if:

- It is in your actual physical possession; or
- It is in your view, after being in your physical possession; or
- It was in your physical possession and you locked it in a tamper-proof container or storage area

All survey participants should receive a copy of the study plan and be knowledgeable of its contents prior to the survey. A presurvey briefing should be held to reappraise all participants of the survey objectives, sample locations and chain-of-custody procedures. After all chain-of-custody samples are collected, a debriefing should be held in the field to check adherence to chain-of-custody procedures, and to determine whether additional evidence samples are required.

SAMPLE COLLECTION

1. To the maximum extent achievable, as few people as possible handle the sample.

2. Water samples are obtained using standard field sampling techniques. When using sampling equipment, it is assumed that this equipment is in the custody of the entity responsible for collecting the samples.

3. The chain-of-custody record tag is attached to the sample container, when the complete sample is collected, and contains the following information: Sample number, time taken, date taken, source of sample (to include type of sample and name of firm), preservative, analyses required, name of person taking sample, and witnesses). The front side of the card (which has been prefilled) is signed, timed, and dated by the person sampling. The tags must be legibly filled out in ballpoint (waterproof) ink. Individual sample containers of groups of sample containers are secured using a tamper-proof seal.

4. Blank samples are also taken. Include one sample container without preservative and containers with preservatives, from all of which the contents will be analyzed by the laboratory to exclude the possibility of container contamination.

5. The Field Data Record logbook should be maintained to record field measurements and other pertinent information necessary to refresh the sampler's memory, if he later takes the stand to testify regarding his actions during the

evidence-gathering activity. A separate set of field notebooks should be maintained for each survey and stored in a safe place, where they can be protected and accounted for at all times. Standard formats have been established to minimize field entries and include the date, time, survey, type of sample taken, volume of each sample, type of analysis, sample number, preservatives, sample location, and field measurements. Such measurements include temperature, conductivity, dissolved oxygen (DO), pH, flow, and any other pertinent information or observations. The entries are signed by the field sampler. The preparation and conservation of the field logbooks during the survey is usually the responsibility of the survey coordinator. Once the survey is complete, field logs should be retained by the survey coordinator, or his designated representative, as a part of the permanent record.

6. The field sampler is responsible for the care and custody of the samples collected until properly dispatched to the receiving laboratory or turned over to an assigned custodian. He should assure that each container is in his physical possession or in his view at all times, or is locked in such a place and manner that no one can tamper with it.

7. Colored slides or photographs are often taken which show the outfall sample location and any visible water pollution. Written documentation on the back of the photo should include the signature of the photographer, time, date, and site location. Photographs of this nature, which may be used as evidence, are handled by chain-of-custody procedures to prevent alteration.

In addition to the above recommendations on sample collection, the EPA (U.S. EPA, 1975) provides procedures on transfer of custody and shipment. A section is also included which presents the EPA's National Field Investigation Center's Laboratory Custody Procedures. These procedures should be closely followed, primarily in compliance monitoring.

QUALITY CONTROL

Decisions made using groundwater data are far reaching. Water quality standards are set to establish satisfactory conditions for a given water use. The laboratory data define whether that condition is being met, and whether the water can be used for its intended purpose. If the laboratory results indicate a violation of the standard, action is required on the part of pollution control authorities. With the present emphasis on legal action and social pressures to abate pollution, the field worker and analyst should be aware of their responsibility to provide results that are a reliable description of the sample. The technician must be aware that his professional competence, the procedures he has used, and the reported values may be used and challenged in court. To satisfactorily meet this challenge, the data must be backed up by an adequate program to document the proper control and application of all of the factors which affect the final result.

Because of the importance of laboratory analyses and the resulting actions which they produce, a program to insure the reliability of the data is essential. It is recognized that all analysts practice quality control to varying degrees, depending somewhat upon their training, professional pride, and awareness of the importance of the work they are doing. However, under the pressure of daily workload, analytical quality control may be easily neglected. Therefore, an established, routine control program applied to every analytical test is necessary in assuring the reliability of the final results.

The need for standardization of methods within a single laboratory is readily apparent. Uniform methods between cooperating laboratories are also important in order to remove the methodology as a variable in comparison or joint use of data between laboratories. Uniformity of methods is particularly important when laboratories are providing data to a common data bank, such as storage and retrieval (STORET), or when several laboratories are cooperating in joint field surveys. A lack of standardization of methods raises doubts as to the validity of the results reported. If the same constituent is measured by different analytical procedures within a single laboratory, or in several laboratories, the question is raised as to which procedure is superior, and why the superior method is not used throughout.

The physical and chemical methods used should be selected by the following criteria:

- The method should measure the desired constituent with precision and accuracy sufficient to meet the data needs in the presence of the interferences normally encountered in polluted waters.
- The procedure should utilize the equipment and skills normally available in the average water pollution control laboratory.
- The selected methods should be in use in many laboratories, or have been sufficiently tested to establish their validity.
- The method should be sufficiently rapid to permit routine use for the examination of large numbers of samples.

The use of EPA methods in all EPA laboratories provides a common base for combined data between Agency programs. Uniformity throughout EPA lends considerable support to the validity of the results reported by the Agency.

Regardless of the analytical method used in the laboratory, the specific methodology should be carefully documented. In some water pollution reports it is customary to state that *Standard Methods* (APHA, 1971) have been used throughout. Close examination indicates, however, that this is not strictly true. In many laboratories, the standard method has been modified because of recent research or personal preferences of the laboratory staff. In other cases the standard method has been replaced with a better one. Statements concerning the methods used in arriving at laboratory data should be clearly and honestly made. The methods used should be adequately referenced, and the procedures applied exactly as directed.

Knowing the specific method which has been used, the reviewer can apply the associated precision and accuracy of the method, when interpreting the laboratory results. If the analytical methodology is in doubt, the data user may honestly inquire as to the reliability of the result he is to interpret.

In field operations, the problem of transport of samples to the laboratory, or the need to examine a large number of samples to arrive at gross values will sometimes require the use of rapid field methods. Such methods should be used with caution, and with a clear understanding that the results obtained may not compare in reliability with those obtained using standard laboratory methods. The data user is entitled to know that approximate values may have been obtained for screening purposes only, and that the results may not represent the customary precision and accuracy obtained in the laboratory.

SOIL

PHYSICAL ANALYSES
Water Content

The water content of field samples is commonly obtained by gravimetry, with oven drying (Gardner, 1965). The sample is placed in a drying oven at about 105 °C. The difference in mass before and after drying represents the water content on a dry mass basis. (Gardner also describes a field method for determining water content using gravimetry, with drying by burning alcohol.)

The cost of a soil-moisture drying oven varies between $300 and $450. The cost of conducting a water content analysis varies from about $4 to $6 per sample. Figure 3-28 gives the cost of multiple water content determinations including a discount rate.

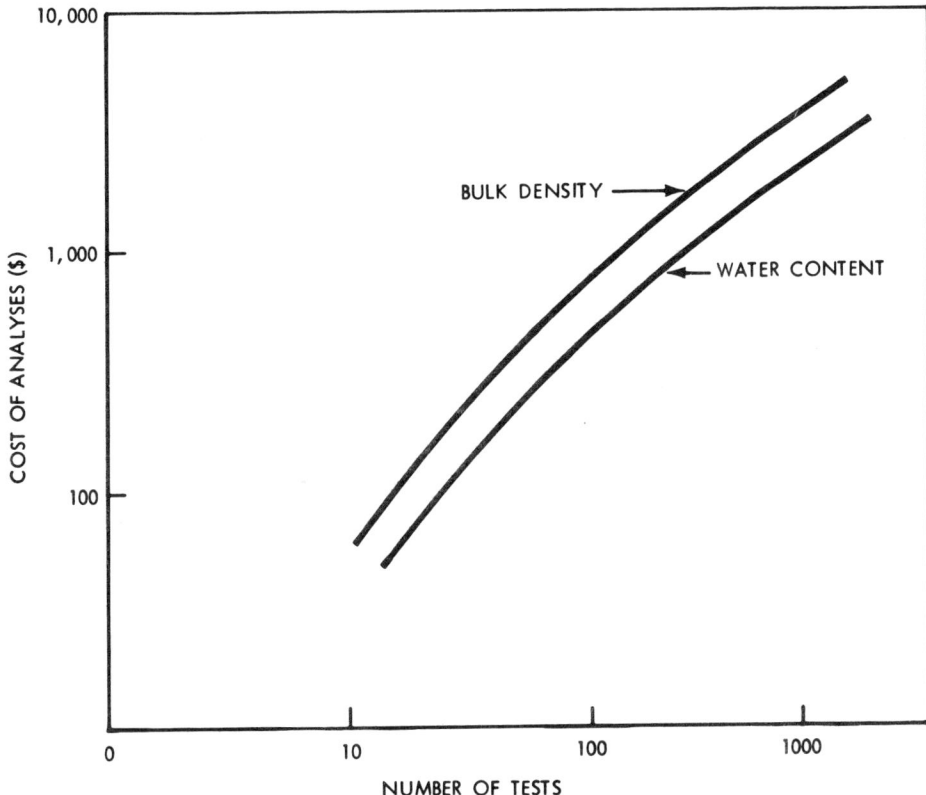

Figure 3-28. Costs for determinations of water content and bulk density in soils, September 1974.

Bulk Density and Porosity

Bulk density is used to calculate soil porosity. Also, together with water content data, bulk density may be used to estimate the amount of water to bring the soil to saturation. Soil cores obtained in the field are oven dried at 105 °C, until constant weight is attained. The bulk density of the sample is the over-dry mass divided by the volume. Blake (1965) presents alternative methods to calculate bulk density, including determination in place by filling the soil cavity with a measured volume of sand. Blake also discusses a double-probe technique for measuring bulk density in situ via gamma radiation.

The cost of a bulk density determination is around $6 to $10 per sample; however, discount rates may be quite high for these determinations. The cost of multiple bulk density tests including a discount rate is given in Figure 3-28.

Particle-Size Distribution

Information on particle-size distribution of drilling samples provides explicit data on the vertical distribution of gravels, sands, silts, and clays. Such data may subsequently be related to gamma logs. According to Day (1965), fractionation and particle-size analyses mainly involve sieving and sedimentation procedures.

Sieving is used to separate particles coarser than 0.05 millimeters. In practice, a sample is prepared in standard fashion and passed through a net of sieves, with openings graded to permit separation of various particle sizes. Either wet sieving or dry sieving is used.

Sedimentation is used principally to determine the percentage of clay in a sample. This fraction is highly active in ion exchange and adsorption. The presence of layers with high clay levels tends to cause lower relative permeability. Sedimentation is accomplished by either the pipette or hydrometer method. Results are usually presented on a curve which plots the percentage of particles, by weight, smaller than a given size, versus the logarithm of the particle size.

The basic equipment used in particle size analysis is a set of sieves and a sieve shaker. Each sieve costs about $12 to $16. A full set of U.S. Standard or Tyler series can cost as much as $225. However, reduced numbers of sieves can be used for soil particle size analysis. The USDA-recommended set of soil sieves costs about $60. The average cost of a mechanical shaker is about $300. Under specialized conditions it may be necessary to determine the sedimentation fraction that passes through even the finest sieves.

The cost of a particle size analysis can vary depending upon the particle size range of interest. The average cost of an analysis of gravel through the clay sizes is $25; the cost of an analysis is $40 if less than a 25 gram sample is available. The normal particle size analysis calls for a sand and gravel determination and this costs about $8 to $10. If an analysis of clay sizes only is required, the cost is about $20 per sample. Volume discount rates may be important in the costs of these tests.

Soil-Water Characteristic Curve

A knowledge of the water content versus pressure relationships of a soil is essential in calculating hydraulic conductivity and the rate of soil-water movement. A soil-water characteristic curve is needed to transform water content data from neutron logging into equivalent soil pressures.

The modified Haines method may be used to obtain the soil-water characteristic curve up to suctions of about 1 atmosphere. This method employs the equipment shown in Figure 3-29. Soil cores are carefully placed on the fritted glass plate and saturated (Nielsen et al., 1973). Excess water is removed, and the hanging water column is adjusted through a desired range of negative pressure. A plot of water content versus soil pressure during a drying cycle, which occurs during the Haines method, is usually different from the curve during wetting, as evidenced in Figure 3-30. This hysteresis phenomenon must be kept in mind during interpretation of field data. Since the fritted glass membrane usually fails at about 0.8 atmosphere negative pressure, other methods are required to obtain the water content versus soil pressure relationship for greater

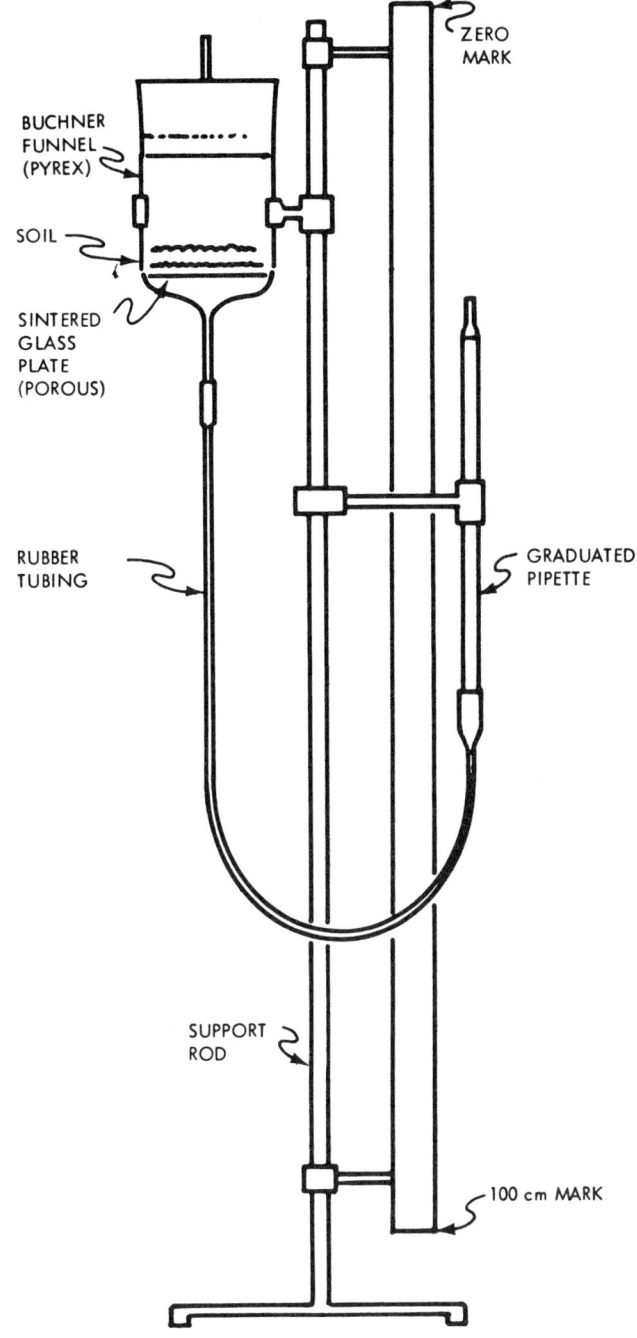

Figure 3-29. Modified Haines Apparatus for obtaining soil-water characteristic curves (after Day et al., 1967).

suction. Klute (1965) describes one such method (the pressure-membrane method), which involves applying air pressure to a soil sample in a weighable cell.

Estimates on the water content of a soil at field capacity may be obtained by the so-called one-third atmosphere technique (Peters, 1965). Alternatively, field capacity can be estimated in situ. A square plot of about 3 meters per side is prepared with a raised border to permit ponding. Water is applied until the desired soil depth is wetted.

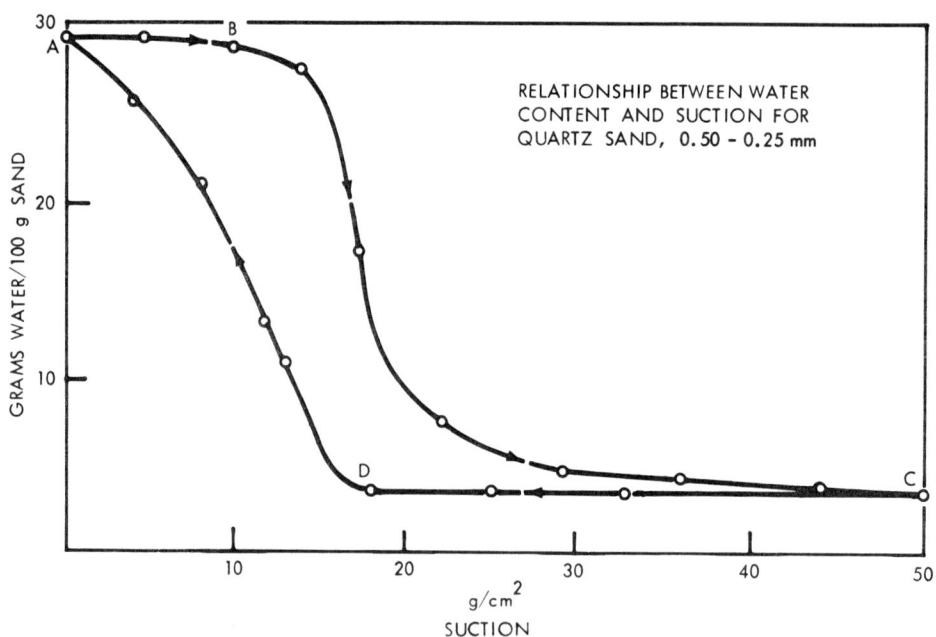

Figure 3-30. Soil-water characteristic curve, showing hysteresis. ABC = drying, CDA = wetting (after Day et al., 1967).

After recession of the water, the soil surface is covered with polyethylene sheeting to prevent evaporation. After allowing about two days for drainage, soil samples are obtained for gravimetric determination of water content. Incidentally, the test pond also can be used to estimate infiltration rates into the soil.

The cost of developing a soil-water characteristic curve is a function of the number of tests to be run. These tests are not routinely run in many laboratories. The average cost of developing a soil-water characteristic is about $30 to $40 per sample. Discounts of over 30 percent can be expected for large numbers of samples. A series of 100 tests, for example, would cost about $2000.

Hydraulic Conductivity
Core samples obtained from the field may be used to evaluate both the saturated and unsaturated hydraulic conductivity for the soil depth sampled. For the saturated case a permeameter may be used, in which a constant positive head is applied to a soil sample, the discharge rate is measured, and the hydraulic conductivity is calculated.

Specific Surface
The cation exchange capacity and specific surface together govern the sorption characteristics of a soil. Unfortunately, the mechanisms for adsorptions directly onto the surface of soil colloids are imperfectly understood.

Mortland and Kemper (1965) discuss the principles of adsorption, relating to the specific surface of clays. In addition, a method for determining specific surface based on sorption of ethylene glycol is presented. Specific surface determinations are not routinely done, and require a specific cost quote.

In addition to the chemical analyses, further soil testing can include a clay mineral identification that costs about $30 per sample. A simple mineral identification costs about $5, while a mineral separation and identification can cost $25. A trace element determination such as zinc, manganese, iron, or copper costs about $5.

The costs of soil-water chemical tests are approximately the same as the costs of water sampling tests.

The range of tests that are performed in a hydrologic laboratory is varied and, usually, quite unfamiliar to the field technician. For this reason an example test schedule and cost quote is provided in Table 3-12.

Unsaturated hydraulic conductivity may also be measured in the laboratory using, for example, the steady-state method of Klute (1965). Soil is placed in a cell containing a porous plate and two tensiometers. After saturating the soil, air pressure is applied, the discharge rate of water leaving the soil is measured, and the tensiometers are read. The hydraulic conductivity at the equivalent negative pressure is calculated from Darcy's equation. With this method a curve may be plotted depicting the relationship between conductivity and negative pressure head. Using a soil-water characteristic curve, the conductivity-pressure relationship may also be obtained.

Special equipment is required to conduct a hydraulic conductivity test in the saturated state. The cost can vary, with the effective stress required, from $30 to $50 for each test. These tests are also subject to large discount rates. The expected cost for 100 tests is about $2400.

CHEMICAL ANALYSIS
Soluble Ions

Soluble salts refer to the soil constituents of an inorganic nature that are appreciably soluble in water. The following steps are usually taken in an analysis for soluble salts: (1) preparing a sample by bringing the soil-water content to some prescribed value; (2) extracting the sample; and (3) measuring or analyzing the salt content of the extract.

The soil-water content at which the salt is extracted is a matter of concern. The saturation extract, for example, is a technique used by agriculturists to relate soluble salts to field moisture range. A problem with diluting the sample, as done by the saturation extract, is that while the concentration of some ions decreases on dilution, the concentration of others may increase (Reitemeier, 1946). For example, as water is added to the soil, precipitated calcium carbonate or gypsum may gradually dissolve, releasing calcium, and possibly magnesium, ions into solution. Concurrently, the additional concentrations of calcium and magnesium may displace sodium on the exchange complex, thereby increasing the sodium level in the soil solution. Overall, calcium, magnesium, and sodium may increase in the soil solution. In contrast, chloride and nitrate may decrease in concentration because of dilution and negative adsorption. Reitemeier (1946) recommended that for arid land soils, dissolved ions should be determined at or near the water content at which the results are to be applied. Consequently, the soil solutes of samples from deeper horizons of the vadose zone should be extracted at the prevailing soil-water pressure.

Pressure membrane apparatus may be used to extract soil solution samples in the dry range (Richards, 1954). Thus, if a soil-water characteristic is available for the sample, together with gravimetric water content data, the equivalent soil-water pressure can be determined. This pressure is then applied to the soil sample in the membrane. A problem with this technique is that solution is extracted from the larger

Table 3-12

HYDROLOGIC LABORATORY TESTS AND COST SCHEDULE, SEPTEMBER 1974

Hydrologic Laboratory Tests	Soil Analyses Costs ($)
Dry Unit Weight	6.00
Electrical Resistivity	20.00
Formation Factor (includes determination of conductivity of interstitial water)	25.00
Hydraulic Conductivity	
Saturation flow, undisturbed samples in standard cylinders at effective stresses 90 psi; tested with formation water	
Vertical flow	30.00
Horizontal flow	35.00
Nonstandard undisturbed samples, vertical or horizontal flow	40.00
Saturated flow, undisturbed samples at simulated overburden loads to 5000 psi and pore pressures to 3000 psi, vertical or horizontal flow; tested with formation water	40.00
Unsaturated flow, undisturbed samples (determination of unsaturated hydraulic conductivity as a function of moisture content over a range of megative heads to -2 atmospheres)	75.00
Moisture Content (as received)	6.00
Moisture-Tension Curve (water content as a function of negative head, by the pressure-plate method)	40.00
Particle-Size Analysis	
Complete (gravel through clay sizes)	25.00
Complete, if <25-gram sample available	40.00
Sand and gravel sizes only	10.00
Clay sizes only (centrifuge method)	20.00
Permeability; single-phase flow of gas or liquid other than water; undisturbed samples	40.00
Pore-Water Extraction -- centrifuge, immiscible displacement, or high-pressure (20,000 psi) squeeze methods	40.00
Porosity and *Specific Yield*	
Total porosity (gravimetric method: includes granular specific gravity and dry unit weight)	12.00
Pore-size distribution curve -- cohesive samples	
Mercury injection method: includes simulated moisture-tension curve, approximate *specific yield*, effective porosity at 2000 psi Hg pressure, total porosity, specific gravity, dry unit weight	35.00
Pore-size distribution curve -- friable samples	
Moisture-tension method, see above: includes *specific yield*, effective porosity at 2 atmospheres moisture tension, total porosity, specific gravity, dry unit weight	50.00
Specific Yield (not including porosity)	
Direct determination, using moisture-tension curve	
Indirect methods	
Centrifuge moisture equivalent (suitable for sands only)	25.00
Mercury-injection method, cohesive samples only (injected porosities at 30 to 50 psi typically correspond to the approximate range of *specific yield*	20.00
Thermal Conductivity	25.00
Specific Gravity (grain density)	8.00
Sample preparation and equipment modification for above tests, if required by unusual sample conditions or test requirements	8.00/hr.

pores, which may not have the same chemical composition as that in the smaller interstices (Rhoades and Bernstein, 1971).

A soil moisture extractor that uses the pressure membrane technique costs about $400. These units, however, require a compressor that costs about $650 and an independent pressure control manifold that costs from $400 to $800, depending upon the number of outlets required.

For surface soils, other factors should be examined in the determination of soluble salts. For irrigated soils, or land disposal operations subject to wetting and drying, the water content of soils may range over large values because of evaporation, so that conversion to the field water content value may not be meaningful (Pratt, 1972). In this case, the saturation extract technique is recommended (Rhoades and Bernstein, 1971). Water content by this method represents about twice that at field capacity; therefore, the salt content extracted from the saturated sample is about one-half the concentration at field capacity. Bower and Wilcox (1965) explain in detail the procedure to obtain a saturated extract.

If detailed information is required on specific ionic constituents, chemical analyses using procedures presented in *Methods of Soil Analysis* (Black, 1969) may be used. The principal ions of importance in most soil studies are: calcium (Ca^{++}), magnesium (Mg^{++}), potassium (K^+), sodium (Na^+), carbonate (CO_3^-), bicarbonate (HCO_3^-), sulfate (SO_4^-), chloride (Cl^-), and boron (B), as well as the nitrogen (N) series of ions. For soluble ions, the procedures given in the water analyses section are applicable.

The cost of an electrical conductivity test on a soil sample is about $5 to $6. The cost of conducting 10, 100, and 1000 tests is about $60, $500, and $4000, respectively. The cost of determining soluble cations such as sodium, potassium, calcium, and magnesium, is about $3 for each chemical parameter tested. To obtain the pH of soil, a saturation extract costs about $2 to $3.

Cation Exchange Capacity
Methods for determining the sum of individual exchangeable ions of a sample are given in Black (1969). This sum is equal to the cation exchange capacity (CEC). Alternatively, the CEC may be obtained directly using methods detailed by Chapman (1965). Briefly, the exchangeable cations in a soil sample are replaced by either ammonium acetate or sodium acetate, and the amounts of ammonium and sodium ions adsorbed are determined. A problem may develop with the use of ammonium ions because this ion becomes strongly adsorbed on some clays. Cation exchange capacity is expressed as milliequivalents (meq) per 100 grams of sample.

The cost of determining the total cation exchange capacity is about $11. The cost of determining the sodium, potassium, magnesium, and calcium exchangeable cations is about $12. The cost of conducting a base exchange analysis, which includes pH of saturated soil paste; electrical conductivity; boron in soil extract; total cation exchange capacity; exchangeable percent of exchange capacity for calcium requirement; nitrates; bicarbonate exchangeable phosphorous; and exchangeable potassium, is about $20 per sample.

WATER

In addition to water sampling techniques and field determinations, the sample containers, preservation technique, and method of analysis are key factors in monitoring groundwater quality. The U.S. Environmental Protection Agency has published *Manual* of *Methods for Chemical Analysis of Water and Wastes* (1974a) and *Methods for Organic Pesticides in Water and Wastewater* (1971). The U.S. Geological Survey has published *Methods of Collection and Analysis of Water Samples* (Rainwater and Thatcher, 1960), and is currently updating and expanding this coverage through its series on *Techniques of Water Resources Investigations, Book 5, Laboratory Analysis* (Barnett and Mallory, 1971; U.S. Geological Survey, 1973).

The American Public Health Association (1971) has published the 13th Edition of *Standard Methods for the Examination of Water and Wastewater.* The major portions treating groundwater pollution monitoring are: Part 100 – "Physical and Chemical Examination of Natural and Treated Waters in the Absence of Gross Pollution;" Part 200 – "Physical, Chemical, and Bioassay Examination of Polluted Waters, Wastewaters, Effluents, Bottom Sediments, and Sludges;" Part 300 – "Examination of Water and Wastewater for Radioactivity;" and Part 400 – "Bacteriological Examination of Water to Determine its Sanitary Quality."

Water testing laboratories can be found at Federal, State, county, and municipal levels. Universities and private companies also provide analysis capabilities. The cost of each analysis varies greatly with the degree of automation, the number and kinds of tests required, and the accuracy of the tests. The Corps of Engineers, the EPA, and other agencies require a certain sensitivity for the tests performed. The costs increase with the sensitivity of the analysis. The costs presented in this section for water-quality analysis represent an average cost for commercial laboratory services. The costs do not include a sample preparation fee, which can vary from $3 to $25. Most commercial laboratories allow a 10 percent discount for invoices over $500 per month, and a 20 percent discount for invoices of $1000 per month or more. In laboratory selection, factors such as location relative to the sampling area, time to complete analyses, and quality assurance programs must be considered. In cost of analysis determinations, special considerations should be given for group rates such as the following October 1974 values:

Major Inorganic Chemicals Group Rate $12		Other Inorganic Chemicals Group Rate $40	
Calcium	Sulfate	Silica	Nitrogen forms
Magnesium	Chloride	Boron	Phosphorous forms
Sodium	Nitrate	Fluoride	Hardness
Potassium	Total dissolved		
Carbonate	solids		
Bicarbonate	pH		
Electrical conductivity			

Drinking Water Trace Elements Group Rate $50		Other Trace Elements Group Rate $55	
Iron	Copper	Vanadium	Aluminum
Manganese	Cyanide	Molybdenum	Cobalt
Arsenic	Lead	Bromide	Lithium
Barium	Selenium	Iodide	Sulfide
Cadmium	Silver	Nickel	Beryllium
Hexavalent chromium	Zinc		

<u>Gases</u>
Group Rate $30

Methane   Carbon dioxide
Hydrogen sulfide   Dissolved oxygen
Residual Chlorine

<u>Insecticides</u>   <u>Herbicides</u>
Group Rate $32   Group Rate $45

Alopin   Dieldrin   2, 4-D   2, 4, 5-T
Chlordane   Endrin   Silvex
DDD   Heptachlor
DDE   Heptachlor Epoxide
DDT   Lindane
 Toxaphene

<u>Radiochemical</u>
Group Rate $100

Gross Alpha, Beta   Radium-226 by Radon
dissolved and suspended   Uranium Fluorometric

DETERMINATIONS

The major constituents of water with respect to groundwater pollution may be grouped into the following physical, inorganic chemical, organic chemical, bacteriological, and radiological classifications:

<u>Physical</u>

Temperature   Odor
Density   Turbidity
 Color

<u>Inorganic Chemical</u>

Major constituents   Trace elements
Other constituents   Gases

<u>Organical Chemical</u>

Carbon   Nitrogen
Biochemical oxygen   Chemical oxygen demand
  demand   Phenolic materials
Methylene blue active   Insecticides and herbicides
  substances

<u>Bacteriological</u>

Coliform group   Pathogenic microorganisms
 Enteric viruses

Radiological

Gross alpha activity	Strontium
Gross beta activity	Radium
Tritium	

Significant removals of the organic chemical, bacteriological, and radiological constituents of percolated wastes ordinarily occur in the topsoil and vadose zone.

PHYSICAL

Measurements of water temperature and notes concerning odor are often made at the sampling point. Density and turbidity determinations are applicable only in special situations. Density is important in the case of disposal of fluids of high density compared to native groundwater, and turbidity is important in well disposal or injection programs.

The cost of a density determination is about $5, and the cost of a turbidity test is about $6. The cost of odor and color determinations are about $1 to $5, respectively.

INORGANIC CHEMICAL

The primary divisions of inorganic chemical constituents with respect to groundwater pollution are the following:

1. Major – the major constituents of groundwater

2. Other – other common constituents of importance in water use and waste disposal

3. Drinking Water Trace – trace constituents of concern for drinking water quality

4. Other Trace – trace constituents of concern in waste disposal

5. Gases – common gases in groundwater and in, or generated by wastes

The constituents within these divisions are:

Major

Calcium	Sulfate
Magnesium	Chloride
Sodium	Nitrate
Potassium	Total dissolved solids
Carbonate	pH
Bicarbonate	Electrical conductivity

Other

Silica	Nitrogen forms
Boron	Phosphorus forms
Fluoride	Hardness

Drinking Water Trace

Iron	Copper
Manganese	Cyanide
Arsenic	Lead
Barium	Selenium
Cadmium	Silver
Hexavalent chromium	Zinc

Mercury

Other Trace

Vanadium	Aluminum
Molybdenum	Cobalt
Bromide	Lithium
Iodide	Sulfide
Nickel	Beryllium

Gases

Methane	Carbon dioxide
Hydrogen sulfide	Dissolved oxygen

Residual chlorine

The procedures for sample collection, treatment, and analytical determination vary widely from wastewaters and effluents to relatively unpolluted or natural groundwaters. This basic difference is recognized in APHA's *Standard Methods* (1971), wherein a major subdivision occurs between natural waters and polluted waters. These differences arise because of factors such as interference effects and biologic activity. In general, wastewaters, particularly those of industrial origin, may often require specialized sample collection and analytical procedures.

Containers

Factors that are pertinent in selecting containers for collecting and storing water samples are resistance to solution and breakage, efficiency of closure, size, shape, weight, availability, and cost. Hard rubber, polyethylene, teflon, and other types of plastics, and some types of borosilicate glass are suitable based on experience within the U.S. Geological Survey and other agencies. Glass bottles may be a problem for analysis of boron, silica, sodium, and hardness. For dissolved oxygen determinations, only glass containers should be used. For silica determinations, only plastic containers should be used.

Before use, all new bottles should be thoroughly cleansed, filled with water, and allowed to soak several days. The soaking removes much of the water-soluble material from the container surface. The source of the sample and conditions under which it was collected should be recorded immediately after collection. In the case of wells, this should include pumping rate, duration of pumping if known, water level, temperature of water, and electrical conductivity. Samples from wells near pollution sources should be accompanied by a description of local conditions, such as "percolation pond empty."

Narrow-mouth, 250-milliliter polyethylene bottles cost about 30 to 40 cents each. The cost of a comparable size glass bottle is about 20 to 30 cents. If the bottles are purchased by the case, the discount rate is about 10 percent for 1 to 4 cases,

15 percent for 5 to 19 cases, and 20 percent for 20 or more cases. Shipping sleeves for use with plastic bottles cost about $18 per 100 sleeves. Glass mailers cost about $35 for a carton of 75.

Preservation of Water and Waste Samples

EPA's *Manual of Methods for Chemical Analysis of Water and Wastes* (U.S. EPA 1974a) is a basic reference for monitoring water and wastes in compliance with the requirements of the Federal Water Pollution Control Act Amendments of 1972. Included is a detailed discussion of sample preservation techniques.

Preservation techniques can only retard the chemical and biological changes that inevitably continue after the sample is taken in the field. Certain changes occur in the chemical structure of the constituents that are a function of physical conditions. Metal cations may precipitate as hydroxides, or form complexes with other constituents; cations or anions may change valence states under certain reducing or oxidizing conditions; and other constituents may dissolve or volatilize with the passage of time. Metal cations may also adsorb onto surfaces (glass, plastic, quartz, among others). Biological changes taking place in a sample may change the valence of an element. Soluble constituents may be converted to organically bound materials, or cellular material may be released into solution.

Methods of preservation are relatively limited, and are intended to (1) retard biological action, (2) retard hydrolysis of chemical compounds and complexes, and (3) reduce volatility of constituents. Preservation methods are usually limited to pH control, chemical addition, refrigeration, and freezing. Refrigeration at temperatures near freezing or below is the best preservation technique available, but it is not applicable to all types of samples. The preservative measures recommended by the EPA (U.S. EPA, 1974a) are given in Table 3-13. When the dissolved concentration is to be determined, the sample is filtered immediately after collection through a 0.45-micron membrane filter, and the filtrate is analyzed by the specified procedure. Specific techniques for monitoring wastewater are given in the EPA's *Handbook for Monitoring Industrial Wastewater* (U.S. EPA, 1973), and American Public Health Association (1971), Part 200.

Sample Collection and Treatment for Groundwaters

Brown et al. (1970) present data that are applicable to groundwater sampling. An analysis will more closely represent the water at the time of collection if separate samples are taken for determination of certain groups of ions. The test procedure should determine whether a given analytical determination should be performed on a settled, filtered, or well-mixed sample.

On-site determinations are mandatory for temperature, pH, dissolved oxygen, and specific conductance. For laboratory determinations the following procedures are recommended by Brown et al.:

> One sample of about 1 liter volume should be collected and immediately filtered through a 0.45-micron membrane filter. The following determinations are made on the filtrate: boron, chloride, fluoride, lithium, nitrate, nitrite, dissolved solids, dissolved phosphorous, potassium, selenium, silica, sodium, and sulfate.
>
> A second sample of about 2 liters should be collected and immediately filtered through a 0.45-micron membrane filter. The filtrate should be acidified with double-distilled, reagent-grade nitric acid to a pH of 3 or less.

Table 3-13

RECOMMENDED SAMPLING AND PRESERVATION TECHNIQUES
FOR INORGANIC CHEMICAL DETERMINATIONS

Measurement	Volume (ml)	Preservative	Holding time
Arsenic	100	HNO_3 to pH<2	6 months
Bromide	100	Cool to 4°C	24 hours
Chloride	50	None required	7 days
Cyanide	500	Cool to 4°C NaOH to pH 12	24 hours
Dissolved oxygen	300	On-site determ.	none
Fluoride	300	Cool to 4°C	7 days
Hardness	100	Cool to 4°C	7 days
Iodide	100	Cool to 4°C	24 hours
Metals, dissolved	200	Filter on site HNO_3 to pH<2	6 months
Metals, total	100	HNO_3 to pH<2	6 months
Mercury, dissolved	100	Filter HNO_3 to pH<2	38 days (glass) 13 days (hard plastic)
Mercury, total	100	HNO_3 to pH<2	38 days (glass) 13 days (hard plastic)
Ammonia nitrogen	400	Cool to 4°C H_2SO_4 to pH<2	24 hours
Nitrate nitrogen	100	Cool to 4°C H_2SO_4 to pH<2	24 hours
Nitrite nitrogen	50	Cool to 4°C	24 hours
pH	25	Cool to 4°C On-site determ.	6 hours
Dissolved orthophosphate	50	Filter on site Cool to 4°C	24 hours
Hydrolyzable phosphorus	50	Cool to 4°C H_2SO_4 to pH<2	24 hours
Total phosphorus	50	Cool to 4°C	24 hours

(continued)

Table 3-13 (Continued)

Measurement	Volume (ml)	Preservative	Holding time
Total dissolved phosphorus	50	Filter on site Cool to 4°C	24 hours
Filterable residue	100	Cool to 4°C	7 days
Non-filterable residue	100	Cool to 4°C	7 days
Total residue	100	Cool to 4°C	7 days
Volatile residue	100	Cool to 4°C	7 days
Selenium	50	HNO_3 to pH<2	6 months
Silica	50	Cool to 4°C	7 days
Specific conductance	100	Cool to 4°C	24 hours
Sulfate	50	Cool to 4°C	7 days
Sulfide	50	2 ml zinc acetate	24 hours
Sulfite	50	Cool to 4°C	24 hours

The following determinations are made on the acidified filtrate: aluminum, arsenic, barium, cadmium, calcium, chromium, cobalt, copper, iron, lead, lithium, magnesium, manganese, molybdenum, nickel, potassium, silver, sodium, strontium, vanadium, and zinc.

A third sample of about 1 liter should be collected and not filtered. The following determinations may be made on this sample: ammonia, organic nitrogen, chemical oxygen demand, cyanide, phosphorus, and turbidity.

Collection of additional samples and individual treatment may be necessary for the nitrogen forms, cyanide, dissolved oxygen, organic phosphorus, and sulfide.

Selection of Specific Determinations

The major cations and anions should usually be determined in a water analysis, along with pH, electrical conductivity, and total dissolved solids at 25 °C. The major cations are Ca^{++}, Mg^{++}, Na^+, and K^+. In areas of polluted groundwater, ammonium ion (NH_4^+) may also be a significant cation. The major anions are CO_3^-, HCO_3^-, Cl^-, SO_4^-, and possibly NO_3^-. Another significant anion may be fluoride (F^-). If the major constituents are determined, the cation-anion balance in equivalents can be checked. Similarly, the calculated total dissolved solids can be compared with the residue. Silicate (SiO_2) is an undissociated form found in many groundwaters that should be determined prior to calculation of total dissolved solids. The electrical conductivity at 25 °C can be compared to the total dissolved solids. The chemical analysis can thus be checked several ways for internal consistency. The determination of major ions also permits interpretation by the use of trilinear diagrams, as well as the application of physical chemistry to groundwater quality problems. The cost of analyses of inorganic chemicals in water is given in Table 3-14. Although costs for dissolved chemicals are of primary interest in groundwater studies, the cost of analysis, which includes suspended chemicals (total), are also given for surface wastewater studies.

Table 3-14

COSTS OF INORGANIC CHEMICAL DETERMINATIONS
FOR GROUNDWATER POLLUTION,[a] OCTOBER 1974

Major parameters (mg/ℓ)	Water analysis cost ($)		
	o.x	o.ox	o.oox
Calcium dissolved	5		
Calcium total	10	12	
Magnesium dissolved	5		
Magnesium total	5	10	20
Sodium dissolved	5		
Sodium total	5	7	10
Potassium dissolved	5		
Potassium total	8	10	15
Carbonate	10		
Bicarbonate	5		
Sulfate	5		
Chloride	5		
Nitrate	2	5	
Dissolved solids, fixed	5		
Dissolved solids, total	1		
Dissolved solids, volatile	5		
pH	3		
Electrical conductivity	3		
Others (mg/ℓ)			
Silica	5	10	15
Boron dissolved	10		
Boron total	15	17	20
Fluoride dissolved	10		
Fluoride total	15	20	
Nitrogen dissolved KJD	10		
Nitrogen NH_4 as N total	5		
Nitrogen NH_4 as N dissolved	5		
Nitrogen NO_2 as N dissolved	5		
Nitrogen NO_2 as N total	5	8	
Nitrogen NO_3 as N dissolved	5	8	
$NO_2 + NO_3$ as N dissolved	5		
$NO_2 + NO_3$ as N total	5		
Nitrogen total KJD as N	10	11	
Phosphate dissolved Ortho as PO_4	3	5	
Phosphate Ortho dissolved as P	3		
Phosphate Ortho total as P	3	5	
Phosphate dissolved as P	5		
Phosphate total HYD as P	10		

(continued)

Table 3-14 (Continued)

Others (mg/ℓ)	Water analysis cost ($)		
	o.x	o.ox	o.oox
Phosphate total as P	10		
Hardness, calcium	3		
Hardness, carbonate	3		
Hardness, magnesium	3		
Hardness, non-carbonate	1		
Hardness, total	3		
Drinking water trace (mg/ℓ)			
Iron	5		
Iron dissolved	5		
Iron +2 and +3	5		
Iron ferrous	5		
Iron total	10	12	15
Manganese	5		
Manganese dissolved	5		
Manganese total	5	7	10
Arsenic dissolved	10		
Arsenic total	15	17	20
Boron dissolved	10		
Boron total	10	12	15
Cadmium dissolved	10		
Cadmium total	10	12	15
Chromium dissolved	8		
Chromium hexavalent	8	9	10
Chromium total	8	9	10
Copper dissolved	5		
Copper total	5	7	10
Cyanide	10	15	
Lead dissolved	10		
Lead total	13	14	15
Selenium dissolved	15		
Selenium total	15	17	20
Silver dissolved	10		
Silver total	10	12	15
Lime dissolved	5		
Lime total	10	12	15
Mercury dissolved	10		
Mercury total	10	12	15

(continued)

Table 3-14 (Continued)

Other traces (mg/ℓ)	Water analysis cost ($)		
	o.x	o.ox	o.oox
Vanadium dissolved	10		
Vanadium total	10	15	20
Molybdenum dissolved	10		
Molybdenum total	10	12	15
Bromine	10	15	
Iodine	10	15	
Nickel dissolved	10		
Nickel total	10	12	15
Aluminum dissolved	10		
Aluminum total	10	15	20
Cobalt dissolved	10		
Cobalt total	10	12	15
Lithium dissolved	12		
Lithium total	12	14	15
Sulfide dissolved	5		
Sulfide total	5		
Beryllium dissolved	10		
Beryllium total	10	12	15
Gases (mg/ℓ)			
Hydrogen sulfide	5	7	
Carbon dioxide	5		
Methane	5		
Dissolved oxygen	5		
Residual chlorine	8		

[a] Significance in reporting data is represented by o.x, o.ox, and o.oox. The sensitivity provided is represented by:

o.x column with sensitivity down to 0.1 milligram per liter (mg/ℓ)
(0.1 mg/ℓ = 0.1 ppm = 100 μ/ℓ),
o.ox column with sensitivity down to 0.01 mg/ℓ
(10 μ/ℓ = 0.01 ppm) and
o.oox column with sensitivity down to 0.001 mg/ℓ
(1 μ/ℓ = 1 ppb).

Specific uses dictate the items of interest in water supply monitoring. In agricultural areas where irrigation water is pumped from groundwater, boron is analyzed, and sodium adsorption ratio and residual bicarbonate are calculated from the analysis of major constituents. In municipal areas using groundwater, fluoride, iron, and manganese, as well as selected trace metals such as arsenic, are analyzed. Hardness is calculated. Many trace metals occur in insignificant concentrations in groundwater, and, with careful judgment, can often be eliminated from the chemical analysis.

In monitoring near waste disposal sites, consideration should be given to constituents in the waste water or percolate. Judgment based on soil chemistry and geochemistry can then be used to predict what constituents should be analyzed. Other forms of specific constituents may be important, such as hydrogen sulfide (H_2S) and nitrite ion (NO_2). Past groundwater quality studies have often reflected a lack of sampling for constituents other than those of interest. It should be remembered that other constituents can often yield significant data. Also, no ion moves by itself; that is, a percolating nitrate ion must be accompanied by some cation, for example. Similarly, thermodynamic or kinetic studies of groundwater quality require knowledge of concentrations of the major constituents, as well as pH and oxidation potential. However, in some cases judgment can be used to effectively utilize partial chemical analyses. In salt-balance studies in irrigated areas, for example, electrical conductivity alone sometimes can be used quite effectively.

ORGANIC CHEMICAL

Goerlitz and Brown (1972) discuss methods for analysis of organic substances in water. Before a proper sampling program for organic substances can be initiated, the nature of the organic compounds must be considered. Organic matter in a body of water may be distributed on the surface, in suspension, adsorbed on suspended sediment, in bed materials, and in solution. Because of this wide, and generally unpredictable, distribution of organic material in a body of surface water, the collection of representative samples requires care and often the use of specialized sampling equipment. In most cases, samples for organic analysis are best collected with the same equipment and technique used for the collection of samples for suspended-sediment measurements (Guy and Norman, 1970). Most surface water investigations are not concerned with surface or bottom material, but rather with the body of the water. However, in groundwater pollution monitoring where percolated water is the prime interest, the bottom material may be of prime importance. The U.S. Environmental Protection Agency (1971) also presents data on organic pesticides sampling analysis. The cost of analysis of organic chemicals in groundwater is given in Table 3-15.

Containers

Organic substances tend to cling to sample containers and special precautions are necessary. Glass bottles are the most acceptable containers for collecting, transporting, and storing samples for organic analysis. Glass appears to be inert relative to organic materials, and can withstand a rigorous cleaning procedure. Because organic materials are so plentiful in the environment, it is extremely difficult to collect samples free from extraneous contamination. Apparatus for containing samples must be scrupulously clean. Boston round-glass bottles of 1-liter capacity, with sloping shoulders and narrow mouths are usually satisfactory. The closure should be inert metal, lined with Teflon.*

---

*Trademark of E.I. duPont de Nemours, Inc.

Table 3-15

COSTS OF ORGANIC CHEMICAL DETERMINATIONS
FOR GROUNDWATER POLLUTION,[a] OCTOBER 1974

Parameter (mg/ℓ)	Water analysis cost ($)		
	0.x	0.0x	0.00x
Carbon in suspension	15		
Carbon dissolved	15		
Carbon total	15	25	
Biochemical oxygen demand	20		
Immediate oxygen demand	10		
Chemical oxygen demand	10		
Methylene blue active substances	10	15	
Nitrogen	10	11	
Phenolic material	15	17	20
Pesticides (insecticides and herbicides), each	40	42	45

[a] Significance in reporting data is represented by o.x, o.ox, and o.oox.

All sample bottles must be cleaned prior to sample collection. The accepted procedure is to wash the bottles in hot detergent solution, rinse them in warm tap water, then rinse them in dilute hydrochloric acid, and finally rinse them in distilled water. The bottles are then put into an oven at 300 $^{\circ}$C overnight. The Teflon cap liners and metal closures are washed in detergent. The caps are rinsed with distilled water and air dried. The liners are rinsed in dilute hydrochloric acid, soaked in redistilled acetone for several hours, and heated to 200 $^{\circ}$C overnight. When the heat treatments are completed, the bottles are capped with the closure and Teflon liners. The cost of glass bottles and mailers has been previously described.

Sample Preservation

Most water samples for organic analysis must be protected from degradation. Icing is the most acceptable method of preserving a sample. The U.S. Environmental Protection Agency (1974a) presents data for organic materials in water and wastes (Table 3-16). Goerlitz and Brown (1972) also recommend preservation techniques for organic substances in water. The procedures are similar, with the following additions:

Chlorophylls	Refrigerate at 4 $^{\circ}$C
Herbicides	Acidify with concentrated H_2SO_4 at a rate of 2 milliliters per liter of sample and refrigerate at 4 $^{\circ}$C
Insecticides	None required for chlorinated compounds

BACTERIOLOGICAL

Coliform organisms have long been used as indicators of sewage pollution, although the group includes bacteria from diverse natural sources. For example, members of the coliform group may come from soil, water, and vegetation, as well as from feces. However, if a water sample contains an appreciable coliform count, the source of the sample is considered to have a disease-producing potential in the absence of contrary information. Standards for drinking-water quality provide criteria as to the number of coliform organisms allowable per 100 milliliters of finished water.

Table 3-16

RECOMMENDED SAMPLING AND PRESERVATION TECHNIQUES
FOR ORGANIC CHEMICAL DETERMINATIONS

Measurement	Volume (ml)	Preservative	Holding time
Biological oxygen demand	1000	Cool to 4°C	6 hours
Chemical oxygen demand	50	H_2SO_4 to pH < 2	7 days
Methylene blue active substances (MBAS)	250	Cool to 4°C	24 hours
Nitrilotriacetic acid (NTA)	50	Cool to 4°C	24 hours
Oil and grease	1000	Cool to 4°C H_2SO_4 to pH < 2	24 hours
Organic carbon	25	Cool to 4°C H_2SO_4 to pH < 2	24 hours
Phenolics	500	Cool to 4°C H_2SO_4 to pH < 4 1.0 g/l $CuSO_4$	24 hours
Kjeldahl nitrogen	500	Cool to 4°C H_2SO_4 to pH < 2	24 hours

Source: U.S. EPA (1974a)

The standard test for presence of members of the coliform group may be carried out by the membrane filter technique or by the multiple-tube fermentation technique described by American Public Health Association (1971).

Fecal coliforms are that fraction of the coliform group present in the gut or the feces of warmblooded animals. They are capable of producing gas from lactose in a suitable culture medium at 44.5 °C. Bacterial organisms from other sources generally cannot produce gas in this manner (American Public Health Association, 1971). The presence of fecal coliform organisms may indicate recent, and possibly dangerous, contamination. The presence of other coliform organisms suggests less recent contamination, or contributions, from sources of nonfecal origin.

Fecal streptococci are being used increasingly as indicators of significant contamination of water because the normal habitat of these organisms is the intestine of man and animals. Fecal streptococcal data verify fecal pollution, and may provide additional information concerning the recency and probable origin of pollution. In combination with data on coliform bacteria, fecal streptococci are used in sanitary evaluation as a supplement to fecal coliforms, where more precise determination of sources of contamination is necessary.

Samples for bacteriologic examination must be collected in bottles that have been carefully cleansed and autoclaved for 20 minutes at 212 °C at 15 psi. Sterilized milk dilution bottles are ideal sample containers. When the sample is collected, ample air space must be left in the bottle to facilitate mixing of the sample by shaking. Care must be taken to avoid contamination of the sample and sample bottle at the time of collection, and in the period prior to analysis.

As soon as possible after collection, preferably within 1 hour and not more than 6 hours, the sample should be filtered and the membrane filter placed on the growth medium. Samples must be kept cool during the time between collection and filtration. If filtration is delayed, the sample should be iced or refrigerated, but not frozen.

The costs of bacterial analysis for groundwater pollution are given in Table 3-17.

Table 3-17

COSTS OF BACTERIAL ANALYSIS FOR GROUNDWATER POLLUTION, OCTOBER 1974

Bacteria	Cost ($)
Standard plate count	7
Coliform, fecal/100 $\mu\ell$	10
Coliform, total/100 $\mu\ell$	10
Enterococcus	10
Klebsiella – aerobacteria	10
Iron bacteria	9
Proteus	10
Pseudomonas aerceginosa	10
Staphylococcus	10
Salmonella	9
Sulfate reducing organisms	10
Streptococci, fecal	10
Streptococci, total	10
Bacteriological identification (90 percent complete)	35

RADIOLOGICAL
Containers
Radioactive elements are often measured in the submicrogram range, and can, therefore, be influenced by any background or residual material that may be in the sample container. Similarly, a radionuclide may be largely or wholly adsorbed on the surfaces of suspended particles. Glass containers tend to have a higher background radioactivity than polyethylene bottles. For most radiochemical analyses (excluding tritium), a polyethylene bottle is recommended.

Preservation
Radiochemical sample containers normally are washed with nitric acid and allowed to fume for several hours before use. After the sample has been taken and separated into suspended and dissolved fractions, a preservative can be added. The kind of preservative is highly dependent upon the kind of radiochemical to be analyzed. Formaldehyde or ethyl alcohol has been suggested as a preservative for highly

perishable samples. Routinely in groundwater, however, hydrochloric and nitric acids are used as general preservatives. Preservatives and reagents should be tested for radioactivity prior to their use.

The costs of radiochemical analyses for groundwater pollution are given in Table 3-18.

Table 3-18

COSTS OF RADIOCHEMICAL ANALYSIS IN GROUNDWATER POLLUTION, OCTOBER 1974

Radiochemical	Cost ($)
Carbon-14 age date	160
Carbon 13/12 ratio	55
Cesium-137 diss.	22
Gamma-scan	70
Gross alpha and beta diss.	25
Gross alpha and beta susp.	25
Lead-210	25
Radium-226, by Radon	30
Strontium-90, diss.	35
Thorium-BTM colorimetric	25
Tritium, liq. scintillation	20
Uranium, diss.dir. fluorometric	18
Uranium, diss.extr. fluorometric	26

SELECTED BIBLIOGRAPHY

L.E. Allison, "A Simple Device for Sampling Groundwaters in Auger Holes," *Soil Science Society of America Proceedings,* Vol. 35, pp. 844-845, 1971.

American Public Health Association, *Standard Methods for the Examination of Water and Wastewater,* American Public Health Association, New York, 1971. (874 pp.)

American Society for Testing Materials, *Manual on Water,* ASTM Special Technical Publication, No. 442, 1969. (360 pp.)

J.T. Anderson, E.E. Hardy, and J.T. Roach, *A Land Use Classification System for Use with Remote-Sensor Data,* U.S. Geological Survey Circular 671, U.S. Government Printing Office, Washington, D.C., 1972.

K.E. Anderson, *Water Well Handbook,* Missouri Water Well Drillers Association, 1967. (281 pp.)

M.A. Apgar and D. Langmuir, "Ground-Water Pollution Potential of a Landfill Above the Water Table," *Ground Water,* Vol. 9, No. 6, pp. 76-93, 1971.

P.R. Barnett and E.C. Mallory, Jr., "Determination of Minor Elements in Water by Emission Spectroscopy," Chapter A2, *Techniques of Water-Resources Investigations, Book 5, Laboratory Analysis,* U.S. Geological Survey, 1971. (31 pp.)

C.M. Barton, "Bore Hole Sampling of Saturated Uncemented Sands and Gravels," *Ground Water,* Vol. 12, No. 3, pp. 170-181, 1974.

R. Bentall, *Methods of Collecting and Interpreting Groundwater Data,* U.S. Geological Survey, Water-Supply Paper 1544-H, pp. 36-46, 1963.

C.A. Black (ed.), "Methods of Soil Analysis," (in two parts), *Agronomy,* No. 9, American Society of Agronomy, Madison, Wisconsin, 1969.

G.R. Blake, "Bulk Density," in "Methods of Soil Analyses," C.A. Black (ed.), *Agronomy,* No. 9, American Society of Agronomy, Madison, Wisconsin, pp. 374-390, 1965.

H.F. Blaney and W.D. Criddle, *Determining Consumptive Use and Irrigation Water Requirements,* U.S. Department of Agriculture Technical Bulletin 1275, 1962. (59 pp.)

H.L. Bohn, B.L. McNeal, and G.A. O'Connor, *Soil Chemistry,* Wiley Interscience, New York, 1979.

H. Bouwer, "Land Treatment of Liquid Waste: The Hydrologic System," *Proceedings of the Joint Conference on Recycling Municipal Sludges and Effluents on Land,* U.S. Environmental Protection Agency, U.S. Department of Agriculture, and the National Association of State Universities and Land-Grant Colleges, Champaign, Illinois, July 9-13, 1973.

C.A. Bower and L.V. Wilcox, "Soluble Salts" in "Methods of Soil Analyses," C.A. Black (ed.), *Agronomy,* No. 9, pp. 933-951, American Society of Agronomy, Madison Wisconsin, 1965.

I. Bowman, *Well-Drilling Methods,* U.S. Geological Survey, Water Supply Paper 257, 1911. (139 pp.)

J.D. Bredehoeft and I.S. Papadopulos, "Rates of Vertical Ground Water Movement Estimated from Earth's Thermal Profile," *Water Resources Research,* Vol. 1, No. 2, pp. 325-328, 1965.

E. Brown, M.W. Skougstad, and M.J. Fishman, "Methods for Collection and Analysis of Water Samples for Dissolved Minerals and Gases," Chapter A1, *Techniques of Water-Resources Investigations, Book 5, Laboratory Analysis,* U.S. Geological Survey, 1970. (160 pp.)

R.H. Brown, A.A. Konoplyantsev, J. Ineson, and V.S. Kovalevsky, "Ground-Water Studies," *An International Guide for Research and Practice,* UNESCO, Paris, 1972.

California Department of Water Resources, *Water Well Standards, State of California,* California Department of Water Resources, Bulletin 74, 1968.

California Office of Emergency Services, *Emergency Plan,* Sacramento, California, 1970. (310 pp.)

M.D. Campbell and J.H. Lehr, "Water Well Technology," Chapter 9, *Formation Identification and Evaluation,* McGraw-Hill Book Co., New York, 1973. (681 pp.)

K. Cartwright, *Temperature Prospecting for Shallow Glacial Aquifers in Illinois,* Circular 433, Illinois State Geological Survey, 1968. (38 pp.)

J.W. Cary, "Soil Water Flowmeters with Thermocouple Outputs," *Soil Science Society of America Proceedings,* Vol. 37, pp. 176-181, 1973.

A. Casagrande, "Seepage Through Dams," *Journal of New England Water Works Association,* June 1937.

D.J. Cederstrom, *Cost Analysis of Ground-Water Supplies in the North Atlantic Region,* U.S. Geological Survey Water Supply Paper 2034, 1970. (47 pp.)

P.B. Chandler, W.L. Dowdy, and D.T. Hodder, *Study to Evaluate the Utility of Aerial Surveillance Methods in Water Quality Monitoring,* California State Water Resources Control Board, Publication No. 41, 1970. (100 pp.)

H.D. Chapman, "Cation Exchange Capacity," in "Methods of Soil Analyses," C.A. Black (ed.), *Agronomy,* No. 9, pp. 891-900, American Society of Agronomy, Madison, Wisconsin, 1965.

E.C. Childs, *An Introduction to the Physical Basis of Soil Water Phenomena,* Wiley-Interscience, 1969. (493 pp.)

R.W. Cruff and T.H. Thompson, *A Comparison of Methods of Estimating Potential Evapotranspiration from Climatological Data in Arid and Subhumid Environments,* U.S. Geological Survey Water Supply Paper 1839-M, 1967. (28 pp.)

S.N. Davis and R.J.M. DeWiest, *Hydrogeology,* John Wiley & Sons, Inc., New York, 1966. (463 pp.)

P.R. Day, "Particle Fractionation and Particle Size Analyses," in "Methods of Soil Analyses," C.A. Black (ed.), *Agronomy,* No. 9, pp. 545-566, American Society of Agronomy, Madison, Wisconsin, 1965.

P.R. Day, D.M. Anderson, and G.H. Bolt, "Nature of Soil Water," in "Irrigation of Agricultural Lands," R.M. Hazan, H.R. Haise, and T.W. Edminster (eds.), *Agronomy,* No. 9, pp. 193-208, American Society of Agronomy, Madison, Wisconsin, 1967.

W.J. Dunlap and J.F. McNabb, *Subsurface Biological Activity in Relation to Ground Water Pollution,* Environmental Protection Technology Series, U.S. Environmental Protection Agency, Corvallis, Oregon, EPA-660/2-73-014, 1973.

B.G. Ellis, "The Soil as a Chemical Filter," *Recycling Treated Municipal Wastewater and Sludge Through Forest and Cropland,* W.E. Sopper and L.T. Kardos (eds.), the Pennsylvania State University Press, 1973.

C.G. Enfield, J.J.C. Hsieh, and A.W. Warrick, "Evaluation of Water Flux Above a Deep Water Table Using Thermocouple Psychrometers," *Soil Science Society of America Proceedings,* Vol. 37, No. 6, pp. 968-970, 1973.

J.G. Ferris, D.B. Knowles, R.H. Brown, and R.W. Stallman, *Theory of Aquifer Tests,* U.S. Geological Survey, Water Supply Paper 1536-E, pp. 69-174, 1962.

R. Follansbee, "Evaporation from Reservoir Surfaces," *American Society of Civil Engineers Transactions,* Paper No. 1871, pp. 704-715, 1933.

Robert H. Forste, *Verification of Groundwater Capital Costs,* Water Resources Research Center, University of New Hampshire, 1973.

W.H. Fuller, *Movement of Selected Metals, Asbestos and Cyanide in Soil: Application to Waste Disposal Problems,* U.S. Environmental Protection Agency, EPA-600/2-77-020, U.S. Government Printing Office, 1977.

M.S. Garber and F.C. Koopman, "Methods of Measuring Water Levels in Deep Wells," Chapter A1, *Techniques of Water-Resource Investigations, Book 8,* U.S. Geological Survey, 1968. (23 pp.)

W.H. Gardner, "Water Content," in "Methods of Soil Analyses," C.A. Black (ed.), *Agronomy,* No. 9, American Society of Agronomy, Madison, Wisconsin, pp. 82-125, 1965.

James P. Gibb, *Cost of Domestic Wells and Water Treatment in Illinois,* Illinois State Water Survey Circular 104, 1971. (23 pp.)

U.P. Gibson and R.D. Singer, *Water Well Manual,* Premier Press, Berkeley, California, 1971. (156 pp.)

D.F. Goerlitz and E. Brown, "Methods for Analysis of Organic Substances in Water," Chapter A3, *Techniques of Water-Resources Investigations, Book 5, Laboratory Analysis,* U.S. Geological Survey, 1972. (40 pp.)

D.H. Griffiths and R.F. King, *Applied Geophysics for Engineers and Geologists,* Pergamon Press, 1969. (223 pp.)

H.P. Guy and V.W. Norman, "Field Methods for the Measurement of Fluvial Sediments," Chapter C2, *Techniques of Water-Resources Investigations, Book 3,* U.S. Geological Survey, 1970. (59 pp.)

D.A. Hackbarth, "Field Study of Subsurface Spent Sulfite Liquor Movement Using Earth Resistivity Measurements," *Ground Water,* Vol. 9, No. 3, pp. 11-16, 1971.

E.A. Hansen and A.R. Harris, "A Groundwater Profile Sampler," *Water Resources Research,* Vol. 10, No. 2, 1974. (375 pp.)

B.B. Hanshaw, W. Back, and M. Rubin, "Radiocarbon Determinations for Estimating Groundwater Flow Velocities in Central Florida," *Science,* Vol. 148, pp. 494-495, 1965.

G.E. Harbeck, Jr., M.A. Kohler, G.E. Koberg, et al., *Water-Loss Investigations: Lake Mead Studies,* U.S. Geological Survey, Professional Paper 298, 1958.

E. Havely and A. Nir, "The Determination of Aquifer Parameters with the Aid of Radioactive Tracers," *Journal of Geophysical Research,* Vol. 61, pp. 2403-2409, 1962.

J.D. Hem, *Study and Interpretation of the Chemical Characteristics of Natural Water,* U.S. Geological Survey Water-Supply Paper 1473, second edition, 1970. (363 pp.)

D. Hillel, *Soil and Water, Physical Principles and Processes,* Academic Press, New York, 1971.

J.W. Holmes, S.A. Taylor, and S.J. Richards, "Measurement of Soil Water," in "Irrigation of Agricultural Lands," R.M. Hazan, H.R. Haise, and T.W. Edminster (eds.), *Agronomy,* No. 9, American Society of Agronomy, Madison, Wisconsin, pp. 275-298, 1967.

C.E. Jacob, "Drawdown Test to Determine the Effective Radius of Artesian Wells," *Transactions of the American Society of Civil Engineers,* Vol. 112, pp. 1047-1070, 1947.

Johnson Division UOP, "Ground Water and Wells," *A Reference Book for the Water-Well Industry,* First Edition, Edward E. Johnson, Inc., St. Paul, Minnesota, 1966. (440 pp.)

Johnson Division UOP, "Ground Water and Wells," *A Reference Book for the Water-Well Industry,* 2nd Ed., Edward E. Johnson, Inc., St. Paul, Minnesota, 1971. (440 pp.)

W.J. Kaufman and G.T. Orlob, "Measure of Groundwater Movement with Radioactive and Chemical Tracers," *Journal of American Water Works Association,* Vol. 48, pp. 559-572, 1956.

W.S. Keys and L.M. MacCary, "Application of Borehole Geophysics to Water-Resources Investigations," Chapter E1, *Techniques of Water-Resources Investigations, Book 2,* U.S. Geological Survey, 1971. (126 pp.)

D. Kirkham and W.L. Powers, *Advanced Soil Physics,* Wiley Interscience, New York, New York, 1972.

A. Klute, "Laboratory Measurement of Hydraulic Conductivity of Unsaturated Soil," in "Methods of Soil Analyses," C.A. Black (ed.), *Agronomy*, No. 9, pp. 253-261, American Society of Agronomy, Madison, Wisconsin, 1965.

M.A. Kohler, *Water-Loss Investigations: Lake Hefner Studies, Technical Report*, U.S. Geological Survey, Professional Paper 269, 1954.

R.L. Lowry and A.F. Johnson, "Consumptive Use of Water for Agriculture," *American Society of Civil Engineers Transactions*, Vol. 107, pp. 1243-1266, 1942.

J.N. Luthin (ed.), *Drainage of Agricultural Lands*, American Society of Agronomy, Madison Wisconsin, 1957.

F.C. McMichael and J.E. McKee, *Wastewater Reclamation at Whittier Narrows*, State of California, Water Quality Publication No. 33, 1966.

B.L. McNeil, "Soil Salts and their Effect on Water Movement," in "Drainage for Agriculture," J. van Schilfgaarde (ed.), *Agronomy*, No. 17, American Society of Agronomy, Madison, Wisconsin, pp. 409-433, 1974.

O.E. Meinzer, "Ground Water," *Hydrology*, O.E. Meinzer (ed.), Dover Publications, Inc., New York, pp. 385-477, 1942.

R.H. Merkel, "The Use of Resistivity Techniques to Delineate Acid Mine Drainage in Ground Water," *Ground Water*, Vol. 10, No. 5, pp. 38-42, 1972.

S.D. Merrill and S.L. Rawlens, "Field Measurement of Soil Water Potential with Thermocouple Psychrometers," *Soil Science*, Vol. 113, No. 2, pp. 102-109, 1972.

M.M. Mortland and W.D. Kemper, "Specific Surface," in "Methods of Soil Analyses," C.A. Black (ed.), *Agronomy*, No. 9, pp. 532-543, American Society of Agronomy, Madison, Wisconsin, 1965.

R.P. Murrmann and F.R. Koutz, "Role of Soil Chemical Processes in Reclamation of Wastewater Applied to Land," *Wastewater Management by Disposal on the Land*, U.S. Army Corps of Engineers, Cold Regions Research and Engineering Lab Specialty Report 171, 1972.

D.L. Myhre, J.O. Sanford, and W.F. Jones, "Apparatus and Techniques for Installing Access Tubes in Soil Profiles to Measure Soil Water," *Soil Science*, Vol. 108, No. 4, pp. 296-299, 1969.

D.R. Nielsen, J.W. Biggar, and K.T. Erh, "Spatial Variability of Field-Measured Soil-Water Properties," *Hilgardia*, Vol. 42, No. 7, pp. 215-260, 1973.

J.O. Osgood, "Hydrocarbon Dispersion in Ground Water: Significance and Characteristics," *Ground Water*, Vol. 12, No. 6, November-December, 1974.

J.D. Oster and L.S. Willardson, "Reliability of Salinity Sensors for the Management of Soil Salinity," *Agronomy*, Vol. 63, pp. 695-698, 1971.

R.R. Parizek and B.E. Lane, "Soil-Water Sampling Using Pan and Deep Pressure-Vacuum Lysimeters," *Journal of Hydrology*, Vol. 11, pp. 1-21, 1970.

H.L. Penman, "Natural Evaporation from Open Water, Bare Soil, and Grass," *Royal Society (London) Proceedings,* Series A, Vol. 193, pp. 120-145, 1948.

D.B. Peters, "Water Availability," in "Methods of Soil Analyses," C.A. Black (ed.), *Agronomy,* No. 9, American Society of Agronomy, Madison, Wisconsin, pp. 279-285, 1965.

C.J. Phene, G.J. Hoffman, and S.L. Rawlens, "Measuring Soil Matrix Potential in-situ by Sensing Heat Dissipation Within a Porous Body: I, Theory and Sensor Construction," and "II, Experimental Results," *Soil Science Society of America Proceedings,* Vol. 35, pp. 225-229, 1971.

J.F. Pickens, J.A. Cherry, G.E. Grisak, W.F. Merritt, and B.A. Risto, "A Multi-Level Device for Ground-Water Sampling and Piezometric Monitoring," *Ground Water,* Vol. 16, No. 5, pp. 322-327, 1978.

C.E. Poulton, "A Comprehensive Remote Sensing Legend System for the Ecological Characterization and Annotation of Natural and Altered Landscapes," *Proceedings of the Eighth International Symposium on Remote Sensing of Environment,* Ann Arbor, Michigan, 1972.

P.R. Pratt, *Nitrate in the Unsaturated Zone Under Agricultural Lands,* U.S. Environmental Protection Agency, Water Pollution Control Research Series, 16060 DOE 04/72, 1972. (45 pp.)

F.H. Rainwater and L.L. Thatcher, *Methods for Collection and Analysis of Water Samples,* U.S. Geological Survey Water-Supply Paper 1454, 1960. (301 pp.)

D.R. Ralston, *Influences of Water Well Design on Neutron Logging,* unpublished M.S. Thesis, University of Arizona, Tucson, Arizona, 1967.

S.L. Rawlens and F.N. Dalton, "Psychrometric Measurement of Soil Water Potential Without Precise Temperature Control," *Soil Science Society of America Proceedings,* Vol. 31, pp. 201-297, 1967.

R.G. Ray, *Aerial Photographs in Geologic Interpretation and Mapping,* U.S. Geological Survey, Professional Paper 373, pp. 201-297, 1960.

R.C. Reeve, "Hydraulic Head," in "Methods of Soil Analyses," C.A. Black (ed.), *Agronomy,* No. 9, pp. 180-196, American Society of Agronomy, Madison, Wisconsin, 1965.

R.F. Reitemeier, "Effect of Moisture Content on the Dissolved and Exchangeable Ions of Soils of Arid Regions," *Soil Science,* Vol. 61, No. 3, pp. 195-214, 1946.

J.D. Rhoades and L. Bernstein, "Chemical, Physical and Biological Characteristics of Irrigation and Soil Water," *Water and Water Pollution Handbook,* L.L. Ciaccio (ed.), Marcel Dekker, Inc., New York, Vol. 1, pp. 142-222, 1971.

L.A. Richards (ed.), *Diagnosis and Improvement of Saline and Alkali Soils,* U.S. Department of Agriculture, Agricultural Handbook 60, 1954.

L.A. Richards, "A Soil Salinity Sensor of Improved Design," *Soil Science Society of America Proceedings,* Vol. 30, No. 11, pp. 333-337, 1966.

C. Rohwer, "Evaporation from Different Types of Pans," *American Society of Civil Engineers Transactions,* Paper No. 1871, 1933.

D.D. Runnells, "Wastewaters in the Vadose Zone of Arid Regions: Geochemical Interactions," *Ground Water,* Vol. 14, No. 6, pp. 374-385, 1976.

E.A. Sammel, "Convective Flow and its Effect on Temperature Logging in Small-Diameter Wells," *Geophysics,* Vol. 33, No. 6, pp. 1004-1012, 1968.

J. Seker, "Hydrologic Significance of Tectonic Fractures Detectable on Airphotos," *Ground Water,* Vol. 4, No. 4, pp. 23-27, 1966.

R.W. Simonson, "What Soils Are," in "Soil," *The Yearbook of Agriculture,* U.S. Department of Agriculture, Washington, D.C., pp. 17-30, 1957.

H.E. Skibitzke, *Electronic Computers as an Aid to the Analysis of Hydrogeological Problems,* International Association of Scientific Hydrology, Publication 52, Brussels, pp. 347-358, 1960.

D.B. Smith, "Flow Tracing Using Isotopes," *Groundwater Pollution in Europe,* J.A. Cole (ed.), Water Information Center, Port Washington, New York, pp. 377-387, 1974.

R.W. Stallman, *Electric Analog of Three-Dimensional Flow to Wells and Its Application to Unconfined Aquifers,* U.S. Geological Survey Water-Supply Paper 1536-4, pp. 205-242, 1963.

R.L. Stollar and P. Roux, "Earth Resistivity Surveys – A Method for Defining Ground-Water Contamination," *Ground Water,* Vol. 13, No. 2, pp. 145-150, 1975.

P.R. Stout, R.G. Burau, and W.A. Allardice, *A Study of the Vertical Movement of Nitrogenous Matter from the Ground Surface to the Water Table in the Vicinity of Grover City and Arroyo Grande – San Luis Obispo County,* Report to Central Coastal Regional Water Pollution Control Board, 1965. (51 pp.)

C.V. Theis, "The Relation Between the Lowering of the Piezometric Surface and the Rate and Duration of Discharge of a Well Using Ground-Water Storage," *Transactions of the American Geophysical Union,* Part 2, pp. 519-524, 1935.

G.W. Thomas and B.J. Barfield, "The Unreliability of Tile Effluent for Monitoring Subsurface Nitrate-Nitrogen Losses from Soils," *Journal of Environmental Quality,* Vol. 3, No. 2, pp. 183-185, 1974.

G.A. Thorley, *Regional Agricultural Surveys Using ERTS-1 Data,* Center for Remote Sensing Research, ERTS-1 Final Report, Berkeley, California, 1973.

C.W. Thornthwaite, "An Approach Toward a Rational Classification of Climate," *Geographic Reviews,* Vol. 38, pp. 55-94, 1948.

D.K. Todd, *Ground Water Hydrology,* John Wiley & Sons, New York, 1959. (336 pp.)

D.K. Todd, R.M. Tinlin, K.D. Schmidt, and L.G. Everett, *Monitoring Groundwater Quality: Monitoring Methodology,* U.S. Environmental Protection Agency, Las Vegas, Nevada, 1976.

A.N. Turcan, *Calculation of Water Quality from Electrical Logs – Theory and Practice,* Louisiana Geological Survey, Water Resources Pamphlet 19, 1966. (23 pp.)

UNESCO, "Measurement of Groundwater Flow Velocity," Part 7.4.2, *Ground-Water Studies, An International Guide for Research and Products,* 1975.

U.S. Department of the Army and Air Force, *Wells,* Technical Manual TM5-197, Washington, D.C., 1957. (264 pp.)

U.S. Environmental Protection Agency, *Procedures Manual for Ground Water Monitoring at Solid Waste Disposal Facilities,* EPA/530/SW-611, Office of Solid Waste, U.S. Government Printing Office, 1977.

U.S. Environmental Protection Agency, *Methods for Organic Pesticides in Water and Wastewater,* National Environmental Research Center, Cincinnati, Ohio, U.S. Government Printing Office, 759-301/2113, 1971. (47 pp.)

U.S. Environmental Protection Agency, *Handbook for Analytical Quality Control in Water and Wastewater Laboratories,* Analytical Quality Control Laboratory, National Environmental Research Center, Cincinnati, Ohio, U.S. Government Printing Office, 479-971, 1972. (54 pp.)

U.S. Environmental Protection Agency, *Handbook for Monitoring Industrial Wastewater,* Technology Transfer Series, U.S. Government Printing Office, 732-349/414, 1973. (91 pp.)

U.S. Environmental Protection Agency, *Manual of Methods for Chemical Analysis of Water and Wastes,* EPA-625-15-75-003, Methods Development and Quality Assurance Research Laboratory, National Environmental Research Center, Cincinnati, Ohio, 1974a. (298 pp.)

U.S. Environmental Protection Agency, "Water Quality and Pollutant Source Monitoring," *Federal Register,* Vol. 39, No. 168, Part III, 1974b.

U.S. Environmental Protection Agency, "Model State Water Monitoring Program," EPA-440/9-74-002, Office of Water and Hazardous Materials Monitoring and Data Support Division, Washington, D.C., 1975. (36 pp.)

U.S. Geological Survey, *Techniques of Water Resources Investigations,* Book 5, *Laboratory Analysis,* 1973.

C.H.M. Van Bavel, "Neutron Scattering Measurement of Soil Moisture: Development and Current Status," *Proceedings of International Symposium on Humidity and Moisture,* Washington, D.C., pp. 171-184, 1963.

J. van Schilfgaarde, "Drainage for Agriculture," *Agronomy,* No. 17, American Society of Agronomy, Madison, Wisconsin, pp. 359-405, 1974.

W.C. Walton, *Groundwater Resource Evaluation,* McGraw-Hill Book Co., New York, 1970. (664 pp.)

W.C. Walton and T.A. Prickett, "Hydrogeologic Electric Analog Computer," *Proceedings of the American Society of Civil Engineers, Journal of the Hydraulics Division,* Vol. 89, pp. 67-91, 1963.

K.K. Watson, "A Recording Field Tensiometer with Rapid Response Characteristics," *Journal of Hydrology*, Vol. 5, pp. 33-39, 1967.

K.K. Watson, "Some Applications of Unsaturated Flow Theory," in "Drainage for Agriculture," J. van Schilfgaarde (ed.), *Agronomy*, No. 17, American Society of Agronomy, Madison, Wisconsin, 1974.

L.S. Willardson, B.D. Meek, and M.J. Huber, "A Flow Path Ground Water Sampler," *Soil Science Society of America Proceedings*. Vol. 36, pp. 965-966, 1973.

L.G. Wilson, "Observations on Water Content Changes in Stratified Sediments During Pit Recharge," *Ground Water*, Vol. 9, No. 3, pp. 29-40, 1971.

L.G. Wilson and K.J. DeCook, "Field Observations on Changes in the Subsurface Water Regime During Influent Seepage in the Santa Cruz River," *Water Resources Research*, Vol. 4, No. 6, pp. 1219-1234, 1968.

L.G. Wilson, *Monitoring in the Vadose Zone - A Review of Technical Elements and Methods*, General Electric Company - TEMPO, GE79TMP-55, April 1980.

W.W. Wood, "A Technique Using Porous Cups for Water Sampling at any Depth in the Unsaturated Zone," *Water Resources Research*, Vol. 9, No. 2, pp. 486-488, 1973.

B.S. Yare, "The Use of a Specialized Drilling and Groundwater Sampling Technique for Delineation of Hexavalent Chromium Contamination in an Unconfined Aquifer, Southern New Jersey Plain," *Ground Water*, Vol. 13, No. 2, pp. 151-154, 1975.

A.A. Zohdy, G.P. Eaton, and D.R. Mabey, "Application of Surface Geophysics to Ground-Water Investigations," Chapter D1, *Techniques of Water-Resources Investigations, Book 2, Collection of Environmental Data*, U.S. Geological Survey, 1974. (116 pp.)

CHAPTER IV

GROUNDWATER DATA MANAGEMENT

SECTION 1 – INTRODUCTION

The development of a management information system (MIS) entails the *identification of system requirements,* system design, organizational design, system procedures design, and, if necessary, programming, implementation, testing, debugging, documenting, and training. The intention of this chapter is to identify the system requirements of a comprehensive groundwater quality monitoring program MIS and to survey the existing capabilities, which may serve to satisfy those requirements.

For those groundwater monitoring agencies whose needs are not met by existing capabilities, this chapter presents generic specifications and guidelines for the structuring of a computerized groundwater surveillance data management system. In addition, the inventory of existing data management capabilities (including generalized data base management packages offered by commercial vendors) presented in this chapter may provide the framework for developing the desired capabilities. The inventory of existing data management systems presented here is not intended to be comprehensive. Rather, existing systems were selected for inclusion on the basis of their significance and relevance.

It is hoped that the discussion herein will convey to the groundwater quality manager the scope and breadth of the field of information management systems, and that it will expose him to the alternatives available to him in structuring an information management capability suitable to his needs.

SCOPE

Effective groundwater quality management requires that relevant information be available to the decision maker in a concise, comprehensive, timely, economical, and reliable manner. Realization of these goals can be accomplished with the assistance of any one of various tools, including file drawers, microfilm, and digital computers. The choice of one of these alternatives will depend, for the most part, on the volume of data involved and the frequency of interaction with the data base.

The discussion here is concerned with the groundwater information management requirements of all levels of governmental monitoring agencies (Federal, State, and local). In recognition of the volume of information likely to be generated by many of these agencies, this report is directed at outlining a comprehensive computer system capability intended to satisfy these requirements. The system described will afford management of ambient groundwater quality information, percolate quality information, compliance monitoring information, and other data relevant to the management of groundwater quality, including citations of groundwater research documentation.

SECTION 2 – GROUNDWATER INFORMATION MANAGEMENT REQUIREMENTS

GENERAL

A complete MIS requirements analysis would call for a survey of the potential users of the system to enable the development of system specifications. Critical factors to be considered by this survey would include the following:

- Information to be managed
- Data volumes
- Frequency of interaction with the data base
- Responsiveness requirements
- Where and how the source information is to be generated
- Required data qualification procedures
- Required output documents

For this chapter, an intensive user survey was superseded by the application of gross, but reasonalby utilitarian, assumptions. The assumptions corresponding to the critical factors listed above are:

- Information to be managed will include monitoring station descriptions (i.e., location, hydrogeology, and local water use), physical and chemical measurements of water samples together with sampling dates, and citations of groundwater research documentation.

- The groundwater surveillance data base will be moderately large (expanding monotonically), consisting of millions of data elements requiring extensive storage capabilities. Once the initial data base is established, input data volume will be relatively low, and output volume in response to user queries somewhat greater.

- Frequency of interaction (updates and queries) with the data base will be moderate.

- Updating and interrogating the groundwater data base will not require quick system response, with several days turnaround generally being adequate. Interrogating information indexing files (water quality data file descriptions and document citations) will require quick system response (i.e., real time), however, to allow for browsing.

- Source information will be generated at locations distributed throughout the U.S., with concentrations in areas of high population density. In general, source information will be generated at locations relatively close to the users of the information.

- Data qualification requirements will include input data editing and provision for specific station, sample, and measurement comments to reflect special conditions.

- Output will be alphanumeric text, tables of primary data computed statistics, and pictorial presentations. Reports will generally be generated on a demand basis, with the possible exception of violation reports associated with compliance monitoring, which may be triggered.

Within the framework established by these assumptions, this section will present a further discussion of the information content of the proposed groundwater MIS, as well as a discussion of the fundamental functions to be performed by this system. These functions are data collection, data communication, data organization and storage, data processing, and information retrieval and display. The nature of these functions will be described, as well as the alternative technologies available to accomplish them. Those technologies which are best suited to a groundwater monitoring program will be identified.

INFORMATION TO BE MANAGED

An effective groundwater monitoring MIS will be capable of maintaining the following types of data:

- Station descriptions
- Quality criteria
- Geologic
- Hydrologic
- Water quality parameter identifiers
- Water quality measurements
- Temporal
- Information qualification data
- Monitoring agency status data
- Information indexing

The individual data elements comprising these information categories will be discussed in the following paragraphs. Each data element will be identified as system specific (i.e., applicable system wide), station specific, sample specific, or measurement specific. Further, those data elements required for retrieval or computational operations will be specified as searchable, indicating that they must be stored as formatted data.

STATION DESCRIPTIONS

Station descriptive data consists of information which specifies the station type (i.e., pumped well, unpumped well, unsaturated zone, information monitoring, compliance monitoring, or other appropriate designation), the party responsible for monitoring the station, a unique station identifier code(s), a unique location (three-dimensional), and directions for locating the station in the field. With the exception of the last item, all of this information should be searchable. Information providing instructions for locating stations in the field can be stored as narrative text along with other special station specific information not required for retrievals or computations (e.g., oil lubricated well subject to bearing leakage, continuous-slot stainless steel well screen, and like information).

The groundwater monitoring station type can be specified as coded information in a field of five characters or more. Station type data would be formatted as follows:

1st character

    Sample extraction method - pump, bail, or probe

2nd character

    Type data - quality, hydrogeologic, and/or DMA status data

3rd character

    Type site - municipal, industrial, or other

4th character

    Location - saturated zone, zone of aeration, or surface

5th character

    Monitoring justification - information, compliance, and/or other

Combinations of attributes can be represented uniquely by coding individual attributes numerically with either a 1, 2, or 4, so that the combination 1 and 4, for instance, could be coded uniquely in one position as a 5.

The designated monitoring agency (DMA) responsible for monitoring the station should be stored as an *agency* code in a searchable field so that, for instance, all stations being maintained by a particular DMA can be retrieved. In addition, the narrative text associated with a station can contain, for example, the names of specific individuals having responsibility for a station, together with their phone numbers.

Each station will require a unique identifier code. This identifier will be maintained permanently within the MIS and provide access to station data, even if and when a station becomes inactive. Provision should be made for storing and retrieving multiple station identifier codes for the case where a DMA uses multiple codes for alternative retrieval schemes, or where more than one DMA is monitoring the same station, and station codes have not been standardized.

Station descriptive data to be maintained by the MIS must include information regarding political jurisdiction (e.g., state, county, city, irrigation district, or park district), as well as a unique areal location. To specify a unique areal location, indication of either the township, range, section, or the familiar conventional geographic coordinate system (latitude/longitude) will be most practicable. The degree of precision associated with the measurement of a station's coordinates should also be stored. Additionally, the depths of both the monitoring station hole and intake screen should be stored as station specific information. It should be noted that in cases where, for example, either a monitoring well is equipped with multiple intake screens or a thief sampler is used, individual sample depths may not correspond to either the well depth or existing water level.

Other major station specific information categories not discussed above are applicable quality criteria, geologic data, and hydrologic data.

QUALITY CRITERIA

Information pertaining to established quality criteria, which a groundwater quality MIS should accommodate as station specific data, includes current and projected land use, current and projected water use, demographic data, economic data, designated protected water uses, applicable permit data (compliance dates and monitoring requirements - parameters and frequency), and water quality criteria (either ambient or discharge limitations).

Demographic and economic data, as well as current and projected land and water use in the neighborhood of a monitoring station, is information typically generated by local planning agencies, which reflects the significance of groundwater pollution in the environs of a monitoring station. This information need not be used for retrieval or computational operations, and, consequently, can be satisfactorily stored in the narrative text associated with each monitoring station.

The development of a comprehensive groundwater quality monitoring program will entail the systematic identification and inventory of principal aquifers and, preferably, the designation of protected uses for these aquifers. In the process of developing the inventory of principal aquifers, full use should be made of the *Catalogue of Aquifer Names and Geologic Unit Codes* compiled by the Office of Water Data Coordination (OWDC), U.S. Department of the Interior (USDI) (Price and Baker, 1974). Aquifer protected use designations would be codified and searchable. Protected use categories would include public water supply and agricultural and industrial use, with allowance made for the possibility of subcategories of the latter two.

Permit data, other than imposed discharge limitations, should not be required for retrieval or computational operations and can, therefore, be stored in the narrative text associated with compliance monitoring stations. Information content would be similar to that contained in National Pollutant Discharge Elimination System (NPDES) applications and permits, including permit numbers, compliance dates, and monitoring requirements. If permit numbers are required for search operations, they can be used as secondary station identifiers.

Permit specified discharge limitations and/or the water quality criteria associated with the designated protected uses established for an aquifer can be stored with the characteristics of each monitoring station as appropriate. Ambient quality criteria to be stored may be those published by EPA in *Proposed Criteria for Water Quality* (EPA, 1973), with provision made for updating these criteria as they are modified. Although it is not likely that they will be needed as record keys, the inclusion of discharge limitations and ambient water quality criteria within the monitoring data base as searchable information will allow efficient generation of exception reports.

GEOLOGIC DATA

In order to uniquely identify the source of groundwater samples, some geologic data is required, in addition to geographical coordinates, to specify the aquifer from which the sample originated. In the case where a monitoring station taps more than one aquifer, aquifer identification is particularly essential, and must be provided as sample specific (i.e., input in conjunction with each set of water quality analysis data), rather than station specific data. The requirement for providing aquifer identification can be satisfied by storing the established aquifer name, if available, or the geologic formation name and age associated with the monitored aquifer (e.g., Mount Simon formation - Cambrian age or glacial drift - Pleistocene age). It should be pointed out that the latter form of identification is not preferred, since aquifers and geologic formations do not necessarily coincide completely. Aquifer identification can be codified and standardized, and search operations facilitated by application of U.S. Geological Survey (USGS) proposed modifications to the stratigraphic coding system developed by the American Association of Petroleum Geologists (Price and Baker, 1974).

Additionally, information regarding the physical properties and chemical constituency of the water bearing materials (aquifer, unsaturated zone, or topsoil) may be necessary, particularly if the synergistic effects between these materials and introduced pollutants are to be modeled. This information may reflect material type and waste attenuation characteristics. If a model is to be computer accessible by the groundwater quality MIS, the information required by the model should be searchable. Otherwise, it can be stored with the narrative text associated with each station description. Frequently, information regarding the characteristics of water bearing materials is generated by drillers, during the installation of a well, and is available in the form of well logs.

HYDROLOGIC DATA

An efficient groundwater quality monitoring system will require a MIS capable of accommodating a wide range of hydrologic information. In general, this type of information has previously been determined, particularly in areas of rigorous groundwater development, and a groundwater quality monitoring program will only demand gathering and storing it. Hydrologic information is necessary to the monitoring program to predict the movement of pollutants, isolate the source of the pollution, and interpret the relationship between groundwater and surface waters.

Most hydrologic information will be station specific and can, therefore, be stored concurrently with the establishment of station descriptions in the data base. In cases where many stations penetrate the same homogeneous medium, it may be possible to store the characteristics of that medium under only one station, together with a list of the other stations common to that medium. Major hydrological data elements will include the following:

- Water bearing material depth, thickness, and areal extent
- Permeability
- Aquifer transmissivity and storage coefficient
- Hydraulic gradient (vector)
- Water table elevation (sample specific)
- Area and magnitude of natural and artificial recharge and discharge
- Station sampling device (e.g., pumped well, suction lysimeter, neutron probe, or other devices) operating characteristics

Hydrologic measurements required for computations, such as to determine hydraulic diffusivity or specific flux, will, of necessity, be stored as searchable information.

WATER QUALITY PARAMETER IDENTIFIERS

The selection of the water quality parameters to be maintained in a groundwater monitoring MIS poses one of the principal design considerations related to the development of the system. This is because of the large number of candidate variables. In many information systems, the data description (i.e., the variable identification) is imbedded in the program logic. However, because of the large number of variables involved in groundwater monitoring, a generalized data storage system is more appropriate. This requires that data identification be independent of the program; the data descriptions must themselves be data inputs to the system. Consequently, the list of water quality parameters maintained can be virtually open ended.

Stipulating the types of quality measurements to be included in a monitoring system is extremely difficult, due to the large number of potential contaminants involved. In 1972, the National Academy of Sciences (NAS) published *Water Quality Criteria – 1972* at the request of and funded by the Environmental Protection Agency (EPA). Subsequently these recommendations were presented nearly intact by the EPA in *Proposed Criteria for Water Quality* (EPA, 1973).

The NAS report propounded criteria, which would serve to preserve water quality for the following purposes:

- Public water supplies
- Agricultural uses
- Industrial uses
- Recreation and aesthetics
- Freshwater aquatic life and wildlife
- Marine aquatic life and wildlife

Normally, only the first three of these would be affected by groundwater quality. The criteria proposed by the NAS for these three use categories, and those imposed by U.S. Public Health Service (USPHS) water standards, would serve as a framework for identifying significant water quality information to be provided by a groundwater information management system (USPHS, 1962). A composite list of the parameters for which the NAS and USPHS have established criteria regarding public, agricultural, and industrial use is presented in Table 4-1.

The set of quality parameters to be examined by any individual groundwater quality monitoring program would, for the most part, be a subset of Table 4-1, which could be considered as a menu of water quality parameters. The sample set to be surveyed at any one groundwater quality surveillance station could be selected, at least partially, from this menu. Additional parameters, not appearing in Table 4-1, might be included as dictated by specific situations.

The justification for presenting Table 4-1 as such a menu rests with the fact that the NAS and the USPHS deem these parameters to be significant, as it is these parameters for which criteria have been developed. The list in Table 4-1 is by no means exhaustive, however. The inadequacy of the list for compliance monitoring purposes is reflected, for example, by *The Toxic Substances List*, published in 1973 by the U.S. Department of Health, Education and Welfare (HEW); National Institute for Occupational Safety and Health. This document identifies 11,000 *toxic*, chemically unique,

Table 4-1

MENU OF CANDIDATE WATER QUALITY PARAMETERS FOR GROUNDWATER MONITORING

Alkalinity (CaCO$_3$)	pH
Ammonia	Phenolic compounds
*Arsenic	i Phosphate
a Aluminum	+Phthalate Esters
*Barium	+Polychlorinated Biphenyls (PCB)
Boron	Radioactivity
*Cadmium	*Selenium
Chloride	*Silver
*Chromium (total)	i Silicon
Color (eg. platinum-cobalt color units)	Sulfate
Copper	i Suspended Solids
*Cyanide	Temperature
Dissolved Oxygen	Total dissolved solids (TDS)
Fluoride	
Foaming agents (MBAS)	Turbidity
Hardness	Viruses
Iron	Zinc
*Lead	*Carbon Chloroform (extractable)
Manganese	a Beryllium
*Mercury	*Total Coliform
*Nitrate-Nitrogen	*Fecal Coliform
+Nitrilotriacetate (NTA)	a Bicarbonates
Odor	a Cobalt
Oil and grease	a Lithium
Organics-Carbon Adsorbable	a Molybdenum
*Pesticides	a Nickel
Insecticides-Chlorinated Hydrocarbons	a Sodium
Insecticides-Organophosphate and Carbamate	a Vanadium
Herbicides-Chlorophenoxy	i Calcium
*Nitrite-Nitrogen	i Potassium

i Industrial impact only
a Agricultural impact only
+ No criteria currently established
* Significant health ramifications

substances (HEW, 1973). It is reasonable to assume that any one of these substances could find its way to a subsurface water reservoir, either by intentional or unintentional introduction, and achieve significance. A groundwater information management system would be required, therefore, to be flexible enough to accommodate a large and inconsistent set of variables.

As stated previously, a centralized groundwater quality MIS is called for to provide support of local efforts. In general, however, a centralized data repository would require more succinct and less detailed information than would be required by decentralized (localized) data banks. Compendiousness can be accomplished by summarization, aggregation, and the use of status indicators. The Council on Environmental Quality has funded (jointly with EPA and USGS) an ongoing study entitled *Comparative Evaluation of Techniques for the Interpretive Analysis of Water Quality*, which will provide methodologies for generating concise data, and will help to satisfy the inherent requirements of the centralized system component.

Water quality parameter identifiers will be codified and system specific. Since water quality parameters are system specific, the system administrator, rather than the DMAs, will have responsibility for depositing and maintaining this type of data in the groundwater MIS. An individual DMA can establish a special parameter identifier by petitioning the system administrator, who will judge the validity, redundancy, and applicability of the new parameter before including it in the data base.

Each water quality parameter identifier entry will consist of two data elements. One will be an alphanumeric descriptor, reflecting the common name of the parameter and the units of measure in which numeric measurements associated with that parameter will be reported. In order for the system to accommodate various units of measure, it will be necessary to assign different parameter identifiers for each one. The second data element comprising a parameter identifier will be the system-administrator-assigned numeric code associated with that identifier. Every effort should be made to organize these codes in a hierarchical fashion so that parameters of a similar nature will be grouped together.

WATER QUALITY MEASUREMENTS

The results of physical and chemical analyses of groundwater, soil, and geologic material samples will be stored as water quality measurement data, which will represent the bulk of the information to be managed by the groundwater MIS. This information will be required for both retrieval and computational operations and must, therefore, by stored as searchable data. Each measurement data element is measurement specific, and must be stored in conjunction with information which specifies the parameter measured (parameter code), the sample analyzed (sample date), and the station sampled (station identifier code). Efficient utilization of the fields set aside for analytical measurements can be realized by also using them to store sample specific data, such as sample depth or sample specific reliability indicators.

TEMPORAL DATA

In order to provide reasonable utility, a water quality information system must be capable of reflecting trends. This would require maintaining water quality data as time series. Water quality data updates need, therefore, to be appending operations, rather than destructive updates. Consequently, a water quality data base can be expected to grow monotonically and linearly, if fluctuations in the number of stations and parameters observed are disregarded.

If water quality data are collected at a constant frequency, it is only necessary to store the data collection rate and initial collection date once for each station (as station specific data). It would also be essential to make provision for entering information regarding interruptions in the period of record.

When data are not collected at a constant frequency, which is most often the case with groundwater monitoring, the date of sampling must be recorded as sample specific data, with each new set of water quality measurements input. Provision for storing dates as searchable information must be incorporated into a groundwater monitoring information system so that any subset of the period of record data set may be retrieved. It should be noted that, in contrast with surface water monitoring, recording calendar dates is usually sufficient to fix the location of a groundwater monitoring sample in time (i.e., clock times are not required). This is due to the far less dynamic nature of groundwater phenomenon.

In situations where significant vertical stratification of water chemistry is present, it will also be necessary to record and store the pumping time, in hours, prior to the collection of either a simple grab, or composite sample. Additionally, in the case of composite samples taken over time, it will be necessary to record and store the duration, in hours, of the composite sampling period.

INFORMATION QUALIFICATION DATA

There is a cogent need for a data qualification capability in any water quality monitoring information system. To accomplish this, the system should include, in addition to data verification, a comprehensive edit function, preferably computerized, which would operate prior to data storage. The edit check can be based on comparison of input data with previous trends, allowable data ranges, and established units of measure. Data failing any one of these checks should not be modified, but flagged and reported as suspect. The capability to compare input data with allowable ranges imposes an additional data requirement, which can be satisfied by storing these ranges as station specific, searchable data.

Improvements in the value of a data base can also be attained by allowing *reliability indicators* to be input, and stored as nonsearchable data. These indicators could be of the type that reflect, for example, station performance anomalies, unusual sampling conditions, unusual methods of measurement, measurement precision, or qualitative judgments of the *goodness* of data. Reliability indicators should be stored either as station specific (in the narrative text), sample specific (as a water quality measurement), or measurement specific (in a special field) as appropriate.

DMA STATUS DATA

A nationwide or statewide groundwater quality monitoring program may involve the periodic inspection of DMA facilities to determine the *operational status* of monitoring programs and equipment. In addition, where a DMA, or other agency, has groundwater pollution control functions, the *readiness status* of a control unit, in terms of its ability to respond to a pollution incident, may also be evaluated. Consequently, a comprehensive groundwater quality MIS should be capable of maintaining this type of information. Most efficiently, a DMA, or pollution control unit, would be regarded as a station by the MIS, an inspection tour regarded as a sampling iteration, and status data regarded as water quality measurements, with parameter codes being established accordingly.

The operational status of a DMA will be estimated based upon its ability to monitor the stations, parameters, and at the frequencies required. An operational status index could be established where, for example:

    1 = 100 percent monitoring effectiveness

    2 = 90 percent monitoring effectiveness

    3 = 80 percent monitoring effectiveness

    4 = 70 percent monitoring effectiveness

    5 = 60 percent monitoring effectiveness, or worse

A *readiness index* could be formulated, which would reflect the ability of a DMA, or other pollution control unit, to respond to a pollution incident. This index would be a function of personnel on hand, personnel training, equipment on hand, and equipment reliability. The readiness index could take the form of a numeric grade where, for example:

    1 = able to respond effectively within 1 day

    2 = able to respond effectively within 2 days

    3 = able to respond effectively within 3 days

    4 = able to respond effectively within 4 days

    5 = able to respond effectively within 5 days, or more

Estimating the operational and readiness ratings of individual DMAs, or pollution control units, would be the responsibility of the national or state groundwater quality monitoring program administrator.

INFORMATION INDEXING

Information indexing allows ready access to abstracts of existing data sets. The groundwater quality MIS should provide indexing of two major categories of data sets: water quality data files present in the MIS data bank, and groundwater research documentation.

Water quality data file abstracts will provide information regarding activities at each station in the monitoring program, and should be accessible by station identifier, geographical coordinates, aquifer code, political jurisdiction, station type, or agency code. Information contained in the water quality file abstract will be station specific, and would include parameters monitored, monitoring frequency, and period of record. All of the information required by the water quality data file index will exist elsewhere in the MIS so that this index can be system generated, and will not require user input.

Research documentation indexing will require special user input. Data elements to be stored, all of which should be searchable, will be document titles, author names, report numbers (access numbers), performing and sponsoring organizations, report dates, textural abstracts, keywords, and geographical area of interest.

SUMMARY

Table 4-2 presents a list of the data elements to be managed by the groundwater quality MIS.

Table 4-2

SUMMARY OF INFORMATION TO BE MANAGED BY GROUNDWATER MIS

1. System Specific Data

    Water Quality Parameter Names
    Units of Measure
    Parameter Codes

2. Station Specific Data

    Station Type
        Sample Extraction Method
        Type Data
        Type Site
        Location (i.e., saturated zone, unsaturated zone, or surface)
        Monitoring Justification
    Responsible Monitoring Agency
    Station Identifier Code(s)
    Geographic Coordinates and Associated Measurement Precision
    Station Location (township, range, section, etc.)
    Station Depth (hole depth and screen depth)
    *Field Location
    *Responsible Individual
    *Station Specific Information Qualification
    Quality Criteria
        *Demographic and Economic Data
        *Land Use
        *Water Use
        Permits
            *Stipulated Monitoring Program (Parameters and Frequency)
            *Compliance Schedules
             Discharge Limitations
        Ambient Criteria
    Political Jurisdiction Code
    Geological Data
        Aquifer Code (may be sample specific)
        Geochemical Information
    Hydrologic Data
        Aquifer Depth, Thickness, and Areal Extent
        Aquifer Transmissivity and Storage Coefficient
        Hydraulic Gradient
        Permeability
        *Area and Magnitude of Natural and Artificial Recharge and Discharge
        *Station Sampling Device Operating Characteristics

3. Sample Specific Data

    Sample Date
    Pumping Duration
    Composite Sample Duration
    Sample Depth
    Water Table Elevation
    *Sample Specific Information Qualification

4. Measurement Specific Data

    Physical-Chemical Analyses
    *Measurement Specific Information Qualification

5. DMA Status Data

    Monitoring Effectiveness Index
    Pollution Control Readiness Index

6. Information Indexing

    Water Quality File Abstracts

        Parameters Monitored
        Monitoring Frequency
        Period of Record

    Research Documentation Citations

        Titles
        Authors
        Report Numbers
        Performing and Sponsoring Agencies
        Report Dates
        Textual Abstracts
        Keywords
        Geographical Area of Interest

---

*Not searchable

DATA COLLECTION

Data collection, in the context of MIS design, is the process of translating information into machine readable form. The primary factors considered in selecting data collection systems are purchase cost, operating cost, reliability, responsiveness, and minimization of the bottleneck created by relatively high internal computer processing speeds and low input speeds.

Total MIS expenditures are particularly sensitive to data collection costs, since data entry typically accounts for 20 to 40 percent of electronic data processing costs (Ferrara and Nolan, 1973). In addition, the data entry process represents the single greatest source of error in a MIS. The significance of the imbalance between input speeds and central processing unit (CPU) speeds can be illustrated by the facts that: a keypunch operator can punch and verify roughly four cards a minute, a card reader can read about 1,000 cards a minute, and a moderately sized CPU can process about 100,000 cards or more a minute (Schwab and Sitter, 1969).

There is a wide variety of available capabilities to provide automated support of the data collection phase of a computerized MIS. These include conventional keypunch, buffered keypunch, key-to-tape, key-to-disc, remote *dumb* terminals, remote *intelligent* terminals, mark sensing, magnetic ink character recognition, and optical character recognition (OCR) devices. These nine options are listed more or less in order of increasing implementation cost and, correspondingly, increasing speed and reliability. The devices listed all have applicability to groundwater data entry. Selection of equipment by each groundwater data depositor will depend primarily upon the magnitude of data flow. If necessary, to further minimize the bottleneck, which can occur at the data input interface, buffered input units and overlapped input systems can be installed at the centralized groundwater computer data bank.

An additional category of devices available to the data depositor, which has particularly attractive applicability to groundwater monitoring, is source data automation. Source data automation is the process of capturing primary data in machine readable form. Examples of such equipment are automatic digital recorders used in conjunction with Keck groundwater level recorders, automatic laboratory chemical analysis equipment, and robot water quality monitoring stations. The advantages of source data automation are that it produces data which are easily converted into other machine usable form, reduces the opportunities for introducing errors, and lower clerical costs.

DATA COMMUNICATIONS

User interaction with a MIS can be segregated into four major activities: (1) file creation; (2) file updating; (3) information requests; and (4) information reception. A computerized MIS accomplishes these functions in one of two modes: (1) batch; or (2) real-time/interactive.

User access to the groundwater surveillance data base should be in the batch mode, whereas access to the information index system component should be in the real-time mode, at least for retrievals. Although user interaction with a batch processing system allows optional use of telecommunication links with the system, telecommunication is mandatory for real-time processing.

A telecommunication link requires a terminal to enter data, modems to encode (in a form acceptable to the transmission channel) and decode data, and a transmission channel. Transmission channels can be ordinary telephone services, such as provided by wide area telephone service (WATS) (best suited to widespread, high volume data flow); dial-up service, such as provided by teletypewriter exchange (TWX) or telephone exchange (TELEX) (best suited to widespread, low volume data flow, which is likely to be the case for groundwater surveillance); or dedicated private line (best suited to high volume data flow concentrated between a few points) (House, 1974). The major factors to be considered in the selection of a transmission service will be responsiveness, reliability, and implementation and operting costs. Data security will not be a significant consideration for groundwater surveillance information.

An ideal groundwater monitoring information system will provide flexible data flow procedures for both data submission to, and data retrieval from, the groundwater surveillance data base. The requirement for flexible data flow procedures is imposed by the desirability of wide system usage and the likelihood that data depositors and data users will have variable transmitting and receiving capabilities. It is important to note that access to the groundwater quality data base should be provided to users with unsophisticated communication capabilities, as well as to those with highly sophisticated capabilities.

All information management systems benefit from the responsiveness of real-time access to computerized data bases. However, provision of real-time capabilities does not always result in the most efficient allocation of data management resources. Since the management of groundwater surveillance data does not necessitate dynamic information flows relative to many other information management functions, real-time systems are not included in the recommendations presented below.

Data collectors should be allowed to submit groundwater data for storage (both file creation and file update) in the following modes:

- Formatted nonmachine readable
- Formatted machine readable (i.e., punch cards, paper tape, or magnetic tape)
- Remote access batch (i.e., teletype of card reader)

Data users should be allowed to request groundwater quality data via the following modes:

- Telephone inquiry
- Letter inquiry
- Teletype batch inquiry

The system should be capable of transmitting data retrievals in any of the following modes:

- Nonmachine readable hardcopy
- Punch cards
- Dial-up remote teletype or remote printer (batch)
- Magnetic tape (To promote intermachine compatibility, options for number of tracks, bits per inch, parity convention and blocked or unblocked output should be provided.)

Figure 4-1 is a diagram showing user access to the proposed groundwater MIS as well as interfile data flow. Unary data is information not subject to update, except where errors necessitate corrections. Multiple data is information subject to update (time series data), and, therefore, multiple data flow channels are likely to support a high volume of data traffic. The content of the files depicted in Figure 4-1 is described in the following section, which discusses data storage requirements.

DATA STORAGE

The development of the data storage component of a MIS entails the selection, or fabrication, of hardware devices, data organization schemes, and data base management software packages. Factors to be considered include cost, storage space, response time, and current and future use of information stored.

Three general classifications of hardware are available for data storage: internal, secondary, and external. Internal storage is best utilized for holding programs and data being immediately executed. Internal storage media include magnetic core, thin films, magnetic rods, and plated wire devices, all of which are characterized by high access speeds and costs. Secondary storage is not an integral part of, but is directly connected (on-line) to the central processing unit (CPU). Secondary storage devices include: magnetic disc, drum, card, and tape peripherals, characterized by moderate access speeds and costs. External storage is not directly connected (off-line) to the CPU. External media include: removable disc packs, magnetic tape, punched cards, and paper tapes, all characterized by low access speeds and costs (Lobel and Farina, 1970).

Figure 4-1 depicts all of the groundwater data files as being in secondary storage and resident in on-line magnetic disc or drum, both of which provide random access. Magnetic cards could also be used, but they are not widely compatible. Although drum storage allows access speeds nearly an order of magnitude greater than disc, disc storage is adequate for storing groundwater data, and will provide significant storage cost savings compared to drum storage. Additional storage cost savings can be realized, if removable disc packs are used (as external storage) and placed on-line only during certain time intervals, and if certain low priority data sets (e.g., seldomly accessed water quality data) are structured for sequential access, and archived on off-line magnetic tapes.

Data files are structured using one or a combination of three basic organizational concepts: sequential, random, and list. Sequential files store records in a specified sequence relative to other records so that the next logical record is also the next physical record. Sequential organization permits rapid access to a series of records logically related to one another, but is cumbersome for updating and retrieving individual records out of sequence.

Random organization requires the establishment of a predictable relationship between a record key and the direct address of the location where the record is stored. In most cases this will require a *dictionary look-up* process as part of each record retrieval. Random access allows rapid retrieval of individual records or data items, where only a small portion of the data file is affected, but is not well suited to retrievals of multiple records.

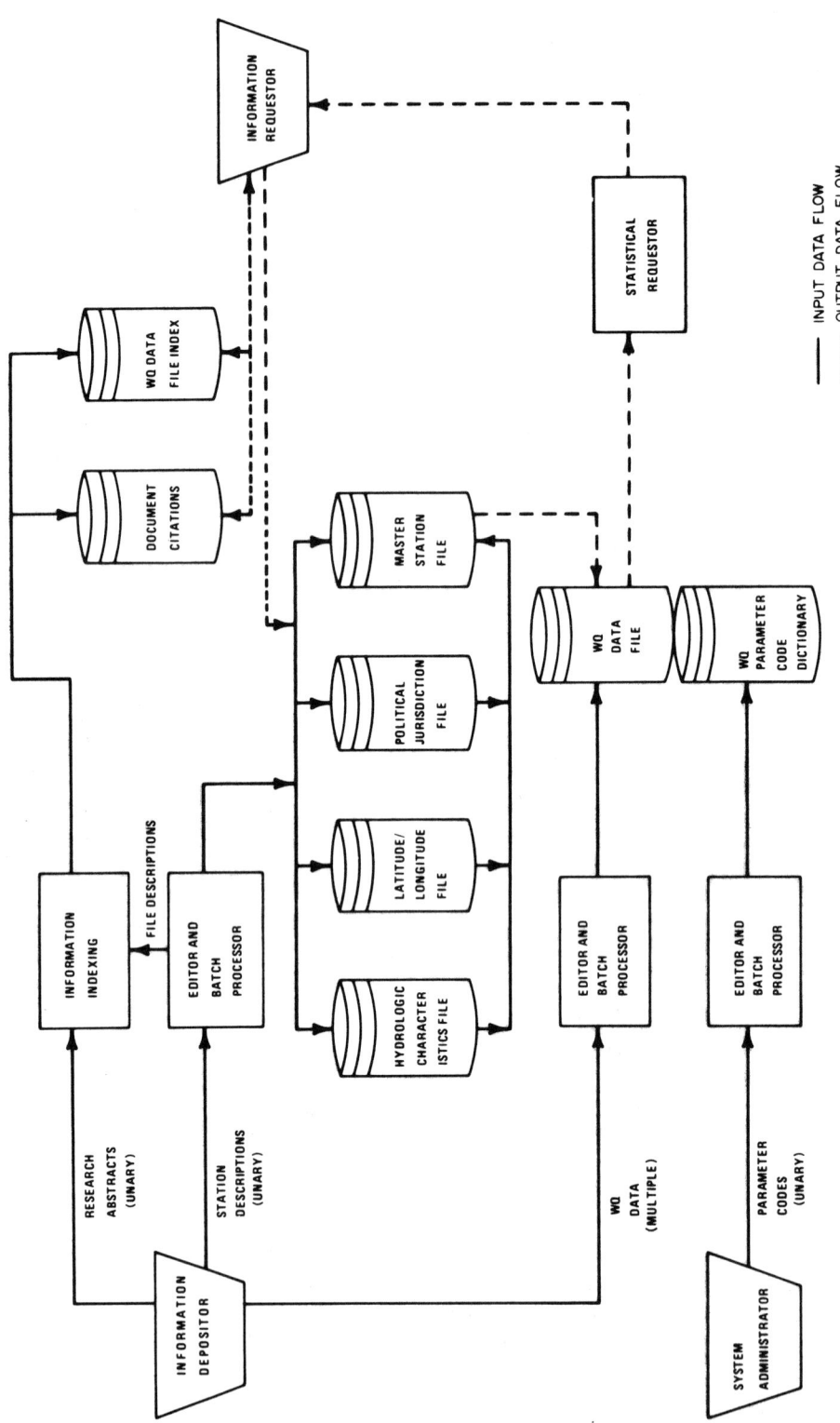

Figure 4-1. User access to groundwater data base.

List structures (simple, inverted, or ring) incorporate pointers in each record, which point to other records that are logically related to the first record. Of particular applicability to the management of groundwater data are inverted list structures, which make every data element available as a record key. For instance, a station type code could be used as the key to a record, which contained pointers to every station of that type. The inverted list approach allows very rapid (and therefore, inexpensive) retrievals, but requires a great deal of storage, and does not foster easy file updates. Therefore, inverted list structures can best be used for files, which are small and frequently accessed, but infrequently updated.

As shown in Figure 4-1, unary station descriptive data should be stored in four separate, directly accessible disc files as follows:

1. The Hydrology file will contain hydrologic characteristics of water bearing media and sampling devices as listed in Table 4-2.

2. The Latitude-Longitude file will list each station number by its latitude and longitude.

3. The Political Jurisdiction file will list each station number by its associated political jurisdiction code (state, county, city, or other political jurisdiction).

4. The Master Station file will contain all station specific data, including station specific narrative text.

The first three files described above can best be structured as inverted lists, since they will likely be frequently accessed and infrequently updated. The Master Station file can be random, using station identifiers as record keys.

The Groundwater Data File, shown in Figure 4-1, will reside in disc storage. This file will contain all sample specific and measurement specific groundwater surveillance data, as well as DMA status data (see Table 4-2). The Groundwater Data File can be organized as a random file, also, using station identifiers as record keys.

The Parameter Code Dictionary should also reside in disc storage, and be structured as a random file, using parameter codes as record keys.

The Groundwater Data File Index should be random, and use station identifiers as record keys. The Document Citation file will actually consist of a number of randomly accessible subfiles. The Master Document Citation file would contain all information regarding each document, and would be accessible by report numbers, which would serve as record keys. Additional files would list report numbers by, and, correspondingly, be keyed by, document title, author, agency, or other pertinent heading.

DATA PROCESSING

Computerized data processing is accomplished either in batches, or on a continuous (real-time) basis. Batch processing requires the accumulation and preprocessing of a group of transactions, all of which will be computer processed at one time. Real-time processing, on the other hand, accepts and processes transactions as they occur. Both processing modes can accept input data from either remote, or local terminals. The basic difference between the two processing methodologies, as seen by the system user, is the difference in response time, with the turn-around time for real-time processing being significantly quicker.

Real-time processing should be implemented only where rapid system response is really needed, since batch processing permits more efficient and economical hardware utilization, by requiring less system redundancy. Therefore, only accession to the groundwater information indexing components of the groundwater MIS requires real-time processing. This requirement is imposed by the users' need to interact intellectually (browse) with the information indexing data base.

Batch processing associated with access to the Groundwater Data File will be composed of editing, sorting, storing, retrieving, and statistical operations. Input editing will examine input data for format errors, check the validity of codes, (parameter codes and aquifer codes, among others), and compare water quality data with acceptable ranges. For compliance monitoring, the input editing module can also be used to compare water quality data with established water quality standards and prepare violation reports as necessary. The sorting and storing processes will organize the data and update the appropriate files. The retrieval commands: access the appropriate data files, organize the requested information, and format output reports. The statistical processor would function in conjunction with the retrieval routines to operate on raw data, as designated by the information requestor. The statistical processor would be required to generate extreme values, first and second moments, regression and correlation coefficients, logarithms, daily loading (for source monitoring), and coordinates necessary to create plots.

DATA RETRIEVAL

Data retrieval is the process of translating information which is meaningful only to machines into a form meaningful to humans. Designing the data retrieval component of an MIS requires identifying the information to be output, specifying the retrieval procedures acceptable to the system, developing the required retrieval software, determining output formats, and selecting hardware.

The data retrieval component of the proposed MIS, which accesses the Groundwater Data File, will be required to yield both alphanumeric and pictorial output. The system should be capable of providing alphanumeric output, which will include the following types of information:

- Measured parameters
- Number of observations
- Beginning and ending sampling dates
- Raw data
- Minimum and maximums
- Arithmetic means
- Standard deviations
- Regression coefficients
- Correlation coefficients
- Percentiles
- Confidence intervals
- Daily loadings
- Logarithms
- Station descriptive paragraphs

User requests for pictorial (graphic) displays may require the following types of plots:

- Physical-chemical variations with time
- Physical-chemical variations with sample depth
- Monitoring stations located geographically
- Physical-chemical variations with distance
- Vertical bar charts
- Circular diagrams
- Radial vector diagrams
- Pattern diagrams
- Trilinear diagrams

The last three information presentation techniques listed above, which may be unfamiliar to some readers, are described by Hem (1959).

The groundwater monitoring MIS can offer the data user the most powerful capabilities, if it can provide a wide range of useful retrieval procedures. A retrieval procedure is characterized by the information required by that procedure as user input to the system to enable the system to locate data and generate output.

The groundwater monitoring MIS user should be able to request data from the system by specifying one or a combination of the following information elements:

- Station number
- Range of station numbers
- Latitude and longitude
- Polygon (specified by the latitude and longitude of its vertices)
- Political jurisdiction
- Sampling date
- Range of sampling dates
- Sampling depth
- Range of sampling depths
- Monitoring agency
- Maximum or minimum parameter values

The user should be able to implement a number of these procedures in conjunction with each other so that Boolean retrieval strategies can be applied. In addition, he should be able to request that various data manipulation and statistical operations be performed, and to dictate, to some extent, the format of the output he receives.

Factors involved in the selection of data retrieval hardware include considerations of speed, cost, flexibility, reliability, noise, number of copies needed, and formatting (i.e., requirements for number of characters per line, number of lines per page, and plot sizes). Retrieval hardware can be categorized according to the following distinctions:

- Impact, non-impact, cathode ray tube (CRT), digital plotter, microfilm, or voice response
- Serial, which produces 10 to 200 characters per second (cps) or parallel, which produces 300 to 10,000 cps (Lorber, 1972)
- Full character or dot matrix

In general, impact printers produce full characters either one at a time (serially) or a line at a time (parallel). Impact printers provide good legibility and multiple copies (a constraining factor for many applications), but, in general, are noisy and subject to relatively frequent breakdowns because of the large number of moving parts they require.

Non-impact printers will best satisfy the requirements of accessing groundwater monitoring data. Non-impact printers can be either serial or parallel printers, which, in a majority of machines, produce dot matrix characters. Ink-jet and electrostatic printers are two types of non-impact printers offering speed, reliability, portability, competitive purchase cost, and quiet operation. The disadvantages normally characteristic of these devices are high operating costs (e.g., electrostatic printers require special paper), the inability to produce multiple copies, and slightly poorer image quality than is provided by impact printers.

CRT displays produce dot matrix characters, either serially or in parallel, as well as graphics. Although CRT displays, themselves, are unable to generate permanent records, they are fast, reliable, and economical to purchase and operate. In addition, these devices afford great flexibility by virtue of the optional peripheral equipment, which may be attached, such as hard copy output, light pens, and information storage capabilities. CRT terminals would be most appropriate for accessing groundwater information indexing files.

Digital plotters capable of producing permanent graphic displays, are available at a wide range of prices and, correspondingly, with a wide range of capabilities. Microfilm systems can receive output directly from a CPU, via either paper to film or CRT to film, and provide the advantages of a compact, inexpensive, external storage medium. Microfilm systems generate output in the form of microfilm (normally 16 mm film), aperture cards (normally 35 mm film), or microfiche (which records many pages of data on one frame of film).

With the exception of voice response units, used most extensively by operations which interface with the public, any of the above mentioned hardware options may find appropriate applications in a groundwater monitoring program. The selection of specific retrieval hardware components will depend upon the requirements of individual data requestors and of interfacing with the central system. The central system should be designed to be flexible so that it represents a minimal constraint on the selection of user output hardware.

SECTION 3 – EXISTING SYSTEMS

GENERAL

This section presents a survey of existing or proposed information management systems relevant to the management of groundwater monitoring information. Table 4-3 lists some of the water resources data management systems currently operational, together with some of their more pertinent characteristics. Table 4-4 presents a selection of computerized information indexing systems, both operational and proposed, which provide data file or research documentation abstracts. Table 4-5 presents several generalized data base management packages, offered by various commercial vendors, which may possess capabilities suited to the needs of specific groundwater data management efforts.

The following discussion describes in further detail some of the more pertinent systems listed in Tables 4-3 and 4-4. Readers with particular interest in one of these systems are referred to the associated users and systems documentation.

STORET

The Storage and Retrieval System (STORET) was developed initially by the U.S. Public Health Service, and is currently operated by the U.S. Environmental Protection Agency, where it is undergoing further development. This system is intended to provide federal assistance to the states in the performance of water quality management, and insure compliance with PL 92-500. Table 4-6 presents a list of those sections of PL 92-500 supported by STORET. Providing the states access to a centralized information retrieval system realizes economies primarily in the areas of system maintenance and user assistance. To date, 42 of the states are utilizing STORET.

The STORET system consisted of two basic files: the Water Quality File (WQF) and the General Point Source File (GPSF). Primarily because of high operating costs, the GPSF was deactivated during February of 1975 and is to be replaced by a less expensive, but also less powerful, generalized information retrieval system called the "Interim Enforcement System." One aspect of this interim measure will be the provision of the capability to store self-monitoring and compliance data in the WQF, with each discharger being treated as a station, and SPDES permit numbers serving as station identification numbers.

The WQF measures the ambient quality of water bodies throughout the nation, whereas the GPSF measured the quality of point source discharges throughout the nation. The software, which updates, manipulates, and retrieves data from these files, is coded in the PL/1 programming language. Updates and retrievals are done in the batch mode, with input provided by card readers or low to medium speed remote terminals. Output reports are generated on a demand basis only.

The WQF contains information, which can be segregated into three categories. The first of these categories consists of information describing the source of water quality samples (i.e., water quality monitoring stations). This descriptive information is required only when the stations are established, in or deleted from the STORET system data base, or when the descriptive information is changed. The input data content and format for station descriptions is presented in Figure 4-2. Header cards 1, 2, 3, 4 and 5 are optional inputs. A detailed description of the procedure for using all of these station storage cards can be found in available STORET literature. Only a brief description of the mandatory agency and station cards is provided here.

Table 4-3

EXISTING ENVIRONMENTAL DATA MANAGEMENT SYSTEMS

System Name	Acronym	Administrator	Information	Storage Location	Groundwater	Computer System
Storage and Retrieval System (1)	STORET	EPA	Water quality	Centralized	43,000 wells	IBM 371/58-OS/MVT
National Water Data Storage and Retrieval System (2)	WATSTORE	USGS	Surface and ground water physical and chemical data	Centralized	25,000+ wells	IBM 370/155
ORSANCO Robot Monitor System (3)	-	Ohio River Valley Water, Sanitation Commission	Surface water quality	Centralized	None	IBM 1130
Groundwater Quality System (4)	-	California	Groundwater quality and hydrographic	Centralized	1,400 wells	CDC 3300
Water Information System for Enforcement (5)	WISE	Michigan DNR	Water quality and discharge inventory	Centralized	22 WQ wells	Burroughs B5500
Tennessee State Groundwater Data Retrieval System (6)	-	Tennessee Department of Conservation	Groundwater yield and quality	Centralized	75,000 wells 800 springs	IBM 370-OS
Well Hydrograph Data Storage and Retrieval System (7)	DSWELL	ERDA Hanford	Well hydrograph	Centralized	300 wells	PDP-9
Groundwater Observation Well Network (8)	GOWN	Canada	Well logs, well data, hydrographs	Centralized	75,000 wells	IBM 360/165
Arizona Water Information System (9)	AWIS	Arizona Water Commission	Water resources	Centralized	2,500+ wells	DEC-10

1. EPA, 1971.
2. Edwards, 1974.
3. Klein et al., 1968.
4. Welsh, 1973.
5. Guenther et al., 1973
6. Wilson et al., 1972.
7. Friedrichs, 1972.
8. Gilliland and Treichel, 1968.
9. Foster and DeCook, 1974.

Table 4-4

COMPUTERIZED INFORMATION INDEXING SYSTEMS

System Name	Acronym	Administrator	File Content	Retrieval Options	Subject	File Size	Computer System
Remote Control System	RECON	ERDA	Document citations	Keywords, publishers countries, authors, etc.	Energy/ Environmental	700,000 citations	IBM 360/75
General Information Processing System	GIPSY	University of Oklahoma	Document citations	Author, any word(s) in abstract, title	Selected water resources abstracts*	80,000 citations*	IBM 360/65*
Environmental Data Index**	ENDEX	NOAA	Data file descriptions	Geographic area (sq), institution, discipline	Environmental	3,000 file references	IBM 360/65
Oceanic and Atmospheric Scientific Information System**	OASIS	NOAA (Environmental Data Service, 1974)	Document citations	Title, keyword, author, publication, etc.	Atmospheric, water and earth resources	10,000,000 citations, 33 files	IBM 360/65 plus others
National Water Data Exchange	NAWDEX	USGS	Type and sources of water data	Station code, WRC Basin code, Lat/Long.	Surface and ground water	Developmental	IBM 370/155
World Science Information System	UNISIST	UNESCO	Type and source of Global Research Documentation	Developmental	Scientific	Developmental	Developmental
International Referral System	IRS	U.N. Environmental Program, Nairobi	Type and source of Global Research Documentation	Developmental	Environmental	Developmental	Developmental
Water Resources Information Program		University of Wisconsin - Madison	Document citations	Free form questions	Water Resources	70,000 citations+	IBM 360/75
Smithsonian Science Information Exchange	SSIE	Smithsonian Science Information Exchange, Inc.	Research in progress	Free form queries	Scientific	170,000 research projects	IBM 370/135

* Department of the Interior, Water Resources Scientific Information Center information base.
** GIPSY is also used to access some modules of the ENDEX and OASIS data bases.

Table 4-5

GENERALIZED DATA BASE MANAGEMENT PACKAGES

System Name	Vendor	Purchase Price	Minimum Core Req'd	Compatability	Applicability
DYL-250/260	Dylakov Computer Systems, Inc.	$8 K+***	32 K	IBM 360/370	Index sequential files, report writing
IMS	IBM	$1316/mo.***	128 K	IBM 360/370	Extremely flexible but complex
MARK IV	Informatics, Inc.	$7.5-35 K**	20 K	IBM 360/370, Univac 70/90/9400	Infrequent, large retrievals from large data base
SYSTEM 2000	MRI System Corp.	$1 -4 K/mo.*	128 K	IBM 360/370, CDC 6000, Univac 1106, 1108, 1110	On-line direct access, inverted files
TOTAL	Cincon Systems, Inc.	$26,500+***	31 K(avg.)	IBM 360/370, CDC Cyber, H 200/2000, Univac 70	Complex interrelationships between data files
ADABAS	Software AG	$120 K***	30 K	IBM 360/370, Univac 9000	Extremely large data bases, many files
PANVALET	Pansophic Systems, Inc.	$5 K***	50 K(avg.)	IBM 360/370	Library maintenance
RAMIS	Mathematica, Inc.	$28 K***	120 K	IBM 360	Hierarchial structured data bases
RSVP	Honeywell Information Systems, Inc.	$ 4 K***	22 K	IBM 360/370	User oriented

* (Gelke, 1972)
** (Steig, 1972)
*** (Datapro Research Corporation, 1974)

Table 4-6

STORET SUPPORTED SECTIONS OF PL 92-500
(after Conger, 1975)

Title I - Research and Related Programs

[1,3] Sec. 104   - Research, Investigations, Training and Information

[1,3] Sec. 104(a)(5) - National Water Quality Surveillance System (NWQSS)

[2,3] Sec. 105   - Grants for Research and Development

[3] Sec. 106   - Grants for Pollution Control Programs

[3] Sec. 107   - Mine Water Pollution Control Demonstrations

Sec. 108   - Pollution Control in the Great Lakes

[3] Sec. 113   - Alaska Village Demonstration Projects

Sec. 114   - Lake Tahoe Study

Title II - Grants for Construction of Treatment Works

[3] Sec. 201   - Construction Grant Facility Plan

[3] Sec. 208   - Areawide Waste Treatment Management Plan

[3] Sec. 209   - Basin Planning

[2] Sec. 210   - Annual Operation and Maintenance Survey

Title III - Standards and Enforcement

[3] Sec. 303   - Water Quality Standards and Implementation Plans

Sec. 303(e)   - River Basin Water Quality Management Plans

Sec. 305(b)   - Water Quality Inventory

[3] Sec. 308   - Inspections, Monitoring and Entry

[3] Sec. 311   - Oil and Hazardous Substance Liability

Sec. 314   - Clean Lakes

[3] Sec. 315   - National Commission on Water Quality

Sec. 316   - Thermal Discharges

Sec. 318   - Aquaculture

Title IV - Permits and Licenses

[3] Sec. 402   - National Pollutant Discharge Elimination System

Sec. 403   - Ocean Discharge Criteria

Sec. 404   - Permits for Dredged or Fill Material

Title V - General Provisions

[2,3] Sec. 516   - Reports to Congress

[3] Sec. 516(b)   - Economics of Clean Environmental Report

---

[1] Requires Federal information management support.

[2] Requires dissemination of information.

[3] Groundwater implications.

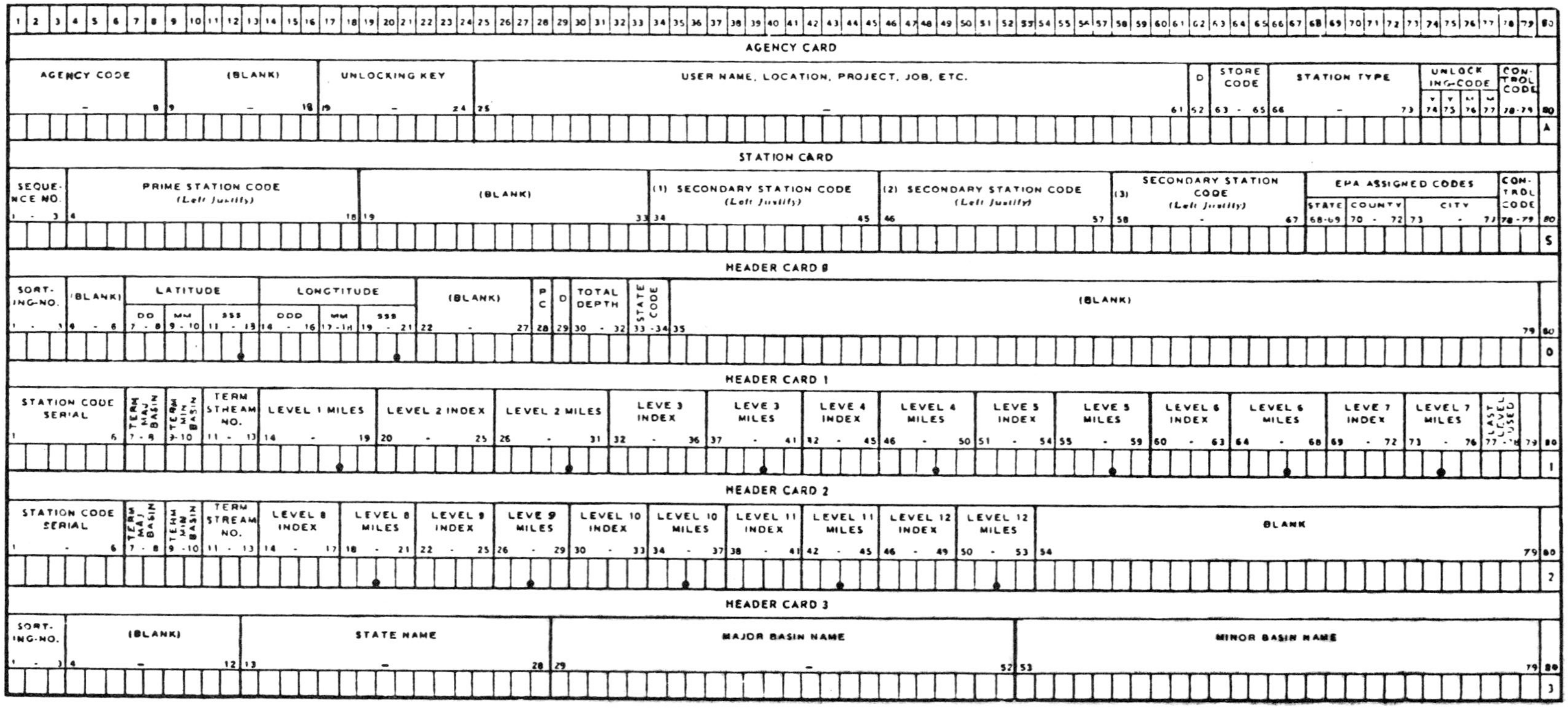

Figure 4-2. STORET system-station storage format.

Figure 4-2 -- Continued

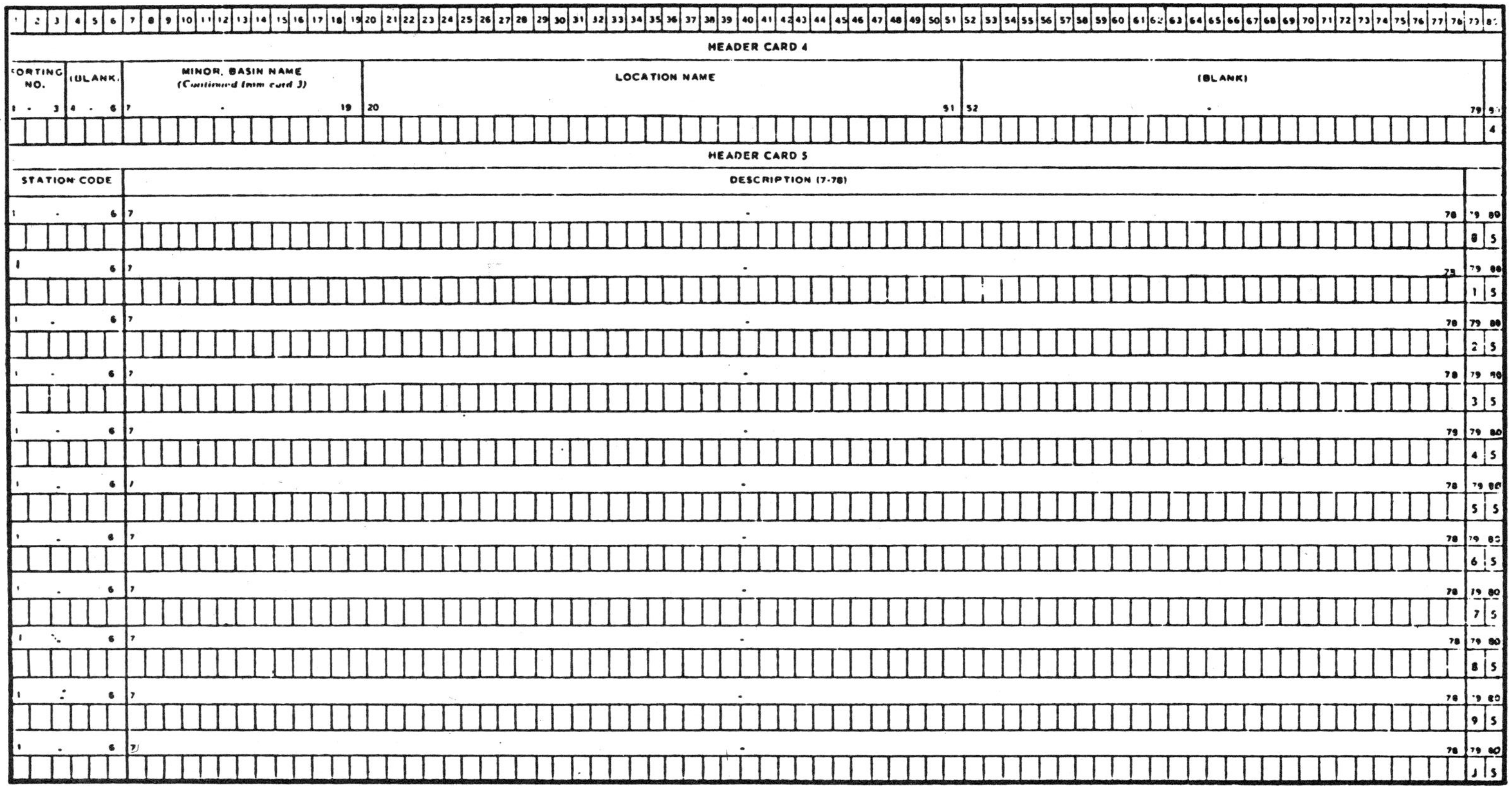

The agency header card contains general information pertaining to a station, or group of stations, involved in a single station storage or retrieval operation. The agency header card is used in the following manner:

- The agency identifier, which associates data with the contributing organization, must be provided in columns 1 through 8.

- An *unlocking key* is an alphanumeric code, which is input via columns 17 through 14 of the agency card, and which is mandatory for all station storage and, if requested by the data contributor, for all retrieval operations.

- Columns 25 through 61 are provided to accommodate the name, location, and telephone number of the individual responsible for storing the station description. Information in this field, though required as input, is not stored as a part of the STORET data base.

- Column 62 is used to record the units in which the sample depths are to be reported, and allows the entering of either an F (feet) or an M (meters).

- Columns 63 through 65 may be used (optionally) to stipulate (by inputting a 1 in the appropriate column) that one desires to store latitude-longitude, RMI code, and/or the state-county-city code as a secondary station number(s).

- Columns 66 through 73 *must* be used to provide a station type code. Station type codes are constructed as shown in Figure 4-3.

- Columns 74 through 77 are used to stipulate the date *after* which data cannot be retrieved without providing an unlocking key.

- Columns 78 and 79 are used as a card use control, and are coded with a CD to change the unlocking date, a CT to change the station type, or left blank for other types of operations.

- For the agency card an A is required in column 80.

The station card is also mandatory and provides a vehicle for inputting information specific to individual stations. In addition to its use in the establishment of a station in the STORET data base, it is also used to delete a station, to change a station location, or to update water quality data. The station card is completed as follows:

- The first field of the card, columns 1 through 3, will contain a sequence number, which corresponds to the entries in the same field of all other location and water quality cards for the same station. This field is not stored, but used for resorting in the event the card deck becomes disarranged.

- The second field, columns 4 through 18, is used to enter the primary station code (alphanumeric) into storage. Only the first 6 characters of this field are used.

- The three fields consisting of columns 34 through 45, 46 through 57, and 58 through 67 are used to store secondary station codes, if required. Secondary station codes are used in the event, for example, that several organizations are storing water quality information derived from the same station, but have assigned different codes to that station.

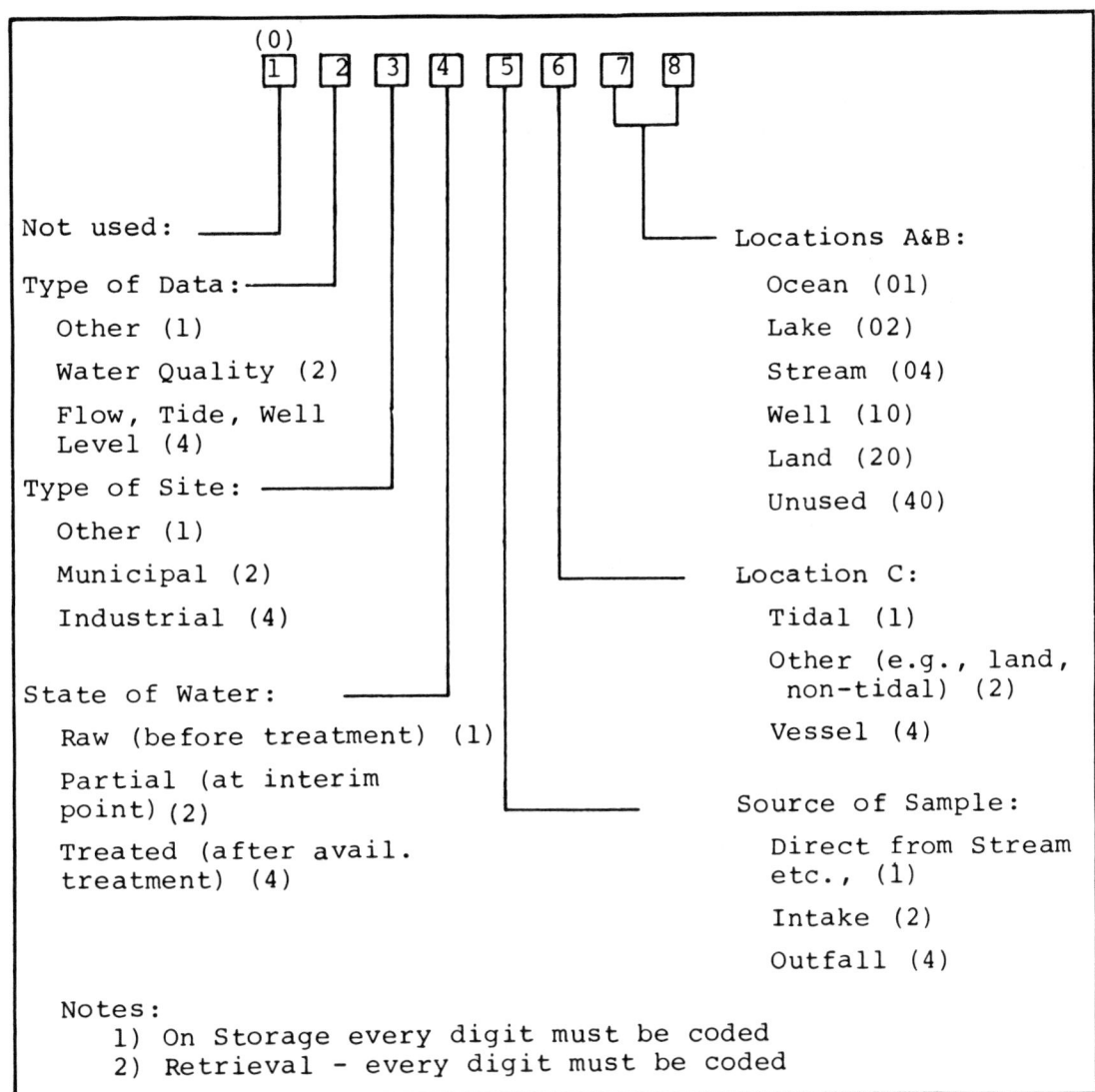

Figure 4-3. STORET system-station type codes.

- The next three fields (columns 68 through 69, 70 through 72, and 73 through 77) are numeric, and are used to store state, county, and city codes, respectively. State and county codes are those adopted by the National Bureau of Standards. City codes are based upon codes adopted by the U.S. Postal Service.

- Columns 78 and 79 are for card use control. This field is coded with an "NS" if an original storage is being executed, with "DD" to delete all data associated with a station, with "DS" to delete both the station and all data associated with the station, with "CN" to change secondary station numbers, with "CC" to change or delete station descriptive data, and with blanks for water quality data updates.

- Column 80 of the station card is coded with an S.

Sampling stations are located areally by stipulating either geographical coordinates (header card 0), hydrologic index (header cards 1 and 2), or both. Locating sampling stations by geographical coordinates allows the retrieval of data from all stations located within a polygon simply by specifying the vertices of that polygon.

Hydrologic indexing, referred to as River Mile Index (RMI) coding, offers an extremely powerful tool, since it defines the hydrologic relationship between a sampling point and the rest of the river system. A complete RMI requires between 15 and 112 numeric characters, and is composed of the following codes:

- Major basin code (2 characters)
- Minor basin code (2 characters)
- Terminal stream number (3 characters)
- Indexes defining direction and level of flow
- Mileages between confluences
- Stream level code (2 characters)

Use being made of the various station locating schemes, as of November 1975, is presented below (Conger, 1974).

	Stations
Total	197,000
RMI	28,000
Geographic	160,000
Both (RMI and Geo.)	15,000
Neither	24,000
Political	193,000

Although RMI coding represents a useful tool, relatively little use is being made of it, undoubtedly because of the level of effort required to generate the code. Most of the stations stored in the WQF using both RMI and geographic coordinates are located in only two areas, the Tennessee and Columbia River basins.

The second category of information stored in the WQF data base is water quality parameter identification. Each water quality measurement stored in the file must be accompanied by a numeric 5-character parameter identifier code. The 5-character water quality parameter identifier code is stored in a 3-byte field in packed decimal format, which allows the storage of 2 numeric characters per byte. The parameter identifier codes are also stored in a cross reference (dictionary) file with the alphanumeric descriptors, which the codes represent.

The WQF can store up to 100,000 parameter identifiers, but only about 2,000 identifiers are currently stored. Eighty-five percent of the water quality data in the WQF is stored under only 187 of the existing identifiers. An effort has been made to commit specific ranges of parameter codes to sets of parameters with similar characteristics. For example, the range of codes 00300-00365 has been dedicated to measurements of oxygen demand.

Of particular interest is the fact that the range of codes from 84,000 to 84,999 has been set aside for identifiers pertinent to groundwater monitoring. To date, the code 84,000 has been designated as a geologic age code, and 84,001 as an aquifer name code. The remainder of the range is uncommitted. Additional parameter codes established specifically to accommodate groundwater monitoring are presented in Table 4-7.

Table 4-7

ESTABLISHED STORET PARAMETER CODES – GROUNDWATER SPECIFIC (EPA, 1971) CELSIUS (BM*)

Code	Output Format**	Parameter Description
72000	xxxxxx.x	Elevation of land surface datum (ft*** above MSL)
72001	xxxxxx.x	Total depth of hole (ft below land surface datum)
72002	xxxxxx.x	Depth to top of water-bearing zone sampled (ft)
72003	xxxxxx.x	Depth to bottom of water-bearing zone sampled (ft)
72004	xxxxxx.x	Pump or flow period prior to sampling (min)
72005	xxxxxxxx	Sample source code (BM* well data)
72006	xxxxxxxx	Sampling condition code (BM* well data)
72007	xxxxxxxx	Formation name code (BM* well data) (AAPG**** code)
72008	xxxxxx.x	Total depth of well (ft below land surface datum)
72009	xxxxxx.x	Elevation of land surface in feet (BM*)
72010	xxxx.xxx	Resistivity (Ω-m) (BM* well data)
72011	xxxx.xxx	Acids, organic (mg/ℓ) (BM* well data)
72012	xxxxx.xx	Specific gravity, temperature, degrees Celsius (BM*)
72013	xxxx.xxx	Specific gravity (BM* well data)
72014	xxxxx.xx	Resistivity, temperature, degrees Celsius (BM*)
72015	xxxxxx.x	Depth to top of sample interval (ft below LSD)
72016	xxxxxx.x	Depth to bottom of sample interval (ft below LSD)
72017	xxxxxxxx	Series code (BM* well data)
72018	xxxxxxxx	System code (BM* well data)
72019	xxxxx.xx	Depth to water level (feet below land surface)
72020	xxxxx.xx	Elevation in feet above MSL
72040	xxxxx.xx	Observed drawdown (ft)
72041	xxxxx.xx	Specific capacity in gpm/ft of drawdown
72042	xxxxxx.x	Pump efficiency (%)
72043	xxxxx.xx	Brake horsepower
72044	xxxxx.xx	Total dynamic pumping head (ft)
72045	xxxxx.xx	Pumping cost in dollars per thousand gallons
72050	xxxxxx.x	Withdrawal of groundwater (millions of gallons/month)
72051	xxxxxxx.x	Withdrawal of groundwater (millions of gallons/year)
84000	xxxxxxxx	Geologic age code (USGS)
84001	xxxxxxxx	Aquifer name code (USGS)

*BM – Bureau of Mines
**Can be modified at retrieval
***See Appendix for conversion to metric units
****American Association of Petroleum Geologists

The third category of information in the WQF is the water quality measurements themselves, together with the depth of the sample and the date and time the sample was taken. The water quality measurements are stored in 4-byte words, in standard IBM 370 floating point format (single precision).

Originally the input module of the STORET system was designed to store only numeric data in the water quality measurement field. System modifications have been accomplished, however, that allow the storage of alphabetic characters, required for aquifer descriptions, in the fields associated with parameter codes 84,000 through 84,999.

The STORET Water Quality File also allows remarks to be input along with water quality measurements. The system accepts remark codes into a 1-character (1-byte) field, one of which has been set aside for each water quality measurement field. The remarks are stored in "Extended Binary Coded Decimal Interchange Code" (EBCDIC) which allows any one of 256 alternative remark codes. These remarks are used, for example, to indicate that the stored data element is not accurate, a field measurement, a lab measurement, a lower limit, or an upper limit.

Recent cost and use data for the WQF are presented below (Notzon, 1975):

Annual operating costs excluding EPA personnel	$1,100,000
Federal, state and local users	240
Cost per user per year	$3,667
Observations stored annually	8-10 million
Observations presently in system	30,000,000
Data acquisition cost	$150-300 million
Annual storage cost per observation	$ .01
Processing cost per observation	$ .011
Retrievals/analysis per year	46,000
Retrieval/analysis cost per job (avg.)	$7.58

The General Point Source File consisted of an inventory of discharges and abatement plans. More specifically, the GPSF contained the following information:

1. Inventory of municipal dischargers

2. Inventory of industrial dischargers

3. Inventory of municipalities

4. Fish kills

5. Agricultural permits

6. Mine drainage permits

7. Deep well injection survey

8. Municipal drinking water supplies

9. Construction needs survey

10. Ocean dumping permits

11. Federal government discharges

12. Grant information

WATSTORE

The National Water Data Storage and Retrieval System (WATSTORE) was implemented in 1971, with the objective of providing the Water Resources Division of the USGS with a comprehensive water data management capability. The system is computerized and operated at the facilities of USGS in Reston, Virginia. Access to WATSTORE is through a telecommunication network, which provides data services to 46 district offices throughout the country. Data are input to WATSTORE by remote entry from laboratories and data centers.

The system data base consists of a "Station Header File," which maintains an index of stations, and provides access to the following files:

- The "Daily Values File" contains physical and chemical data reported daily.
- The "Water Quality File" contains the results of analysis (chemical and physical) of all samples taken. This includes groundwater samples taken on an infrequent and irregular basis.
- The "Peak Flow File" contains annual maximum discharge and stage values for surface water sites.
- The "Groundwater Site Inventory File" contains physical, topographic, aquifer hydraulic, and text data pertinent to groundwater monitoring sites. Parameters maintained in this file are presented in Table 4-8.

WATSTORE retrieval capabilities enable the output of text, tabular, and graphic reports. Retrieval options include individual station, station type (e.g., wells), specific periods, polygon, political, aquifer code (for groundwater sites), and individual parameter retrievals. In addition, data for a particular parameter, which falls within a specified range, may be retrieved.

The WATSTORE system is designed to recognize the possibility that a groundwater monitoring station (well) can penetrate more than one aquifer, and that samples can be drawn from individual aquifers separately, with the use of screen plugs. Therefore, WATSTORE allows for the storage of aquifer identifiers along with the water quality analysis data for each sample. The aquifer identifier are stored as 8-character codes based on the stratigraphic coding system proposed by the American Association of Petroleum Geologists.

Table 4-8

PARAMETERS MAINTAINED IN WATSTORE GROUNDWATER SITE INVENTORY FILE (BAKER, 1975)

Site Id
Site Type
Record Classification
Source Agency
Project Number
District
State
County
State County*
Latitude
Longitude
Coordinate Accuracy
Local Number
Land Net Location
Location Map Id
Location Map Scale
Altitude
Altitude Method
Altitude Accuracy
Topographic Setting
OWDC Hydrologic Unit
Date Constructed
Date Const. Acc.*
Site Use
Water Use
Second Water Use
Third Water Use
Hole Depth
Well Depth
Well Depth Source
Water Level
Water Level Date
WL Date Accuracy*
Water Level Source
Water Level Method
Site Level Status
Pump
Geohydro Data Source
Last Update*
Verified*
Hydraulic Data
  Hydraulic Seq. No.
  Hydraulic Unit Id.
  Test Interval Top
  Test Interval Bottom
  Hydraulic Unit Type
  Hydraulic Remarks
  Coefficients
    Coef. Seq. No.
    Transmissivity
    Horizontal Cond.
    Vertical Cond.
    Storage Coef.
    Leakage
    Diffusivity
    Specific Storage
Quality Network(QN)
  QN Begin Year
  QN End Year
  QN Data Source
  QN Frequency
  QN Analysis Type

Lift
  Lift Type
  Lift Date
  Lift Date Accuracy*
  Intake Setting
  Power Type
  Horsepower
  Major Pump
    Manufacturer
    Serial Number
    Power Company
    Account
    Meter
    Consumption
    Pump Maintainer
  Standby
    Standby Power Type
    Standby Horsepower
Geohydrologic Units
  Geohydro Top
  Geohydro Bottom
  Geohydro Unit
  Lithology
  Lithology Modifier
  Aquifer
    Aquifer Date
    Aquifer Date Acc.
    Aquifer Static Level
    Aquifer Contribution
Remarks
  Remarks Date
  Remark

Pump Production
  Pump Seq. No.
  Pump Meas. Date
  Pump Date Acc.*
  Pump Discharge
  Pump Discharge Source
  Pump Discharge Method
  Pump Prod. Level
  Pump Static Level
  Pump Level Source
  Pump Level Method
  Pump Period
Owners
  Ownership Date
  Ownership Date Acc.*
  Last Name
  First Name
  Middle Initial
Minor Repairs
  Repair Seq. No.
  Repair Nature
  Repair Date

Construction
  Const. Sequence No.
  Date Completed
  Const. Date Accuracy*
  Contractor
  Const. Data Source
  Const. Method
  Finish
  Seal Type
  Seal Bottom
  Development Method
  Development Duration
  Special Treatment
  Holes
    Hole Top
    Hole Bottom
    Hole Diameter
  Casings
    Casing Top
    Casing Bottom
    Casing Diameter
    Casing Material
    Casing Thickness
  Openings
    Opening Top
    Opening Bottom
    Opening Type
    Screen Material
    Opening Diameter
    Opening Width
    Opening Length
Site Visits
  Inventory Date
  Inventory Person

Other Ids
  Other Id
  Other Id Assigner
Field Water Quality
  FWQ Sample Date
  FWQ Date Acc.*
  FWQ Geohydro Unit
  FWQ Parameter
  FWQ Measurement
Logs
  Log Type
  Log Top
  Log Bottom
  Log Source
Well Group(WG)
  Number Wells
  WG Deepest
  WG Shallowest
  WG Method
Pond Tunnel Drain
  PTD Length
  PTD Width

Table 4-8 (Continued)

```
Level Network (LN)          Repair Date Acc.*        PTD Depth
  LN Begin Year             Repair Contractor        Cooperator Data
  LN End Year               Performance Changes        Cooperators Id
  LN Data Source          Springs                      Contractor Reg. No.
  LN Frequency              Spring Name                Inspection Status
Pumpage Network (PN)        Spring Type                Reason Unapproved
  PN Begin Year             Permanence                 Date Inspected
  PN End Year               Discharge Sphere           Cooperator Remarks
  PN Data Source            Improvements             Laterals
  PN Frequency              Number Spring Openings     Lateral Number
  PN Data Method            Flow Variability           Lateral Depth
Flow Data                   Flow Var. Accuracy         Lateral Length
  Flow Seq. No.           Other Data                   Lateral Diameter
  Flow Meas. Date           Type Data                  Lateral Mesh
  Flow Date Acc.*           Data Location
  Flow Discharge
  Flow Discharge Source
  Flow Discharge Method
  Flow Prod. Level
  Flow Static Level
  Flow Level Source
  Flow Level Method
  Flow Period
```

*System-Generated

The 8-character code consists of three parts. The first 3 characters are numeric, and identify the geologic age (Erathem, System and Series, respectively) of the aquifer, as shown in Table 4-9. The next 4 characters constitute an alphanumeric mneumonic code, which specifies the name of the rock-stratigraphic unit. The rock-stratigraphic unit name code is generated by the use of an algorithm, which specifies the order in which characters are to be eliminated from the original term until only 4 remain. The last character of the 8-character is optional, and provides for modifiers of the rock-stratigraphic unit name. For example, the complete code for the Pliocene Upper Pico Formation in California is 121PICOU.

The WATSTORE system currently stores data for several hundred different water quality parameters. The list of water quality parameters is open ended, and is expended as necessary. The water quality parameters stored in WATSTORE are coded with a 5-character code established in cooperation with the EPA STORET User Assistance Branch so that the parameter codes are the same in both systems. WATSTORE is equipped with a module which generates STORET input corresponding to WATSTORE data file updates. The input formats for storing data in the WATSTORE Water Quality File are presented in Figure 4-4.

NAWDEX

The National Water Data Exchange (NAWDEX) is a developmental computerized information indexing capability being implemented by the Water Resources Division of the U.S. Geological Survey. This effort resulted from a determination by the U.S. Department of the Interior that accessibility to water data on a national scale required upgrading.

Table 4-9

USGS NUMERIC CODES FOR GEOLOGIC AGE IDENTIFICATION
(Price and Baker, 1974)

Age	Code	Age	Code
Unknown Age	000	Paleozoic (cont'd)	
		Middle	324
Cenozoic	100	Des Moinesian	325
Quaternary	110	Atokan	326
Holocene	111	Lower	327
Pleistocene	112	Morrowan	328
Tertiary	120	Mississippian	330
Pliocene	121	Upper	331
Miocene	122	Chesterian	332
Oligocene	123	Meramecian	333
Eocene	124	Lower	337
Paleocene	125	Osagean	338
		Kinderhookian	339
Mesozoic	200	Devonian	340
Cretaceous	210	Upper	341
Upper	211	Chautauquan	342
Gulfian	212	Senecan	343
Lower	217	Middle	344
Comanchean	218	Erian	345
Coahuilan	219	Lower	347
Jurassic	220	Ulsterian	348
Upper	221	Silurian	350
Middle	224	Upper	351
Lower	227	Cayugan	352
Triassic	230	Middle	354
Upper	231	Niagaran	355
Middle	234	Lower	357
Lower	237	Ordovician	360
		Upper	361
Paleozoic	300	Cincinnatian	362
Permian	310	Middle	364
Upper	311	Champlainian	365
Ochoan	312	Lower	367
Guadalupian	313	Canadian	368
Lower	317	Cambrian	370
Leonardian	318	Upper	371
Wolfcampian	319	Middle	374
Pennsylvania	320	Lower	377
Upper	321		
Virgilian	322	Precambrian	400
Missourian	323		

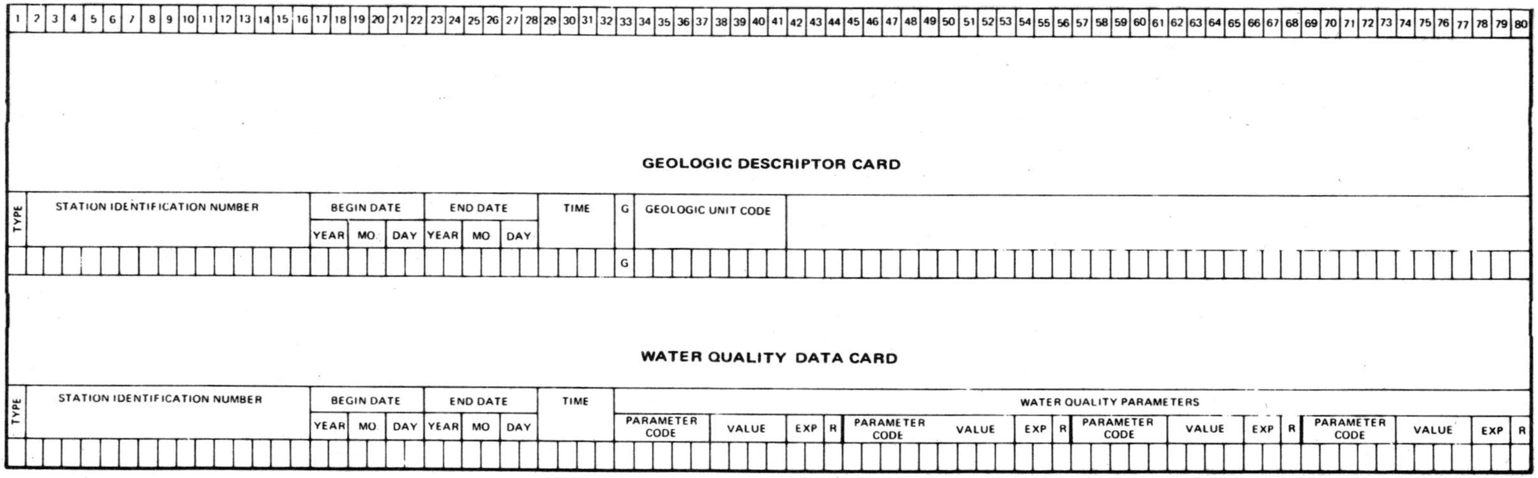

Figure 4-4. WATSTORE Water Quality File – data storage format.

NAWDEX will consist of a centralized data inventory file and communications links, not necessarily automated, with management information systems maintained by the various data depositors that subscribe to NAWDEX. The centralized data file will contain monitoring stations descriptions as well as source and types (parameters and monitoring frequency) of available water data. Access to this file is provided by requiring the user to stipulate his interest in either surface or groundwater and geographical area of interest (e.g., hydrologic basin code). The system allows additonal information, as available, from the data requestor to further narrow the file search (Planning Research Corporation, 1974).

SECTION 4 – PROPOSED MODIFICATIONS TO EXISTING SYSTEM

1. The STORET parameter code dictionary should be appended to include those groundwater monitoring related parameters listed in Table 4-10.

2. The STORET system should be modified to accept multiple remark codes with individual measurements. It is recognized that a modification of this type would represent a major commitment of resources.

3. The STORET groundwater data file should be developed separately from the existing STORET surface water data file (i.e., the WQF). This will promote faster updates of the groundwater data file, and avoid degradation of update times for the surface water data file.

4. The STORET groundwater data file should be maintained on a detachable magnetic disc, and placed on-line on the basis of some constant schedule (e.g., Tuesdays and Fridays from 2:00 p.m. to 6:00 p.m.). The periods during which the file will be on-line can be determined by performing a survey of potential users.

5. Some groundwater data should be archived off-line on magnetic tape. The data set to be archived can be defined either on the basis of its age (e.g., data over two years old), or on the basis of its activity level (e.g., stations not accessed or updated within the preceding year).

6. The proposed STORET groundwater data file should be allowed to accept compliance monitoring data, as well as background information monitoring data. Discharge permit numbers may be used as station identifier codes. The fact that a monitoring station is generating compliance data can be indicated in the station type code. In addition, the groundwater data file should be able to accept DMA status data, with the DMA treated as a station, and the DMA code used as a station code.

7. For the groundwater data file, the eight character STORET station type code should be modified and interpreted as follows:

    - Column 1, which is not currently used, should be allowed to accept a code to indicate the sample extraction method employed at the subject station (i.e., pump = 1, bail = 2, and probe = 4).

    - In column 2, a 1 would indicate DMA status data, a 2 would indicate water quality data, and a 4 would indicate hydrogeologic data.

    - In column 5, and 1 would indicate information monitoring, a 2 would indicate compliance monitoring, and a 4 would indicate other.

    - In columns 7 and 8, a 10 would indicate monitoring directly in the saturated zone, a 20 would indicate surface monitoring, and a 40 would indicate monitoring of the zone of aeration.

8. The STORET groundwater data file should store water quality criteria (ambient or effluent) as sample data. The date of enactment of the criteria would be stored in the STORET sample date field, and some exclusive value, such as 8888 for ambient criteria and 9999 for effluent limitation, would be stored in the STORET sample time field.

9. STORET retrieval options should be expanded to allow more extensive Boolean retrieval strategies. It is recognized that these additions would require setting up new index and cross-reference files and, correspondingly, entail a significant additional commitment of resources.

10. STORET user assistance capabilities and policies should be expanded to allow non-machine compatible user interface with the data base on a routine basis.

11. Either the General Information Processing System (GIPSY) or the Remote Control System (RECON) document citation retrieval systems should be modified to accommodate polygon type retrievals. This would allow the groundwater investigator to provide geographic delimiters and receive research documentation abstracts regarding his geographical area of interest.

Table 4-10

PROPOSED ADDITIONAL STORET PARAMETER CODES

Code	Parameter Description
84100	Horizontal permeability (gpd/ft^2)
84105	Vertical permeability (gpd/ft^2)
84107	Specific yield (dimensionless)
84110	Effective porosity (%)
84112	Void ratio
84115	Soil bulk density (g/ℓ)
84117	Soil moisture content (%)
84120	Soil exchangeable sodium (%)
84123	Soil specific gravity (g/cm^3)
84130	Soil gradation - % clay or silt fines
84131	Soil gradation - % fine sand
84132	Soil gradation - % medium sand
84133	Soil gradation - % coarse sand
84134	Soil gradation - % fine gravel
84135	Soil gradation - % coarse gravel
84136	Soil gradation - % cobbles
84138	Coefficient of soil uniformity
84140	Coefficient of curvature of soil gradation plot
84142	Capillary head (ft)
84200	Hydraulic gradient
84205	Hydraulic gradient direction (degrees from North)
84210	Transmissivity (gpd/ft)
84215	Storage coefficient (dimensionless)
84220	Leakage - downward (gpd/mi.^{2})
84222	Leakage - upward (gpd/mi.^{2})
84225	Diffusivity (gpd/ft)
84230	Specific flux (gpd/ft^2)
84300	Highest use made of aquifer (protected use)
84500	Monitoring agency status index
84505	Pollution control readiness index
84600-84610	Alphanumeric, sample specific comments - 10 fields, 4 characters each

SECTION 5 – CONCLUSIONS AND RECOMMENDATIONS

A nationwide groundwater monitoring program will produce a large volume of highly diversified information. The best use of this information can be realized only if efficient information management is exercised as an integral element of the overall monitoring program.

The prevalent proximity of the groundwater data user to the source of the data, as well as the specialized needs of individual users, indicates that decentralized (localized) groundwater data management systems are appropriate. A centralized (Federal) data management system is called for as well, however, for the coordination of the national effort (making data available for multiple users and uses and minimizing redundant data collection and analysis activities), the provision of interim groundwater data management support, the achievement of economies of scale, and the encouragement of local compliance with national groundwater monitoring requirements. Consequently, the development of comprehensive groundwater information management capabilities should be undertaken at the Federal, State, and, where necessary, local levels. Whereas the volume of data likely to be involved at the Federal level dictates the need for a computerized system, this is not necessarily so below the Federal level.

A comprehensive groundwater data management capability is composed of three major components: maintaining the data generated by groundwater surveillance, indexing that data so that it can be accessed expeditiously, and maintaining concise citations of relevant groundwater research documentation. At the Federal level, these capabilities can be provided adequately by existing or proposed computerized information management systems, with only minor modifications. Below the Federal level, it may be necessary to develop a computerized capability to maintain groundwater surveillance data, a task already accomplished by some States (e.g., California, Colorado, Tennessee, and Texas). Data indexing and the management of document citations are capabilities which can be provided to agencies below the Federal level by existing Federal systems.

The management of groundwater surveillance data at the Federal level can be satisfactorily achieved by application of the STORET system currently operated by the EPA. Suggestions for modifications to this system aimed at improving its effectiveness for managing groundwater data are presented in Section 4 of this chapter. The STORET system is also available to State and local users, whose participation is encouraged by the EPA. A system designed for a broad-based user population is characteristically not responsive to unique individual requirements, however, and State and local users should consider the merits of developing computerized systems designed specifically for their needs. In addition, it should be noted that STORET is not now used on a major scale for groundwater analyses, and a major new STORET user community will require a further evaluation and commitment of resources by the EPA.

Groundwater data indexing capabilities, which allow the data user to expeditiously locate pertinent groundwater data and examine its nature, prior to accessing the data itself, can be provided to Federal, State, and local users, by the National Water Data Exchange (NAWDEX) proposed, and currently being developed by the U.S. Geological Survey. The community of water data collectors and users should support and coordinate with this effort.

The Water Resources Scientific Information Center, U.S. Department of the interior, provides computerized storage of and access to document citations through use of the Remote Control System (RECON) and the General Information Processing System (GIPSY). These capabilities are available to all categories of groundwater investigators, and are generally sufficient to meet their needs.

LIST OF ABBREVIATIONS AND ACRONYMS

AWIS	Arizona Water Information System
BM	Bureau of Mines
CPU	Central processing unit
CRT	Cathode ray tube
DMA	Designated monitoring agency
DNR	Department of Natural Resources
DSWELL	Well Hydrograph Data Storage and Retrieval System
EBCDIC	Extended Binary Coded Decimal Interchange Code
EMSL	Environmental Monitoring and Support Laboratory
ENDEX	Environmental Data Index
EPA	Environmental Protection Agency
ERDA	Energy Research Development Agency
GIPSY	General Information Processing System
GOWN	Groundwater Observation Well Network
GPSF	General Point Source File
HEW	Department of Health, Education, and Welfare
IRS	International Referral System
MIS	Management information system
NAS	National Academy of Sciences
NAWDEX	National Water Data Exchange
NPDES	National Pollutant Discharge Elimination System
NOAA	National Oceanic and Atmospheric Administration
NWQSS	National Water Quality Surveillance System
OASIS	Oceanic and Atmospheric Scientific Information System
OCR	Optical character recognition
ORSANCO	Ohio River Valley Water, Sanitation Commission
OWDC	Office of Water Data Coordination
RECON	Remote Control System
RMI	River Mile Index
SSIE	Smithsonian Science Information Exchange
STORET	Storage and Retrieval system
TELEX	Telephone Exchange
UNESCO	U.N. Educational, Scientific and Cultural Organizaiton

LIST OF ABBREVIATIONS AND ACRONYMS (Continued)

UNISIST	World Science Information System
USDI	U.S. Department of the Interior
USGS	U.S. Geological Survey
USPHS	U.S. Public Health Service
WATS	Wide Area Telephone Service
WATSTORE	National Water Data Storage and Retrieval System
WISE	Water Information System for Enforcement
WQF	Water Quality File

SELECTED BIBLIOGRAPHY

C. Baker, Written Communication, U.S. Geological Survey, Water Resources Division, Reston, Va., April 3, 1975.

C.S. Conger, Personal Communication, U.S. Environmental Protection Agency, Data Processing and User Assistance Branch, Washington, D.C., November 4, 1974.

C.S. Conger, Written Communication, U.S. Environmental Protection Agency, Data Processing and User Assistance Branch, Washington, D.C., March 4, 1975.

Datapro Research Corporation, *A Buyer Guide to Data Base Management Systems,* Delran, N.J., December, 1974. (12 pp.)

Melvin D. Edwards, *The Processing and Storage of Water Quality in the National Water Data Storage and Retrieval System,* U.S. Geological Survey, Water Resources Division, Reston, Va., 1974. (85 pp.)

Environmental Data Service, *User's Guide to OASIS - Oceanic and Atmospheric Scientific Information System,* National Oceanic and Atmospheric Administration, Washington, D.C., 1974.

R. Ferrara, and R.L. Nolan, "New Look at Computer Data Entry," *Journal of Systems Management,* Association for Systems Management, pp 24-33, February, 1973.

K.E. Foster, and J. DeCook, *Implementation of Arizona Water Information System (AWIS) Remote Terminal Accessible Hydrologic Data Sets on DEC-10 Computer,* University of Arizona, Tucson, Arizona, 1974. (21 pp.)

D.R. Friedrichs, *Information Storage and Retrieval System for Well Hydrograph Data - User's Manual,* Battelle Pacific Northwest Laboratories, Richland, Washington, 1972. (23 pp.)

J.A. Gilliland, and A. Treichel, "GOWN – A Computer Storage System for Groundwater Data," *Canadian Journal of Earth Sciences,* Vol. 5, pp. 1518-1524, September 1968.

G. Guenther, D. Mincavage, and F. Morley, *Michigan Water Resources Enforcement and Information System,* U.S. Environmental Protection Agency, Office of Research and Monitoring, Socioeconomic Environmental Studies Series, EPA-R5-73-020, Washington, D.C., 1973. (161 pp.)

J.D. Hem, *Study and Interpretation of the Chemical Characteristics of Natural Water,* U.S. Geological Survey, Water Supply Paper 1473, 1959. (269 pp.)

W.C. House, ed., *Data Base Management,* Mason and Lipscomb Publishers, Inc., New York, New York, 1974. (470 pp.)

W.L. Klein, D.A. Dunsmore, and R.K. Horton, "An Integrated Monitoring System for Water Quality Management in the Ohio Valley," *Environmental Science and Technology,* Vol. 2, American Chemical Society, pp. 764-771, October, 1968.

Jerome Lobel, and M.V. Farina, "Selecting Computer Memory Devices," *Automation*, Penton Publishing Co., Cleveland, Ohio, pp. 66-70, October 1970.

Matthew Lorber, "Evaluating Computer Output Printers," *Automation*, Penton Publishing Co., Cleveland, Ohio, pp. 64-67, March 1972.

E.M. Notzon, Written Communication, U.S. Environmental Protection Agency, Monitoring and Data Support Division, Washington, D.C., October 2, 1975.

Planning Research Corporation, *Support in the Implementation of a National Water Data Exchange*, Second Quarterly Progress Report (September-November, 1974), PRC-D-1863, December 1974. (61 pp.)

W.E. Price, and C.H. Baker, *Catalog of Aquifer Names and Geologic Unit Codes Used by the Water Resources Division*, U.S. Department of the Interior, Geological Survey, Water Resources Division, Reston, Va., 1974. (306 pp.)

B. Schwab, and R. Sitter, "Economic Aspects of Computer Input-Output Equipment," *Financial Executive*, Financial Executives Institute, pp. 75-87, September 1969.

C.R. Showen, and O.O. Williams, *Index to Water Quality Data Available from the U.S. Geological Survey in Machine-Readable Form to December 31, 1972 - Western Region*, PB-232-794, U.S. Geological Survey, Water Resources Division, Washington, D.C., 1973. (520 pp.)

D.B. Steig, "File Management Systems Revisited, "*Datamation*, Barrington, Illinois, pp. 48-51, October 1972.

P.L. Taylor, Written Communication, U.S. Environmental Protection Agency, Data Reporting Branch, Washington, D.C., November 27, 1974.

U.S. Department of Health, Education and Welfare, *The Toxic Substances List, 1973 Edition*, National Institute for Occupational Safety and Health, Rockville, Maryland, 1973. (1001 pp.)

U.S. Environmental Protection Agency, *Storage and Retrieval of Water Quality Data. Training Manual*, PB-214 580, Washington, D.C., 1971. (302 pp.)

U.S. Environmental Protection Agency, *Proposed Criteria for Water Quality*, Vol. 1, Washington, D.C., 1973. (425 pp.)

U.S. Public Health Service, "Drinking Water Standards," Federal Register, Government Printing Office, Washington, D.C., pp. 2152-2155, March 6, 1962.

J.M. Wilson, M.J. Mallory, and J.M. Kernodle, *Summary of Groundwater Data for Tennessee through May 1972*, Miscellaneous Publication Number 9, State of Tennessee, Department of Conservation, Division of Water Resources, Nashville, Tennessee, 1972.

Larry Welke, "A Review of File Maintenance Systems," *Datamation*, Barrington, Illinois, pp. 52-54, October 1972.

J.L. Welsh, "Ground-Water Quality Data for Planning, Monitoring, and Surveillance," *Proceedings at the Ninth Biennial Conference on Ground Water*, Goleta, California, September 1973.

CHAPTER V

MONITORING DISPOSAL WELLS

SECTION 1 – INTRODUCTION

As of mid-1973, at least 278 industrial wastewater injection wells had been constructed, and 61 percent of them were operating (Warner and Orcutt, 1973). This is a relatively small number of waste disposal units, but the number has continued to increase at a rate of about 30 wells annually, and could increase even more rapidly in response to the objective of eliminating discharge to surface waters and demands of new technologies such as geothermal energy production, desalination, and radioactive waste disposal. Regardless of the number of industrial wastewater injection wells, they have been an object of unusual attention by regulatory agencies and environmentalists.

This attention is reflected by inclusion of specific references to disposal wells in Public Law 92-500, the Federal Water Pollution Control Act Amendments of 1972. A provision of that Act is the requirement that the Administrator of the Environmental Protection Agency (EPA) shall, in cooperation with the States or other Federal agencies, establish a system for the surveillance of the quality of surface and groundwaters. The enactment of Public Law 93-253, the Safe Drinking Water Act, further requires the Administrator to propose and promulgate, for State underground injection programs, minimum monitoring requirements to assist in preventing underground injection, which endangers drinking water sources.

This chapter provides technical information concerning data needed for monitoring, and the methods and tools available for monitoring of wastewater injection wells and examples of their application. However, the material presented cannot be expected to satisfy the monitoring requirements of all aspects of underground fluid injection that will likely be included in the rules and regulations promulgated in response to P.L. 93–523. The definition of the term underground injection is sufficiently broad in P.L. 93-253 to include subsurface emplacement of fluids by many means, such as ponds, pools, lagoons, and pits. This chapter relates specifically to the subsurface emplacement of fluids through cased disposal wells.

Monitoring of groundwater is often thought of as the observation of groundwater quality by sampling of wells and springs. In this chapter, monitoring is meant to include the full spectrum of consideration given to determining the effects of wastewater injection systems, from planning of the system through well construction, testing, operation, and, finally, abandonment. The policy of the EPA is consistent with this approach (see Appendix; also, Hall and Ballantine, 1973). The Ohio River Valley Water Sanitation Commission (ORSANCO), 1973, also has established a basis for injection well monitoring. Both the EPA policy statement and the ORSANCO publication provide suggestions for a suitable data base for monitoring. ORSANCO also suggests a series of administrative procedures, which, if followed, assure the early involvement of regulatory agencies in monitoring, and provide for their continued surveillance of injection systems throughout construction, use, and abandonment.

This chapter is intended to complement existing documents, such as those mentioned above, by providing a more extensive discussion of the data that characterize the subsurface environment, how these data are obtained, and how they are used to predict and interpret injection well response. The surveillance of operating injection wells is treated in more detail here than in earlier publications.

SECTION 2 – THE SUBSURFACE ENVIRONMENT

In devising a monitoring program for a wastewater injection system, the first consideration is definition of the regional and local subsurface environment. Factors in such an appraisal are stratigraphic and structural geology, lithology, fluid properties, mechanical properties of injection and confining units, hydrodynamics, and subsurface resources. Other publications (Warner, 1965 and 1968) have reviewed, in general, the relation of the subsurface environment to wastewater injection. The purpose of the following discussion is to provide more specific detail and examples of the methodology for applying these concepts to monitoring. It will be attempted, insofar as possible, to avoid repetition of material that has been previously presented.

STRATIGRAPHIC GEOLOGY

Regional stratigraphy is determined by use of outcrop and borehole data, which have been interpreted and are usually presented in the form of columnar sections, isopach maps, facies maps, and cross sections.

The basic data unit used in studies of stratigraphic geology is the columnar section, which is a graphic representation of the sequence, thickness, lithology, and relationship of the rock units at a location. A generalized columnar section may be prepared, which shows these parameters for a region. Figure 5-1 is a generalized columnar section for northeastern Illinois. Columnar sections are prepared by using cores, cuttings, and geophysical logs from boreholes and, where outcrops are present, from them. Some possible injection horizons in Figure 5-1 are the St. Peter, Ironton, Galesville, and Mt. Simon Formations. Of these, the Mt. Simon is the deepest, and can be seen to be overlain by the Eau Claire Formation, which may contain confining shale beds. On the other hand, the St. Peter Formation is shallower, and is overlain by limestones and dolomites, which are less dependable as aquitards. The St. Peter, therefore, has a lesser potential for wastewater injection.

Isopachous maps indicate, by contour lines, the varying thickness of a stratigraphic unit. Figure 5-2 is an isopachous map of the Mt. Simon Formation in Illinois, showing that this sandstone unit varies in thickness from 0 to over 2,000 feet within that State. Other factors being equal, locations where the Mt. Simon Formation is thickest have greatest potential for wastewater injection.

The facies of a stratigraphic unit are its laterally varying aspects, such as lithology, fossil content, and so forth. For example, the Eau Claire Formation, which overlies the Mt. Simon Formation, consists of a mixture of siltstone, shale, dolomite, and sandstone in northeastern Illinois (Figure 5-1), but passes by facies change eastward into sandstone in central Ohio and dolomite in eastern Ohio (Figure 5-3).

Some types of facies maps are ratio maps, percentage maps, and isolith maps. These facies maps are different ways of showing the relative amounts of the various lithologies in a rock unit or units. The ratio and percentage maps show contours of the ratios or percentages of the aggregate thicknesses of lithologic classes.

Figure 5-4 is a lithologic ratio map, showing the relative ratios of sandstone, shale, and dolomite in post-Mt. Simon, pre-Knox rocks in Ohio. This figure shows that this group of rocks changes from a sandy facies in western Ohio to a dolomite facies in eastern Ohio. The rocks depicted in Figure 5-4 are equivalent to the Eau Claire Formation in Figure 5-1. So, in eastern Ohio, the Eau Claire Formation is almost

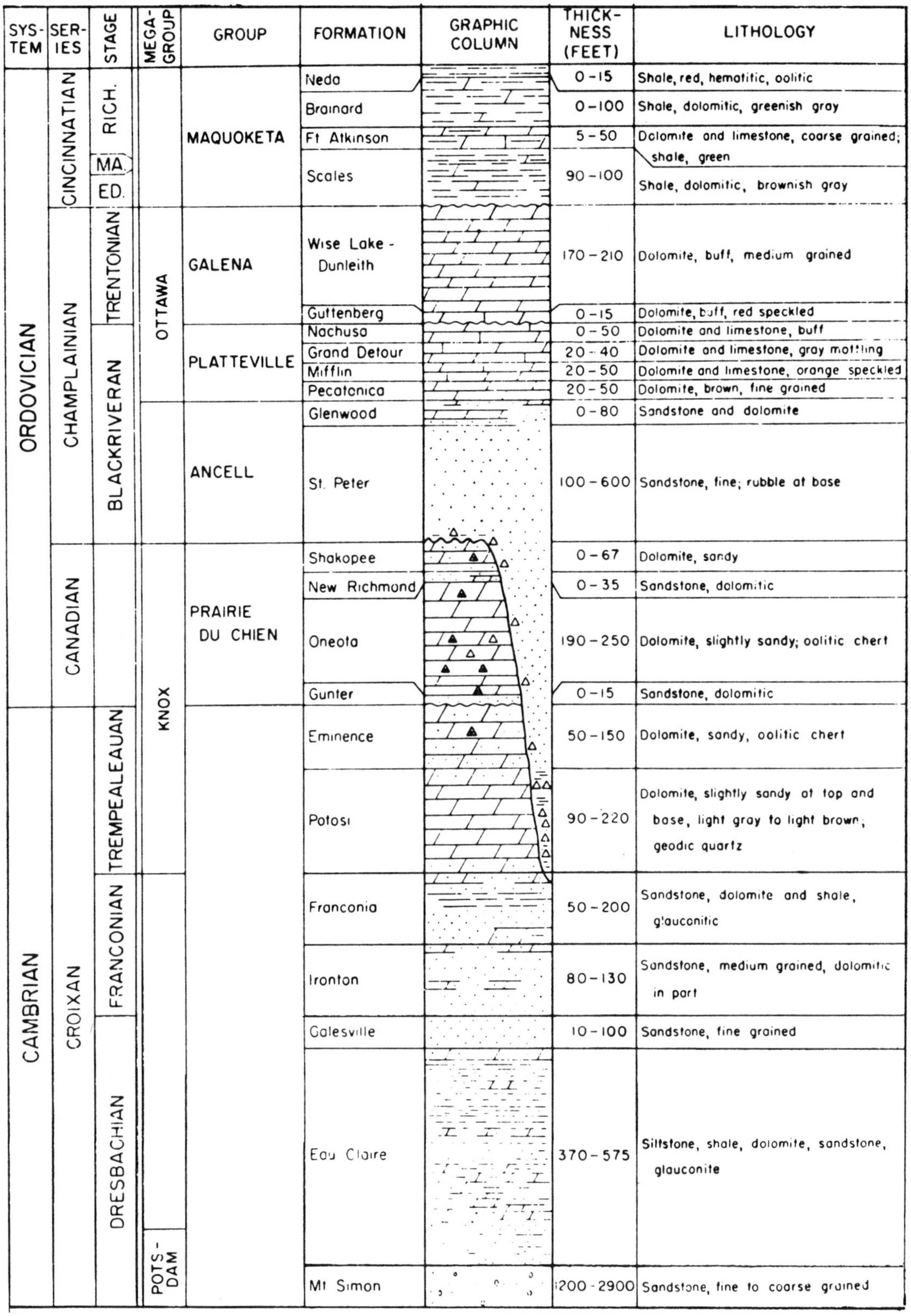

SYS-TEM	SER-IES	STAGE	MEGA-GROUP	GROUP	FORMATION	GRAPHIC COLUMN	THICK-NESS (FEET)	LITHOLOGY
ORDOVICIAN	CINCINNATIAN	RICH.		MAQUOKETA	Neda		0-15	Shale, red, hematitic, oolitic
		MA. ED.			Brainard		0-100	Shale, dolomitic, greenish gray
					Ft. Atkinson		5-50	Dolomite and limestone, coarse grained; shale, green
					Scales		90-100	Shale, dolomitic, brownish gray
	CHAMPLAINIAN	TRENTONIAN	OTTAWA	GALENA	Wise Lake - Dunleith		170-210	Dolomite, buff, medium grained
					Guttenberg		0-15	Dolomite, buff, red speckled
				PLATTEVILLE	Nachusa		0-50	Dolomite and limestone, buff
					Grand Detour		20-40	Dolomite and limestone, gray mottling
					Mifflin		20-50	Dolomite and limestone, orange speckled
					Pecatonica		20-50	Dolomite, brown, fine grained
		BLACKRIVERAN		ANCELL	Glenwood		0-80	Sandstone and dolomite
					St. Peter		100-600	Sandstone, fine; rubble at base
	CANADIAN		KNOX	PRAIRIE DU CHIEN	Shakopee		0-67	Dolomite, sandy
					New Richmond		0-35	Sandstone, dolomitic
					Oneota		190-250	Dolomite, slightly sandy; oolitic chert
					Gunter		0-15	Sandstone, dolomitic
CAMBRIAN	CROIXAN	TREMPEALEAUAN			Eminence		50-150	Dolomite, sandy, oolitic chert
					Potosi		90-220	Dolomite, slightly sandy at top and base, light gray to light brown; geodic quartz
		FRANCONIAN			Franconia		50-200	Sandstone, dolomite and shale, glauconitic
					Ironton		80-130	Sandstone, medium grained, dolomitic in part
					Galesville		10-100	Sandstone, fine grained
		DRESBACHIAN			Eau Claire		370-575	Siltstone, shale, dolomite, sandstone, glauconite
			POTS-DAM		Mt. Simon		1200-2900	Sandstone, fine to coarse grained

Figure 5-1. Generalized columnar section of Cambrian and Ordovician strata in northeastern Illinois (Buschbach, 1964, p. 16).

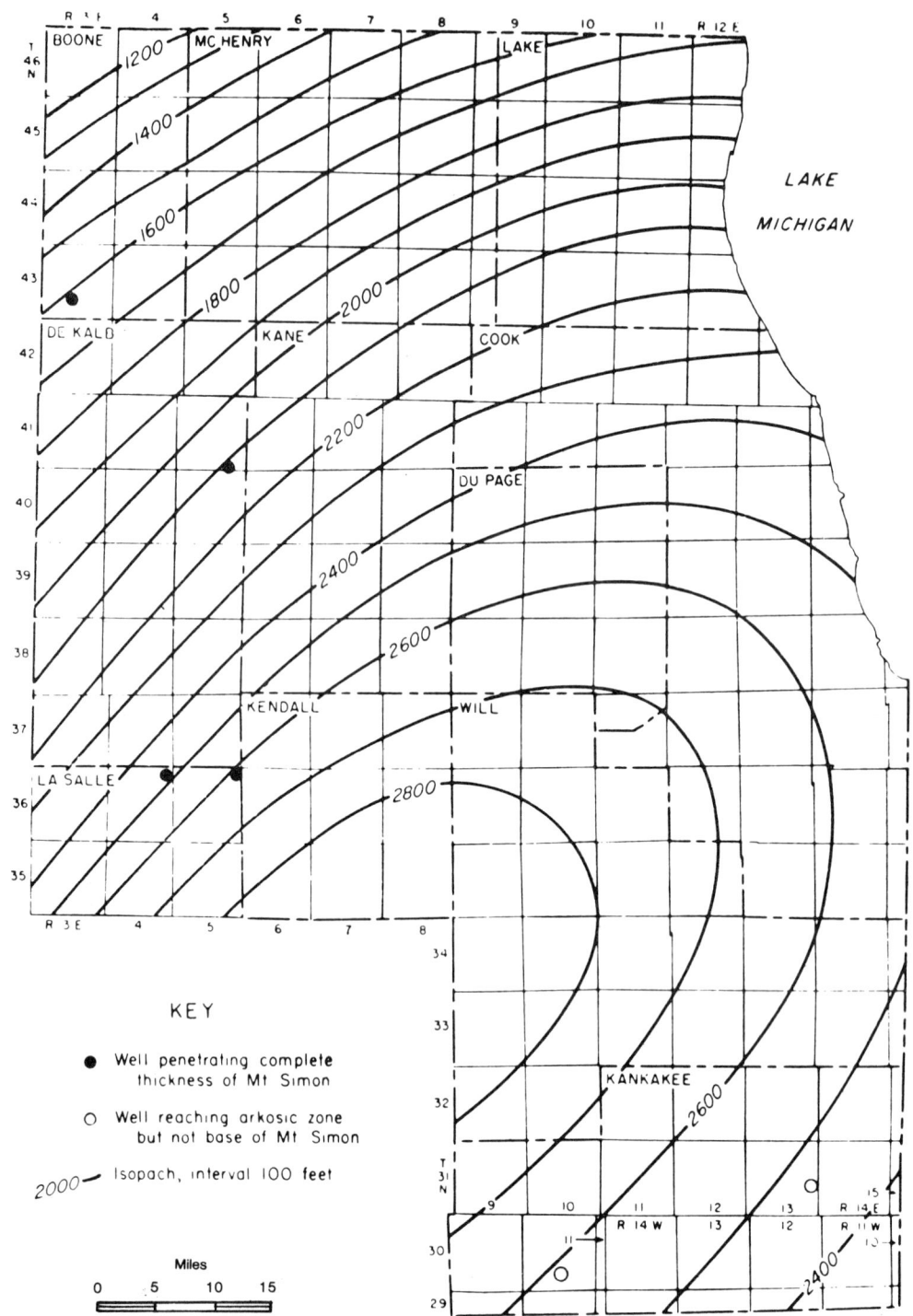

Figure 5-2. Isopach map of Mt. Simon Formation in northeastern Illinois (Buschbach, 1964).

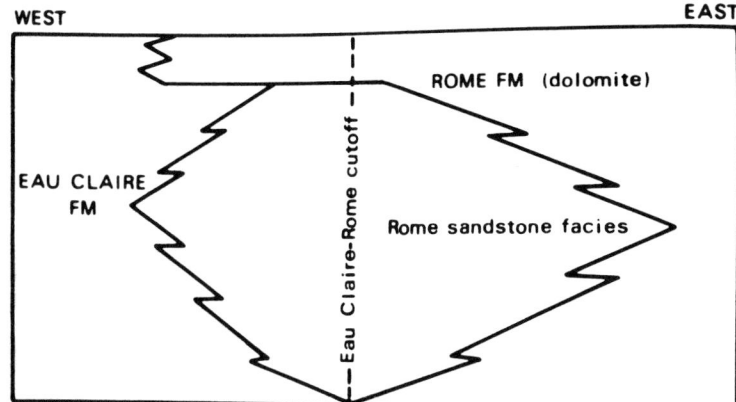

Figure 5-3. Schematic east-west section of the Eau Claire and equivalent Rome strata (Janssens, 1973, p. 10).

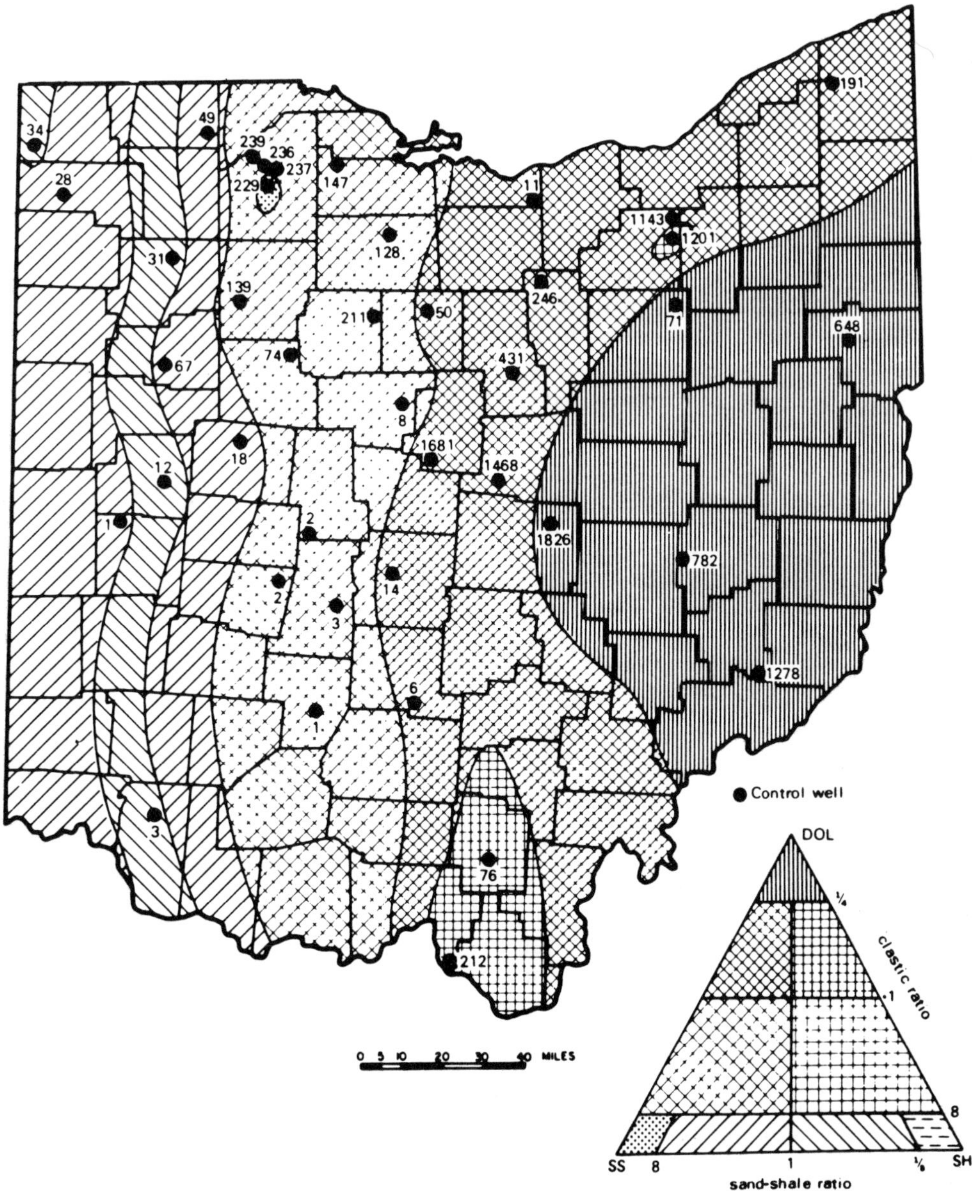

Figure 5-4. Lithologic ratio map of post-Mt. Simon, pre-Knox rocks (Janssens, 1973, p. 19).

entirely dolomite, rather than the mixed lithology shown in Figure 5-1. Without further information, Figures 5-3 and 5-4 indicate that the Eau Claire Formation becomes less promising as a confining unit for the Mt. Simon Formation as it is traced eastward from Illinois into Ohio.

Figure 5-5 is an east-west cross section of Paleozoic rocks extending from east-central Illinois to northwestern Pennsylvania. This cross section shows the facies changes in the Eau Claire described above and shown in earlier figures. The cross section also shows that the Mt. Simon Formation is about 1,500 feet thick in east-central Illinois, but thins to about 100 feet across northern Ohio and into northwestern Pennsylvania. Thus, much of the same information conveyed in the previous figures is summarized in a readily understandable form in such a cross section.

Local stratigraphy is first projected from regional data before drilling of a well, then determined in detail for the well when it is drilled. As previously mentioned, the means of displaying the stratigaphy of a well is the columnar section.

STRUCTURAL GEOLOGY

Structural geology means the folding, faulting, and fracturing of rocks, and the geographic distribution of these features. One means of showing regional structural geologic features is a map which includes areas or lines of major features. Figure 5-6 is such a map for the Ohio River Basin. Another type of map is the structural contour map. Figure 5-7 is a structural contour map on the top of the Mt. Simon Formation in Illinois. Such a map allows an estimate of the approximate depth to the mapped unit, and shows the location of known faults and folds that may influence decisions concerning the location and monitoring of an injection well.

LITHOLOGY

Lithology refers to the composition and texture of a rock. The generalized columnar section in Figure 5-1 contains brief, highly generalized lithologic descriptions of rock units in northeastern Illinois. The descriptions prepared for individual wells are very detailed. An example of a description of a core from the top of the Mt. Simon Formation, in one well, is shown in Table 5-1.

Such detailed descriptions are prepared from cores, cuttings, and geophysical logs, and are necessary for determining the rock-unit characteristics in a test well. From such descriptions, and other data, injection intervals, confining beds and casing points are selected, and other engineering decisions made.

FLUIDS

CHEMISTRY

Judgment as to whether wastewater may or may not be permitted to be injected into a rock unit depends, in part, on the chemistry of the aquifer water. The chemistry of aquifer water is also important because of the possibility of reactivity with injected wastewater.

Policy concerning the minimum salinity of water in aquifers approved for wastewater injection varies by State. In the Ohio Valley region, Illinois agencies have determined that groundwater containing less than 10,000 mg/ℓ total dissolved solids should be protected. In New York, waste injection is prohibited in aquifers with a dissolved solids content of 2,000 mg/ℓ, or less. In Florida, the limiting value is 1,500 mg/ℓ.

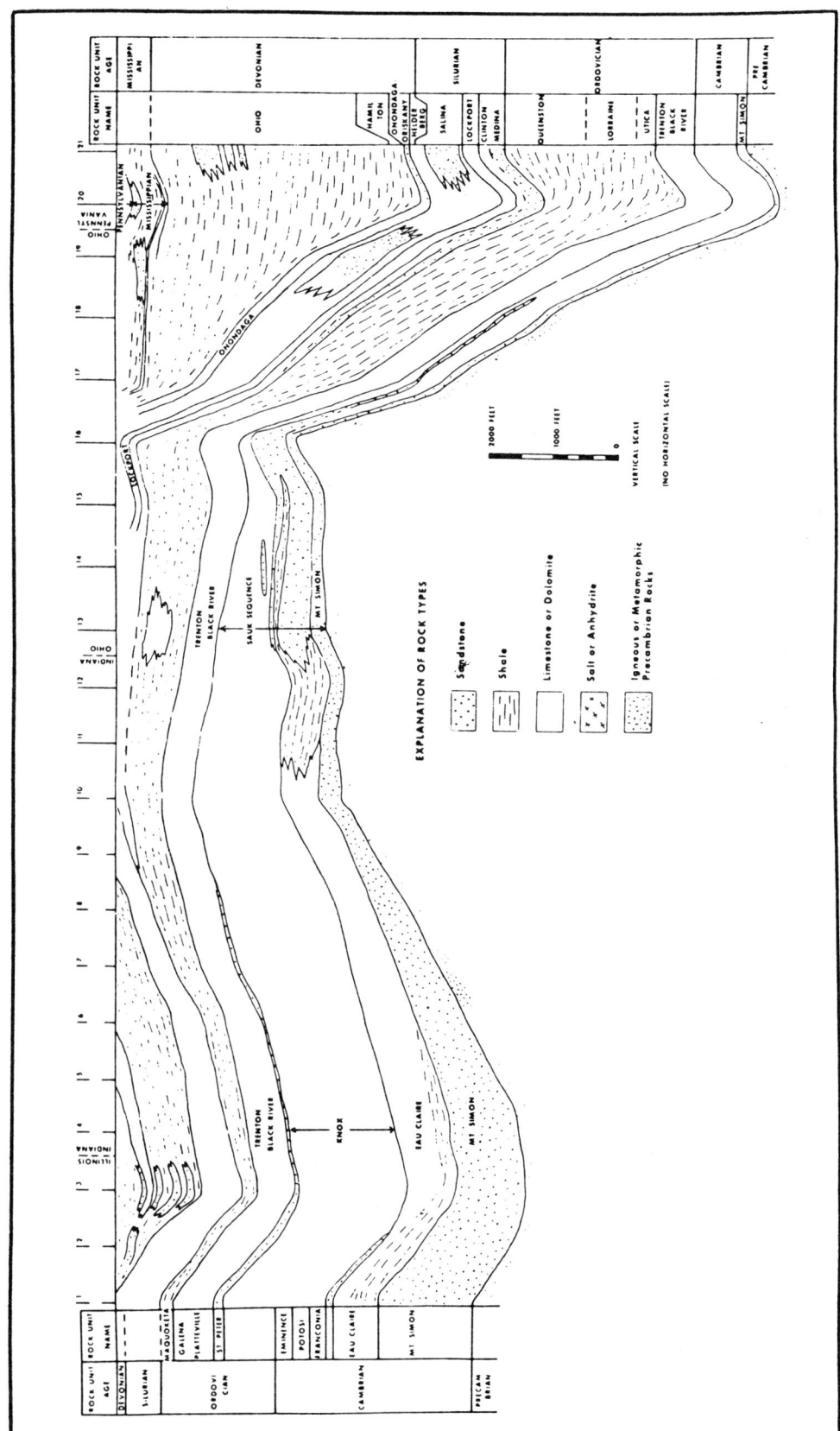

Figure 5-5. East-west cross section of Paleozoic rocks in the northern Ohio River Valley -- modified after cross sections in American Association of Petroleum Geologists cross section Publication 4, 1966 (Ohio River Valley Water Sanitation Commission, 1973, p. 51).

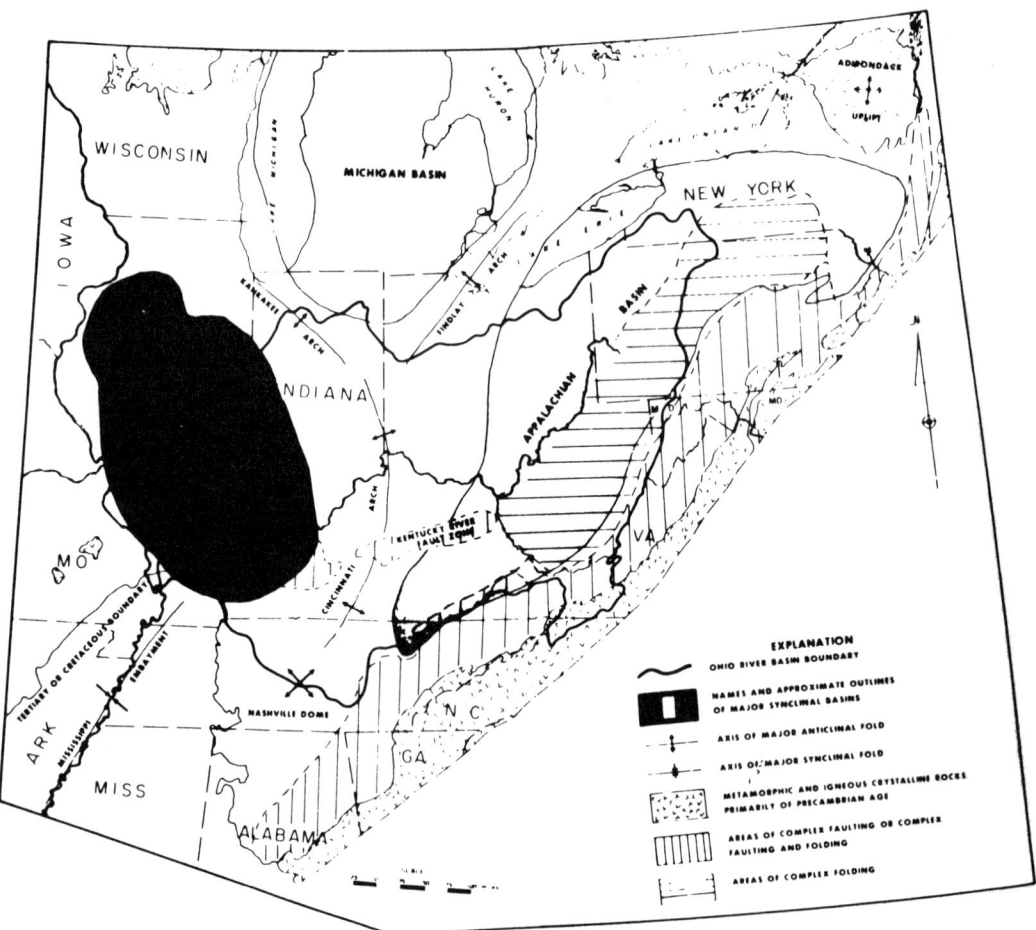

Figure 5-6. Map of the Ohio River Basin and vicinity showing some major geologic features. Data modified from published maps (Ohio River Valley Sanitation Commission, 1973, p. 24).

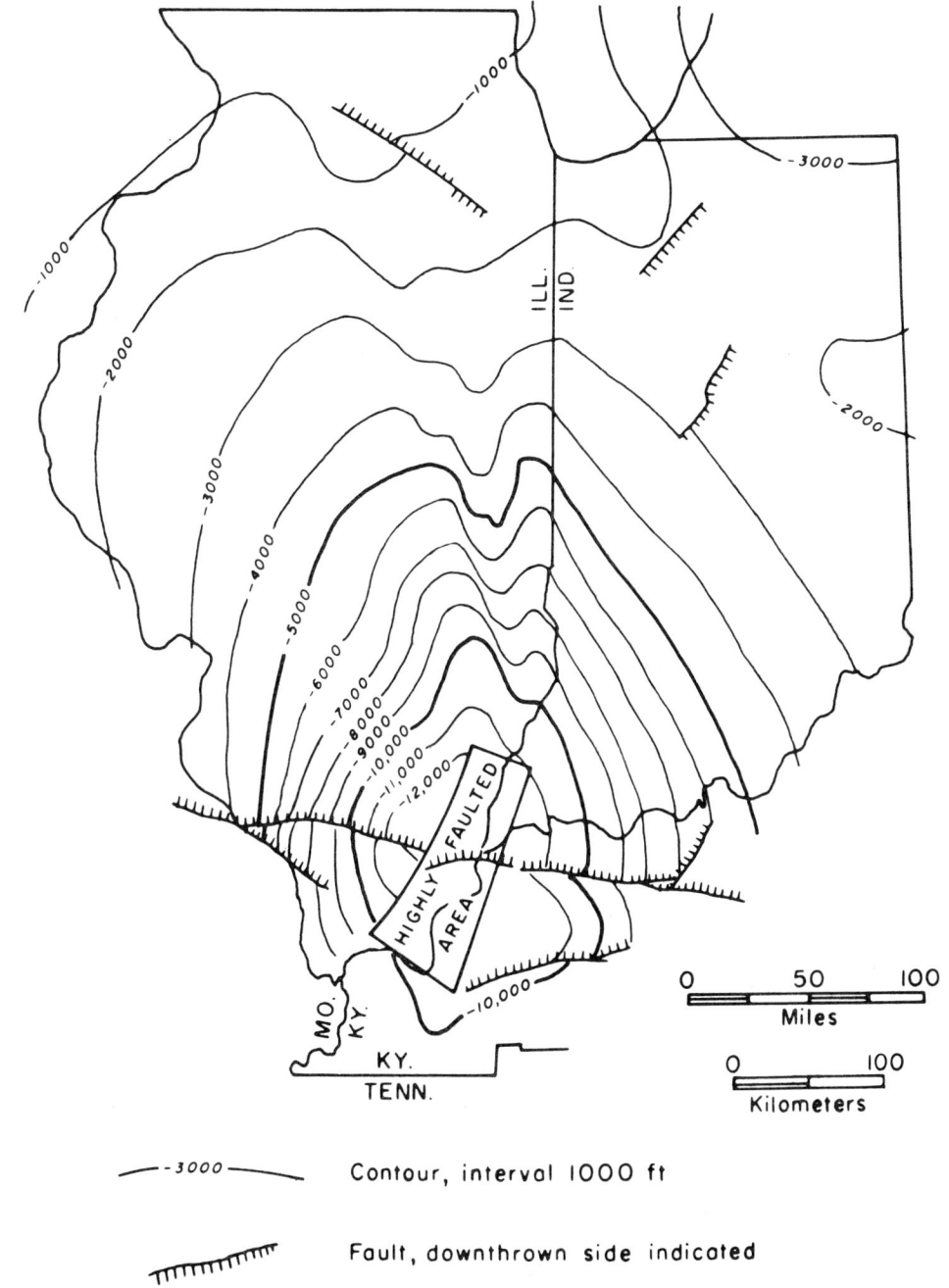

Figure 5-7. Structure on top of Mt. Simon Formation (Bond, 1972, p. 36).

Table 5-1

TYPICAL DESCRIPTION OF A CORE
FROM THE TOP OF THE MT. SIMON FORMATION IN ILLINOIS

Depth in Well	Lithologic Description
3019.4 - 3020.5	Sandstone; grayish-white; medium to very coarse grained; grains are broken, pitted, and chipped; very cohesive and hard; very tight; semi-quartzitic.
3020.5 - 3021.8	Sandstone; as above, very poor sorting; medium to very coarse, rounded grains, with abundant fine-grained matrix; glassy; slightly pyritic; cohesive and hard; not as tight as above zone; limited mud invasion.
3021.8 - 3023.8	Sandstone; good sorting; very fine to fine, sub-angular grains; slightly pyritic; cohesive and firm; limited mud invasion; very few shale laminations.

The problem of potential reactivity between wastewater and aquifer minerals and water is summarized by Warner (1968). Several recent papers concerning this topic are contained in the *Proceedings of the Symposium on Underground Waste Management and Environmental Implications* (Cook, 1972).

In order to evaluate the details of the chemistry of aquifer water, it is necessary to obtain samples after a well is drilled; samples from previously drilled wells may provide a good indication of what will be found. Geophysical logs are also useful for estimating the dissolved solids content of aquifer water in intervals that are not sampled, as will be discussed later.

In Illinois, the Mt. Simon Formation has been found to contain water ranging in dissolved solids content from less than 1,000 mg/ℓ in the northern part of the State, to over 300,000 mg/ℓ in the southern part. Such information can be displayed in the form of an isocon map (Figure 5-8). Most of the dissolved solids are sodium chloride, but significant amounts of calcium, magnesium, and sulfate are also present (Table 5-2).

VISCOSITY

Viscosity is the ability of a fluid to resist flow, and is an important property in determining the rate of flow of a fluid through a porous media. The common unit of viscosity is the poise, or the centipoise, which is one one-hundredth of a poise. Figure 5-9 shows the variation in viscosity of water with temperature and salinity. Both temperature and dissolved solids content can have a significant effect. In most cases, the effects will be offsetting in subsurface waters, since temperature and dissolved solids content both tend to increase with increasing depth. The viscosity of some wastewaters may be unusually high as a result of the presence of dissolved organic chemicals. Pressure in the range of interest has an insignificant effect on viscosity.

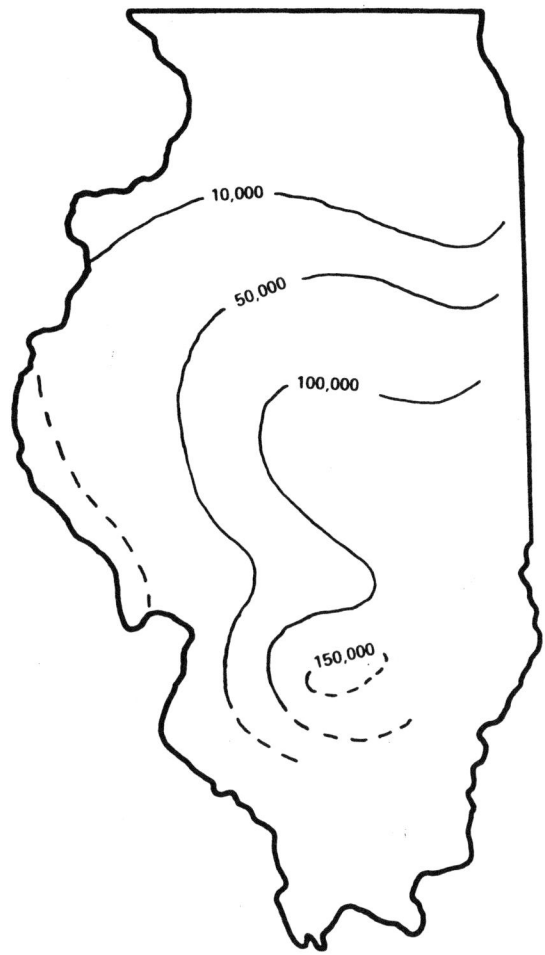

Figure 5-8. Isocon map, showing the dissolved solids content in parts per million of water in the upper 100 feet of the Mt. Simon Formation in Illinois.

Table 5-2

ANALYSIS OF WATER FROM THE MT. SIMON FORMATION
IN THE VICINITY OF BLOOMINGTON, ILLINOIS

Analysis	Result	
Specific gravity	1.050	
pH	6.6	
Hydrogen sulfide	0.0	mg/ℓ
Carbonate alkalinity	0.0	mg/ℓ
Bicarbonate alkalinity	68	mg/ℓ
Chlorides	39,250	mg/ℓ
Total hardness	17,900	mg/ℓ
Calcium	5,200	mg/ℓ
Magnesium	1,190	mg/ℓ
Sulfates	1,700	mg/ℓ
Manganese	1.3	mg/ℓ
Total iron	27.0	mg/ℓ
Total dissolved solids (calculated)	65,460	mg/ℓ

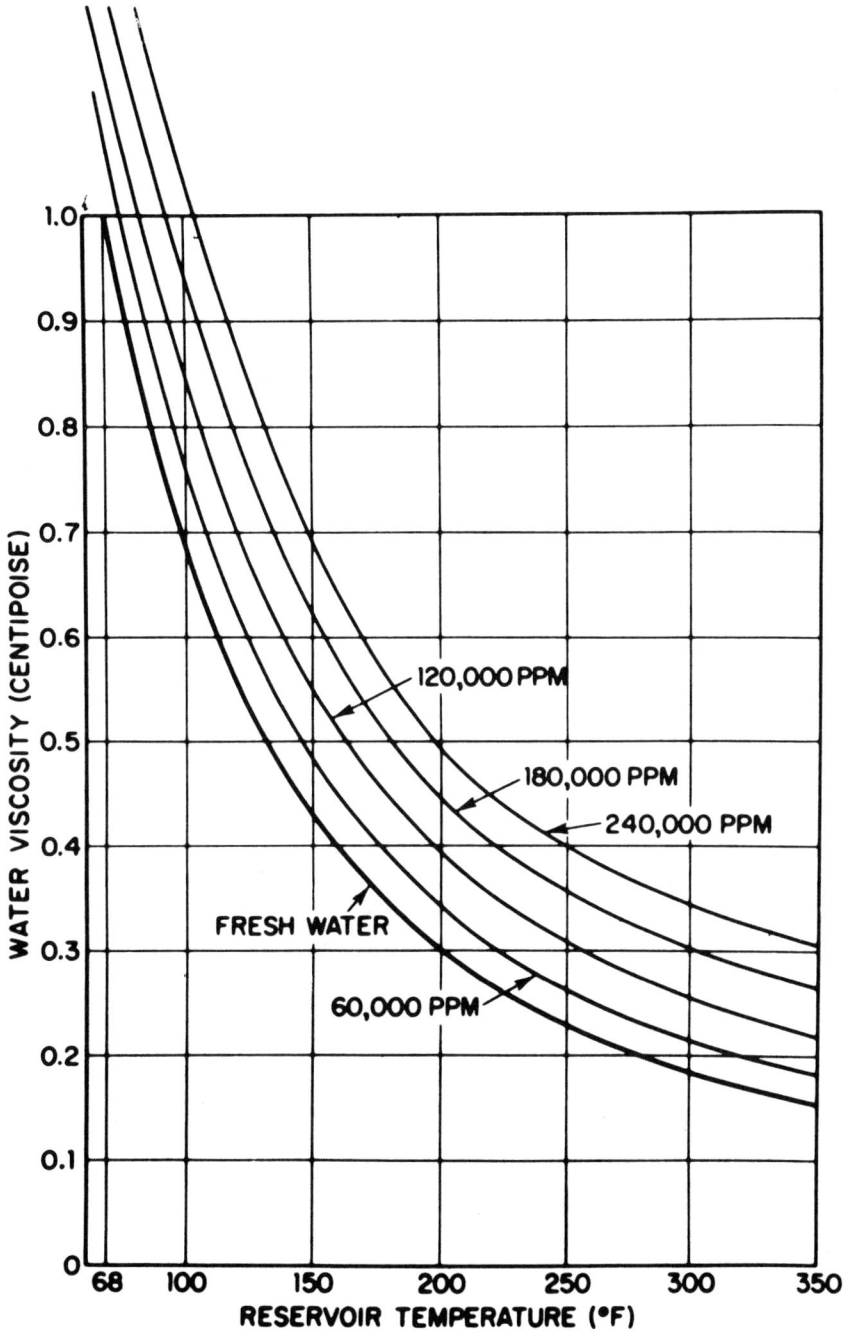

Figure 5-9. Water viscosity as a function of temperature and salinity (ppm NaCl) (Pirson, 1963, p. 40).

DENSITY

The density of a fluid is its mass per unit volume. The density of a liquid increases with increased pressure, and decreases with increased temperature. However, the density of water changes very little within the range of pressures and temperatures of interest. For example, the density of water decreases only 0.04 gm/cm^3 between 60 °F and 210 °F (Figure 5-10), and increases only about 0.04 gm/cm^3 from 0 to 14,000 psi (Figure 5-11). A more important influence on the density of water is the total dissolved solids content. Figure 5-12 shows the effect of various amounts of sodium chloride on specific gravity (or density).* Since natural brines may differ significantly from sodium chloride solutions, it may be desired to develop empirical relationships between density and dissolved solids as was done by Bond (1972) for the Illinois basin (Figure 5-13).

PRESSURE

A knowledge of fluid pressure in the unit proposed for wastewater injection is important. Fluid pressure can be measured directly in the borehole at the depth of the injection horizon, usually by performing a drill-stem test, which will be described later. Fluid pressure at the injection horizon can also be measured indirectly by determining the static water level in the borehole, then computing the pressure of the fluid column at the depth of interest.

Figure 5-14 shows how fluid pressure increases with depth in a well bore filled with freshwater with a specific gravity of 1.0. When the average specific gravity of the

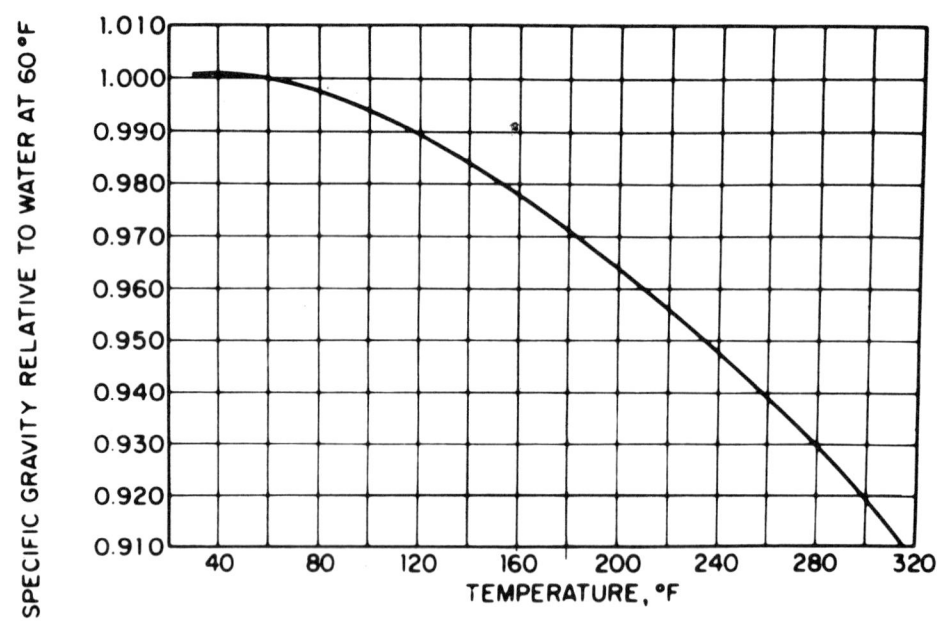

Figure 5-10. Specific gravity of distilled water as a function of temperature (Pirson, 1963, p. 39).

---

*Specific gravity is the ratio of the mass of a body to that of an equal volume of pure water, so for practical purposes, the numerical values of density and specific gravity are equal. Specific gravity, however, is dimensionless.

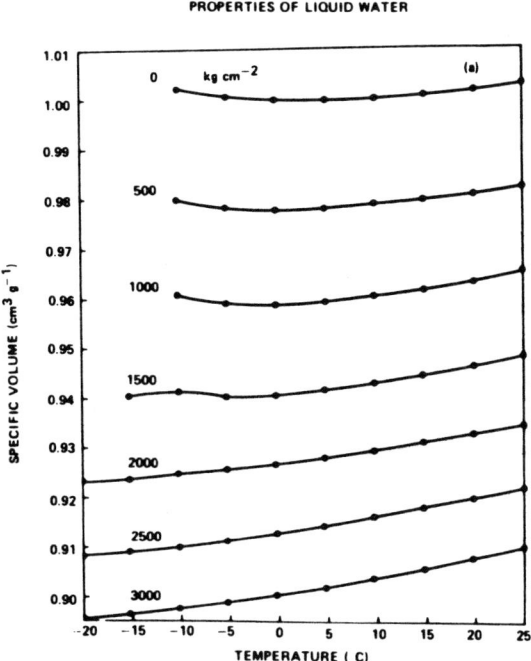

Figure 5-11. Specific volume of water as a function of temperature and pressure (Eisenberg and Kauzmann, 1969, p. 186).

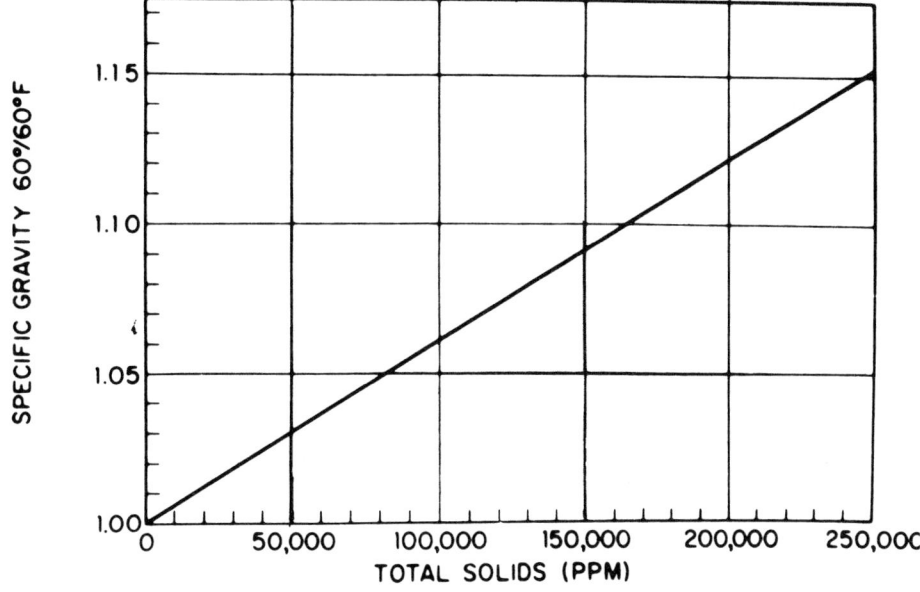

Figure 5-12. Specific gravity of formation waters (D_ω) versus total solids in ppm (data for NaCl solutions) (Pirson, 1963, p. 39).

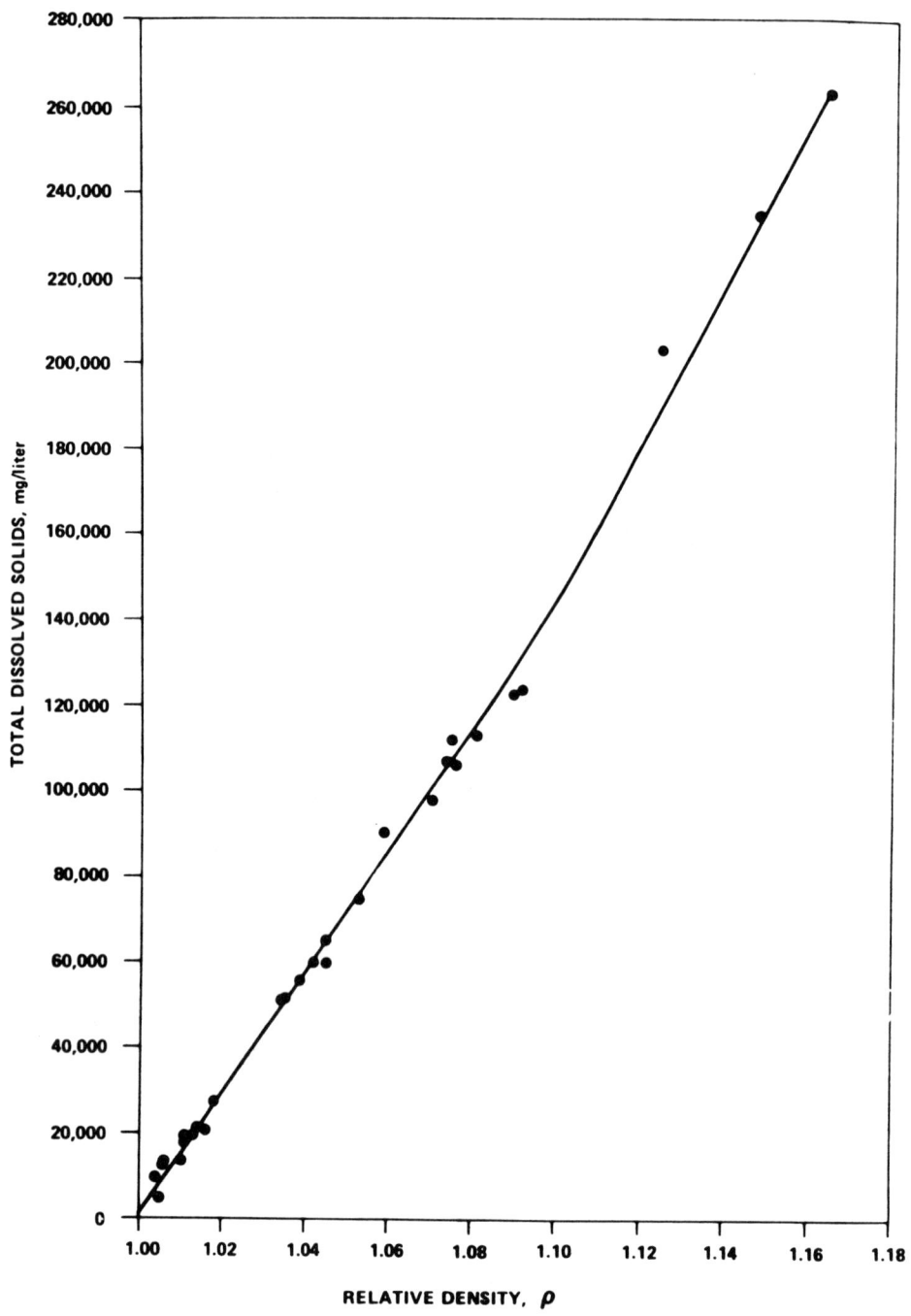

Figure 5-13. Relation between relative density and dissolved solids content of brines in deep aquifers of the Illinois basin (Bond, 1972).

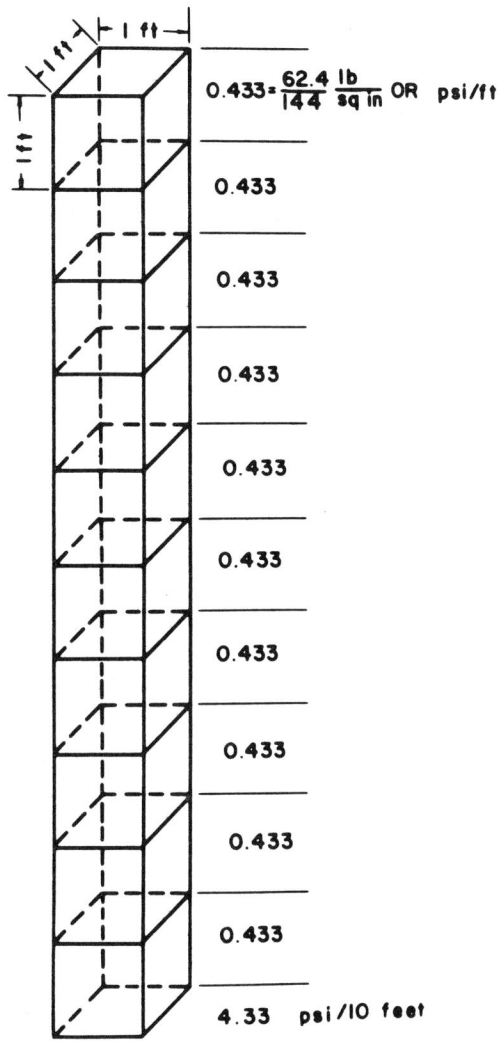

Figure 5-14. Hydraulic pressure gradient in a column of water (Katz and Coats, 1968, p. 11).

water, or wastewater, is other than 1.0, the rate of pressure increase varies accordingly. For example, if a well bore is filled with formation water with a dissolved solids content of 65,000 mg/ℓ and a specific gravity of 1.05, fluid pressure increases at a rate of 0.455 psi/ft, and would be 455 psi at the bottom of a 1,000-ft-deep water-filled well. The fluid pressure must be added to the pump pressure in injection calculations to determine the total pressure.

Although instances of truly anomalous formation pressure are likely to be relatively rare at sites selected for wastewater injection, the existence of unusually high or low pressures, and the possible reasons for their existence, should be recognized. Some causes of anomalous pressure are:

1. Compaction of sediments
2. Tectonic forces
3. Osmotic effects
4. Massive extraction or injection of fluids

Abnormally high pressures can result from 1, 2, and 3, and from massive injection. Abnormally low pressures can result from osmotic effects and extraction of fluids. Abnormally high pressures resulting from compaction of sediments are common in deep wells of the Gulf Coast (Dickinson, 1953). Berry (1973) concluded that abnormally high pressures in the California Coast Ranges are a result of tectonic forces. Hanshaw (1972) discussed natural osmotic effects and their relation to subsurface wastewater injection.

COMPRESSIBILITY

The compressibility of an elastic medium is defined as:

$$\beta = \frac{-\partial V}{V \partial p} \quad (F/L^2)^{-1} \tag{1}$$

where

β = compressibility of medium (pressure^{-1})
V = volume
p = pressure

The compressibility of water varies both with temperature and pressure, as is shown in Figure 5-15. For problems in wastewater injection, β will generally be within the range of 2.8 to 3.3 x 10^{-6} psi^{-1}, and 3.0 x 10^{-6} psi^{-1} is a reasonable value to assume in most cases.

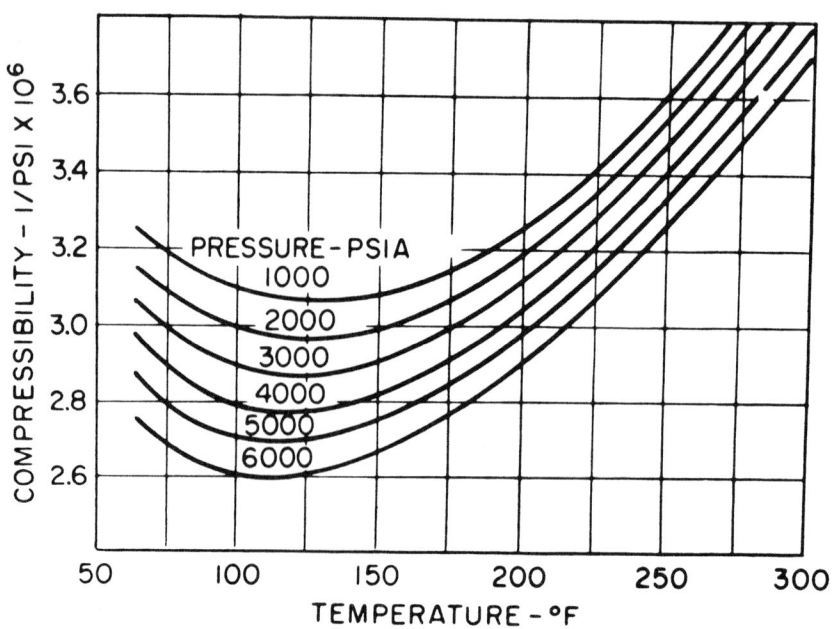

Figure 5-15. Compressibility of water (Katz and Coats, 1968, p. 93).

MECHANICAL PROPERTIES OF INJECTION AND CONFINING UNITS

POROSITY

Porosity is defined as:

$$\phi = \frac{V_v}{V_t} \quad \text{(dimensionless)} \tag{2}$$

where ϕ = porosity, expressed as a decimal fraction
V_v = volume of voids
V_t = total volume of rock sample.

Porosity is also commonly expressed as a percentage. Porosity may be total porosity or effective porosity. Total porosity is a measure of all void space. In comparison with total porosity, effective porosity is based on the total volume of interconnected voids. Effective porosity better defines the hydraulic properties of a rock unit, since only *interconnected* porosity is available to fluids flowing through the rock. In the remainder of the chapter, reference to porosity implies effective porosity, unless otherwise stated.

Porosity may also be classified as primary or secondary. Primary porosity includes: original intergranular or intercrystalline pores and the porosity associated with fossils, bedding planes, and so forth. Secondary porosity results from fractures, solution channels, and from recrystallization and dolomitization. Intergranular porosity occurs principally in unconsolidated sands and sandstones, and can be measured reasonably well in the laboratory using core samples taken from wells. Porosity contributed by fractures and solution channels is difficult to measure in the laboratory. Various borehole geophysical methods that will be discussed later can be used to determine the porosity of strata in place. Porosity values in reservoir formations range from a maximum of about 0.40 in unconsolidated sands to as little as 0.02 in dense limestones. Porosity in the Mt. Simon Formation of Illinois ranges from about 0.20 to 0.02, as shown in Figure 5-16.

PERMEABILITY

Permeability is the capacity of a rock to transmit fluid. Permeability is quantified by the coefficient of permeability, or hydraulic conductivity. When both the properties of the fluid and the porous medium are considered, the coefficient of permeability $\overline{K}$ is defined by Darcy's law as:

$$\overline{K} = \frac{Q\mu}{A\rho g} \frac{dL}{dh} \quad (L^2) \tag{3}$$

where Q = flow rate through porous medium
A = cross-sectional area through which flow occurs
μ = fluid viscosity
ρ = fluid density
L = length of porous medium through which flow occurs
h = fluid head loss along L
g = acceleration of gravity.

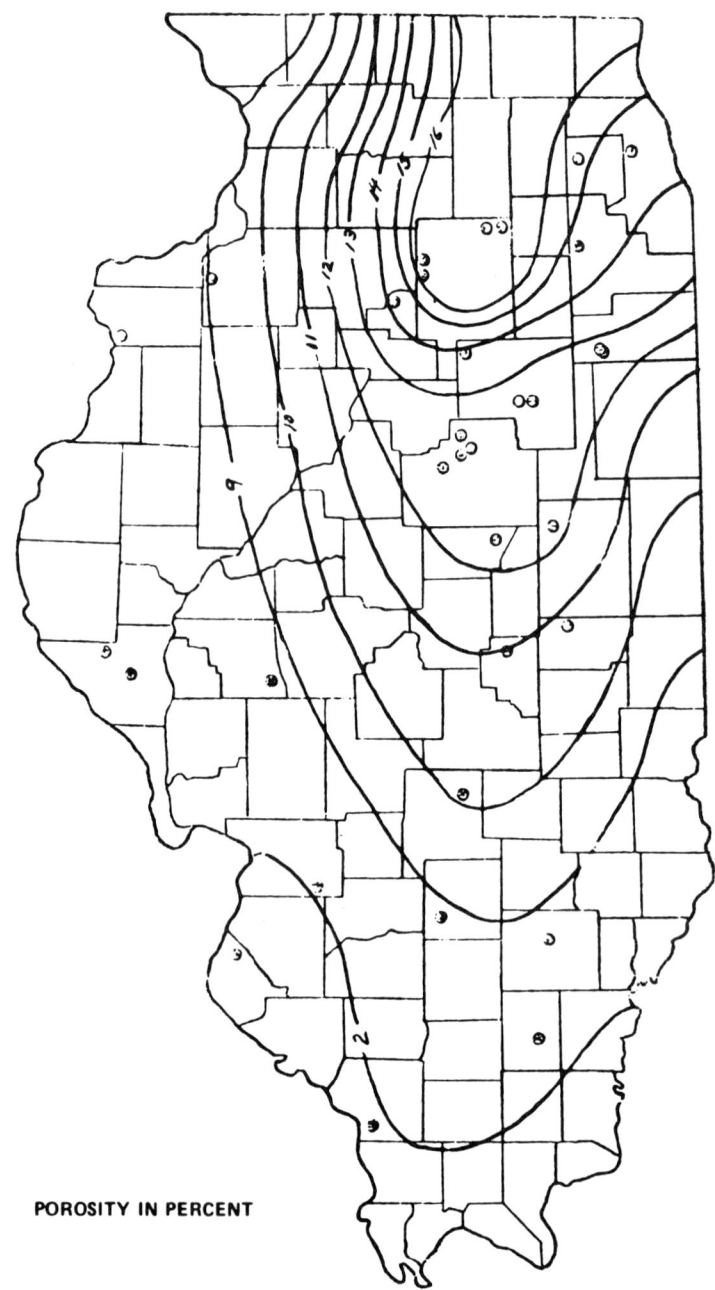

Figure 5-16. Map showing distribution of the average porosity of the Mt. Simon Formation in Illinois.

If cgs units are used, the coefficient of permeability from Equation 3 will be expressed in cm^2. The unit of permeability used in oil field work is the darcy, which is defined by:

$$\bar{K} = \frac{Q\mu}{A}\frac{dL}{dp} \quad (L^2) \tag{4}$$

where $\quad p_c = \rho gh$ = pressure

$$1 \text{ darcy} = \frac{1 \text{ cm}^3/\text{s} \times 1 \text{ cP} \times 1 \text{ cm}}{1 \text{ cm}^2 \times 1 \text{ atm}}$$

A still simpler form of Darcy's law is used in groundwater studies, where the density and viscosity of water do not vary greatly:

$$K = \frac{Q}{A}\frac{dL}{dh} \quad (L/T) . \tag{5}$$

The constant K is referred to as hydraulic conductivity, and is usually expressed in centimeters per second or in U.S. Geological Survey units, which are gallons per day x feet^2 (meinzers). A table for conversion of permeability units is given below (Table 5-3).

Table 5-3

TABLE OF EQUIVALENCY OF PERMEABILITY VALUES
IN VARIOUS UNITS
(Davis and Deweist, 1966, p. 165)

1 darcy	= 9.87 x 10^{-9} cm^{-2} = 1.062 x 10^{-11} ft^2
10^{-10} cm^2	= 1.012 x 10^{-12} darcys
0.1 cm/day	= 1.15 x 10^{-6} cm/s $\approx$ 1.18 x 10^{-11} cm^2 for water at 20 °C
1.0 cm/s	$\approx$ 1.02 x 10^{-5} cm^2 for water at 20 °C
1 darcy	$\approx$ 18.2 meinzer units for water at 60 °F
1 meinzer	= 0.134 ft/day = 4.72 x 10^{-5} cm/s $\approx$ 5.5 x 10^{-2} darcys for water at 60 °F

Permeability values for the formations used for wastewater injection range from several darcys to less than a millidarcy (one millidarcy = 10^{-3} darcy). Average permeability values for the Mt. Simon Formation in Illinois range from more than 100 millidarcys in the north to less than 1 millidarcy in the south. The permeability of shale beds in the Eau Claire Formation, overlying the Mt. Simon Formation, is consistently less than 0.001 millidarcy.

A useful constant in hydrogeologic work is the coefficient of transmissivity (transmissibility), which is the permeability, or hydraulic conductivity, multiplied by the thickness of the aquifer. When the unit of permeability is the darcy, transmissivity is in darcy-feet per centipoise.

COMPRESSIBILITY

The compressibility of an aquifer includes the compressibility of the aquifer skeleton and of the contained fluids. Thus, the total compressibility of an aquifer is

$$C = \phi\beta + \alpha \quad (F/L^2)^{-1} \tag{6}$$

where
- C = compressibility of aquifer (pressure^{-1})
- ϕ = porosity
- β = compressibility of water
- α = compressibility of aquifer skeleton.

The compressibility of water has previously been discussed. The compressibility of aquifer skeletons varies greatly, from as little as 1×10^{-8} psi^{-1} in consolidated rocks, to as much as 1×10^{-5} psi^{-1} in unconsolidated materials.

The coefficient used in analysis of reservoir response to injection or pumping is the storage coefficient (storativity), which is defined by:

$$S = \phi\gamma b \left(\beta + \frac{\alpha}{\phi}\right) \quad \text{(dimensionless)} \tag{7}$$

where ϕ, β, and α are as previously defined, and
- S = storage coefficient
- γ = ρg = specific weight of water per unit area
- b = aquifer thickness.

The storage coefficient is the volume of water an aquifer releases or takes into storage per unit surface area per unit change in hydraulic head. The storage coefficient may be estimated from the equation above, or determined from aquifer tests that will be described later. Values of S are reported to range from 5×10^{-5} to 5×10^{-3} for confined aquifers. As an estimate of the value of S for the Mt. Simon Formation in northern Illinois assume that $\phi = 11\%$, $b = 1,700$ ft, $\gamma = 0.45$ psi/ft, $\beta = 3.0 \times 10^{-6}$ psi^{-1}, and $\alpha = 6.7 \times 10^{-6}$.* Then, from the equation above, $S \approx 5.4 \times 10^{-3}$. This is a high value, but the aquifer is very thick. If the compression of the water alone were considered, then S would be 2.5×10^{-4}. The Illinois State Water Survey (1973) estimated an average storage coefficient of 1×10^{-4} for the Mt. Simon Formation in northern Illinois, which is probably too low, if the entire thickness of the formation is considered.

TEMPERATURE

The temperature of the aquifer and its contained fluids is important because of the effect that temperature has on fluid properties. The temperature of shallow groundwater is generally about $2°$ to $3°$ greater than the mean annual air temperature. In Illinois, this is from about 60 °F in the south to 50 °F in the north. Below 30 to 60 ft, the temperature increases approximately $1°$ to $2°$F per 100 ft of depth. Figure 5-17 is

---

*Testing of the Mt. Simon Formation, in a gas storage field in northern Illinois, yielded a value of compressibility of the formation and its contained water of about 7×10^{-6}. Since the water only occupies 11% of the rock, the rock skeleton compressibility at that location is 6.7×10^{-6}.

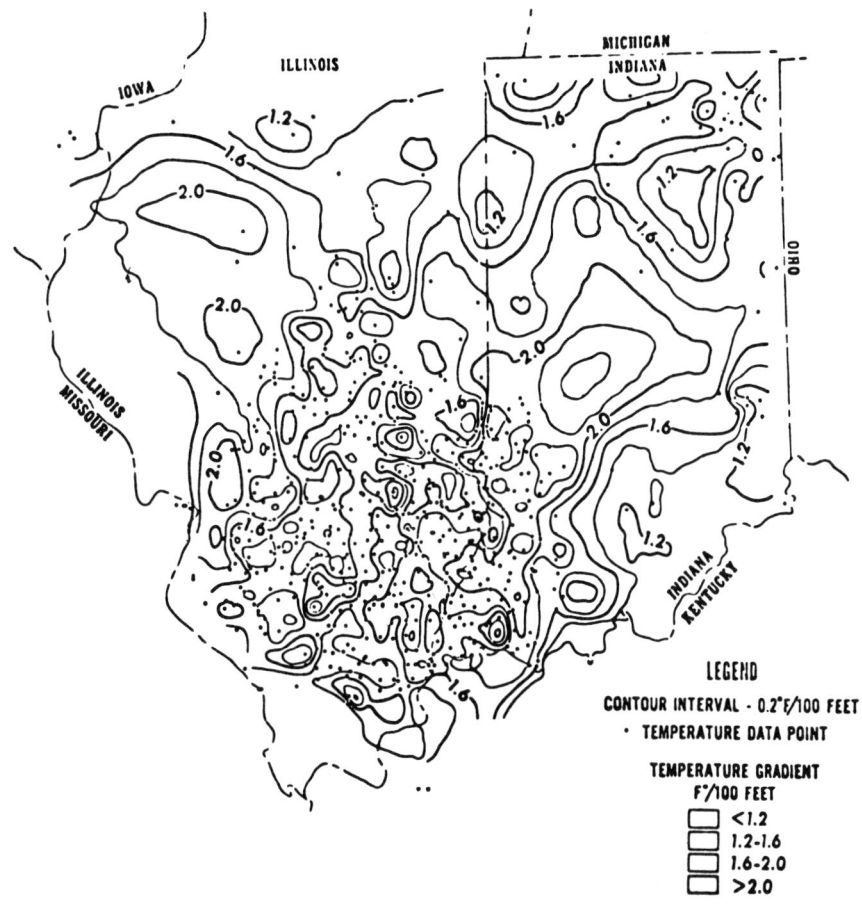

Figure 5-17. Reproduction of portfolio map No. 10, American Association of Petroleum Geologists Geothermal Survey of North America (Gould, 1974).

a geothermal gradient map of Illinois and Indiana. At a depth of 3,000 feet, in northern Illinois, the calculated temperature would be about 86 °F. The measured temperature at 3,000 ft near Pontiac, Illinois, was 90 °F. Geothermal gradient maps for the United States have been prepared by the American Association of Petroleum Geologists, Tulsa, Oklahoma, and can be obtained from that organization. Figure 5-17 is a modification of one of the AAPG maps.

STATE OF STRESS

In order to predict the pressure at which hydraulic fracturing or fault movement would be expected to occur, it is necessary to estimate the state of stress at the depth of the injection horizon. On the other hand, determination of the actual fracture pressure allows computation of the state of stress (Kehle, 1964).

The general equation for total normal stress across a plane in a porous medium is:

$$S_t = p_o + \sigma_i \quad (F/L^2) \tag{8}$$

317

where S_t = total stress

p_o = fluid pressure

σ_i = effective or intergranular normal stress.

Effective stress, as defined by Equation 8, is the stress available to resist hydraulic fracturing, or the stress across a fault plane that acts to prevent movement on that fault. The equation shows that, if total stress remains constant, an increase in fluid pressure reduces the effective stress, and a decrease in fluid pressure increases effective stress. When the effective stress is reduced to zero by fluid injection, hydraulic fracturing occurs. Fault movement will occur before normal stresses across the fault plane are reduced to zero, since there must be some shear stress acting on the fault blocks to cause them to move.

In a sedimentary rock sequence, the total normal vertical stress increases with depth of burial, under increasing thicknesses of rock and fluid. It is commonly assumed, and the validity of the assumption can easily be verified, that the normal vertical stress increases at an average of about 1 psi/ft of depth. The lateral stresses may be greater or less than the vertical stress, depending on geologic conditions. In areas where crustal rocks are being actively compressed, lateral stresses may exceed vertical ones. In areas where crustal rocks are not in active compression, lateral stresses should be less than the vertical stress. The basis of estimating lateral stress prior to drilling of a well is hydraulic fracturing data from nearby wells and/or knowledge of the tectonic state of the region in which the well is located. The tectonic state of various regions is only now being determined. For example, Kehle (1964) concluded, as a result of hydraulic fracturing data from four wells, that the stresses at the well locations in Oklahoma and Texas were representative of an area that was tectonically in a relaxed state. In contrast, Sbar and Sykes (1973) characterized much of the eastern and north-central United States as being in a state of active tectonic compression. Further discussion concerning the state of stress and hydraulic fracturing will be presented in the section on hydraulic fracturing.

HYDRODYNAMICS

Hydrodynamics, as the term has been adopted for use in subsurface hydrology, refers to the state of potential for flow of subsurface fluids, particularly in deep sedimentary basins. As examples of its application, recent publications by Bond (1972) and Clifford (1973) discuss the flow potential in deep aquifers of Illinois, Indiana, and Ohio, as determined from pressure, water level, and water density measurements made in deep wells.

The potential for flow in deep aquifers that are used for wastewater injection is important, because it can be used to estimate natural groundwater flow rates and directions. Figure 5-18 is a map showing the potentiometric surface of the Mt. Simon Formation in Ohio and Indiana. The arrows indicate the directions of regional groundwater flow in the Mt. Simon Formation, as indicated by the potentiometric contours. Bond (1972 and 1973) discusses some of the difficulties in interpretation and application of potentiometric data.

RESOURCES

An objective in the monitoring of subsurface wastewater injection is to verify that fresh groundwater, oil or gas, coal, or other subsurface resources are not being jeopardized. Therefore, the occurrence and distribution of all significant subsurface

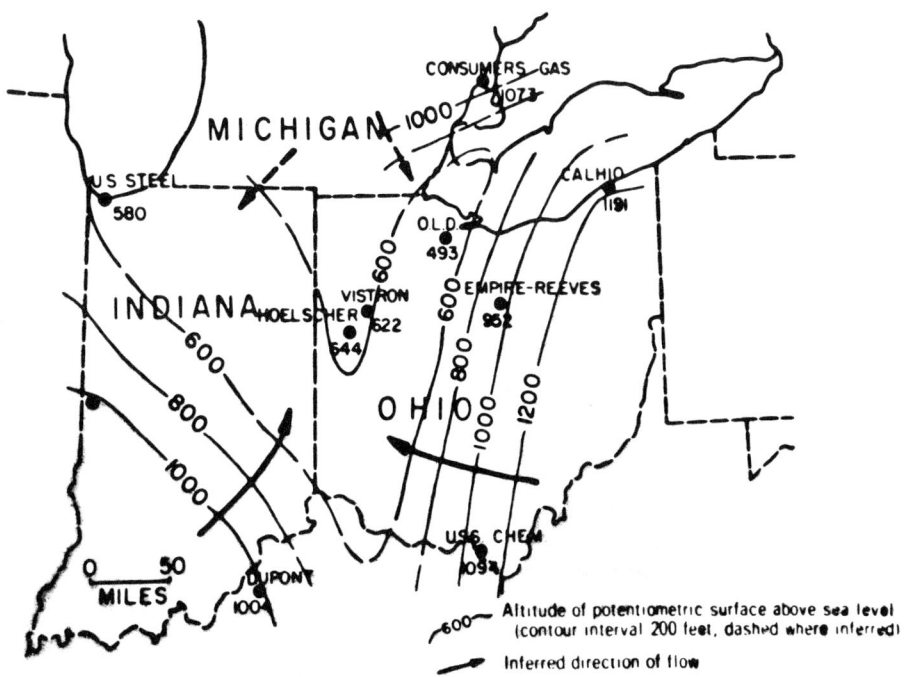

Figure 5-18. Potentiometric surface of the Mt. Simon Formation in Ohio and vicinity (Clifford, 1973).

resources must be determined. This determination is made by reference to published reports and consultation with public officials, companies, and individuals familiar with subsurface resources of the area. Also, the actual drilling of the well will show the location and nature of resources present in the subsurface at the well site.

In reviewing the occurrence of subsurface resources, the locations, construction, use, and ownership of all wells, both shallow and deep, within the area of influence of the injection well, should be determined. The plugging record for all abandoned deep wells should be obtained to verify the adequacy of such plugging. In States where oil has been produced for many years, there are often areas where wells are known to have been drilled, but for which no records are available, as well as wells which are located, but for which plugging records are not available, or for which plugging is known to have been inadequate. Documenting the status of deep wells near the injection well may be the most important step in monitoring of injection wells in areas that are, or have been, active oil or gas provinces, because these wells provide the greatest hazard for escape of wastewater or formation water from otherwise well-confined aquifers.

SECTION 3 - ACQUISITION OF SUBSURFACE DATA

PRIOR TO DRILLING

In order to estimate the performance of injection wells and evaluate the subsurface environment prior to construction, the types of information described in Section 2, "The Subsurface Environment," are estimated from sources such as the figures and tables from that section. The information in those figures and tables has, of course, come from previously drilled wells; and if it has not been compiled on maps, cross sections, and tables, it may be necessary to do so before it can be used. Basic information for previously drilled wells is available in most States through State geological surveys, oil and gas agencies, and water resources agencies. In addition, private companies acquire and sell well logs and other subsurface data. In some cases it may be necessary to go to individual oil companies or consultants for subsurface data not publically available. Companies and individuals are usually cooperative in releasing information not considered confidential.

DURING WELL CONSTRUCTION AND TESTING

ROCK SAMPLES

Most deep wells drilled today are drilled by rotary drilling rigs. Rotary drilling rigs use two basic types of drilling bits, rock bits and core bits. Rock bits grind the strata into small chips that are usually carried from the hole by a viscous drilling mud, but sometimes by water or air. The chips are periodically collected, usually after each 5 or 10 feet of new hole, washed, and examined with a low-powered binocular microscope. The methods for collection, examination, and description of such samples are presented in a reference edited by Haun and LeRoy (1958). Figure 5-19 is an example of a sample log prepared by examination of cuttings. Soft, unconsolidated clays will not yield chips, but will break down into mud, and unconsolidated, or soft, sandstones into individual grains, when drilled. Samples are of only limited value in such areas.

Cores taken with rotary core bits and barrels give a much more accurate picture of the subsurface formations than cuttings, but core samples are very expensive (>$50/ft) in deep wells, and can usually only be afforded in limited numbers. In deep wells, core samples are commonly about 4 inches in diameter. Cores are described just as are cuttings, but since a continuous sample of the formation is available, a detailed foot by foot description can be prepared (Table 5-1). Whole-core samples can be analyzed for porosity and permeability in the laboratory, or small cores can be taken from the large core and analyzed. The latter procedure is the most common. Table 5-4 shows typical laboratory core data from the Mt. Simon Formation in Illinois.

FORMATION FLUIDS

Samples of water from subsurface formations can be obtained from deep wells, before they are completed, from cores, by formation testing devices, and by swabbing.

When cores are taken, as previously described, the water in the cores can be carefully extracted, and its chemistry analyzed. Contamination is a serious problem, since the core has been exposed to infiltration by drilling mud and mud filtrate.

Drill-stem testing is a technique whereby a zone in an open borehole can be isolated by an expandable packer, or packers, and fluid from the formation allowed to flow through a valve into the drill pipe.

Figure 5-19. Sample log (Moore, 1951).

Table 5-4

LABORATORY CORE ANALYSIS DATA
FROM THE MT. SIMON FORMATION IN ILLINOIS[a]

Sample Number	Depth (feet)	Permeability (millidarcys)		Porosity (percent)
		Horizontal	Vertical	
408	3154.5	6.9	0.11	6.4
409	3155.5	<0.10	0.17	6.4
410	3156.6	<0.10	<0.10	9.7
411	3157.5	0.17	0.31	8.6
412	3158.5	0.26	0.72	8.3
413	3159.5	<0.10	<0.10	8.1
414	3160.5	1.9	0.12	9.6
415	3161.5	<0.10	<0.10	8.7
416	3162.5	2.3	0.98	8.1
417	3163.5	0.43	0.46	6.2
418	3164.5	12.	0.12	8.2
419	3165.5	3.1	1.1	14.7
420	3166.5	0.31	0.44	10.7
421	3167.5	7.8	0.79	10.0
422	3168.5	8.5	5.4	9.9
423	3169.5	5.0	3.2	7.2
424	3170.5	6.2	3.6	6.9
425	3171.5	3.4	1.2	8.3
426	3172.5	10.	2.5	12.2
427	3173.5	1.4	0.46	8.9
428	3174.5	11.	2.0	8.0
429	3175.5	8.5	1.5	8.2
430	3176.5	2.6	0.91	7.7
431	3177.5	0.74	<0.10	5.9

Note:
[a] Mt. Simon Core No. 15 3148.0 - 3178.0

The basic drill-stem test tool assembly consists of:

1. A rubber packing element, or packer, which can be expanded against the hole to segregate the annular sections above and below the element

2. A tester valve to (a) control flow into the drill pipe, that is, to exclude mud during entry into the hole, and allow formation fluids to enter during the test; and an equalizing or bypass valve to (b) allow pressure equalization across the packer(s) after completion of the flow test.

Figure 5-20 illustrates the procedure for testing the bottom section of a hole. While going in the hole, the packer is collapsed, allowing the displaced mud to rise as shown by the arrows. After the pipe reaches bottom, and the necessary surface

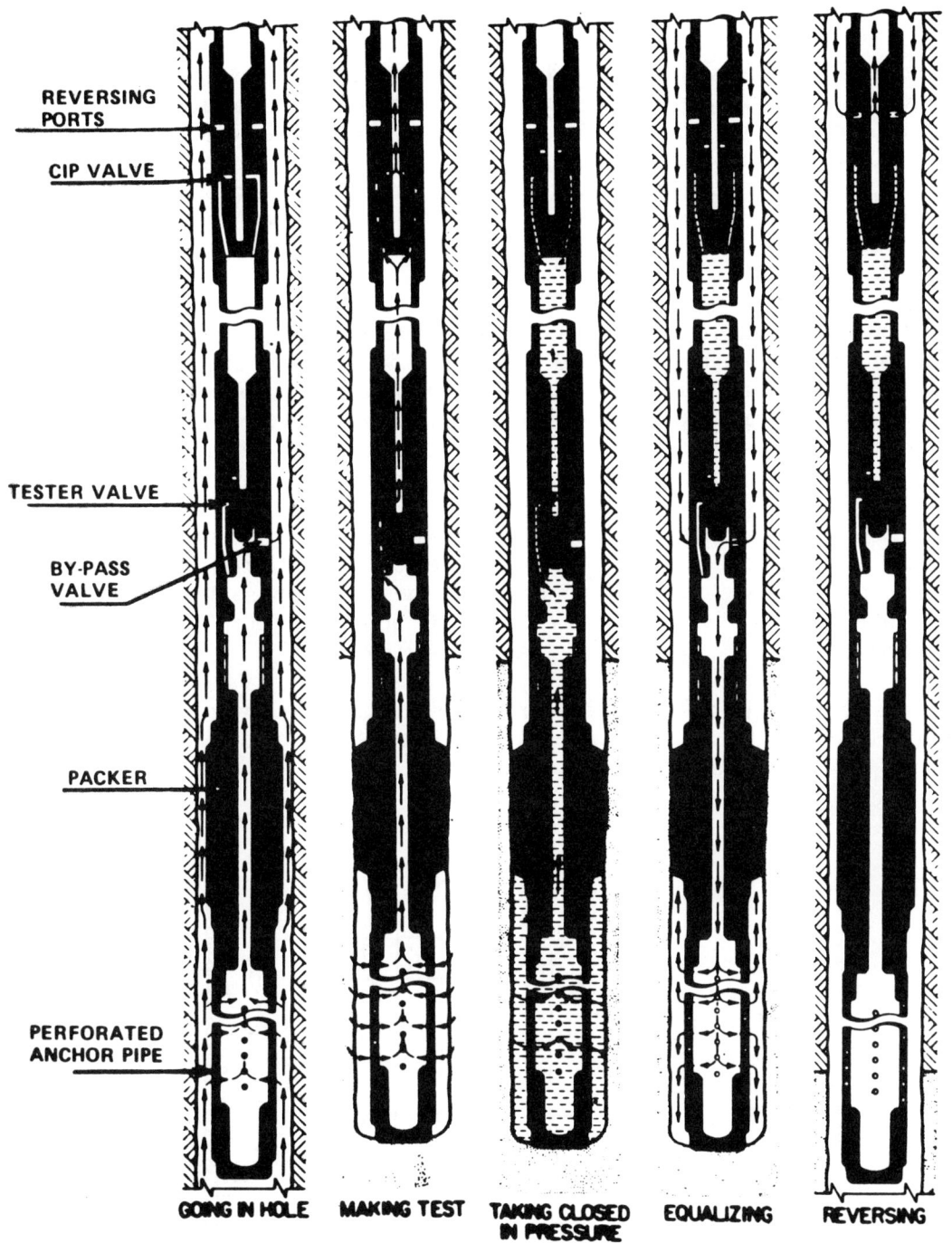

Figure 5-20. Fluid passage diagram for a conventional bottom section, drill stem test (Gatlin, 1960).

preparations have been made, the packer is set (compressed and expanded); this isolates the lower zone from the rest of the open hole. The compressive load is furnished by a slacking off of the desired amount of drill string weight, which is transferred to the anchor pipe below the packer.

The tester valve is then opened and the isolated section is exposed to the low pressure inside the empty, or nearly empty, drill pipe. Formation fluids can then enter the pipe, as shown in the second picture. At the end of the test, the tester valve is closed, trapping any fluid above it, and the bypass valve is opened to equalize the pressure across the packer. Finally, the setting weight is taken off, and the packer is pulled free. The pipe is then pulled from the hole until the fluid-containing section reaches the surface. As each successive pipe section is removed, its fluid content may be examined.

Although the above is a very common type of test, there are many other variations of procedure, as indicated in Figure 5-21. The straddle packer test is necessary when isolation from formations both above and below the test zone is necessary. Such a situation arises when it is desired to test a zone previously passed by. Straddle testing is less desirable than conventional testing, from both a cost and an operational hazard standpoint. Two packers are more apt to become stuck than one, since any material which sloughs or caves from the test zone may accumulate between the packers. Also, two positive, pressure-tight packer-formation seals are required for a successful test. Consequently, this procedure is not preferred, and is applied only when necessary. This should not be construed to mean that these disadvantages prevent one from making such tests, but rather that the additional problems the tests entail should be recognized.

Formation testing devices are available, which can be lowered into the borehole on a wire line. In this case, the sample is limited to the amount that can be contained in the testing device (up to about 5 gallons).

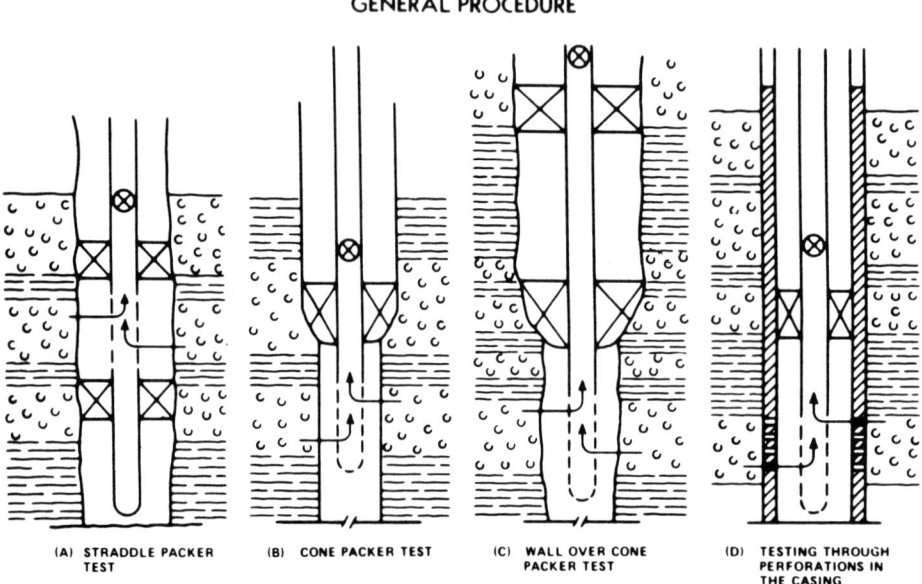

Figure 5-21. Schematic illustration of various drill stem test conditions (Kirkpatrick, 1954).

Swabbing is a method of producing fluid that is similar to pumping a well. In swabbing, fluid is lifted from the borehole through drill pipe, casing, or tubing, by a swab that falls freely downward through the pipe and its contained fluid, but which seats against the pipe walls on the upstroke, drawing a volume of fluid above it as it is raised. Swabbing may be used in conjunction with drill-stem testing to increase the volume of fluid obtained. The advantage of swabbing is that it can be continued until all drilling mud has been drawn from the pipe, and the formation and the chemistry of the water obtained reaches a steady state. This procedure helps to insure that a representative sample of formation water is obtained.

BOREHOLE GEOPHYSICAL LOGS

After a well has been drilled, a variety of borehole logging tools are available that can be used to produce a record of the nature of the formations penetrated and their contained fluids. In borehole logging, a probe is lowered into a well at the end of a wire cable, and selected geophysical properties are measured and recorded at the surface as a function of depth.

Current methods of well logging are too numerous to discuss in detail here. A broad classification of the more commonly used methods is shown in Table 5-5, together with their main applications. Because the variety of available logging methods is so

Table 5-5

WELL LOGGING METHODS AND THEIR APPLICATIONS
(MODIFIED AFTER JENNINGS AND TIMUR, 1973)

	Method	Property	Application
ELECTRICAL	Spontaneous Potential (SP)	Electrochemical and electrokinetic potentials	Formation water resistivity (R_w); shales and nonshales; bed thickness; shaliness
	Nonfocused Electric Log	Resistivity	a. Water and gas/oil saturation b. Porosity of water zones c. R_w in zones of known porosity d. True resistivity of formation (R_t) e. Resistivity of invaded zone
	Focused Conductivity Log	Resistivity	a, b, c, d Very good for estimating R_t in either freshwater or oil base mud
	Focused Resistivity Logs	Resistivity	a, b, c, d Especially good for determining R_t of thin beds Depth of invasion
	Focused and Nonfocused Microresistivity Logs	Resistivity	Resistivity of the flushed zone (R_{xo}) for calculating porosity Bed thickness
ELASTIC WAVE PROPAGATION	Transmission	Compressional and shear wave velocities	Porosity; lithology; elastic properties, bulk and pore compressibilities
		Compressional and wave attenuations	Location of fractures; cement bond quality
	Reflection	Amplitude of reflected waves	Location of vugs, fractures; orientation of fractures and bed boundaries; casing inspection

(continued)

Table 5-5 (Cont'd)

	Method	Property	Application
RADIATION	Gamma Ray	Natural radioactivity	Shales and nonshales; shaliness
	Spectral Gamma Ray	Natural radioactivity	Lithologic identification
	Gamma-Gamma	Bulk density	Porosity, lithology
	Neutron-Gamma	Hydrogen content	Porosity
	Neutron-Thermal Neutron	Hydrogen content	Porosity; gas from liquid
	Neutron-Epithermal Neutron	Hydrogen content	Porosity; gas from liquid
	Pulsed Neutron Capture	Decay rate of thermal neutrons	Water and gas/oil saturations; reevaluation of old wells
	Spectral Neutron	Induced gamma ray spectra	Location of hydrocarbons; lithology
OTHER	Caliper	Borehole diameter	Calculation of cement volume; location of mud cake
	Dipmeter	Azimuth and inclination of bedding planes	Dip and strike of beds
	Deviation Log	Azimuth and inclination of borehole	Borehole position
	Gravity Meter	Density	Formation density
	Ultra-Long Spaced Electric Log	Resistivity	Salt flank location
	Nuclear Magnetism	Amount of free hydrogen; relaxation rate of hydrogen	Effective porosity and permeability of sands; porosity for carbonates
	Production or Injectivity	Temperature, flow rate, fluid specific gravity, pressure	Downhole production or injection
	Temperature Log	Temperature	Formation temperature

great, the suite used in logging a well must be carefully selected to provide the desired information at an acceptable cost. Local practice in the particular geographic area is a valuable guide, since it represents the cumulative experience obtained from logging many wells. Some of the objectives in logging injection wells are: determination of lithology; bed thickness; amount, location, and type of porosity; and salinity of formation water. In order to achieve these objectives, a chosen suite of logs should include a gamma ray log, a focused resistivity log, and one or more porosity measuring logs, selected from among the various radiation and elastic wave logs. Some other frequently used logs include the spontaneous potential (SP) and nonfocused electric logs, the caliper log, and the temperature log.

Figures 5-22, 5-23, and 5-24 are intervals from a Laterlog*-gamma ray-neutron log, a sonic log, and a temperature-caliper log run in a wastewater disposal well in northern Illinois. On the Laterlog-gamma ray-neutron log, the contact between the Eau Claire Formation and the Mt. Simon Formation is shown at 3108 feet, where it was picked by the Illinois Geological Survey. However, it is apparent from the gamma ray log that, for engineering purposes, the shale confining interval terminates at 2900 ft and that sandstones usable for injection begin at 2900 ft. From the sonic log, it can be seen that the first sandstone interval from 2900 to 2940 ft has an average interval transit time of about 72 μs/ft. Using tables provided by the logging company (Schlumberger, 1972a) and a matrix velocity of 19,500 ft/s, the average porosity of this sandstone body

---

*Laterlog is a trade name of Schlumberger, Ltd., for a resistivity log.

is estimated as 15%. The temperature log shows a temperature of about 83.5 °F from 2900 to 2940 ft and from the Laterlog (Figure 5-22), the resistivity of this interval is about 40 Ω-m. From the Archie equation (Schlumberger, 1972) the formation factor F is 45, and the resistivity of the formation water is 0.625 Ω-m. A sodium chloride water with a resistivity of 0.625 Ω-m has a dissolved solids content of about 8,000 ppm at 83.5 °F. Actually the formation water salinity is probably about twice the calculated value because the Laterlog yields incorrectly high resistivities, when run in low-salinity mud, as is the case here. An induction log would yield more accurate results in such a situation. This example illustrates some of the principal uses of borehole geophysical logs in conjunction with the evaluation of geological conditions in wastewater injection wells. Further uses will be covered in Section 5, on well monitoring. Keys and Brown (1973) give a more complete discussion of the application of borehole geophysical logs to wastewater injection than is possible here.

TESTING OF INJECTION UNITS AND CONFINING BEDS

Examination of the records of many of the wastewater injection wells that have been constructed up to the present time shows that, with few exceptions, the maximum amount of usable geologic and engineering information has not been obtained during the testing of wastewater injection wells. This is regrettable, because such tests provide the best basis for analyzing reservoir conditions prior to injection, predicting the long-term behavior of the well and the reservoir, detecting and understanding changes in well performance that may occur during operation, and analyzing the history of a well from its records.

The methods for testing of pumping or injection wells and the techniques for analysis of test data are discussed in numerous textbooks and in hundreds of other publications concerning groundwater and petroleum engineering. Because the number of published articles and the scope of their content are so extensive, only a few selected references are mentioned, and a few examples discussed here to establish the reasons for and methods of well testing.

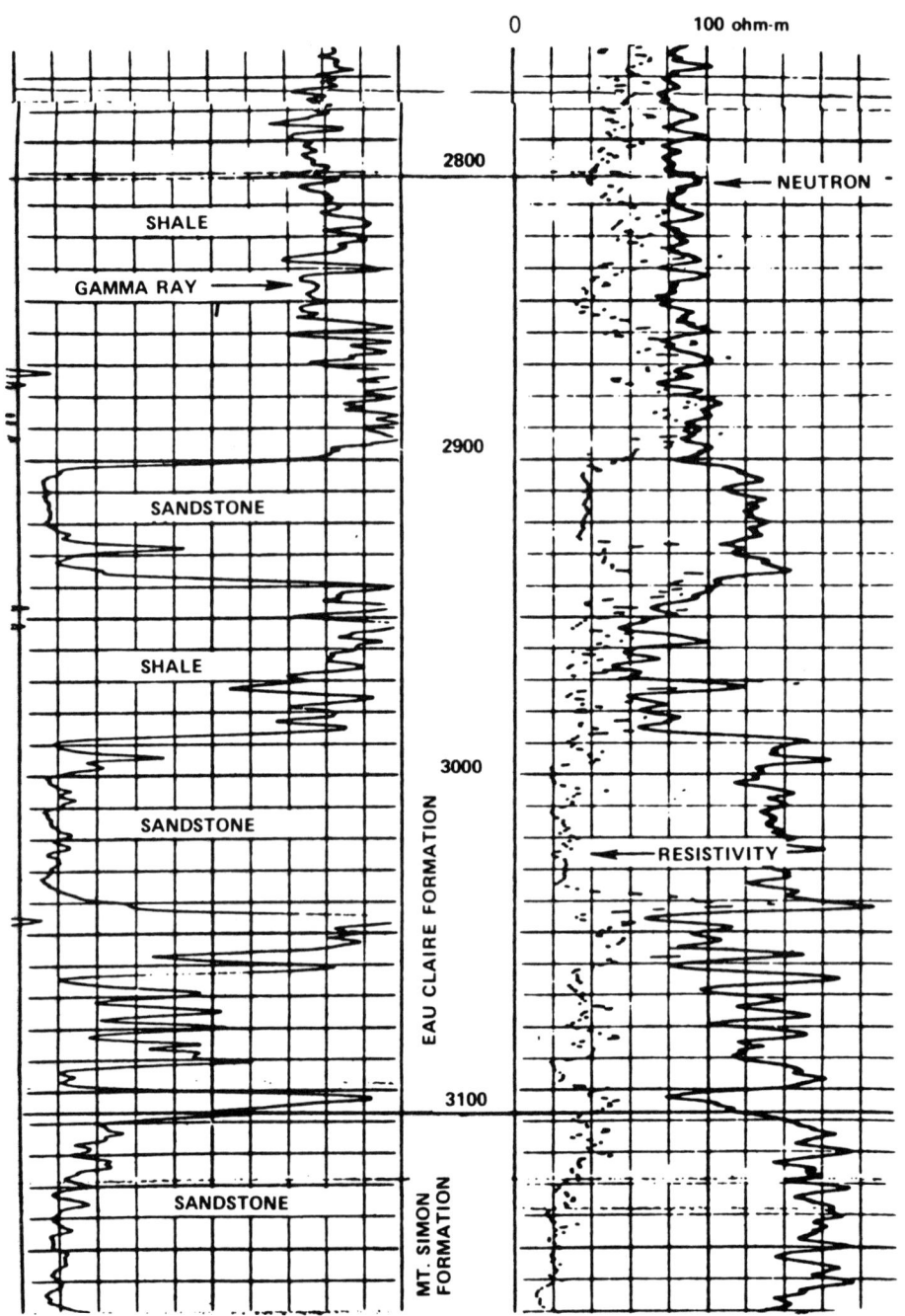

Figure 5-22. Portion of a Laterlog-gamma ray-neutron log from a deep well in northern Illinois.

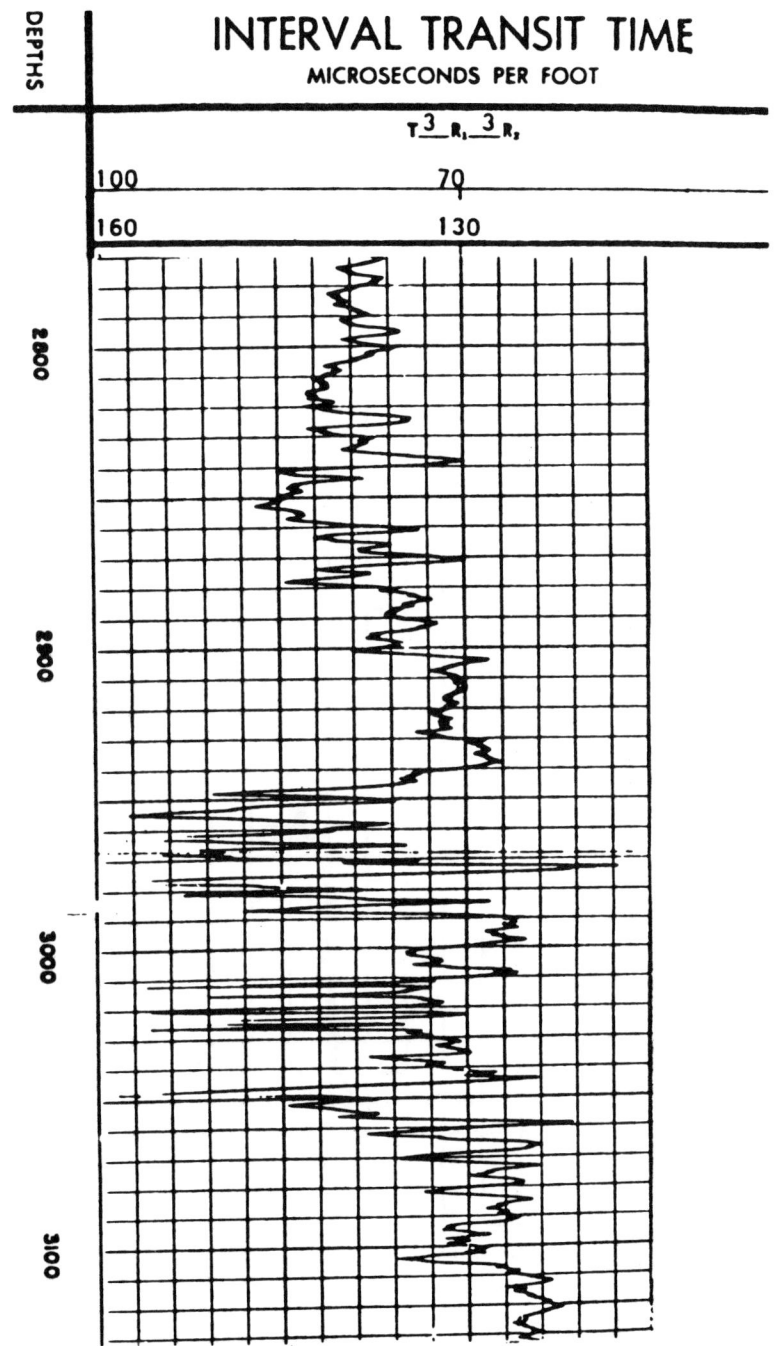

Figure 5-23. Portion of a sonic log from a deep well in northern Illinois.

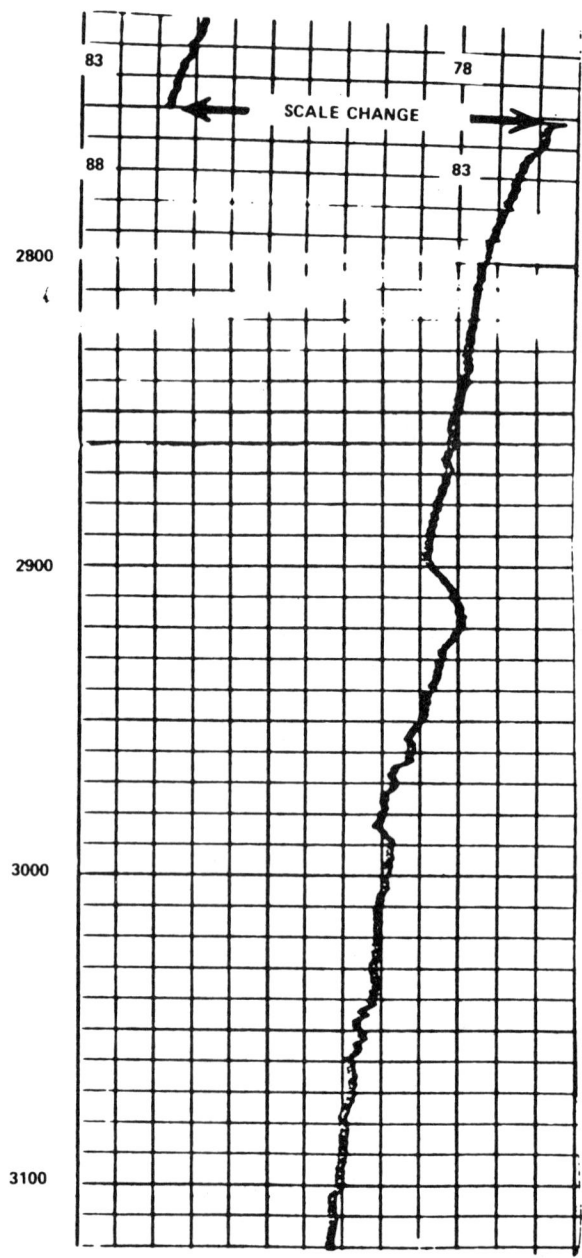

Figure 5-24. Portion of a temperature log from a deep well.

A well can be tested by pumping from it, or injecting into it. Measurements of reservoir pressure or water level can be made during pumping or injection or, alternatively, after pumping or injection has ceased, and the reservoir is adjusting to its original condition. Furthermore, reservoir pressure or water level can be measured in the principal well or in adjacent observation wells. Any one of these approaches will yield much of the same information.

DRILL STEM TESTING

In the case of the usual deep, and rather expensive, wastewater injection well, there will be no observation well, and testing will be in the well itself. In the sequence of well construction and testing, the first type of formation test likely to be made is the drill stem test (DST). As has previously been mentioned, this test is analogous to a pumping test of limited duration. Quantitative analysis is usually made using data obtained during the period of pressure buildup, following the period in which the reservoir is allowed to flow.

Figure 5-25a is a schematic DST pressure record, with a description of the sequence of events in a successful test. Figure 5-25b is a schematic representation of a test in which no fluid was produced. Conditions that may be encountered in a DST are widely variable, and considerable experience may be required in order to interpret an unusual test. The companies that provide the testing services also provide assistance in test interpretation.

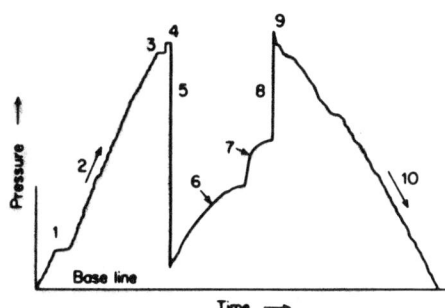

1. Putting water cushion in drill pipe
2. Running in hole
3. Hydrostatic pressure (weight of mud column)
4. Squeeze created by setting packer
5. Opened tester, releasing pressure below packer
6. Flow period, test zone producing into drill pipe
7. Shut in pressure, tester closed immediately above packer
8. Equalizing hydrostatic pressure below packer
9. Released packer
10. Pulling out of hole

Figure 5-25a. Normal sequence of events as recorded on the chart during a successful drill stem test (Kirkpatrick, 1954).

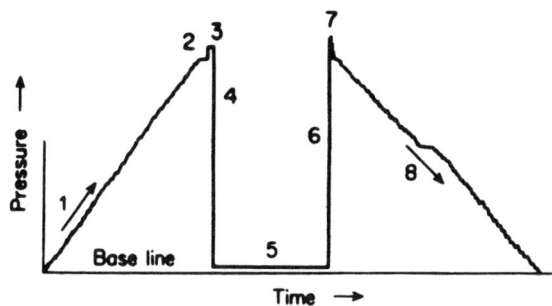

1. Running in hole
2. Hydrostatic pressure (weight of mud column)
3. Squeeze created by setting packer
4. Opened tester, releasing pressure below packer
5. Flow period, test zone open to atmosphere
6. Closed tester and equalizing hyd. pressure below packer
7. Pulled packer loose
8. Pulling out of hole

Figure 5-25b. Sequence of events as recorded during a drill stem test when no fluids were produced (Kirkpatrick, 1954).

If a test is successful, pressure buildup data from the test are taken from the DST chart and tabulated. These data are then plotted as shown in Figure 5-26. A series of calculations of formation properties are then made. The properties that are routinely calculated and are of importance here are:

1. Static bottom hole pressure

2. Transmissivity

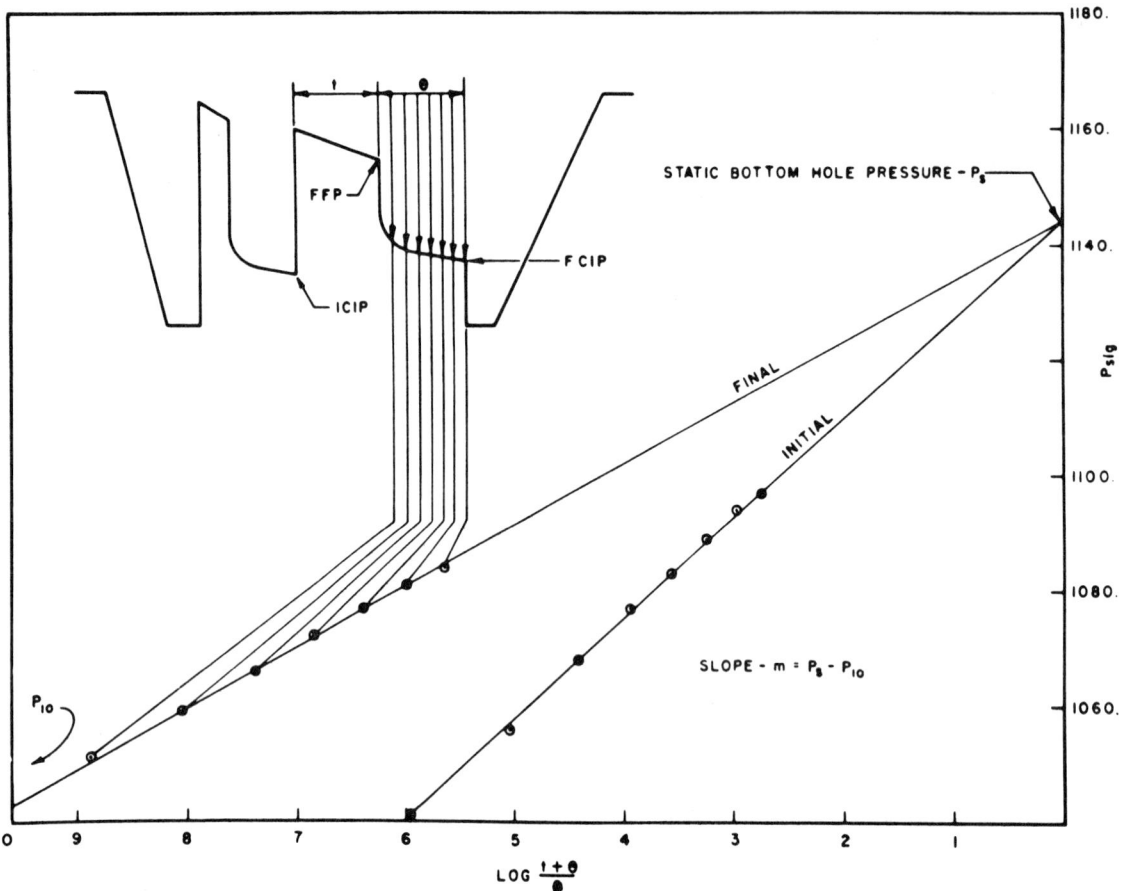

Figure 5-26. Example of a plot of data from a drill stem test with dual closed-in periods (Murphy, undated).

3. Average effective permeability
4. Damage ratio
5. Approximate radius of investigation

The static bottom hole pressure, as determined from a successful test, is assumed to closely represent the formation pressure at the elevation of the pressure recording device. Transmissivity is average permeability multiplied by the thickness of the test interval. The damage ratio is an indication of the amount of plugging of pores in the formation during drilling of the well. In addition to this routine information, drill stem tests may indicate the presence of and distance to nearby faults or facies changes that act as barriers to flow or channels for rapid flow.

For detailed presentations of drill stem test analysis, the reader is referred to Gatlin (1960), Lynch (1962), Matthews and Russell (1967) and Pirson (1963). Also, literature, such as that by Murphy (undated), is readily available from companies that provide drill stem testing services.

As an example of DST analysis, data from testing of the Mt. Simon Formation in a well in Ohio were selected. Figure 5-27 is a plot of the pressure buildup data for that test. Extrapolation of the data to the logarithm of $(t + \theta)/\theta = 0$ shows that the static formation pressure is 2750 psig. The gage was at a depth of 5886 ft in the well, so the fluid pressure gradient is 0.467 psi/ft of depth.

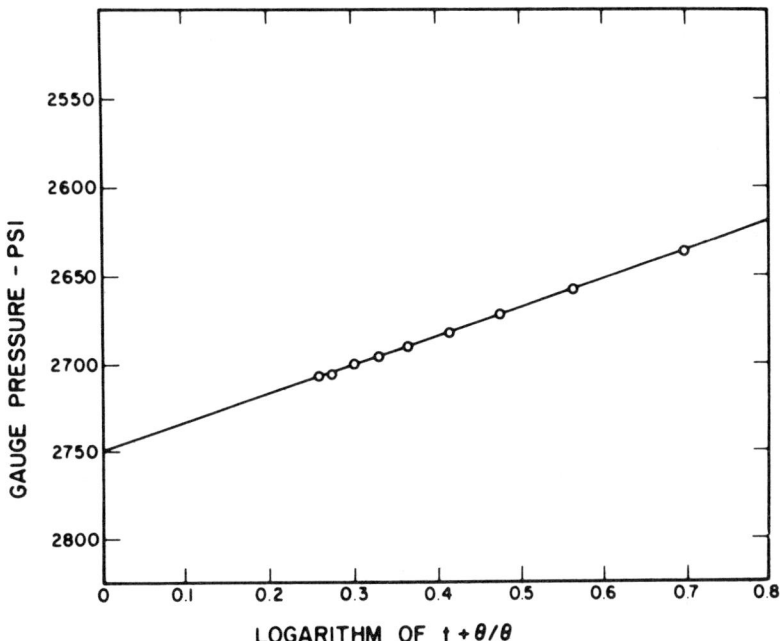

Figure 5-27. Plot of extrapolated pressure from drill stem test data from an injection well in Ohio.

For the remaining calculations, the following values from the test are needed (dimensionalized in oil field units):

P_f = final flow pressure = 1061 psig

t = final flow time = 62 min

m = $P_s - P_{10}$ = 163 psi/log cycle

Q = average flow rate = 347 bbls/day

µ = water viscosity = 1.065 cP

b = formation thickness = 105 ft

Then,

T = transmissivity = $162.6 \frac{Q}{m}$ (millidarcy-ft/cP)   (9)

$\bar{K}$ = average permeability = $\frac{T\mu}{b}$ (millidarcys)   (10)

DR = damage ratio = $\frac{0.183(p_s - p_f)}{m}$ (dimensionless)   (11)

r = radius of investigation $\cong (\bar{K}t)^{1/2}$.   (12)

The transmissivity is computed to be 345 millidarcy-ft/cP, the average permeability 3.5 millidarcys, the damage ratio 1.9, and the radius of investigation 14.73 ft. These calculations reveal that the Mt. Simon Formation, at this location, has a very low capacity to accept injected fluids. The capacity could theoretically be improved nearly 100% by removing formation damage; reservoir stimulation by hydraulic fracturing would also help, but the reservoir is not promising. No hydrologic boundaries were encountered within the radius of investigation, which was only about 14 ft. Further well testing and core analysis results to confirm these findings are discussed in the material that follows.

INJECTIVITY TESTS

After an injection well has been drilled, and possible injection intervals identified by coring, geophysical logging, and drill stem testing, injection tests will usually be run. For initial injection testing, truck-mounted pumps are often rented, and treated water used for injection rather than wastewater. Frequently, more than one possible injection interval is present, and tests are performed on the intervals individually or on more than one at a time. The common practice when performing an injection test is to begin injection at a fraction of the final estimated rate, to inject at this rate for at least several hours, then to repeat this process at increasingly greater rates until a limiting rate or pressure is reached. Injection is then stopped, and the reservoir allowed to return to its original pressure state. Pressures may or may not be recorded during this fall-off period.

Regardless of the sequence in which a test is performed, if pressure, time, and flow data are accurately recorded, and the test is run long enough, it is theoretically possible to analyze the test. However, the simpler the test, the simpler and probably more reliable the interpretation. Tests performed on more than one interval at a time are particularly difficult to interpret, and should be avoided if possible or, alternatively, both single and multiple zone tests performed.

Figure 5-28 is a plot of the data from a constant-rate injectivity test of the Mt. Simon Formation. The test was run at 75 gpm for about 25 hr. The equation used to determine formation transmissivity from Figure 5-28 is:

$$T = \frac{2.30 Q}{4\pi \Delta h} \quad (L^2/T). \tag{13}$$

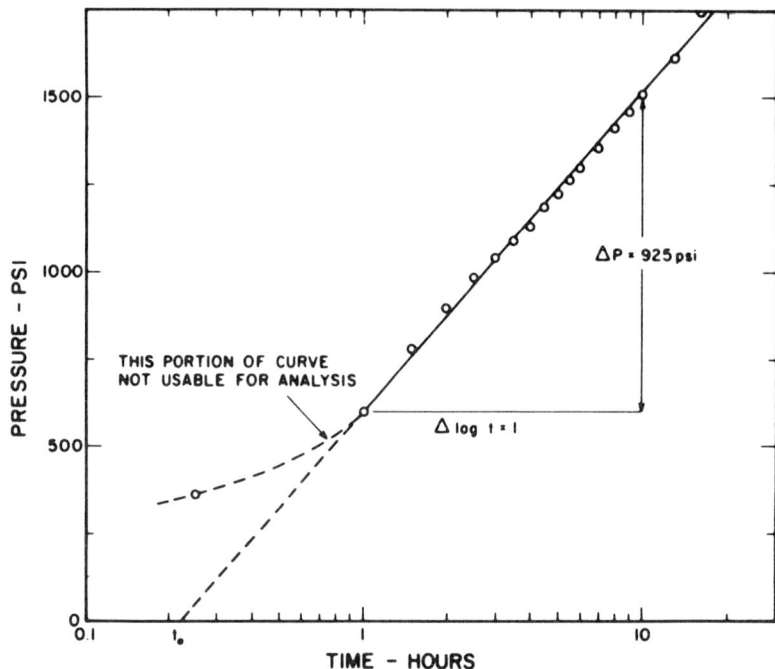

Figure 5-28. Plot of pressure buildup data from an injectivity test of the Mt. Simon Formation in Ohio.

Alternatively, Equation 9 can be used. Any consistent units can be used in Equation 13, whereas Equation 9 is dimensionalized for oil field units, as previously indicated.

Using Equation 9

$$T = \frac{162.6 \times 2571 \text{ bbl/day}}{925 \text{ psi/log cycle}} = 452 \text{ millidarcy-ft/cP.}$$

Using Equation 13

$$T = \frac{2.30 \times 14{,}434 \text{ ft}^3/\text{day}}{4\pi \times 2136 \text{ ft/log cycle}} = 1.24 \text{ ft}^2/\text{day}$$

or

$$T = 9.3 \text{ gal/day ft.}$$

This test was run on the same well for which the drill stem test analysis was given, but the well bore was cleaned up and acidized before the injectivity test, thus leading to a slightly higher transmissivity.

The injectivity test can further be used to determine the formation storage coefficient from

$$S = \frac{2.25\, T t_o}{r^2} \quad \text{(dimensionless)} \tag{14}$$

where t_o = intercept of extrapolated test curve with time axis
r = radius of well bore.

In Figure 5-28, $t_o = 2.2$ hr and

$$S = \frac{2.25 \times 1.24 \text{ ft}^2/\text{day} \times 0.0092 \text{ days}}{(0.396)^2} = 0.16.$$

As was previously discussed, storage coefficient values for confined aquifers are generally at least three orders of magnitude lower than the calculated value of 0.16. As a better estimate, Equation 7 yields a value for the storage coefficient of 3.34×10^{-4}. It is believed that the discrepancy in this case results from the fact that the well was hydraulically fractured during an earlier injection test, leading to a greatly enlarged effective well bore. As an estimate of the degree of enlargement, Equation 14 is rearranged and solved for r, using the calculated storage coefficient, yielding:

$$r = \sqrt{\frac{2.25\, T t_o}{S}}$$

$$r = \sqrt{\frac{2.25 \times 1.24 \text{ ft}^2/\text{day} \times 0.009 \text{ days}}{3.34 \times 10^{-4}}} = 8.7 \text{ ft.}$$

This is a reasonable value and will be used in later calculations.

If early time data are available, an alternative form of analysis that involves curve matching can be employed. Figure 5-29 is such a plot of recovery data for an injection well at Mulberry, Florida. The details of the analysis of this test are given by

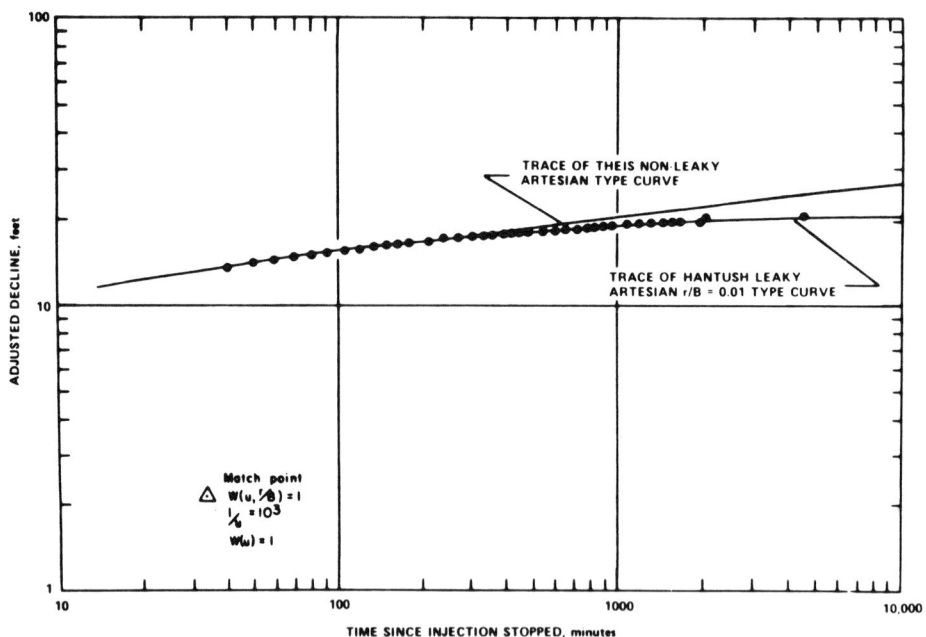

Figure 5-29. Plot of recovery data and matching-type curves for an injection test of a well at Mulberry, Florida (Wilson et al. 1973).

Wilson et al. (1973). The most interesting aspect of this example is that the test data indicate an observable amount of leakage through the confining beds. Witherspoon and Neumann (1972) discuss in some detail the theory and procedure for analysis of leaky confining beds, and give two field examples from gas storage projects.

Readers wishing to pursue the subject of aquifer testing further are referred to the same references previously given for drill stem test analysis, particularly to the Society of Petroleum Engineers Monograph prepared by Matthews and Russell (1967). Additionally, publications in the groundwater field by Lohman (1972) and Kruseman and De Ridder (1970) are excellent recent summaries of this subject, as is the reference by Witherspoon et al. (1967), which was prepared for the underground gas storage industry.

SECTION 4 - PREDICTION OF AQUIFER RESPONSE

FLOW THEORY

The basic equation used to describe the flow of fluids in porous media is Darcy's law, alternate forms of which are given by Equations 3, 4, and 5 in Section 2. Darcy's law alone can be used for calculations of steady flow. Steady flow occurs when the same quantity of fluid is entering an aquifer as is leaving it, so that no change in volume of the aquifer or its contained fluid is occurring with time.

When flow is unsteady or, as stated in oil field terminology, when formation pressures are transient, Darcy's law must be combined with the continuity equation so that time and the compressibility of the aquifer and aquifer fluids may be taken into account. The appropriate partial differential equation and its derivation may be found in most modern texts on hydrogeology and petroleum reservoir engineering, along with numerous solutions.

The solution first formulated and still most widely used is that for a well pumping from, or injecting into, an aquifer under the following conditions:

1. The aquifer is, for practical purposes, infinite in a real extent
2. The aquifer is homogeneous, isotropic, and of uniform thickness over the area of influence
3. Natural flow in the aquifer is at a negligible rate
4. The aquifer is sufficiently confined so that flow across confining beds is negligible
5. The well penetrates the entire thickness of the aquifer
6. The well is small enough that the storage in the well can be neglected, and water removed from storage in the aquifer is discharged instantaneously

This is a formidable list of assumptions, which are obviously not completely met in any real situation. However, if one reviews the characteristics of aquifers such as the Mt. Simon Formation, it can be concluded that they probably comply with the assumptions sufficiently for practical purposes.

The equation that describes the response of such an aquifer to a single injection well is then:

$$\Delta h = \frac{Q}{4\pi T}\left(-0.577216 - \log_e u + u - \frac{u^2}{2 \cdot 2!} + \frac{u^3}{3 \cdot 3!} - \ldots\right) \; (L) \qquad (15)$$

where

$$u = \frac{r^2 S}{4Tt} \quad \text{(dimensionless)}$$

and Δh = hydraulic head change at radius r and time t
  Q = injection rate
  T = transmissivity
  S = storage coefficient
  t = time since injection began
  r = radial distance from well bore to point of interest

One can easily enter the desired values into this series solution. Tables with the series evaluated are available in the previously referenced publications on aquifer testing.

For large values of time, small values of radius of investigation, or both, Equation 15 can be reduced to:

$$\Delta h = \frac{2.30 Q}{4\pi T} \log \frac{2.25 T t}{r^2 S} \quad (L) \quad . \tag{16}$$

Equations 15 and 16 are not dimensionalized; therefore, any consistent units can be used.

Two very important characteristics of the equations presented above are that individual solutions can be superimposed, and that hydrologic boundaries such as faults can be simulated by a properly located imaginary well. The fact that solutions can be superimposed allows the effects of multiple wells to be easily analyzed. Because the effect of boundaries is analogous to that of properly located pumping or injection wells, the existence of boundaries can be detected by observing aquifer response to injection or pumping or, conversely, the effects of known or suspected boundaries can be estimated.

REGIONAL FLOW

As examples of the application of Darcy's law to analysis of regional flow, the velocity of natural flow in the Mt. Simon Formation in Ohio and the lower Floridan aquifer in Florida will be considered.

From Figure 5-18 (Section 2) it can be seen that, at the location of the Empire-Reeves injection well, the hydraulic gradient is 8 ft/mi toward the northwest. At this location, the Mt. Simon Formation has a permeability of 24 millidarcys (from a drill stem test) and a porosity of 10.4% (Clifford, 1973). Rearranging Darcy's law:

$$\bar{v} = \frac{Q}{A} = K \frac{dh}{dL} \quad (L/T) \tag{17}$$

where $\bar{v}$ = apparent velocity through entire area A.

Then,

$$v = \frac{\bar{v}}{\phi} = \frac{Q}{A\phi} = \frac{K}{\phi} \frac{dh}{dL} \tag{18}$$

where v = average velocity of flow through pores

ϕ = porosity.

From the data given above, converted to consistent units, and entered into Equation 18

$$v = \frac{21.3639 \text{ ft/yr}}{0.104} \times \frac{8 \text{ ft/mi}}{5{,}280 \text{ ft/mi}}$$

$$= 0.31 \text{ ft/yr} \quad .$$

This evaluation shows that water in the Mt. Simon Formation in north-central Ohio is moving northwest at a rate of 0.31 ft/yr. The source of the hydraulic gradient and the fate of the moving water are not understood. Furthermore, there are complications in the analysis itself, as pointed out by Bond (1973). However, in spite of such uncertainties, it can be indisputably concluded that water in the Mt. Simon Formation is moving at a negligible rate, if at all, at this location. This fact is sufficient for a practical analysis of the monitoring needed at such a wastewater injection site.

As a further example, Figure 5-30 shows the potentiometric surface for the lower Floridan aquifer in northwest Florida. There the hydraulic gradient was estimated to be about 1.33 ft/mi toward the southwest, in the vicinity of the Monsanto Company injection well prior to its operation. The permeability is about one darcy, and the porosity is estimated to be 10% (Goolsby, 1972 and 1972). The velocity of natural flow in the lower Floridan aquifer is then estimated to be

$$v = \frac{890 \text{ ft/yr}}{0.10} \times \frac{1.33 \text{ ft/mi}}{5,280 \text{ ft/mi}} = 2.24 \text{ ft/yr} \ .$$

This analysis is more easily interpreted than the previous one for Ohio, because it is well known that the source of hydraulic head lies to the north of the injection well site, and that the discharge area lies to the south as shown in Figure 5-31. The velocity of flow is again very low; it appears that more than 200,000 years would be required for injected waste to reach the subsea discharge point 100 miles to the south.

PRESSURE EFFECTS OF INJECTION

Wastewater injected into deep aquifers does not move into empty voids; rather it displaces existing fluids, primarily saline water. The displacement process requires exertion of some pressure, in excess of the natural formation pressure. The pressure increase is greatest at the injection well, and decreases in approximately a logarithmic manner away from the well. The amount of excess pressure required and the distance to which it extends depend on the properties of the formation and the fluids, the amount of fluid being injected, and the length of time that injection has been going on. The pressure or head changes resulting from injection are added to the original regional hydraulic gradients to obtain a new potentiometric surface map that depicts the combined effects of regional flow and the local disturbances.

By use of the theory described, potentiometric surface maps can be produced to show the anticipated situation at any time in the future. If observation wells exist, the actual potentiometric surface at any time can be constructed from the water levels or pressures recorded in the wells.

Figure 5-32 shows the theoretical potentiometric surface map for the lower Floridan aquifer in northwestern Florida in 1971, after wastewater injection had been in progress near Pensacola for about eight years. The estimated pressure effects of injection can be seen by comparing Figure 5-30 with Figure 5-32. The comparison indicates that changes in hydraulic head may extend out for 30 miles or more from the injection site. Although Figure 5-32 is titled a theoretical potentiometric surface map, it is, in fact, partially substantiated by observation wells. If more observation wells were available, the map would be constructed entirely from observed data.

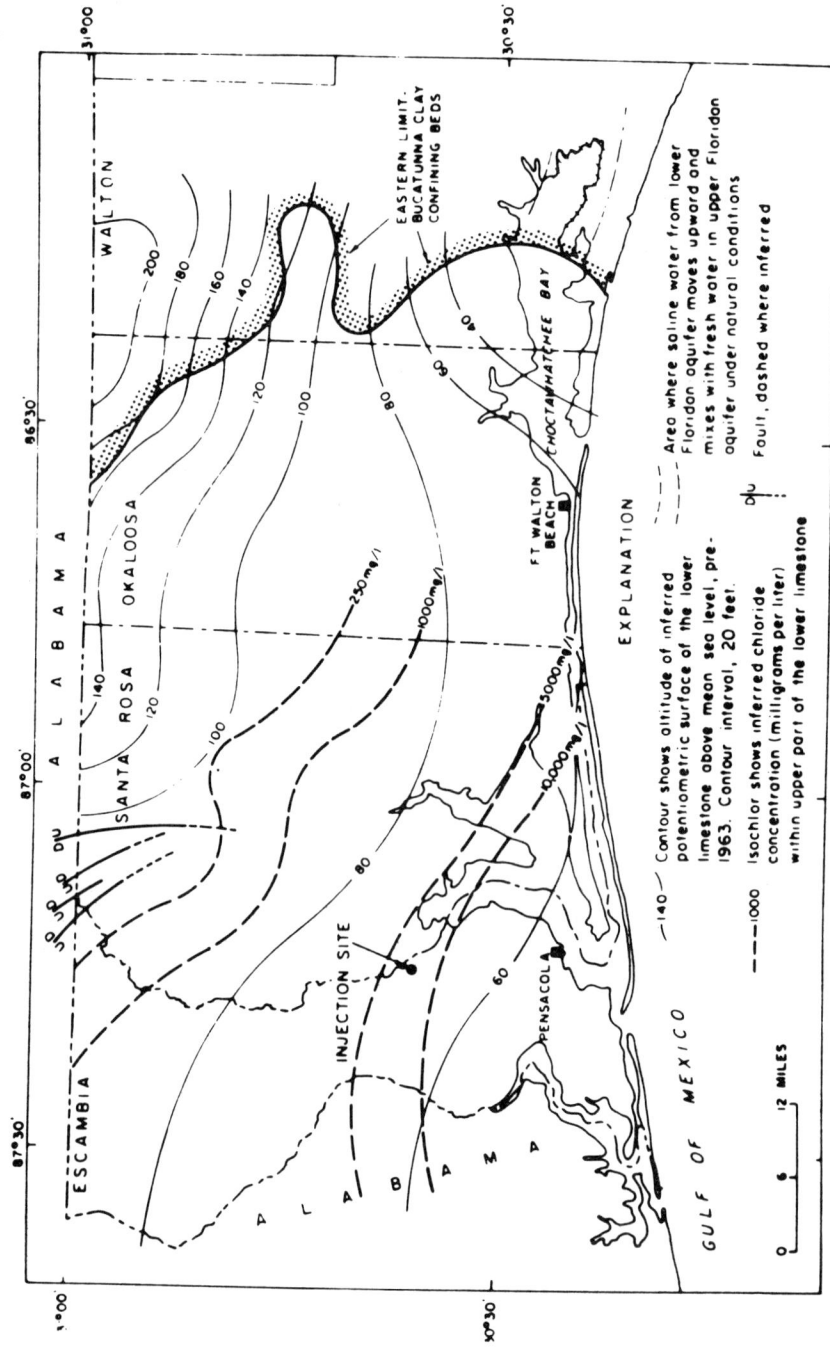

Figure 5-30. Hydrogeology of the lower Floridan aquifer in northwest Florida (Goolsby, 1972).

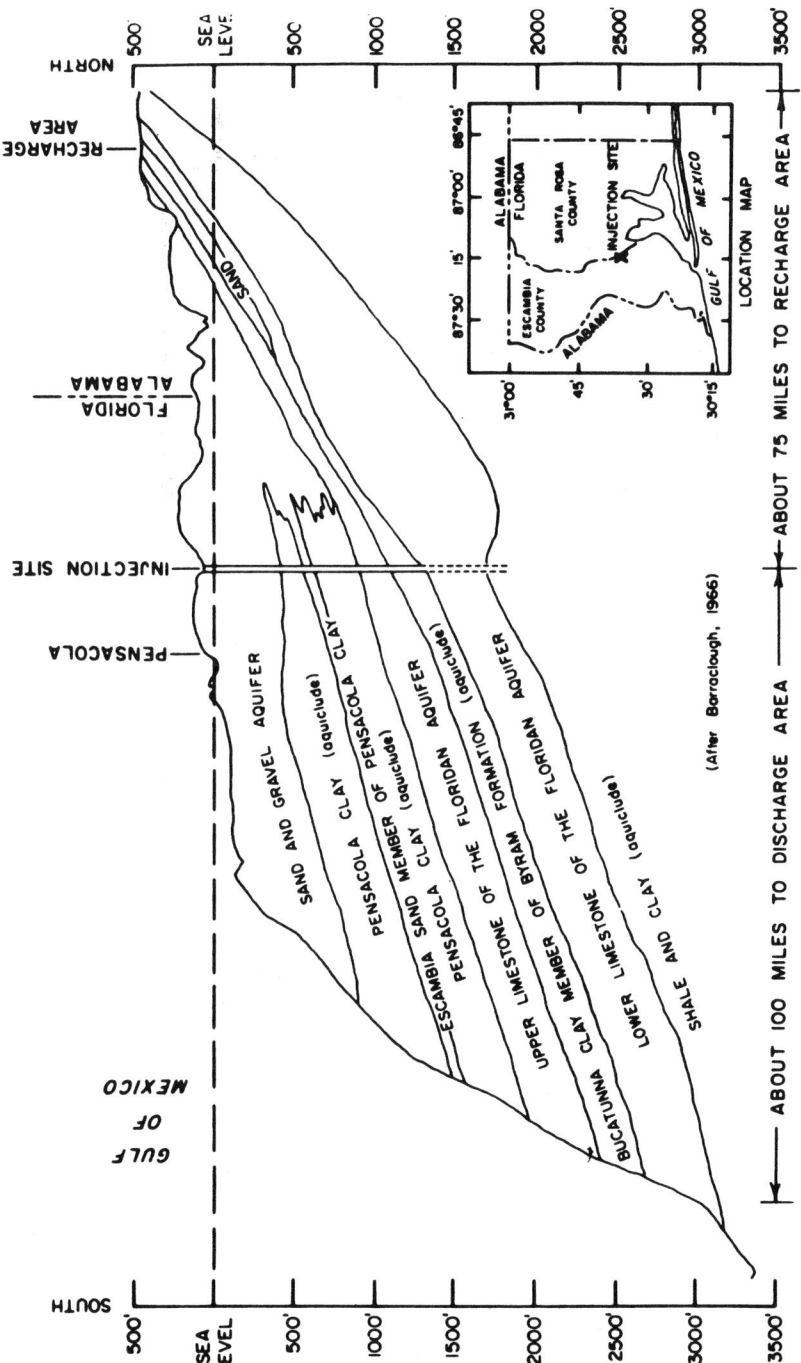

Figure 5-31. Generalized north-south geologic section through southern Alabama and northwestern Florida (Goolsby, 1972).

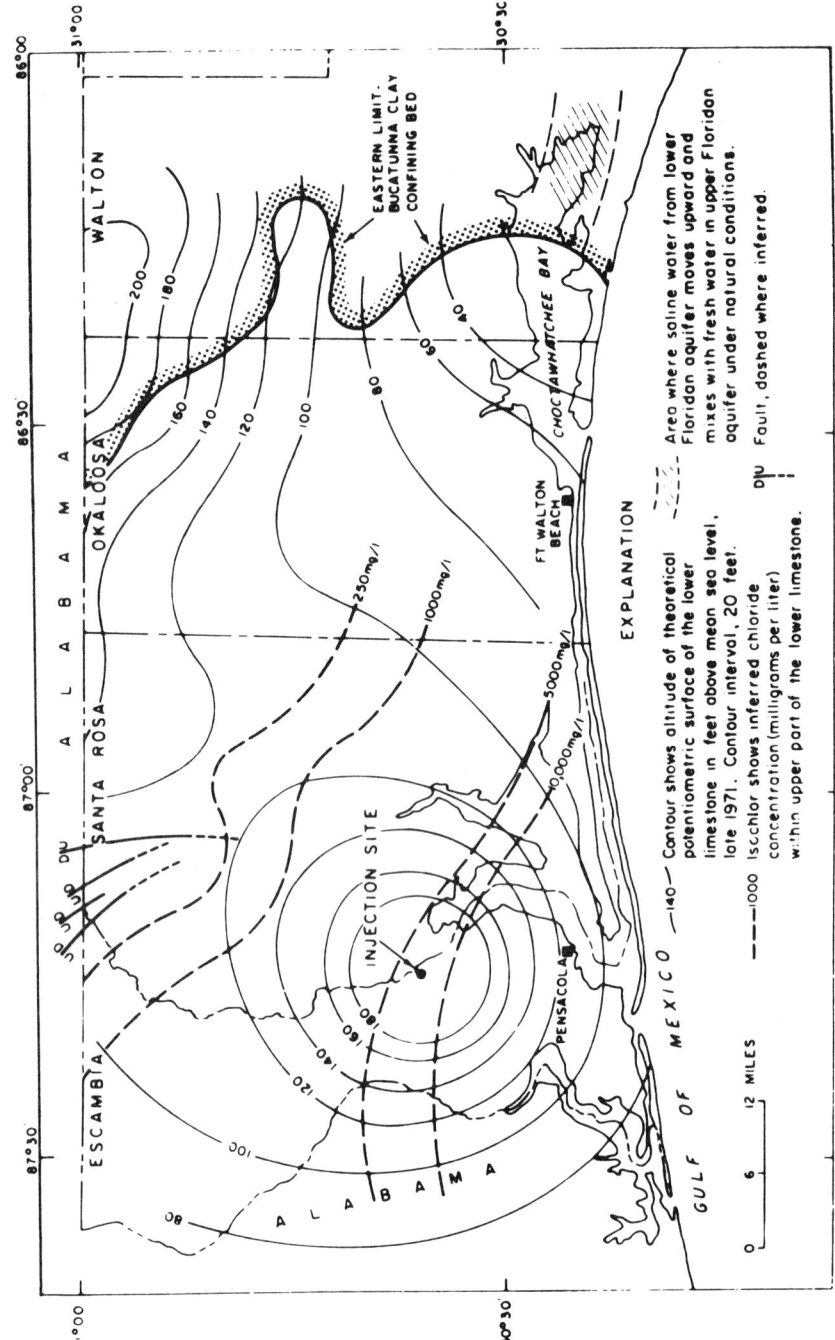

Figure 5-32. Theoretical potentiometric surface of lower limestone of Floridan aquifer in late 1971 (Goolsby, 1972).

As an example of the development of such a theoretical potentiometric surface map, one point on Figure 5-32 will be determined. The point will be one at a radial distance of 6 miles northeast of the injection well site, which places it at a potential of about 77 ft on Figure 5-30 and 180 ft on Figure 5-32, showing a head increase of about 103 ft. From Goolsby (1972), the following data were obtained or estimated:

$$Q = 2.427 \times 10^6 \text{ gal/day} = 3.244 \times 10^5 \text{ ft}^3/\text{day}$$
$$T = 6,300 \text{ gal/day} \cdot \text{ft} = 842 \text{ ft}^3/\text{day} \cdot \text{ft}$$
$$t = 3,000 \text{ days}$$
$$r = 6 \text{ mi} = 31,680 \text{ ft}$$
$$S = 2 \times 10^{-4} \text{ (dimensionless)}$$

Therefore, from Equation 16, the head increase in 3,000 days 6 miles northeast of the injection site is:

$$\Delta h = \frac{2.30 \times 3.244 \times 10^5 \text{ ft}^3/\text{day}}{4\pi \times 842 \text{ ft}^3/\text{day} \cdot \text{ft}}$$

$$\times \log \frac{2.25 \times 842 \text{ ft}^3/\text{day} \cdot \text{ft} \times 3,000 \text{ days}}{(31,680 \text{ ft})^2 \times 2 \times 10^{-4}}$$

$$= 70.50 \log 28.31 = 102.4 \text{ ft}$$

The calculated increase of 102.4 ft compares very well with the 103 ft obtained from Goolsby's maps. As many points as desired can be calculated to produce the contour map. Rather than calculating the pressure at a point (actually on a circle with radius r), even head increments can be selected, and the radii to them calculated, which simplifies the contouring process.

The well in Ohio for which core data are available, and for which a drill stem test and an injection test were presented, will also be used as an example. The core data yielded a transmissivity of 954 millidarcy-ft/cP, the drill stem test 345 millidarcy-ft/cP, and the injection test 452 millidarcy-ft/cP. The value from the injection test will be used, because it is considered the most reliable. Pressure buildup will be calculated at the well, which, as previously explained, appears to have an effective radius of 8.7 ft. The storage coefficient calculated from Equation 7 is 3.34×10^{-4}. The information of interest is, for injection rates of 25, 50, or 75 gpm, how long wastewater can be injected before a limiting allowable surface pressure increase of 1800 psi has been reached. Rearranging Equation 16, and entering the values given above, which have been converted to consistent units:

$$\log t = \frac{4\pi T \Delta h}{2.30 Q} - \frac{\log 2.25 T}{r^2 S}$$

$$\log t = \frac{(4\pi)(1.24 \text{ ft}^2/\text{day})(4157 \text{ ft})}{(2.30)(14,437 \text{ ft}^3/\text{day})} - \log \frac{(2.25)(1.24 \text{ ft}^2/\text{day})}{(8.7 \text{ ft})^2 (3.34 \times 10^{-4})}$$

$$\log t = 1.95 - \log 110.36 = -0.092$$
$$t = 0.81 \text{ days} = 19.4 \text{ hr.}$$

This value could also have been obtained by extrapolating to 1800 psi the line in Figure 5-28 (Section 4), but only for the same injection rate and radius of investigation, and not for other rates and radii.

As the injection rate is changed, the amount of time required for the pressure to increase to a particular level changes proportionately, so that for an injection rate of 50 gpm, $t = 27$ hr, and for an injection rate of 25 gpm, $t = 54$ hr.

For this well, the calculations simply confirm what could already have been intuitively deduced; the fact that the Mt. Simon Formation will not be a suitable injection unit at this location. Similar calculations could have been made from core data and from the drill stem test, and this conclusion reached prior to injection testing.

In comparison with the Ohio example, a well in northern Illinois had the following characteristics:

$$b = 1734 \text{ ft}$$
$$\bar{K}_{av.} = 36 \text{ millidarcys}$$
$$T = 62.42 \text{ darcy} \cdot \text{ft}$$
$$Q = 100 \text{ gpm}$$
$$r_{well} = 4.4 \text{ in.}$$
$$S = 5.46 \times 10^{-3}$$

Using these data, we can calculate what the injection pressure increase at the well will be after five years of continuous operation.

$$\Delta p = 0.4333 \text{ psi/ft} \left[\frac{2.30 \times 19{,}248 \text{ ft}^3/\text{day}}{4\pi \times 167 \text{ ft}^2/\text{day}} \right.$$
$$\left. \times \log \frac{2.25 \times 1825 \text{ days} \times 167 \text{ ft}^2/\text{day}}{(0.36 \text{ ft})^2 \times 5.46 \times 10^{-3}} \right] = 81 \text{ psi}.$$

This calculation shows that the pressure increase will be negligible. In actual operation, the injection pressure has averaged 120 to 300 psi; the difference between predicted and observed performance is not of concern in this case, unless the observed pressure continues to increase, indicating possible progressive plugging of the formation.

MULTIPLE WELLS

As previously mentioned, estimating the combined pressure effects of multiple wells is made easy by virtue of the principle of superposition. It is only necessary to estimate the separate effects of two or more wells at the point of interest, then add them to obtain their combined effect. For example, referring to the last Mt. Simon well discussed above, what would the combined effects of two wells spaced 1,000 ft apart be on each other after five years? Assume both wells have the same characteristics:

$$\Delta p = 81 \text{ psi} + 0.433 \text{ psi/ft} \left[\frac{2.30 \times 19{,}248 \text{ ft}^3/\text{day}}{4\pi \times 167 \text{ ft}^2/\text{day}} \right.$$

$$\left. \times \log \frac{2.25 \times 1825 \times 167 \text{ ft}^2/\text{day}}{(1000 \text{ ft})^2 \times 5.46 \times 10^{-3}} \right]$$

$$\Delta p = 81 \text{ psi} + 19 \text{ psi} = 99 \text{ psi}.$$

HYDROLOGIC DISCONTINUITIES

Another common situation is one in which a barrier to flow, a fault, or facies change is present within the area of influence of an injection well. Faults may also act as channels for escape of fluid from the injection horizon.

In predicting aquifer response in the presence of such features, the image-well concept is used. Assume the presence of a fault or lithologic change that acts as an impermeable barrier, 500 ft in any direction from the Mt. Simon Formation injection well that is discussed above. Assume also, according to image-well theory, an imaginary injection well with all of the same properties as the real injection well is placed 1,000 ft from the real well, on the opposite side of the fault, on a line that passes through the real well, and is perpendicular to the fault. Figure 5-33 shows the potentiometric surface and flow lines that would develop in such a situation; the pressure effect of the barrier would be the same as that calculated above for an actual injection well 1,000 ft from the first well.

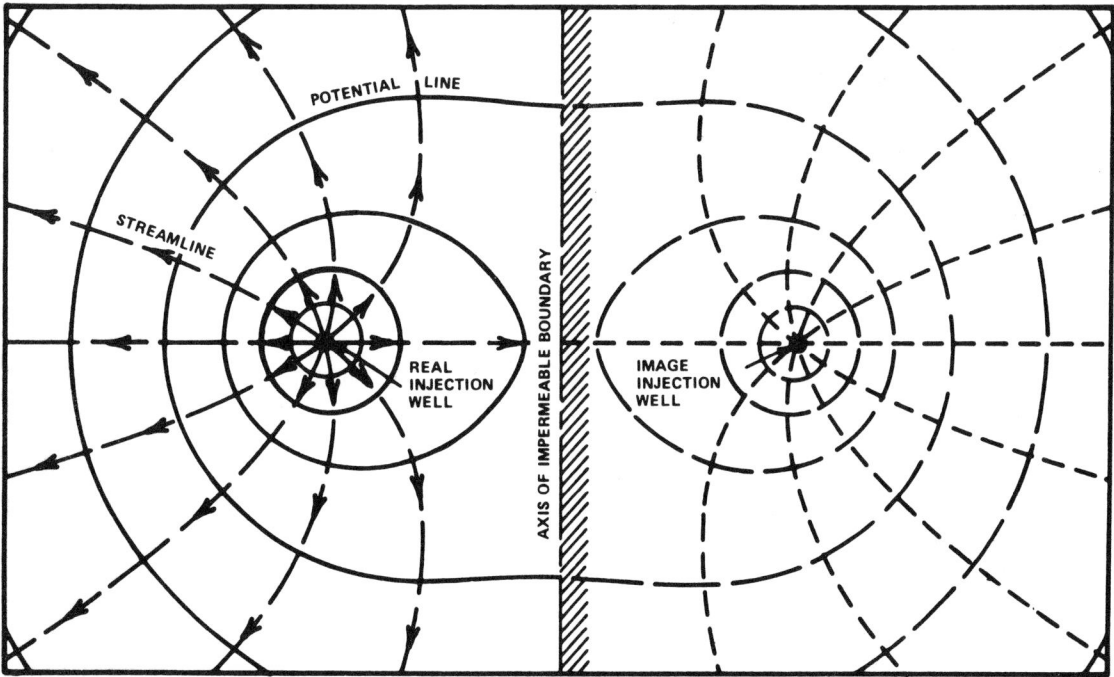

Figure 5-33. Generalized flow net showing the potential lines and stream lines in the vicinity of an injection well near an impermeable boundary (Ferris et al., 1962).

If the hydrologic discontinuity were a leaky fault rather than a sealed one, the opposite effect would occur; the pressure at any time would be reduced as if a discharging well were present.

The equations and examples given are for the most basic hydrogeologic circumstances, but many injection wells can be treated this way because these are the conditions sought when choosing an injection site and receiving aquifer. However, cases of virtually any complexity can be analyzed by use of the appropriate solution to the basic flow equations; where analytical solutions are not possible, numerical models can be developed. The limitations to an analysis are usually pragmatic rather than theoretical — lack of data, limitations of time and funds, or the fact that a simplified estimate is sufficient for the circumstances.

RATE AND DIRECTION OF FLUID MOVEMENT

As with pressure response to injection, the rate and direction of movement of the injected fluid depend on the hydrogeology of the site; therefore, the same factors previously listed require consideration. In addition, the properties of the formation water and the injected wastewater assume major importance.

Broad flow patterns in an aquifer with a significant existing potentiometric gradient can be deduced from a map of the regional potentiometric surface with the effects of the injection system superimposed.

Figure 5-34 is a duplication of Figure 5-32, with flow lines added to show how the flow directions of aquifer water and injected wastewater can be deduced from the potentiometric surface map. The wastewater will never actually travel as far northward as the map indicates, but displaced aquifer water will be forced in this direction, ahead of the small cylinder of wastewater that surrounds the well. The extent of this wastewater cylinder will be discussed next.

A good estimate of the minimum distance of wastewater flow from an injection well can be made by assuming that the wastewater will uniformly occupy an expanding cylinder, with the well at the center. The equation for this case is:

$$r = \sqrt{\frac{V}{\pi b \phi}} \quad (L) \tag{19}$$

where
- r = radial distance of wastewater front from well
- V = Qt = cumulative volume of injected wastewater
- b = *effective* aquifer thickness
- ϕ = average *effective* porosity.

For a Mt. Simon injection well with the following characteristics:

Q = 100 gpm

t = 5 yr

b = 1618 ft

ϕ = 13.5%

$r = \sqrt{\dfrac{35{,}128{,}993 \text{ ft}^3}{\pi \times 1618 \text{ ft} \times 0.135}}$

    = 226 ft.

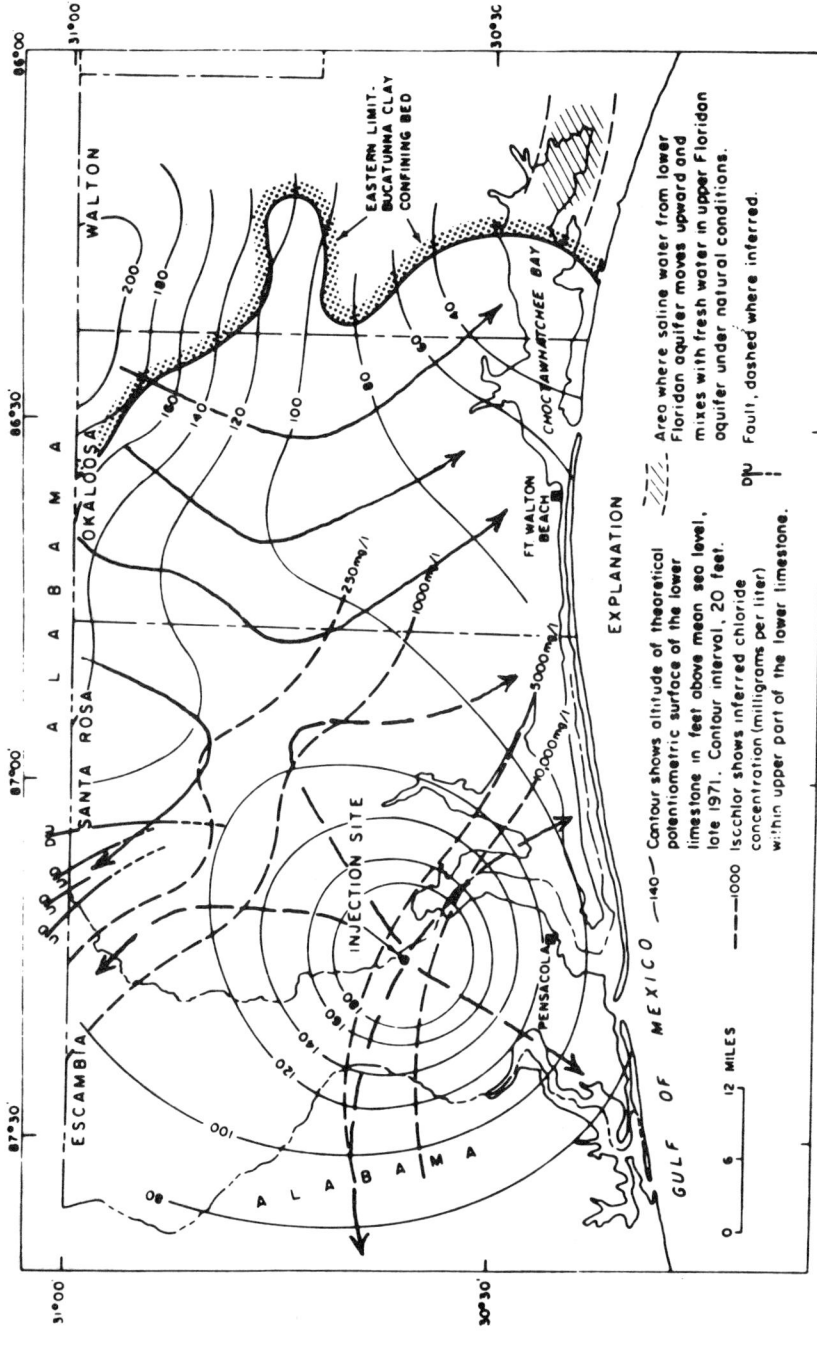

Figure 5-34. Theoretical potentiometric surface of lower limestone of Floridan aquifer in late 1971, with flow lines showing the directions of aquifer water and wastewater movement. Solid flow lines show the direction of flow of diverted aquifer water, dashed flow lines show direction of flow of injected wastewater and displaced aquifer water (modified after Goolsby, 1972).

It is noted that effective aquifer thickness and average effective porosity should be used. The effective aquifer thickness is, for example, that part of the total aquifer that consists of sandstone in the case of a mixed sandstone-shale lithology. The effective porosity has been previously defined as that part of the porosity in which the pores are interconnected.

In most situations the minimum radial distance of travel will be exceeded, because of dispersion, density segregation, and channeling through high permeability zones. Flow may also be in a preferred direction, rather than radial, because of hydrologic discontinuities (e.g. faults), selectively oriented permeability paths, or natural flow gradients.

An estimate of the influence of dispersion can be made with the following equation:

$$r' = r + 2.3\sqrt{Dr} \quad (L) \qquad (20)$$

where
- r' = radial distance of travel with dispersion
- D = dispersion coefficient; 3 ft for sandstone aquifers and 65 ft for limestone or dolomite aquifers.

Equation 20 is obtained by solving equation (10.6.65) of bear (1972) for the radial distance at which the injection front has a chemical concentration of 0.2% of the injected fluid.

The detailed development of dispersion theory is presented by bear (1972). The dispersion coefficients given are high values for sandstone and limestone aquifers obtained from the literature. No actual dispersion coefficients are known to have been obtained for any existing injection well.

Then, for the above example, which is a sandstone:

$$r' = 226 \text{ ft} + 2.3\sqrt{3 \text{ ft} \times 226 \text{ ft}}$$
$$= 286 \text{ ft}.$$

It is clear that, in this example, the distance of wastewater travel from the well is negligible, and could not possibly be of concern, if actual conditions comply even generally with those that were assumed. This conclusion has been found to apply to many of the wells that have been constructed to date. Since almost no attempts have been made to determine the actual wastewater distribution around existing injection wells, there is little evidence for comparison with theory. However, if such a calculation were in error by several hundred percent, there would still be no cause for concern, since the injection well, in this and many other cases, is tens of miles from the nearest other well penetrating or known resource in the injection zone.

To proceed beyond the calculations that have been shown may not be necessary or, in many cases, meaningful. However, it may be possible, if necessary, to account for some of the additional complications that are mentioned. For example, Bear and Jacobs (1964), in one of a series of reports, considered the flow of water from a groundwater recharge well in an aquifer of uniform flow, when the densities and viscosities of the injected and interstitial fluids are the same. Gelhar and others (1972) developed analytical techniques for describing the mixing of injected and interstitial waters of different densities.

So far, the travel of the injected wastewater has been treated as though it were an inert fluid and would not react with the aquifer water or minerals, be affected by bacterial action, or decompose or radioactively decay. If the wastewater is not inert, changes in chemical composition with time and distance may also need to be considered. Bredehoeft and Pinder (1972) discuss the methodology for a unified approach to this type of problem, and Robertson and Barraclough (1973) presented an example of a case in which radioactive decay, dispersion, and reversible sorption were considered. However, no procedure exists at this time for simultaneously considering the full range of practical possibilities that may be involved in wastewater movement.

In spite of the degree of sophistication used in development of theories for rate and direction of travel of injected fluid from an injection well, nonuniform distribution of porosity and permeability will preclude making accurate estimates in many cases. In general, wastewater flow in unfractured sand or sandstone aquifers would be expected to more closely agree with theory than flow in fractured reservoirs, or in carbonate aquifers with solution permeability. However, even in sand aquifers, flow can be expected to be nonideal, as shown by tests reported by Brown and Silvey (1973). Particularly great deviations from predictions may occur in limestone or dolomite aquifers. Figure 5-35 is an example of this. The radial zones around Well No. 1 show the predicted extent of waste travel, using Equations 19 and 20. The irregular boundary shows the probable actual extent of wastewater spread, as indicated by evidence from Wells 2 and 3. In this case, the wastewater apparently traveled selectively in a single, thin, porous, and permeable interval, rather than throughout the several zones indicated by testing results. Accurate prediction of the rate and direction of movement in such a case may well be technically infeasible even in the future because the amount of information needed will seldom, if ever, be available.

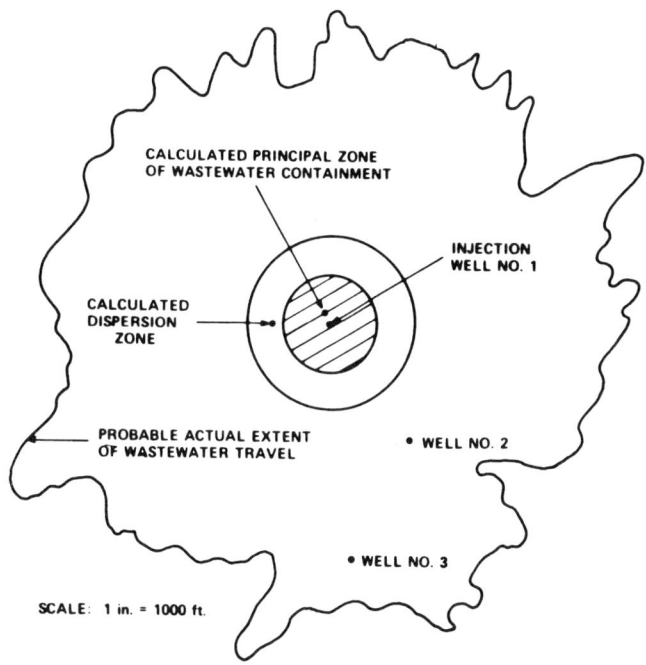

Figure 5-35. Predicted and probably actual extent of wastewater travel for a well completed in a carbonate aquifer.

HYDRAULIC FRACTURING

Hydraulic fracturing may be deliberately accomplished to increase formation permeability, or it may occur during injection testing or wastewater injection, if the fracture initiation pressure is exceeded. Regulatory policy may or may not allow short-term hydraulic fracturing operations for well stimulation, but continuous injection at pressures above the fracture point are prohibited by most, if not all, agencies. This is because of the danger of damage to well facilities, and of the uncertainty about where the fractures and injected fluids are going, as fractures continue to be extended.

Figure 5-36 is a schematic diagram of bottom-hole pressure versus time during hydraulic fracturing. Before injection begins, the pressure is that of the formation fluid (p_o) and the column of fluid in the well bore. Pressure is increased until fracturing occurs; then, as fluid continues to be pumped into the well, the pressure stabilizes at p_f, the flowing pressure, during which the fractures continue to be extended. When injection is ceased, and the well shut in, the pressure quickly stabilizes to a constant value, the instantaneous shut-in pressure. This pressure is considered to be equal to the least principal earth stress in the vicinity of the well.

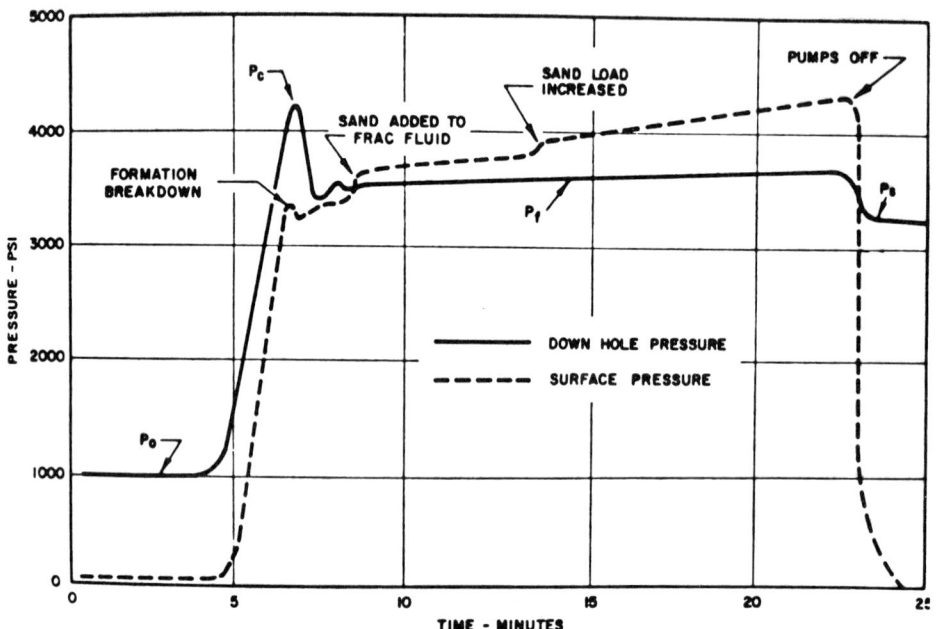

Figure 5-36. Schematic diagram of pressure versus time during hydraulic fracturing (Kehle, 1964).

In estimating the fluid pressure at which hydraulic fracturing will occur, one of two conditions is usually assumed:

1. The least principal stress is less than the vertical lithostatic stress caused by the rock column. In this case, fractures are assumed to be vertical.

2. The vertical lithostatic stress is the least principal stress. In this case, fractures will be horizontal.

In the first case, the minimum bottom-hole pressure required to initiate a hydraulic fracture can be estimated from (Hubbert, 1972):

$$p_i \cong \frac{S_z + 2p_o}{3} \quad (F/L^2) \tag{21}$$

where p_i = fracture initiation pressure
 S_z = total lithostatic stress
 p_o = formation fluid pressure.

The fracture gradient, that is, the injection pressure required per foot of depth, can be estimated by entering representative unit values into Equation 21. The unit values for S_z and p_o are, respectively, 1.0 and 0.46 psi/ft. This yields a p_i gradient of 0.64 psi/ft as a minimum value for initiation of hydraulic fractures. This situation implies a minimum lateral earth stress. As the lateral stresses increase, the bottom-hole fracture initiation pressure also increases up to a limiting value of 1.0 psi/ft. Actually, fracture pressures may exceed 1.0 psi/ft when the rocks have significant tensile strength, and no inherent fractures pass through the well bore. In any particular case, injection tests can be run to determine what the actual fracture pressure is, then operating injection pressures held below the instantaneous shut-in pressure. In the absence of any specific data, arbitrary limitations of from 0.5 to 1.0 psi/ft of depth have been imposed on operating injection wells. Regional experience should be used as a criterion in establishing an arbitrary limit, since regional tectonic conditions and fluid pressure gradients dictate what a safe limit will be.

GENERATION OF EARTHQUAKES

As a matter of background, it is widely, but not universally, accepted that a series of earthquakes that began in the Denver area in 1962 was initiated by injection of wastewater into a well at the Rocky Mountain Arsenal. Since the association of seismic activity with wastewater injection at Denver, apparently similar situations have been observed at Rangely, Colorado and Dale, New York. The former related to water injection for secondary recovery of oil, and the latter to disposal of brine from solution mining of salt. On the other hand, there are presently about 160 operating industrial wastewater injection wells and tens of thousands of oil field brine disposal wells that have, apparently, never caused any noticeable seismic disturbance, so these three examples would have to be considered very rare.

It has been erroneously stated by many that the seismic events have been stimulated by *lubrication* of a fault zone by injected fluids. What happened, if injection was involved, is that the water pressure on a fault plane was increased, thus decreasing the friction on that plane and allowing movement, and consequent release of stored seismic energy.

Based on this interpretation of the mechanism of earthquake triggering by fluid injection, some of the conditions that would have to exist in order to have such earthquakes would be:

1. A fault with forces acting to cause movement of the blocks on either side of the fault plane, but which are being successfully resisted by frictional forces

2. An injection well that is constructed close enough, vertically and horizontally, to the fault so that the fluid pressure changes caused by injection will be transmitted to the fault plane

3. Injection at a sufficiently great rate and for a sufficiently long time to increase fluid pressure on the fault plane to the point that frictional forces resisting movement become less than the forces tending to cause movement. (At this time, movement will occur and stored seismic energy will be released. That is, an earthquake will occur.)

As has been discussed earlier in the section on state of stress, relatively little is known about stress distribution in the earth's crust, and even less is known about stress distribution along fault systems. In the absence of this information, only qualitative estimates of the probability of earthquake stimulation can be made. In the great majority of cases, the potential for earthquake stimulation will be nonexistent or negligible because only very limited areas in the country are susceptible to earthquake occurrence. The susceptible areas are delineated by records of earthquakes that have occurred in the past and by tectonic maps that show geologic features associated with belts of actual or potential earthquake activity.

In a case where subsurface stresses are known, or are determined by hydraulic fracturing or other means, and where the location and orientation of the fault plane are known, then a quantitative estimate of the pressure required to cause fault movement can be made. Raleigh (1972) provides an example of such a calculation from the Rangely, Colorado, oil field.

SECTION 5 - SURVEILLANCE OF OPERATING WELLS

INJECTION WELL MONITORING

The principal means of surveillance of wastewater injection presently practiced is monitoring at the injection well of the volume, chemistry, and biology of the injected wastewater, and of the well-head and annulus pressures (Figure 5-37). To some, this apparently seems inadequate. However, if all of the necessary evaluations have been made during the planning, construction, and testing of the well, this may be a satisfactory program, when combined with periodic inspection of surface and subsurface facilities. This is because, as pointed out by Talbot (1972), the greatest risk of escape of injected fluids is normally through the injection well itself, rather than from leakage through permeable confining beds, fractures, or unplugged wells.

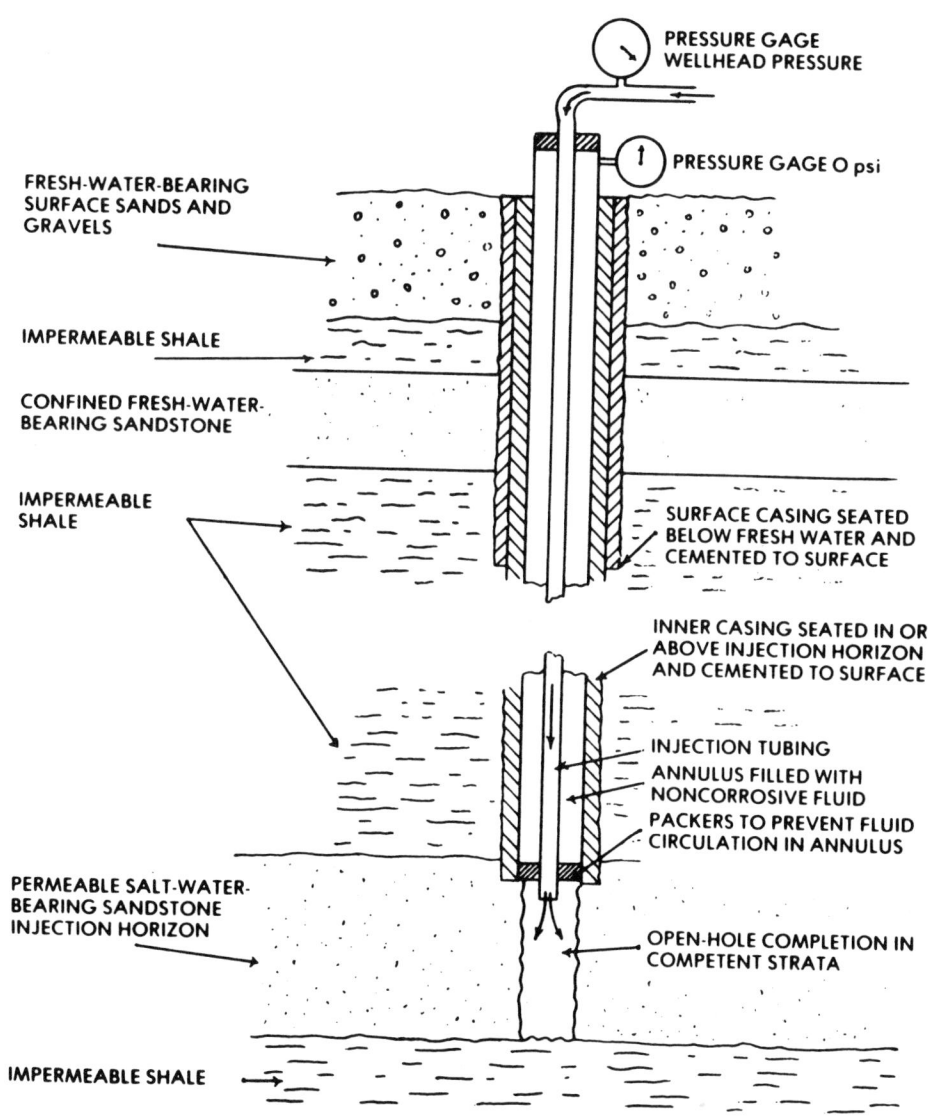

Figure 5-37. Schematic diagram of an industrial waste injection well completed in competent sandstone (modified after Warner, 1965).

The purpose of monitoring the volume of injected wastewater is to allow for estimates of the distance of wastewater travel and interpretation of pressure data, as well as to provide a permanent record of the volume of emplaced wastewater. Also, a record is needed as evidence of compliance with restrictions, for interpretation of well behavior, and as a precaution in the event that a chemical parameter should deviate from design specifications. Some characteristics that have been monitored continuously are suspended solids, pH, conductance, temperature, density, dissolved oxygen, and chlorine residual. Complete chemical analyses are frequently made on a periodic basis on composite or grab samples. Because bacteria may have a damaging effect on reservoir permeability, periodic biological analysis of some wastewaters may be desirable to insure that organisms are not being introduced.

Injection pressure is monitored to provide a record of reservoir performance, and as evidence of compliance with regulatory restrictions. Injection pressures are limited to prevent hydraulic fracturing of the injection reservoir and confining beds, or damage to well facilities. As with flow data, injection pressure should be continuously recorded.

Pressure fall-off data collected after any extended period of continuous operation can be used to check the performance of the reservoir as compared with its original condition. However, it should be noted that the time scale of continuous recorders is not usually adequate for providing data during the early period of a pressure fall-off test, so the continuously recorded data will probably need to be supplemented with additional observations in order to have a complete record of the test.

Figure 5-28 is an example of the pressure response that would ideally be expected during a period of continuous injection. Pressure increase through time should be linear on a semilogarithmic scale, after an early period of adjustment.

In contrast with this ideal behavior, Figure 5-38 shows the injection pressure history of a wastewater injection well completed in a carbonate reservoir. Two marked periods of pressure decline are shown, one in 1967-1968 and one in 1970. The explanation for this is believed to be that the wastewater being injected, initially an acid solution, reacted with the carbonate reservoir to increase the permeability, and thus decrease the injection pressure. The period of gradual pressure increase during 1969-1970 is probably the normal buildup following this initial period of permeability increase. In 1970, the wastewater composition was changed to include a second acid

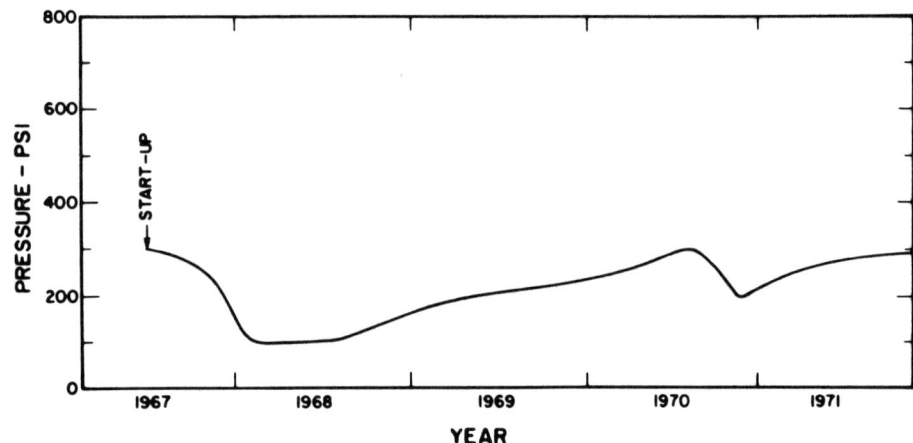

Figure 5-38. Pressure history of a well injecting into a carbonate aquifer.

stream. This new stream apparently caused additional permeability increase and a temporary reduction in injection pressure, after which the expected pressure buildup resumed.

Figure 5-39 shows the plots of two pressure fall-off tests performed in an injection well of the Monsanto Company, Pensacola, Florida. This well is also constructed in a carbonate aquifer. One test was made in November 1967, before injection of an acidic wastewater stream began. The other test was performed in January 1969, after the acidic wastewater had been injected for nine months. The second test shows a much slower rate of fall off, indicating an increased permeability in the vicinity of the well bore, caused by reaction of the acidic wastewater with the carbonate aquifer. This conclusion is substantiated by an increase in the injection index for this and another well during the same time period, as shown in Figure 5-40.

Some other possible causes of deviation from the ideal response are the presence of hydrologic barriers of conduits, leaky confining beds, permeability reduction from suspended solids, chemical reactions, and other phenomena. The variety of factors that may influence well behavior indicates the need for maintaining an accurate, detailed well history so that the probable cause of any unusual performance can be deduced, and the appropriate action taken.

Pressure in the casing-tubing annulus is monitored to detect any changes that might indicate leakage through the injection tubing or the tubing-casing packer. When a packer is used, the casing-tubing annulus pressure should be zero, except perhaps for some pressure resulting from expansion of the injection tubing. In cases where a packer is not used, pressure will be exerted directly on the fluid in the annulus, and indication of leakage would be a significant change in the annulus pressure.

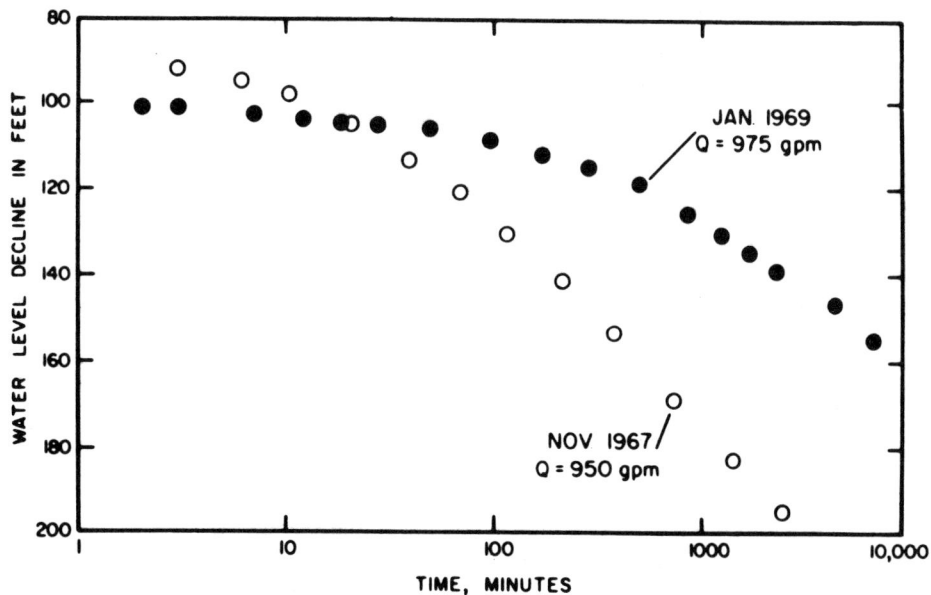

Figure 5-39. Semilogarithmic plot of two pressure fall-off tests measured in an injection well of the Monsanto Company, Pensacola, Florida (Goolsby, 1971).

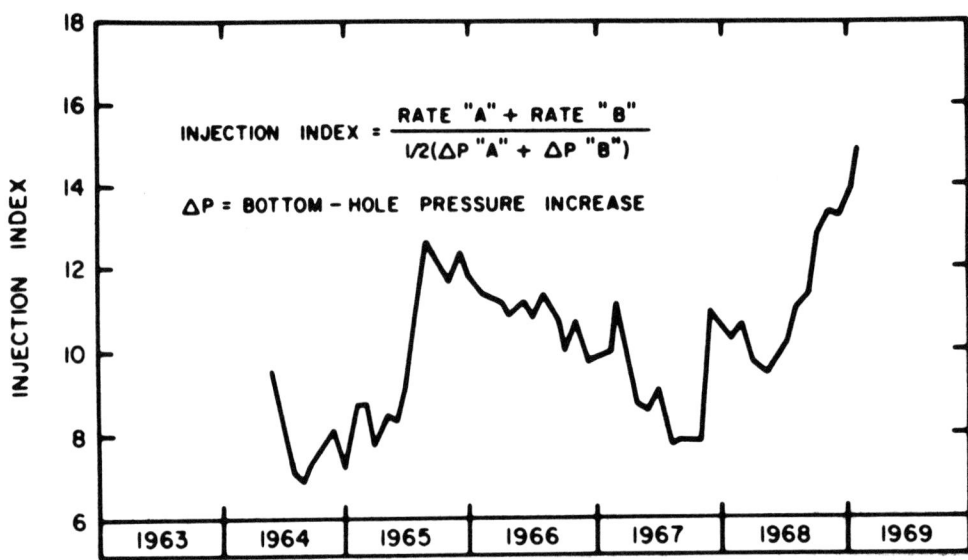

Figure 5-40. Monthly average injection index of two injection wells of the Monsanto Company, Pensacola, Florida (Goolsby, 1971).

Other methods of monitoring the injection well also deserve mention. The corrosion rate of well tubing and casing may be monitored by use of corrosion coupons inserted in the well. A conductivity probe may be used to detect a change in the chemistry of the fluid in the casing-tubing annulus. In wells with packers and in wells without tubing to detect shifts in the interface between the injected fluid and the casing-tubing fluid, the conductivity probe can be used to detect tubing leaks. Another technique used to monitor the casing-tubing annulus is continuous cycling of the annulus fluid, and analysis of the return flow for evidence of contamination by wastewater.

PERIODIC INSPECTION AND TESTING

Sufficient incidents have occurred in the past to emphasize the need for periodically inspecting or testing the subsurface facilities of injection wells, particularly when chemically reactive wastes are being injected. One such incident was the rather spectacular failure of a wastewater injection well at the Hammermill Paper Mill, Erie, Pennsylvania. In that instance, the well casing parted as a result of corrosion, and a portion of it was, reportedly, lifted from the hole by fluid pressure. Substantial loss of wastewater into Lake Erie and abandonment of the well resulted. Other cases have been reported in which portions of tubing or casing have failed by corrosion, and caused temporary or permanent shutdown of the well. There may also be reason to examine the well bore to check for the location of zones of wastewater entrance, enlargement due to chemical reaction, the location and orientation of induced fractures, buildup or precipitates or filtered solids, or other potential problems. Examples are available of wells that have been abandoned or modified because of borehole enlargement that led to collapse of the borehole, damage to the casing, or cement near the bottom of the casing string.

Methods of inspection of casing, tubing, cement and the well bore are:

- Pulling of tubing and visual or instrumental inspection
- Utilization of magnetic logs for inspection of casing or tubing in place

- Utilization of caliper or televiewer logs for inspection of casing, tubing, or the well bore
- Pressure testing of casing
- Utilization of cement bond logs for inspection of casing cement
- Utilization of injectivity or temperature profiles, or other appropriate logs for inspection of casing, cement, or the well bore

The process of pulling and inspecting tubing is self-explanatory. Mechanical methods are available, for example, for inspection of lined steel tubing for flaws in the lining. Individual joints of tubing can be pressure tested at the surface for leakage.

Magnetic down-hole casing, or tubing, inspection services are provided by oil field service companies. These logs indicate, by virtue of the electromagnetic response of steel pipe, the relative pipe thickness. Thin areas may indicate corrosion or other damage. If such a log is run early in the life of the pipe, logs run after the well has been in operation are much more easily interpreted. Figure 5-41 shows the response of a pipe inspection log and photographs of the casing that was pulled after running the log. Figure 5-42 is a portion of a pipe inspection log from a wastewater injection well, which indicates possible corrosion in the interval from 1480 to 1510 ft; regular deflections on the log represent casing joints. Corrosion could be either on the inside, or the outside of the casing.

Caliper logs provide a record of the inside diameter of pipe or borehole walls, and may show intervals of pipe corrosion, borehole enlargement, or borehole plugging at the formation face. Figure 5-43 shows portions of a caliper log run before injection,

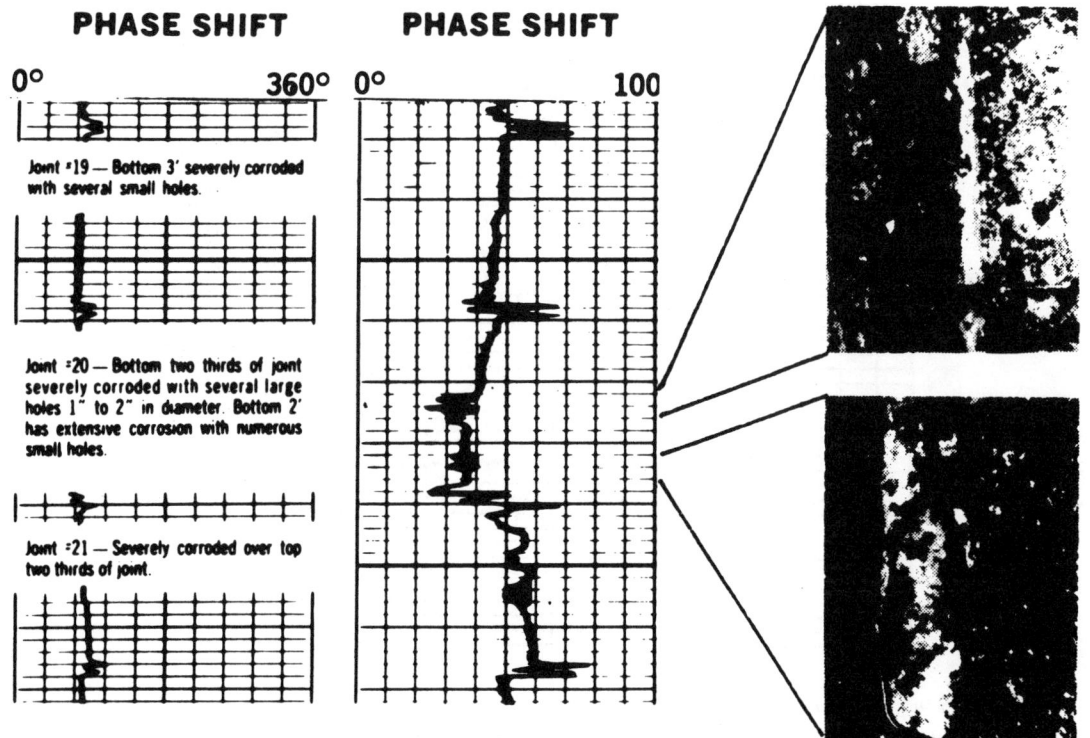

Figure 5-41. Pipe inspection log and photographs of casing pulled after log was run to verify the log (Schlumberger, 1970).

357

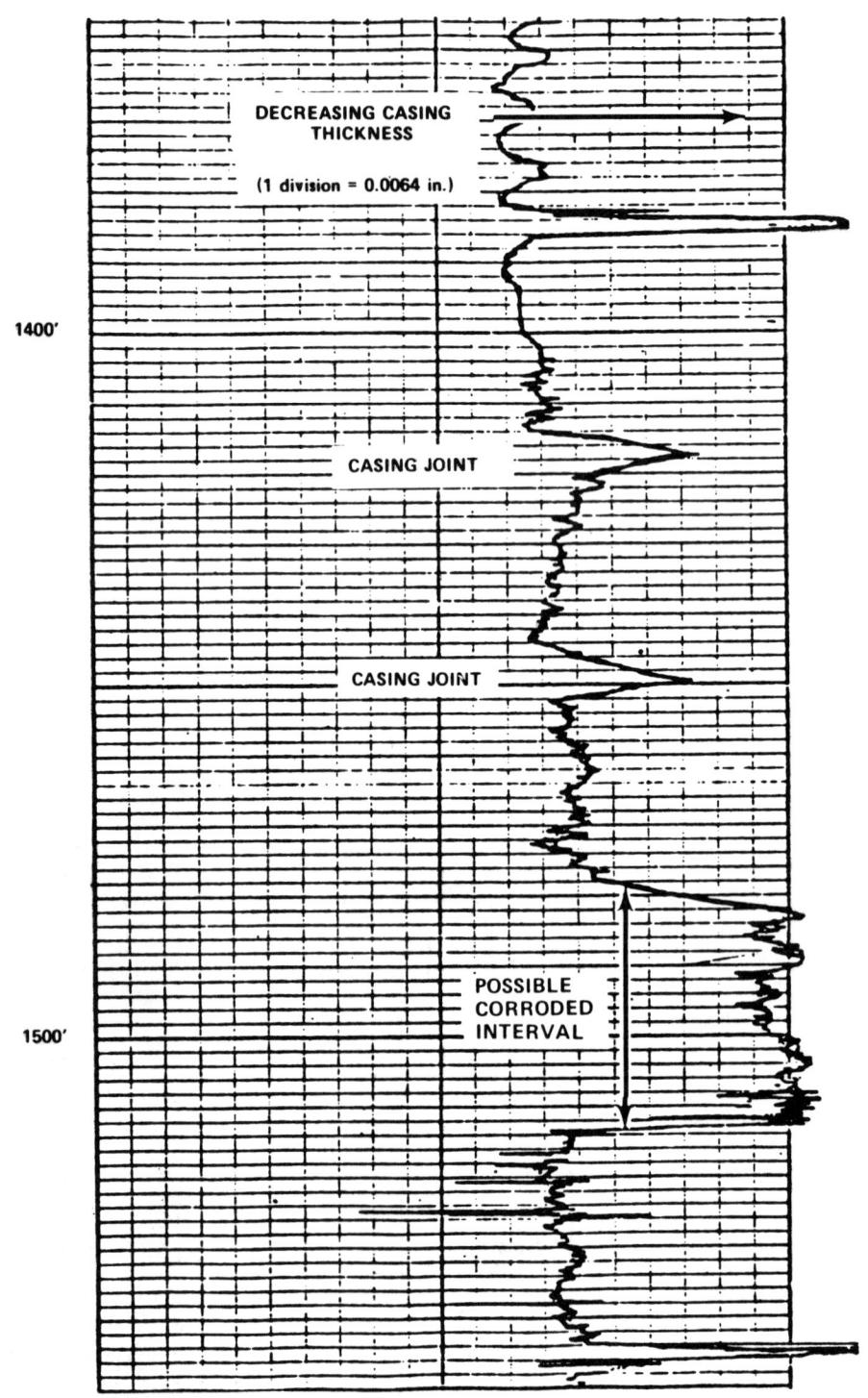

Figure 5-42. Portion of a casing inspection log run in a wastewater injection well showing possible corrosion in the interval from 1480 to 1510 ft.

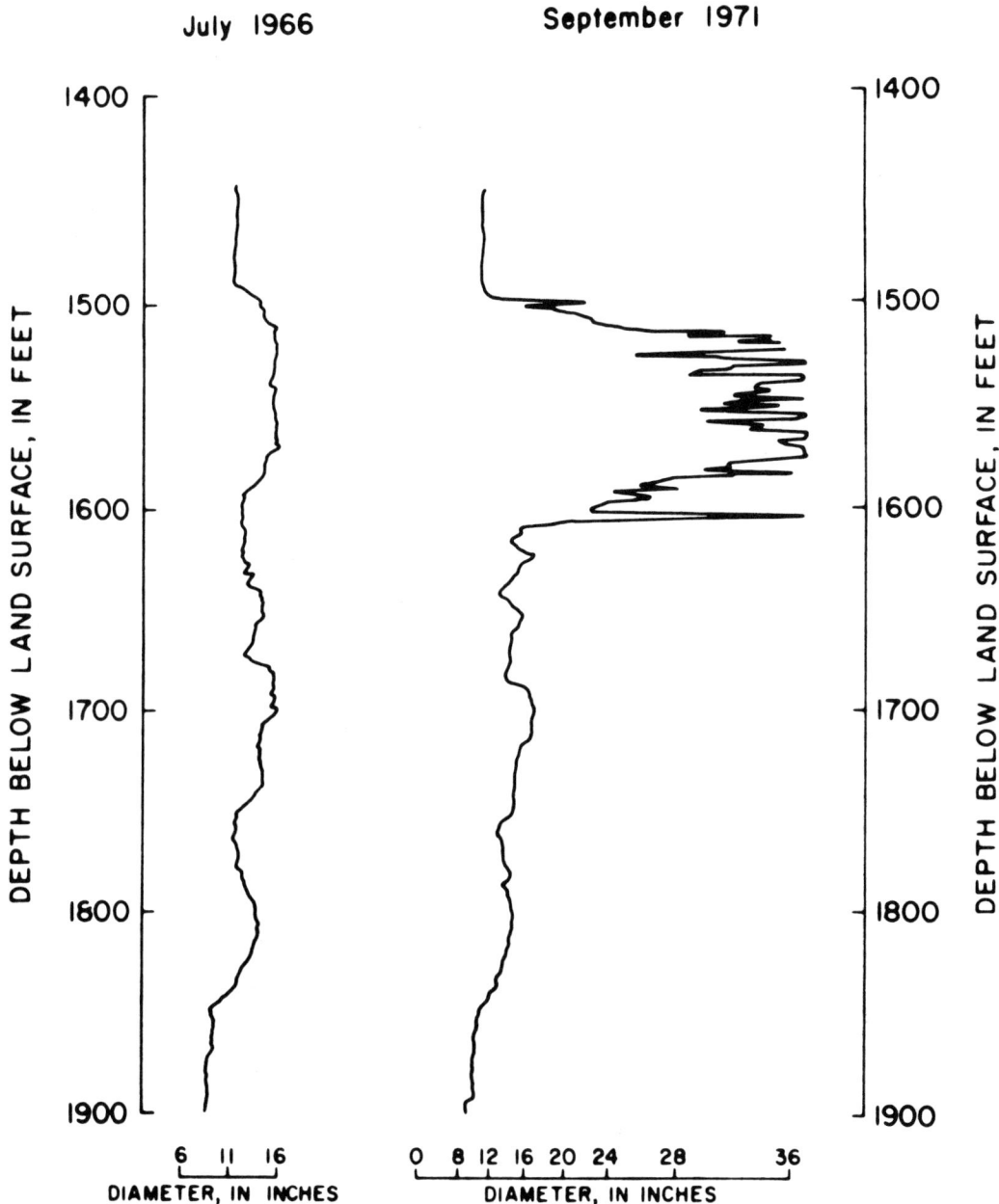

Figure 5-43. Preinjection and postinjection caliper logs from a wastewater injection well at Belle Glade, Florida, showing solution of the limestone aquifer in the 1500- to 1600-ft interval by acidic wastewater (Black, Crow, and Eidsness, 1972).

and after five years of injection of an acidic wastewater into a limestone aquifer. The log indicates considerable borehole enlargement as a result of dissolution of the limestone by the injected acidic waste in the interval from 1500 to 1600 ft. It would be reasonable to conclude that most of the wastewater entered that interval.

Borehole televiewers provide an image of the pipe of borehole wall, as produced by the reflection of sound waves emitted from a sonde. The combination sound source and receiver is highly directional, and is rotated rapidly as the tool is moved up the hole. Thus, the hole is continuously scanned. The resulting information is displayed on an oscilloscope, and a film made of the scope display. The picture obtained depicts the well bore as though it were split open and laid out for inspection. Figure 5-44 illustrates the detail with which the borehole televiewer can indicate casing damage. In Figure 5-45, vertical fractures in the borehole wall of a well in Oklahoma are shown.

Pressure testing can be used to detect casing leaks, and is required by law in many oil-producing States, as a method of testing the integrity of casing in new wells at the time that the casing is cemented into the borehole. In such tests, a cement plug is left at the bottom of the casing during cementing, and allowed to harden. The interior of the casing is subjected to a specified amount of fluid pressure (0.2 psi/ft of casing in

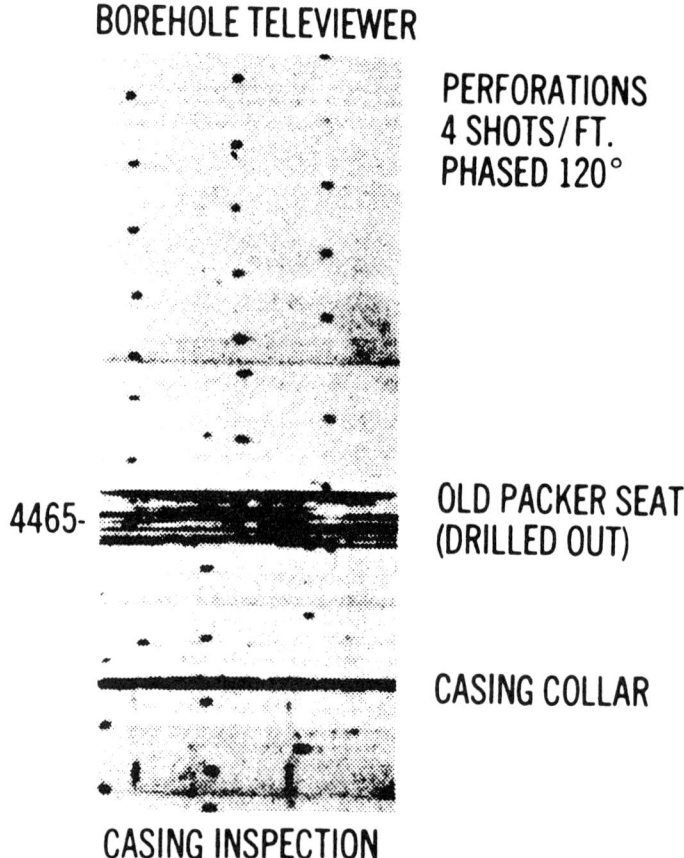

Figure 5-44. Borehole televiewer log of a section of casing showing casing perforations, packer seat, and casing collar (Schlumberger, 1970).

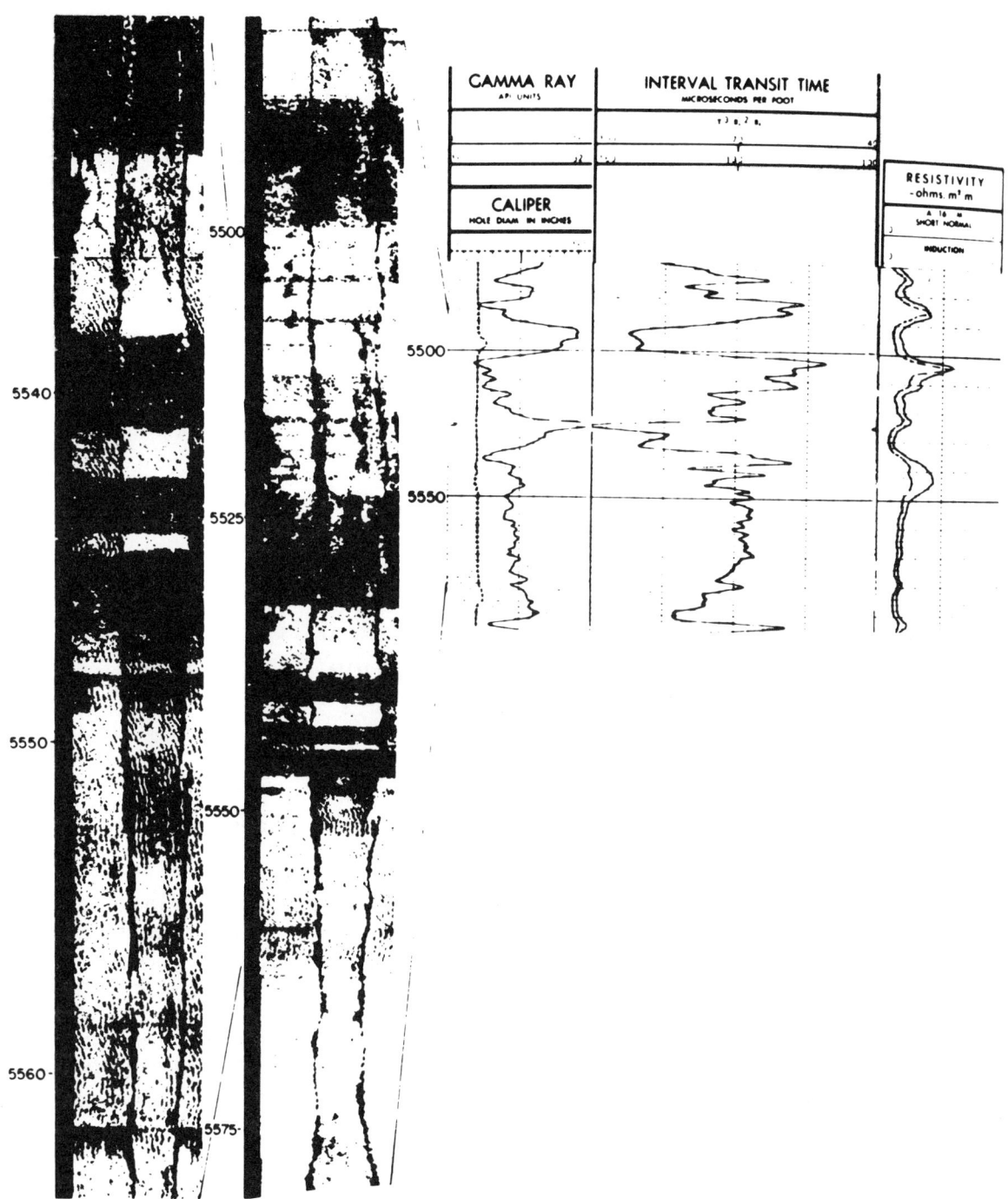

Figure 5-45. Borehole televiewer log showing vertical fractures in the borehole wall of a well in Oklahoma (Zemanek et al., 1970).

Texas).* Rapid decline in pressure indicates leakage from the casing. Such a test could also be performed periodically in operating wells by setting temporary plugs or using packers.

The cement bond log is used to determine the quality of the casing-cement bonding, detect channels in the cement behind the casing, or detect damage to cement from high-pressure injection of chemical reaction. The cement bond log is a continuous measurement of the amplitude of elastic waves after they have traveled through a short length of pipe, cement, and perhaps formation (Figure 5-46). The amplitude of the

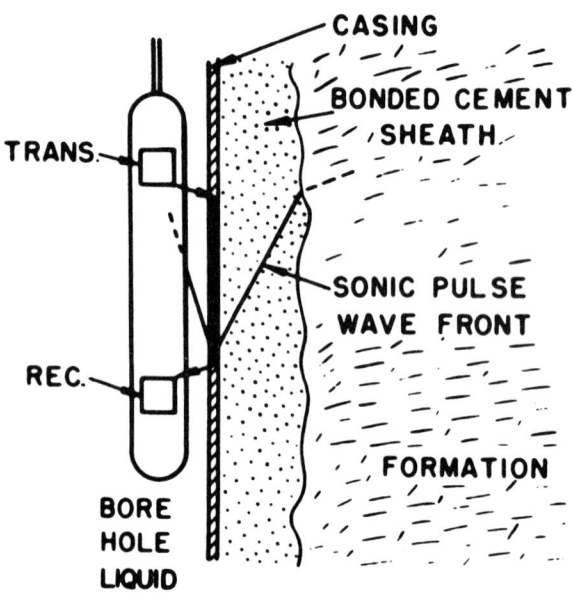

Figure 5-46. Schematic diagram of a cement bond logging tool in a borehole (Grosmangin et al., 1960).

elastic wave is maximum in uncemented casing, and will generally be lower as the degree of bonding and integrity of the cement improves. Thus, the relative amplitudes of the waves in different portions of a well can be interpreted to indicate the condition of the cement and degree of bonding. Complications that occur in the interpretation of cement bond logs are discussed by Fertl et al. (1974). Figure 5-47 shows portions of a cement bond log from an acid wastewater injection well. It appears that the casing in the vicinity of 1900 to 2000 ft is not bonded. The interval from 2700 to 2800 ft, near the base of the casing, shows progressively better bonding between the casing and cement.

Some other possible inspection methods are radioactive tracer injectivity profiles, flow-meter injectivity profiles, and temperature profiles. The objective of these methods is to determine where injected fluid is going. Radioactive tracer injectivity profiles accomplish this through injection of a radioactive tracer, and logging of the borehole with a gamma ray detector. The detector measures concentrations of tracer, which indicate paths of tracer flow. Flow-meter injectivity profiles are similar,

---

*Texas Railroad Commission rules.

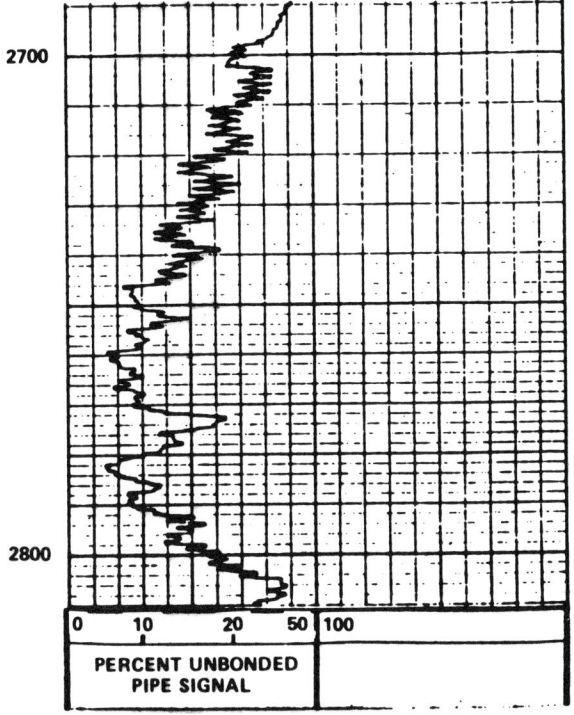
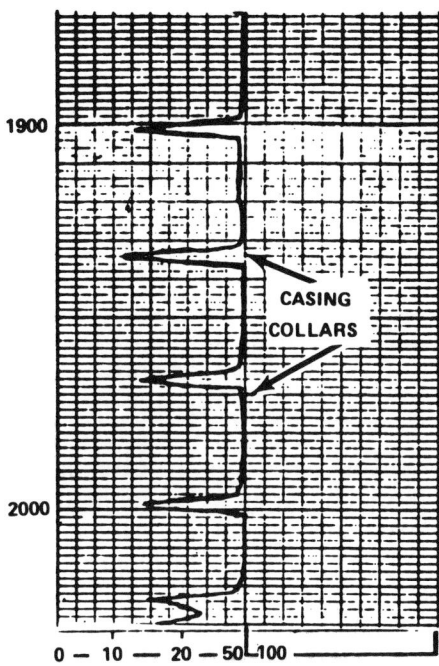

Figure 5-47. Portions of a cement bond log from an acid wastewater injection well.

except that flow paths of injected fluid are indicated by a flow meter, rather than by an injected tracer. Temperature profiles may indicate anomalies at points where injected fluids enter the receiving formation, or where they escape through casing or tubing leaks. Such anomalies would obviously be most likely to be detectable in wells where significant temperature contrasts exist between injected fluids and the aquifers.

Repetitive running of resistivity or radioactive logs may also be used to locate the zones that are accepting injected wastewater. Resistivity logs are limited to the uncased portion of a well, but radioactive logs have been used to locate a freshwater-saline water interface behind casing (Keys and MacCary, 1973).

MONITORING WELLS

The subject of monitoring wells has been a controversial one in regulation of wastewater injection. Such wells are routinely used in shallow groundwater studies, but are less frequently used in conjunction with wastewater injection, for reasons that will be examined.

At least three hydrogeologically different types of monitor wells can and have been constructed, each with different objectives as shown below:

Well Type	Objective
1. Constructed in receiving aquifer – nondischarging	A. Obtain geologic data B. Monitor pressure in receiving aquifer C. Determine rate and direction of wastewater movement

Well Type	Objective
	D. Detect geochemical changes in injected wastewater
	E. Detect shifts in freshwater-saline water interfaces
2. Constructed in, or just above, confining unit – nondischarging	A. Obtain geologic data B. Detect leakage through confining unit
3. Constructed in a freshwater aquifer above receiving aquifer	A. Obtain geologic data B. Detect evidence of freshwater contamination

Monitor wells constructed in the receiving aquifer are normally nondischarging because a discharging well would defeat most of the purposes of this type of monitor well. Also, the produced brines would have to be eliminated. Although it is not normally necessary to monitor pressure in the receiving aquifer, except at the injection well, special monitor wells may be desired, where pressure at a distance from the injection well is of concern because of the presence of known, or suspected, faults or inadequately plugged abandoned wells. The pressure response in a monitor well at such locations would indicate the extent of danger of flow through such breaches in the confining beds, and possibly also indicate whether leakage was occurring.

Constructing a monitor well, or wells, in the receiving aquifer is the only direct means of verifying the rate and direction of wastewater movement. More than one well will frequently be necessary to meet this objective, because monitor wells of this type only sample wastewater plumes that pass directly through the well bore; nonuniformity in aquifer porosity and permeability can cause the wastewater to arrive very rapidly, or perhaps not at all, at a particular well. A single well might be satisfactory, where aquifer and fluid properties are such that it is judged most unlikely that wastewater movement will be radial and reasonably uniform, or where the objective is to detect wastewater arrival at a particular point of interest. These same comments apply to wells intended to detect geochemical changes in injected wastewater. A difference is that a well for monitoring geochemical changes would be placed near enough to the injection well so that the wastewater front will arrive within a relatively short time, whereas, a well for detecting wastewater arrival at a point of concern might be beyond the expected ultimate travel distance of the wastewater.

A well intended to detect a shift in a freshwater-saline water interface should be located either within that interface, or in the freshwater portion of the aquifer just beyond the interface. Because movement of this interface will be in response to increased aquifer fluid pressure, rather than to actual displacement by the wastewater front, detection of its movement should be possible with a small number of observation wells, perhaps even a single properly located one. It is possible to estimate rates of movement for a particular case and determine if a monitor well is likely to be able to detect such a shift. Monitoring would be for confirmation of the calculations and to allow for revisions in regulation if unexpected results occur.

Negative factors should be considered in any case where deep monitor wells are contemplated: monitor wells in the receiving aquifer may be of limited usefulness, and they provide an additional means by which injected wastewater could escape from the receiving aquifer. In a number of cases, multiple injection wells have been constructed at a site, one or more of which may be standby injection wells. Standby wells can be used for monitoring of aquifer pressure, and for sampling of aquifer water. However, if they have been operated or even extensively tested, their use for monitoring may be impaired.

Some examples of the use of observation wells in the receiving aquifer are given by Goolsby (1971 and 1972), Kaufman et al (1973), Leenheer and Malcolm (1973), Peek and Heath (1973), and Hanby and Kidd (1973).

For detection of leakage, the idea of using nondischarging monitor wells completed in the confining beds, or in a confined aquifer immediately above the confining beds, has been widely discussed but has been little used. This type of well has the potential for acting as a very sensitive indicator of leakage by allowing measurement of small changes in pressure (or water level) that accompany leakage. A well of this type is best suited for use where the confining unit is relatively thin and well defined and the engineering properties of the two aquifers are within a range such that pressure response in the monitored aquifer will be rapid if leakage occurs. Use of the concepts outlined by Witherspoon and Neuman (1972) allow evaluation of the possibilities of success of this monitoring method in a specific situation. In many actual cases, confining beds are several hundred to several thousand feet thick, and do not contain aquifers suitable for such monitoring. In other cases, the physical circumstances are amenable to such monitoring, but several thousands of feet of interbedded aquitards and saline water aquifers are present. In these cases, slow vertical leakage across the aquitard immediately over the injection interval is not significant because it can be predicted that there will be no measurable influence at the stratigraphic level, where freshwater or other resources occur.

Two good examples of the usefulness of monitoring an aquifer immediately above the confining beds are provided by Kaufman et al (1973) and Leenheer and Malcolm (1973). In the case described by Kaufman et al, wastewater leakage from the lower Floridan aquifer through 150 ft of confining beds into the upper Floridan aquifer was detected by geochemical analysis of water from a monitor well constructed in the upper Floridan aquifer. No pressure effects were noticed in this instance. Leenheer and Malcolm summarized a case history in which leakage through the confining beds was detected first, by pressure increase in an overlying aquifer, and later, confirmed by chemical analysis, which showed wastewater contamination of water in the aquifer.

The type of monitor well most commonly in use is that completed in a freshwater aquifer above the injection horizon for detecting freshwater contamination. In a number of locations, this type of monitoring is provided by wells that are a part of the plant's water supply system. In other cases, the wells have been constructed particularly for monitoring, and are not used for water supply. Wells for detection of freshwater contamination should be discharging wells because they sample an area of aquifer within their cone of depression. As previously mentioned, nondischarging wells are of limited value for detection of contamination because they sample only that water that passes through the well bore. Wells for monitoring freshwater contamination should be located close to the anticipated sources of contamination, which are:

- The injection well itself
- Other nearby deep wells, active or abandoned
- Nearby faults or fracture zones

No example is known to the writer where monitor wells of this type have detected wastewater contamination of a water supply aquifer.

In the preceding discussion, it has been implied that separate wells would need to be constructed for surveillance of aquifers and aquicludes at different depths. This is not necessarily the case. Talbot (1972) shows how the injection well itself can be adapted for monitoring of overlying aquifers, and how monitor wells may be constructed for a surveillance of more than one aquifer. Wilson et al (1973) describe a case where the injection well was modified as shown in Figure 5-48 for monitoring of two aquifers overlying the injection zone.

Since the objectives for each of the types of monitoring wells discussed are worthwhile ones, why are monitor wells not more widely used? The answer is that the potential benefits are often judged to be small in comparison with the costs and negative aspects. Therefore, such wells may not be voluntarily constructed by the operating companies or required by the regulatory agencies. In particular, monitor wells constructed in the receiving aquifer are often difficult to justify because such wells are the most expensive form of surveillance, and may yield very little information that is important for regulation. It can reasonably be concluded that monitor wells should not be arbitrarily required, but should be used where the local circumstances justify them.

OTHER MONITORING METHODS

A method of monitoring not yet mentioned is the sampling of springs, streams, or lakes that could be affected by injection. There are few instances where such monitoring would be applicable; but, for example, where springs originate along a fault within the area of pressure influence of the injection well, an increase in discharge rate or change in water quality could be an indication of leakage of formation water along the fault, in response to the increased pressure from injection. Also, springs and gaining streams act similarly to discharging wells in that they provide a composite sample of groundwater over their area of influence; thus, they might reveal leakage from unknown fracture zones or abandoned wells that connect a shallow groundwater aquifer with the injection interval. In a similar way, lakes may be collecting points for groundwater seepage or streams, and may reflect quality changes in shallow groundwater aquifers.

Surface geophysical methods offer some limited possibilities for monitoring of wastewater injection. Barr (1973) discussed the feasibility of monitoring the distribution of injected wastewater with seismic reflection. Monitoring by seismic reflection depends on the existence of a sufficient density contrast between injected and interstitial water, and no field trials of monitoring by seismic reflection have been reported. Electrical resistivity surveying could be useful for monitoring the movement of freshwater-saline water interfaces, or for detecting saline water pollution of freshwater aquifers (Swartz, 1937; Warner, 1969).

Monitoring for earthquake occurrence is accomplished by use of a network of seismometers placed in the vicinity of the injection well and in the vicinity of nearby faults along which seismic events might be triggered. Examples of this form of monitoring are described by Raleigh (1972) and Hanby and Kidd (1973). In a case where earthquake stimulation is considered a possibility, seismic monitoring should begin before the well is operated to obtain background data.

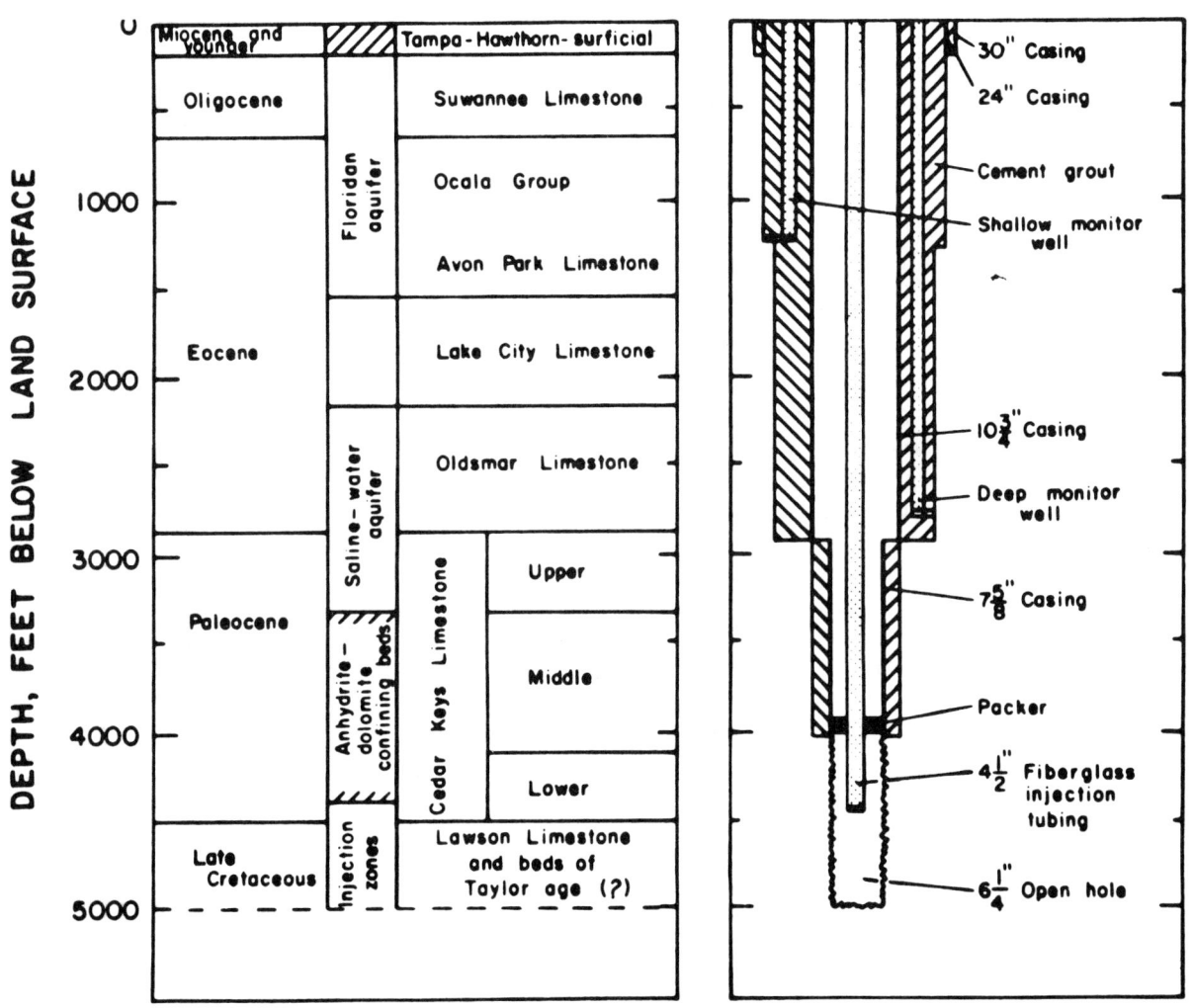

Figure 5-48. Geologic column and construction of a wastewater injection well at Mulberry, Florida, where two aquifers above the injection zone are monitored through the injection well (Wilson et al, 1973).

SECTION 6 - CONCLUSIONS AND RECOMMENDATIONS

The subsurface environment is a complex one characterized by the rocks and their structure, lithology, contained fluids and other resources, and mechanical properties. The static and dynamic states of the rocks and fluids are also characteristic of a region and a specific location.

An estimate of the characteristics of the subsurface environment can be made prior to drilling of a well based on projections of data from outcrops, previously drilled wells, and possible surface geophysical studies. A much more accurate knowledge of the local subsurface environment is obtained when a well is drilled and tested. Data obtained from a well are based on rock and fluid samples, geophysical logs, and pumping or injection tests.

When the characteristics of the subsurface environment have been estimated or determined, the response to wastewater injection can be predicted. Such predictions are essential to monitoring because they provide a baseline of expected performance, including rate of pressure buildup and rate and direction of travel of injected wastewater.

The principal means of injection well monitoring surveillance is the injection well itself. This provides more protection than is commonly realized, because the well is, in most cases, the most likely source of escape of injected wastewater. Periodic inspection and testing of injection well facilities complements continuous monitoring of well performance, and should prove helpful in detecting deterioration of these facilities prior to failure.

Monitoring wells can be used for several purposes; they may be constructed in the injection aquifer, in or just above the confining beds, or in freshwater aquifers. Local geology and hydrology, the waste being injected, and economics are factors in determining if monitor wells are needed, and, if so, how many and where.

Other types of monitoring include surface geophysics; sampling of springs, streams, and lakes; and monitoring to record any seismic events, which might be related to operation of the injection well.

Monitoring of subsurface wastewaters injection should be thought of as the full spectrum of consideration given to determining the effects of a wastewater injection system, from planning of the system through well construction, testing, operation, and abandonment.

Policy guidelines of the environmental Protection agency and of The Ohio River Valley Water Sanitation Commission should be used as a basis for injection well monitoring. These sources also provide suggestions for a suitable data base for monitoring and ORSANCO outlines a series of administrative procedures that should be followed.

This chapter has provided a discussion of the tools and methods for obtaining the needed data base and the use of the resulting subsurface data for prediction and interpretation of well behavior during operation. It also discusses the surveillance of operating wells in some detail. The maximum use should be made of the methods and tools that are available, consistent with the practicalities of available resources. Because of the obvious complexity of many of the tools and methods, regulatory agencies should not hesitate to request the assistance of other public agencies and private consultants in monitoring injection systems.

SELECTED BIBLIOGRAPHY

F.J. Barr, Jr., "Feasibility Study of a Seismic Reflection Monitoring System for Underground Waste-Material Injection Sites," in *Underground Waste Management and Artificial Recharge*, Jules Braunstein, ed., pp. 207-218, 1973.

Jacob Bear, *Dynamics of Fluids in Porous Media*, Elsevier Publishing Co., New York, 1972. (764 pp.)

J. Bear and M. Jacobs, *The Movement of Injected Water Bodies in Confined Aquifers*, Underground Water Storage Study Report No. 13, Technion, Haifa, Isreal, 1964.

F.A.F. Berry, "High Fluid Potentials in California Coast Ranges and their Tectonic Significance," *American Association of Petroleum Geologists Bulletin*, Vol. 57, No. 7, pp. 1219-1249, 1973.

Black Crow and Eidsness, Inc., *Engineering Report on Modification to Deep-Well Disposal System: Effect of Monitoring Wells and Future Monitoring Requirements for Sugar Cane Growers Cooperative of Florida, Belle Glade, Palm Beach County, Florida*, Engr. Rept. Proj. No. 387-71-01, 1972. (40 pp.)

D.C. Bond, *Hydrodynamics in Deep Aquifers of the Illinois Basin*, Illinois State Geological Survey Circular 470, 1972. (72 pp.)

D.C. Bond, "Deduction of Flow Patterns in Variable-Density Aquifers from Pressure and Water-Level Observations," in *Underground Waste Management and Artificial Recharge*, Jules Braunstein, ed., Am. Assoc. Petroleum Geologists, Tulsa, Oklahoma, pp. 357-378, 1973.

J.D. Bredehoeft and G.F. Pinder, "Application of Transport Equations to Groundwater Systems," in *Underground Waste Management and Environmental Implications*, T.D. Cook, ed., Am. Assoc. Petroleum Geologists Memoir 18, pp. 191-199, 1972.

D.L. Brown and W.D. Silvey, "Underground Storage and Retrieval of Fresh Water from a Brackish-Water Aquifer," in *Underground Waste Management and Artificial Recharge*, Jules Braunstein, ed., Am. Assoc. of Petroleum Geologists, Tulsa, Oklahoma, pp. 379-419, 1973.

T.C. Buschbach, *Cambrian and Ordovician Strata of Northeastern Illinois*, Illinois Geol. Survey Report of Investigation 218, 1964. (90 pp.)

M.J. Clifford, "Hydrodynamics of the Mount Simon Sandstone, Ohio and Adjoining Areas," in *Underground Waste Management and Artificial Recharge*, Jules Braunstein, ed., Am. Assoc. of Petroleum Geologists, Tulsa, Oklahoma, pp. 349-356, 1973.

T.D. Cook, ed., *Underground Waste Management and Environmental Implications*, Am. Assoc. of Petroleum Geologists Memoir 18, 1972. (412 pp.)

S.H. Davis and R.J.M. De Weist, *Hydrogeology*, Wiley and Sons, Inc., New York, New York, 1966. (463 pp.)

George Dickinson, "Geological Aspects of Abnormal Reservoir Pressures in the Gulf Coast Louisiana," American Association of Petroleum Geologist Bulletin, Vol. 37, No. 2, pp. 410-432, 1953.

D. Eisenberg and W. Kauzmann, *The Structure and Properties of Water*, Oxford University Press, New York, New York, 1969. (296 pp.)

J.G. Ferris, et al, *Theory of Aquifer Tests*, U.S. Geological Survey Water Supply Paper 1536-E, 1962. (174 pp.)

W.H., Fertl, et al, "A Look at Cement Bond Logs," *Journal of Petroleum Technology*, Vol. 26, pp. 607-617, June 1974.

Carl Gatlin, *Petroleum Engineering Drilling and Well Completions*, Prentice-Hall, Inc., Englewood Cliffs, N.J., 1960.

L.W. Gelhar, and others, *Density Induced Mixing in Confined Aquifers*, U.S. Environmental Protection Agency Water Pollution Control Research Series Publication 16060 ELJ 03/72, 1972.

D.A. Goolsby, "Hydrogeochemical Effects of Injecting Wastes into a Limestone Aquifer Near Pensacola, Florida," *Ground Water*, Vol. 9, No. 1, pp. 13-19, 1971.

D.A. Goolsby, "Geochemical Effects and Movement of Injected Industrial Waste in a Limestone Aquifer," in *Underground Waste Management and Environmental Implications*, American Association of Petroleum Geologists Memoir 18, Tulsa, Oklahoma, pp. 355-367, 1972.

H.R. Gould, *History of the AAPG Geothermal Survey of North America*, unpublished paper presented at the 1974 American Association of Petroleum Geologists Annual Meeting, San Antonio, Texas, 1974.

M. Grosmangin, et al, "A Sonic Method for Analyzing the Quality of Cementation of Borehole Casings," *Journal of Petroleum Technology*, pp. 165-171, February 1961.

C.W. Hall and R.K. Ballentine, "U.S. Environmental Protection Agency Policy on Subsurface Emplacement of Fluids by Well Injection," in *Underground Waste Management and Artificial Recharge*, Jules Braunstein, ed., American Association of Petroleum Geologists, Tulsa, Oklahoma, pp. 783-793, 1973.

K.P. Hanby and R.E. Kidd, "Subsurface Disposal of Liquid Industrial Wastes in Alabama—A current Status Report," in *Underground Waste Management and Artificial Recharge*, Jules Braunstein, ed., American Association of Petroleum Geologists, Tulsa, Oklahoma, pp. 72-90, 1973.

B.B. Hanshaw, "Natural Membrane Phenomena and Subsurface Waste Emplacement," in *Underground Waste Management and Environmental Implications*, T.D. Cook, ed., American Association of Petroleum Geologists Memoir 18, Tulsa, Oklahoma, pp. 308-315, 1972.

J.D. Haun and L.W. Le Roy, eds., *Subsurface Geology in Petroleum Exploration*, Colorado School of Mines, Golden, Colorado, 1958.

M.K. Hubbert and D.G. Willis, "Mechanics of Hydraulic Fracturing," in *Underground Waste Management and Environmental Implications,* T.D. Cook, ed., American Association of Petroleum Geologists Memoir 18, Tulsa, Oklahoma, 1972. (411 pp.)

Illinois Water Survey, *Feasibility Study of Desalting Brackish Water from the Mt. Simon Aquifer in Northeastern Illinois,* Urbana, Illinois, 1973. (120 pp.)

A. Janssens, *Stratigraphy of the Cambrian and Lower Ordovician Rocks in Ohio,* Ohio Division of Geological Survey Bulletin 64, 1973. (197 pp.)

H.Y. Jennings and A. Timur, "Significant Contributions in Formation Evaluation and Well Testing," *Journal of Petroleum Technology,* Vol. 25, pp. 1432-1446, December 1973.

D.L. Katz and D.L. Coats, *Underground Storage of Fluids,* Ulrich's Books, Inc., Ann Arbor, Michigan, 1968. (575 pp.)

Kaufman, et al, "Injection of Acidic Industrial Waste in a Saline Carbonate Aquifer," in *Underground Waste Management and Artificial Recharge,* Jules Braunstein ed., American Association of Petroleum Geologists, Tulsa, Oklahoma, pp. 526-551, 1973.

R.O. Kehle, "The Determination of Tectonic Stresses through Analysis of Hydraulic Well Fracturing," *Journal of Geophysical Research,* Vol. 69, No. 2, pp. 259-273, 1964.

W.S. Keys and R.F. Brown, "Role of Borehole Geophysics in Underground Waste Storage and Artificial Recharge," in *Underground Waste Management and Artificial Recharge,* Jules Braunstein, ed., American Association of Petroleum Geologists, Tulsa, Oklahoma, pp. 147-191, 1973.

W.S. Keys and L.M. MacCary, *Location and Characteristics of the Interface between Brine and Fresh Water from Geophysical Logs of Boreholes in the Upper Brasos River Basin, Texas,* U.S. Geological Survey Prof. Paper 809-B, 1973. (23 pp.)

C.V. Kirkpatrick, "Formation Testing," *The Petroleum Engineer,* pp. B-139, 1954.

G.P. Kruseman and N.A. DeRidder, *Analysis and Evaluation of Pumping Test Data,* International Institute for Land Reclamation and Improvement, Bulletin 11, Wageningen, The Netherlands, 1970. (200 pp.)

J.A. Leenheer and R.L. Malcolm, "Case History of Subsurface Waste Injection of an Industrial Organic Waste," in *Underground Waste Management and Artificial Recharge,* Jules Braunstein, ed., American Association of Petroleum Geologists, Tulsa, Oklahoma, pp. 565-584, 1973.

S.H. Lohman, *Ground-Water Hydraulics,* U.S. Geological Survey Professional Paper 708, 1972. (70 pp.)

E.J. Lynch, *Formation Evaluation,* Harper and Row, New York, New York, 1962. (422 pp.)

C.S. Matthews and D.G. Russell, *Pressure Buildup and Flow Tests in Wells,* American Institute of Mining, Metal, and Petroleum Engineers, Society of Petroleum Engineers, Monograph Vol. 1, 1967.

C.A. Moore, ed., *Second Symposium on Subsurface Geological Techniques*, University of Oklahoma Extension Division, Norman, Oklahoma, 1951.

W.C. Murphy, *The Interpretation and Calculation of Formation Characteristics from Formation Test Data*, Halliburton Services, Duncan, Oklahoma, undated.

Ohio River Valley Water Sanitation Commission, *Underground Injection of Wastewater in the Ohio Valley Region*, Cincinnati, Ohio, 1973. (63 pp.)

H.M. Peek and R.C. Heath, "Feasibility Study of Liquid-Waste Injection into Aquifers Containing Salt Water, Wilmington, North Carolina," in *Underground Waste Management and Artificial Recharge*, Jules Braunstein, ed., American Association of Petroleum Geologists, Tulsa, Oklahoma, pp. 851-878, 1973.

S.J. Pirson, *Handbook of Well Log Analysis*, Prentice-Hall, Inc., Englewood Cliffs, N.J., 1963. (326 pp.)

C.B. Raleigh, "Earthquakes and Fluid Injection," in *Underground Waste Management and Environmental Implications*, T.D. Cook, ed., American Association of Petroleum Geologists Memoir 18, pp. 273-279, 1972.

J.B. Robertson and J.T. Barraclough, "Radioactive- and Chemical-WasteTransport in Groundwater at National Reactor Testing Station, Idaho: 20-year Case History and Digital Model," in *Underground Waste Management and Artificial Recharge*, Jules Braunstein, ed., American Association of Petroleum Geologists, Tulsa, Oklahoma, pp. 291-322, 1973.

M.L. Sbar and M.L. Sykes, "Contemporary Compressive Stress and Seismicity in Eastern North America: An Example of Intra-Plate Tectonics," *Geological Society of American Bulletin*, Vol. 84, No. 6, pp. 1861-1882, 1973.

Schlumberger, *Schlumberger Engineered Production Services*, Schlumberger, Houston, Texas, 1970.

Schlumberger, Limited, *Log Interpretation Volume I–Principles*, Schlumberger Limited, New York, New York, 1972.

Schlumberger, Limited, *Log Interpretation Charts*, Schlumberger Limited, U.S.A., 1972a.

J.H. Swartz, "Resistivity Studies of some Saltwater Boundaries in the Hawaiian Islands," *Transactions of the American Geophysical Union*, Vol. 18, pp. 387-393, 1937.

J.S. Talbot, "Requirements for Monitoring of Industrial Deep Well Disposal Systems," in *Underground Waste Management and Environmental Implications*, T.D. Cook, ed., American Association of Petroleum Geologists Memoir 18, pp. 85-92, 1972.

D.L. Warner, *Deep-Well Injection of Liquid Waste*, U.S. Dept. of Health, Education, and Welfare, Public Health Service Publication No. 99-WP-21, 1965. (55 pp.)

D.L. Warner, "Subsurface Disposal of Liquid Industrial Wastes by Deep-Well Injection," in *Subsurface Disposal in Geologic Basins–A Study of Reservoir Strata*, J.E. Galley, ed., American Association of Petroleum Geologists Memoir 10, pp. 11-20, 1968.

D.L. Warner and D.H. Orcutt, "Industrial Wastewater-Injection Wells in the United States—Status of Use and Regulation, 1973," in *Underground Waste Management and Artificial Recharge*, Jules Braunstein, ed., American Association of Petroleum Geologists, Tulsa, Oklahoma, pp. 687-697, 1973.

W.E. Wilson, et al., "Hydrologic Evaluation of Industrial-Waste Injection at Mulberry, Florida," in *Underground Waste Management and Artificial Recharge*, Jules Braunstein, ed., American Association of Petroleum Geologists, Tulsa, Oklahoma, pp. 552-564, 1973.

Witherspoon, et al., *Interpretation of Aquifer Gas Storage Conditions from Water Pumping Tests*, American Gas Association, Inc., New York, New York, 1967. (273 pp.)

P.A. Witherspoon and S.P. Neuman, "Hydrodynamics of Fluid Injection," in *Underground Waste Management and Environmental Implications*, T.D. Cook, ed., American Association of Petroleum Geologists Memoir 18, Tulsa, Oklahoma, 1972.

Joe Zemanek, et al., "Formation Evaluation by Inspection with the Borehole Televiewer," *Geophysics*, Vol. 35, No. 2, pp. 254-269, 1970.

CHAPTER VI

ILLUSTRATIVE EXAMPLES

SECTION 1 - INTRODUCTION

Chapter II presents the following 15-step methodology for monitoring groundwater quality degradation resulting from man's activities:

Step 1 — Select Area or Basin for Monitoring
Step 2 — Identify Pollution Sources, Causes, and Methods of Waste Disposal
Step 3 — Identify Potential Pollutants
Step 4 — Define Groundwater Usage
Step 5 — Define Hydrogeologic Situation
Step 6 — Study Existing Groundwater Quality
Step 7 — Evaluate Infiltration Potential of Wastes at the Land Surface
Step 8 — Evaluate Mobility of Pollutants From the Land Surface to Water Table
Step 9 — Evaluate Attenuation of Pollutants in the Saturated Zone
Step 10 — Prioritize Sources and Causes
Step 11 — Evaluate Existing Monitoring Programs
Step 12 — Establish Alternative Monitoring Approaches
Step 13 — Select and Implement the Monitoring Program
Step 14 — Review and Interpret Monitoring Results
Step 15 — Summarize and Transmit Monitoring Information

Application of these 15 steps by a Designated Monitoring Agency (DMA), at the Statewide or local level, will result in the selection of the area to be monitored, identification and prioritization of pollution sources, and causes for monitoring. Chapter III summarizes specific techniques for monitoring groundwater quality, and gives detailed estimates of their costs.

In many instances, government and industrial organizations divide monitoring responsibilities among their various departments according to activities such as agriculture, mining, public health, and other interest areas. As a consequence, classes of pollution sources (e.g., irrigation return flow, waste-water treatment and disposal, and solid waste disposal) have become institutionalized along similar lines. Individuals with a responsibility for monitoring such class-specific problems are not likely to be interested in the complete 15-step, areawide monitoring methodology, but, instead, will want to know how to deal with a particular pollution source. Sections 2 and 3 of this chapter have been specially prepared to illustrate the application of site-specific monitoring methodology.

Section 2 presents a critique of five actual groundwater pollution case histories, given with the monitoring techniques employed, as well as a retrospective view of these techniques and their efficacy. The five case histories are as follows: Brine Disposal in

Arkansas; Plating Waste Contamination in Long Island, New York; Landfill Leachate Contamination in Milford, Connecticut; Pollution Potential of an Oxidation Pond Near Tucson, Arizona; and Multiple-Source Nitrate Pollution in the Fresno-Clovis, California, Metropolitan Area.

Section 3 presents illustrative examples of how to apply those steps of the methodology applicable in site-specific situations to the following pollution sources: Agricultural Return Flow; Septic Tanks; Percolation Ponds; and Solid Waste Landfills. Steps 1, 10, 14, and 15 are omitted from the discussion because the area will already have been specified, and the priority of the sources for monitoring established. In addition, the review, interpretation, and transmission of the monitoring results will be a function of the goals of the DMA, which are not specified in this instance. The costs for the monitoring methods selected in each example were obtained from Everett et al. (1976).

SECTION 2 – GROUNDWATER POLLUTION CASE HISTORIES AND EVALUATION OF MONITORING TECHNIQUES

BRINE DISPOSAL IN ARKANSAS

BACKGROUND

Disposal of the salt water produced along with the oil from oil wells has long been a pollution problem. Prior to the 1960s the salt water was commonly placed in *evaporation* pits for disposal. Pits dug into an impermeable formation, or lined, would generally function as intended; however, in most cases, considerable saltwater infiltrated downward to pollute the groundwater. In older oil fields, use of such pits was widespread, and, even though many pits have been filled and are no longer visible, plumes of polluted groundwater are still present.

Today, oil-field brines are more commonly disposed of through wells. Saltwater disposal wells can be classified in increasing order of safety from pollution as follows: (1) disposal through the annulus between the surface casing and production casing, (2) disposal using a converted abandoned oil well, and (3) disposal in a well specifically designed and drilled for saltwater disposal. The most common causes of pollution associated with saltwater disposal wells are (1) corroded or broken casing allowing the saltwater to escape into a fresh-water aquifer, and (2) excessive injection pressure resulting in upward movement of brines outside an improperly cemented casing or through fractures in containing formations.

The groundwater pollution discussed in this subsection was caused by disposal of oil-field brine first, through an evaporation pit and later, through a faulty disposal well (Fryberger, 1972). The scope of the original report includes the following:

- A history of the cause of the pollution
- Determination of the extent of pollution
- Evaluation of chemical changes in the polluting brine
- Cost-benefit evaluation of potential remedial measures

DESCRIPTION OF AREA
Location

The polluted area is in Miller County, in the southwest corner of Arkansas (Figure 6-1). The site selected for detailed investigation is 2-1/2 miles southwest of Garland City, on the floodplain of the Red River, and about 2-1/2 miles west of the present river channel.

Geologic Setting

The floodplain of the Red River is 9 miles wide in the project area, and is characterized by oxbow lakes and poorly drained bayous, typical of a mature, meandering, aggrading river. Clean, highly permeable sand was deposited by the river during much of its early depositional history. In the polluted area, the alluvial sand extends to a depth of 40 feet, but elsewhere the alluvium extends up to 90 feet (Ludwig, 1973). Alluvial clays and silts extending from ground surface to a depth of about 12 feet overlie the alluvial sands in the polluted area. The bedrock underlying the alluvium consists of sedimentary formations of Eocene and Cretaceous ages.

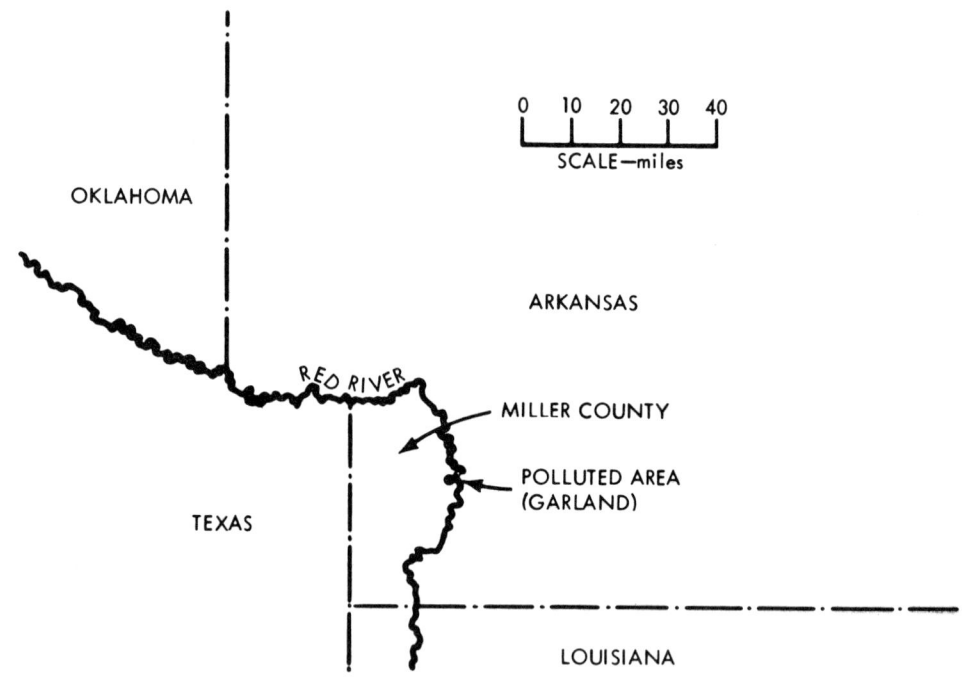

Figure 6-1. Location of polluted areas.

The static water level in the alluvium is about 8 feet below ground level in the polluted area. This alluvial aquifer is the most commonly tapped water source in the County for municipal, domestic, and agricultural uses.

History

In 1967 a farmer owning land adjacent to the polluted area complained to State agencies that his 1,000 gallons per minute irrigation well had turned salty.

During the summer of 1967, interested State agencies, with overlapping jurisdiction (Arkansas Soil and Water Conservation Commission, Arkansas Geological Commission, Pollution Control Commission, and the Oil and Gas Commission), conducted a preliminary investigation to determine the source of the pollution. This investigation consisted of sampling the sand-water mix brought to the surface by a continuous auger-type drill in the area around the nearby evaporation pit. These preliminary test hole samples strongly suggested that the evaporation pit was the source of the polluting brine.

In addition, under a reconnaissance study being conducted simultaneously by the U.S. Geological survey, groundwater samples were obtained over a 20-square mile area along the Red River in Miller County. This more general study delineated two other polluted areas where chlorides exceeded 500 milligrams per liter. Pollution of all three of the areas is believed to be caused by improper disposal of oil-field brines through evaporation pits, two of which had been abandoned and filled in. The operator of the remaining evaporation pit was ordered to fill the pit and dispose of the brine using some other means. An abandoned oil well next to the pit was then used by the operator as a disposal well.

Funded by an EPA grant, the Arkansas Soil and Water Conservation Commission conducted a detailed investigation of the polluted area using numerous test/observation wells. During this investigation it was noted that periods of water-level rises in an observation well 500 feet from the disposal well correlated with periods of saltwater injection into the disposal well. Further investigation revealed that, when the disposal well was first put into operation, injection pressures of 300 to 400 pounds per square inch (psi) were required to pump the brine into the disposal formation at a depth of about 2500 feet. However, at the time of observation, no injection pressure was required to inject the brine at the same flow rate. Based on these facts and other observations, it was determined that most of the brine then being pumped was escaping from the disposal well through a corroded or faulty surface casing, and was being injected into the fresh-water alluvial aquifer. A new injection well was constructed nearby for disposal of the brine. The continued pollution through the original faulty disposal well could have been avoided had the appropriate State agency required monitoring of a fluid-filled annulus outside of the tubing to detect leaks.

The investigation of the Arkansas Soil and Water Conservation Commission was completed with preparation of the report, *Rehabilitation of a Brine-Polluted Aquifer*, Fryberger (1972).

MONITORING METHODS
Objectives

Objectives of most investigative projects can be divided into two categories. The first includes general objectives that are common to most investigations of a similar nature; the second includes special objectives not common to all projects of a similar nature.

The general objectives of this investigation were to (1) delineate the lateral and vertical areal extent of pollution and the gradation in concentration away from the source, (2) positively identify the source, and (3) determine the rate and direction of movement of the pollution plume by determining the transmissivity of the aquifer and the slope of the water table.

The special objectives of the investigation were to (1) determine the chemical changes taking place in the aquifer as a result of the polluting brine mixing with the native groundwater and the formation of solids in the aquifer, (2) determine the present and future monetary loss caused by the pollution, and (3) determine the technical and economic feasibility of rehabilitation of the aquifer by removing the saltwater.

Further discussion of the monitoring program is limited to the three general objectives and to the first special objective.

Exploration Alternatives

The selection of specific monitoring methods from the wide choices available is primarily dependent on the objectives and the geologic conditions in the project area, and is secondarily dependent on relative reliability and costs of the alternate choices.

Three alternate general exploration methods for collecting the data required to meet the objectives of the investigation were considered. These were: (1) drill test holes, install permanent casing, and obtain water samples, (2) conduct a surface resistivity survey (supplemented by drilling and sampling), and (3) use rotary drill holes (uncased), and run electric logs on the holes (supplemented by casing and sampling).

The first alternate, drilling test holes, installing permanent casing, and obtaining water samples, was chosen as the general exploration method because it was the only monitoring method of the three that provided the means to achieve all of the objectives. It offered the advantages of:

- Ability to determine both lateral and vertical distribution of brine
- Ability to obtain water samples for chemical analysis to determine chemical changes in the injected brine
- Ability to obtain water levels and formation samples to determine rate of movement of the plume
- Provision for a permanent monitoring system for determining actual plume movement with time

Although the cost of drilled and cased monitoring holes is higher than the alternate choices, it was not considered prohibitive, because of the shallow depths and easy drilling provided by the soft geologic formations.

Under different geologic conditions it may have been advantageous to use the alternate methods to augment, or replace, some of the cased drill holes. For instance, if the polluted aquifer was deeper and presented more costly drilling conditions, then to reduce costs, down-the-hole electric logging methods to determine relative water salinity could be used in place of water sampling at some of the sampling points. In addition, under vertically and laterally uniform geologic conditions, surface resistivity could be used to replace some test holes to help delineate the lateral extent of the pollution at less cost than the test holes (Oklahoma Water Resources Board, 1974).

Drilling Alternatives

Having selected cased holes and water sampling as the exploration method, the investigators considered alternative methods of setting the casing in order to best fit the objectives and the requirements of geologic conditions, cost, and reliability of data. The methods considered were: use of drive points, continuous-flight auger drilling, mud/water rotary drilling, cable tool drilling, and air rotary drilling. The drilling method selected for this project was a combination of two alternatives, continuous-flight auger drilling and drive points.

Two advantages of continuous-flight auger drilling are: (1) there is no drilling fluid to contaminate the native groundwater, and (2) the drilling rigs are relatively mobile. Disadvantages include: (1) formation samples are mixed, therefore obtaining an accurate detailed geologic log is not possible, (2) the casing cannot be set because the unconsolidated, loose formations, such as saturated sands, cave in, and (3) drilling is possible only in relatively soft formations.

Drive points, which are pipes with a short, pointed well screen on the bottom, are driven through the formation to the desired depth. Although practical only for shallow, soft formations, drive points have the advantages of low cost and no need for using foreign matter, such as drilling fluid; therefore, water samples can be obtained quickly without fear of contamination. However, because no samples of the formation material are obtained in the process, detailed geologic logs cannot be constructed. Also, determination of aquifer transmissivity and rate of movement of the plume is much less reliable without formation samples.

In sinking the well shafts, a continuous-flight auger was used to drill a hole down through the 10 to 15 feet of clayey soil overlying the sand aquifer. This method was fast and economical for that specific part of the work, but not suited for drilling the entire hole because the loose sand aquifer would not stand open. Drive points with 2-inch pipe were set into the open hole, drilled through the remaining clay, and mechanically driven to the desired depth in the sand aquifer. The drive points could not be driven through the clay because excessive force would have been required, and the clay would have sealed off the drive point well screen. This approach made it feasible to obtain water samples at intermediate depths as the drive point was being driven to its final depth. The bottom of the soft alluvial sand was detected when the drive point reached the hard underlying bedrock.

Water samples were obtained by pumping with a centrifugal pump attached directly to the top of the drive point pipe. If the water table had been lower than about 20 feet, other pumping methods, such as a submersible or turbine pump, would have been required, and a larger-diameter casing would have been necessary.

The primary disadvantage of the method selected is the lack of aquifer formation samples obtained. This made it more difficult to determine aquifer transmissivity necessary to calculate the rate of movement of the plume. Pumping tests and sample descriptions from other sources, not in the immediate area, had to be used for transmissivity estimates.

CRITIQUE OF MONITORING PROJECT
Results Obtained

Twenty-eight sampling sites were located around the pollution source. At many of the sampling sites water samples were obtained at intermediate depths, in order to provide data on the vertical distribution of the brine.

Figure 6-2 shows the distribution of the sampling sites relative to the pollution source, the lateral distribution of chloride concentration at the bottom of the aquifer, and the location of the vertical section through the area that is depicted in Figure 6-3.

Generally, the monitoring methods selected for this project were successful. The vertical and horizontal distribution of the chlorides within the polluting plume were determined with sufficient accuracy. A survey of water level elevations in all the test wells, plus adjacent domestic wells, provided accurate data to calculate the slope of the water table. Good water samples were obtained from the test wells to meet both the general and special objectives, and additional water samples were obtained from adjacent domestic wells to determine background quality.

Recommended Changes

In retrospect, two changes would be desirable, if the project were to be repeated. One change involves a general objective, and the other a special objective.

First, a rotary drill using clear water for a drilling lubricant would be used in place of an auger and drive points. Rotary drilling would permit sampling of the aquifer formation in order to better evaluate the aquifer transmissivity for calculating the rate of movement of the polluting plume. This procedural change would entail (1) drilling and sampling the hole, (2) setting casing with a short screen on the bottom to the desired depth, and (3) pumping for a sufficient time to clear all of the drilling fluid out of the formation. The length of pumping time required would be determined by monitoring the pumped water in the field, using a portable specific conductivity meter.

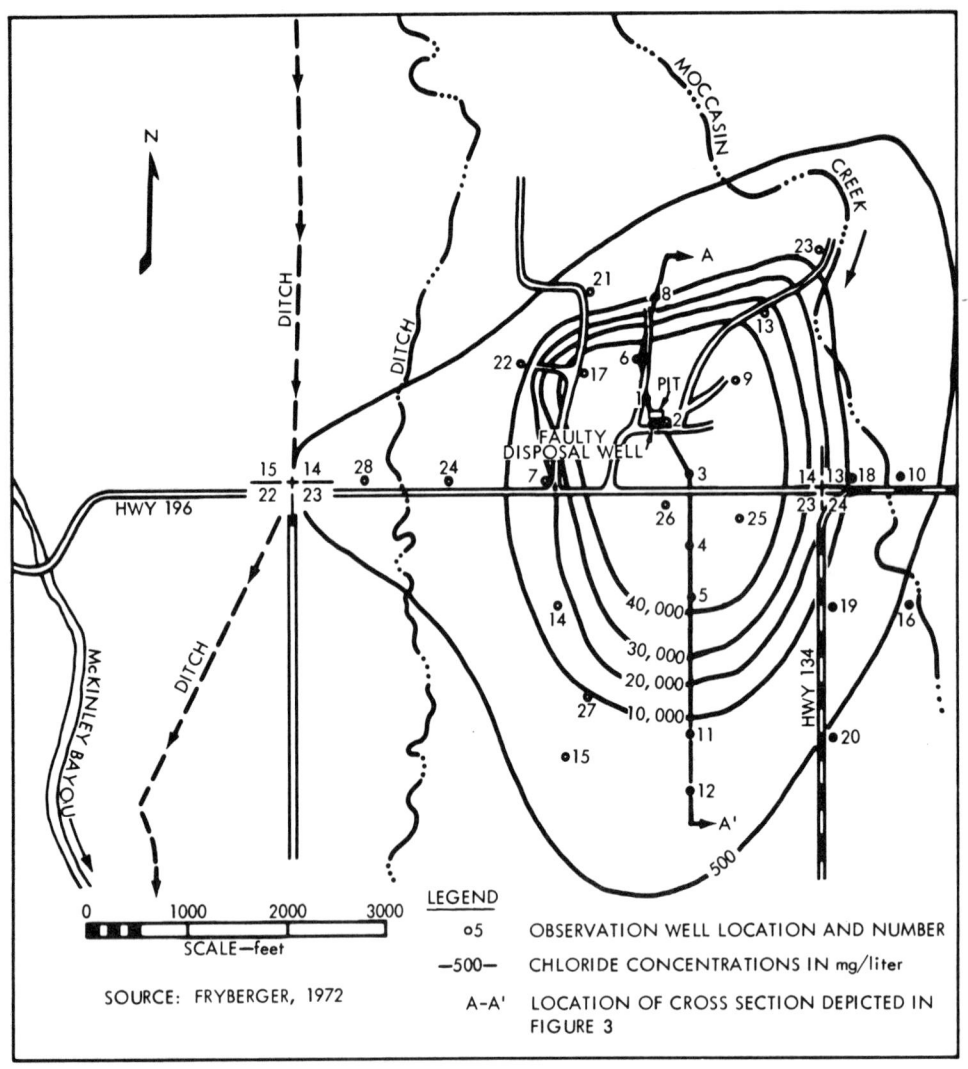

Figure 6-2. Contours of chloride concentration at bottom of alluvium.

Second, certain key chemical analyses such as for carbon dioxide (pH, CO_2) and bicarbonate (HCO_3), would be conducted, either in the field or, if in a laboratory, shortly after samples were obtained, to provide better quality data for the objective of determining chemical changes in brine as it mixed with the native water and aquifer solids.

PLATING WASTE CONTAMINATION IN LONG ISLAND, NEW YORK

BACKGROUND

In 1942, the Nassau County Department of Health, Long Island, N.Y., undertook a routine survey of the water-supply system in a former aircraft plant at South Farmingdale. The survey revealed that water from a well near a basin used to dispose of wastes from metal-plating operations at the plant contained about 0.1 miligrams per liter of chromium.

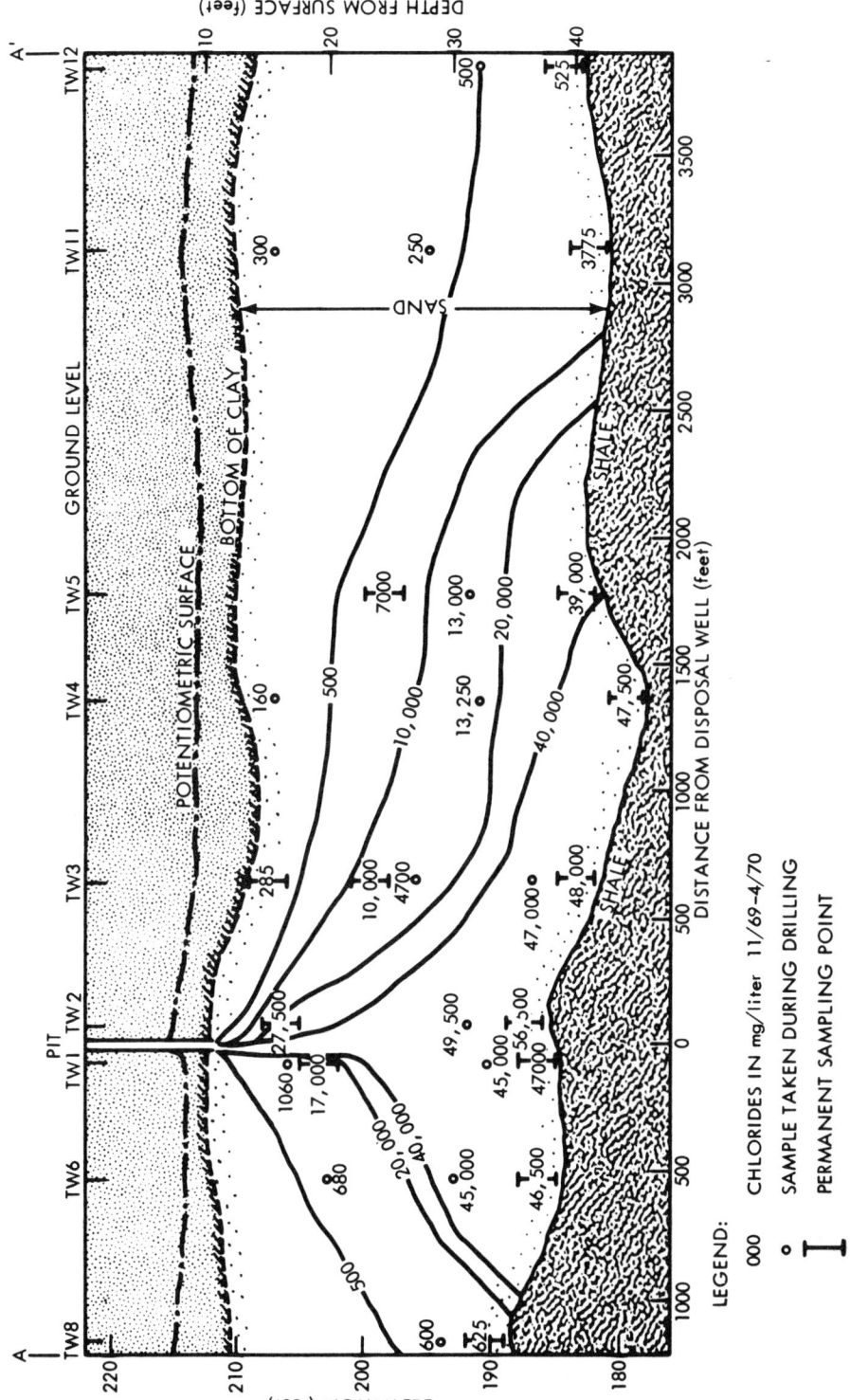

Figure 6-3. Section A-A' showing brine distribution.

The plating-waste contaminants were derived from chemical solutions used chiefly in anodizing and other metal plating processes, starting in about 1941. Until several years after World War II, large quantities of virtually untreated plating-waste effluents were recharged into the groundwater through disposal basins. Only scanty records were kept of the quantities, but it is estimated that during the early 1940s, as much as 200,000 to 300,000 gallons of effluent, containing about 52 pounds of chromium and smaller amounts of other metals, was recharged daily into the upper glacial aquifer. After the war, the quantities of effluent were reduced substantially, and the character of the waste changed to some extent.

It was not determined, initially, whether the chromium was in the nontoxic trivalent ion or the toxic hexavalent ion, but the plant management was advised by the Health Department to prohibit the use of water from the contaminated well from drinking purposes and to initiate treatment for removal of chromium from the plating wastes.

Nothing further was done until 1945, when the Department of Water Supply, Gas, and Electricity of the City of New York installed a series of shallow test wells in an area several hundred feet south of the aircraft plant. This work was undertaken because the City was concerned over potential contamination of its auxiliary groundwater system at Massapequa, several miles to the south. The chromium content of the water from the test wells, which penetrated the water table for only a short distance, ranged from zero to a trace, and, consequently, City officials decided that no real threat to the water system existed.

In 1946, after the U.S. Public Health Service established a limit of 0.05 milligram per liter for hexavalent chromium in drinking water, the New York State Department of Health requested that the new owners of the metal-plating facilities present plans for the removal of chromium from the plating wastes before disposal into the groundwater reservoir. In 1948, the New York State Department of Health analyzed another set of samples from the test wells drilled in 1945, and also took samples from a shallow domestic well about 1500 feet south of the disposal basins. The results showed about 1 to 3.5 milligrams per liter of hexavalent chromium, 0 to 0.24 milligrams per liter of cadmium, and 0.06 to 0.16 milligrams per liter of copper and aluminum. In conformance with recommendations of the County and State Health Departments, a waste-treatment unit for chromium was placed in operation in 1949, but discharge of the effluent containing cadmium and other metals continued at the disposal basins.

Hydrogeologic Setting

The groundwater reservoir in the South Farmingdale area consists of about 1300 feet of saturated, consolidated deposits resting on crystalline bedrock. In general, this sedimentary sequence is divided into three principal aquifers, or water-bearing units. The upper glacial aquifer extends from the water table, at depths of less than 15 feet below land surface, to the top of the second aquifer, referred to as the Magothy aquifer. The upper glacial aquifer consists mainly of beds and lenses of fine-to-coarse sand and gravel. The Magothy aquifer, whose upper surface ranges from about 80 to 140 feet below land surface, consists chiefly of beds and lenses of fine sand, sandy and silty clay, and clay. The third aquifer, the Lloyd sand member of the Raritan formation, lies more than 1,000 feet below land surface. Only the upper glacial aquifer was affected by the contamination.

The general direction of groundwater movement in the upper glacial aquifer is toward the south, and the shallow groundwater in the area of the study eventually discharges into Massapequa Creek. Water enters the upper glacial aquifer by direct infiltration of precipitation and lateral subsurface inflow.

Dimensions of Plume

Figures 6-4 and 6-5 show a plan view and vertical sections of the plume of hexavalent chromium and cadmium contamination, as defined by the test drilling and sampling. The plume was as much as 4,300 feet long and 1,000 feet wide in 1962. In the vertical dimension, it extended from the water table to depths of as much as 50 to 70 feet below land surface. Maximum concentrations of hexavalent chromium, which were as much as 40 milligrams per liter in 1949, had decreased to about 10 milligrams per liter in 1962. Concentrations of cadmium were as high as 3 milligrams per liter in 1962, and as high as 10 milligrams per liter in 1964, at a spot not previously tested.

MAPPING OF PLUME

Aware of the potential danger to public water supplies, the Nassau County Departments of Health and Public Works began a systematic investigation of the plume in 1949 with the drilling of about 40 test wells along several streets south of the disposal basins (Figure 6-4). These wells were 1-1/4-inch driven points, which were sampled by hand pumping at 5-foot intervals. Drilling was continued to depths (generally 50 to 60 feet) at which field testing showed no evidence of chromium contamination. Maps and profiles of the plume were prepared for the first time from the results of this drilling (Suter et al., 1949).

In 1953, 1958, and 1962, additional test wells were installed to map changes in the boundaries of the plume and changes in concentrations of the cadmium and chromium. In the 1962 investigation, which was the most detailed of the series, about 100 sampling wells were installed, and several test holes were drilled in which cores were taken to define the lithology and hydraulic coefficients of the geologic units in more detail. Extensive sampling of Massapequa Creek and underlying beds also was undertaken to determine the concentration and load of heavy metals at various points in and beneath the stream. Spectrographic analyses of several water samples were made to determine the presence of metals other than cadmium and chromium. Detailed maps and cross sections of the plume again were prepared, a water budget calculation was made, and the pattern of movement of contaminated water from the shallow aquifer into the stream was delineated (Perlmutter and Geraghty, 1963; Perlmutter et al., 1963).

STEPS THAT COULD HAVE BEEN TAKEN

The basic objectives of the investigation were somewhat limited, and were focused mainly on defining threats to drinking-water supplies, especially the ground-water facility operated by the City of New York and private wells in the area. Although a relatively large expenditure was made on the study, several approaches were not tried that might have yielded better information, and might have led to a better understanding of the occurrence and behavior of the plume. These other approaches are: surface resistivity surveys, more comprehensive chemical analyses, pumping tests, and contaminant transport modeling.

Surface Resistivity Survey

Surface resistivity surveys might have proven of value by detecting changes in subsurface conductivity caused by differences in the conductivities of the native and contaminating fluids. Such surveys are simple to conduct, do not require extensive test drilling, and offer a quick way of mapping the broad dimensions of a plume. Although they do not provide quantitative information on the mineral composition of the plume, they may be useful in selecting areas within which drilling should be undertaken. The success of such surveys depends on the degree of contrast in conductivities and the depths of the fluids to be detected.

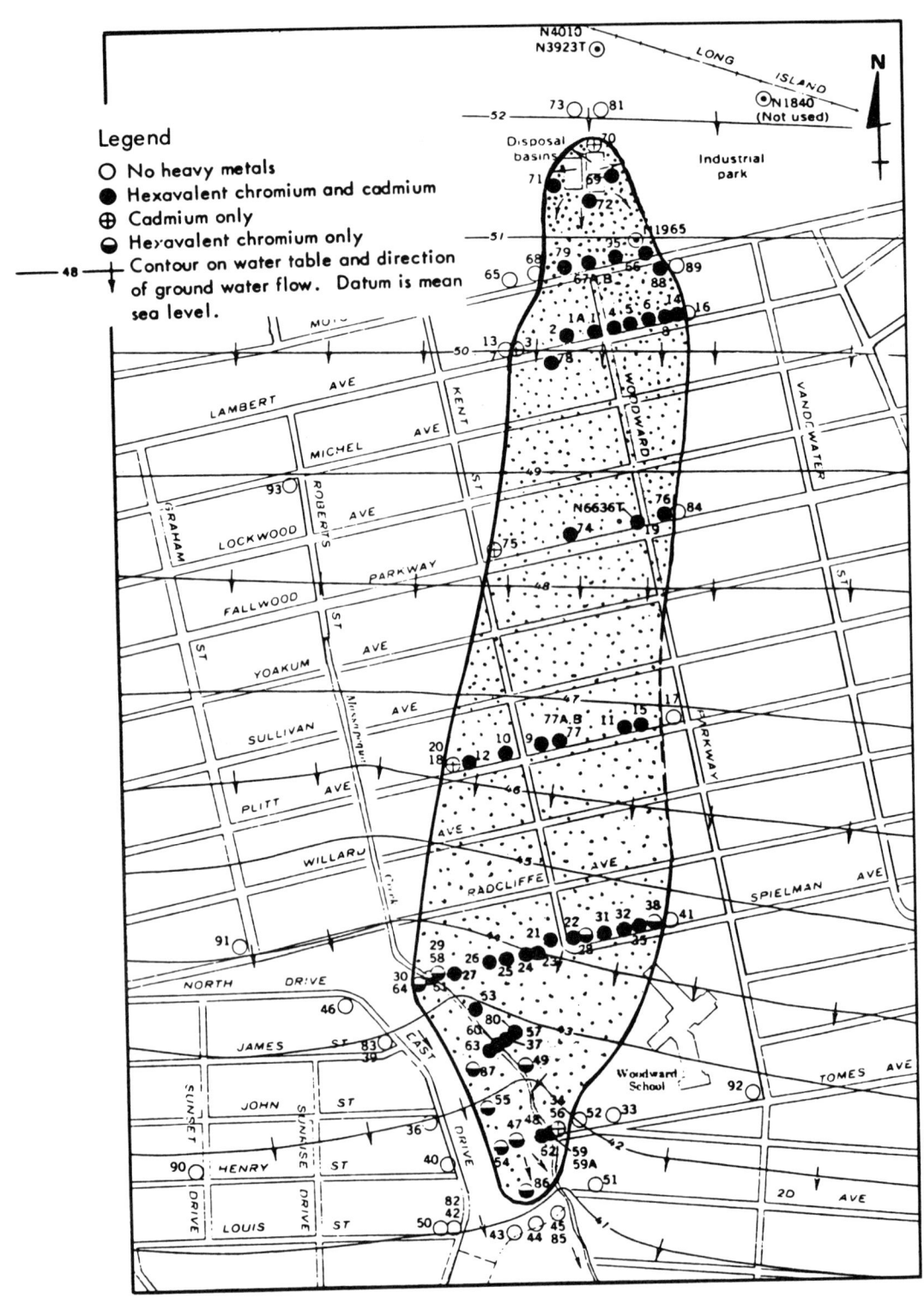

Figure 6-4. Map showing plating-waste plume, water-table contours, and selected test wells (modified after Perlmutter and Lieber, 1970).

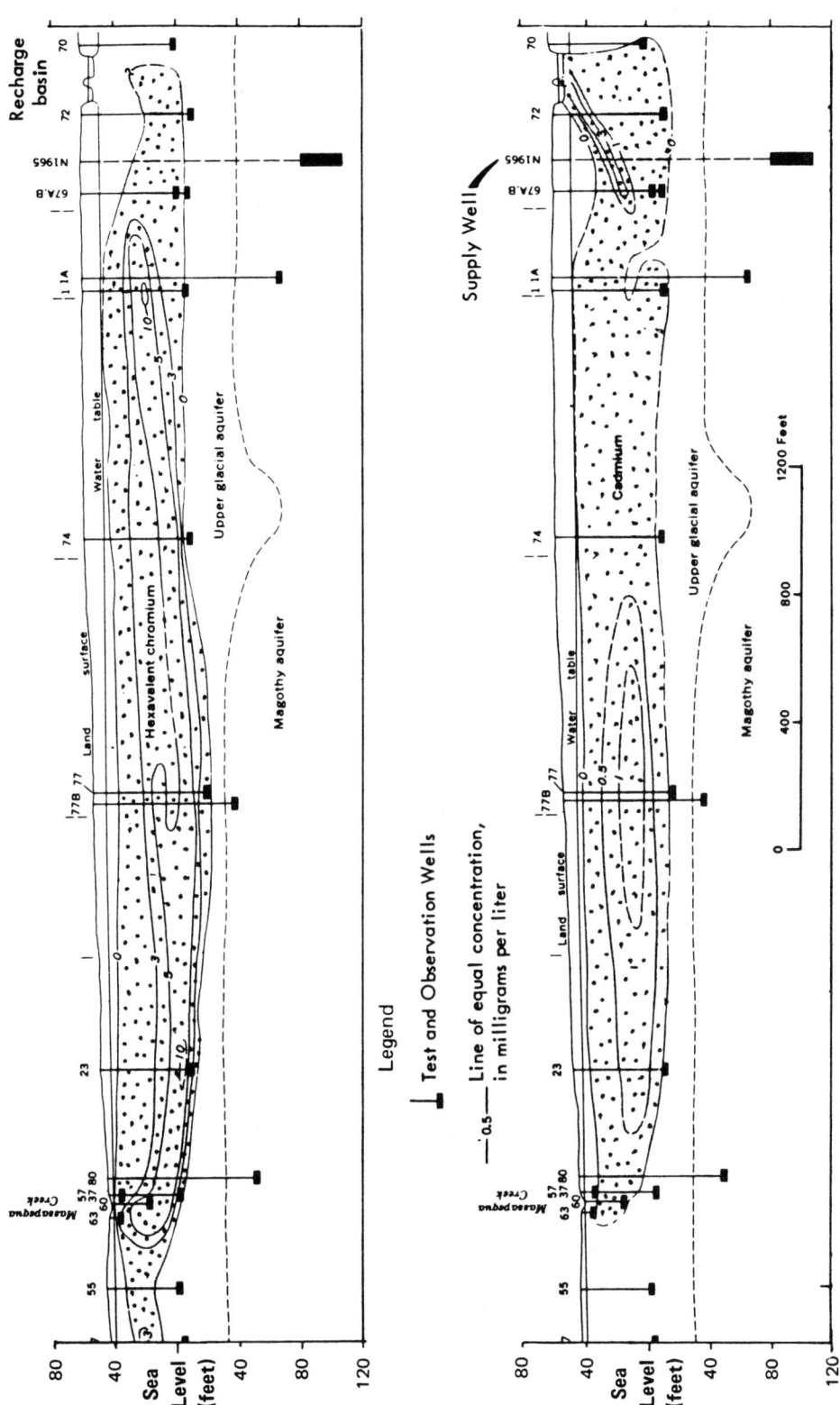

Figure 6-5. Vertical profiles of hexavalent chromium and cadmium along the center line of the plating-waste plume (modified after Perlmutter and Lieber, 1970).

Comprehensive Chemical Analyses

Some of the samples taken during the investigation should have been analyzed for more chemical constituents, especially other heavy metals and trace elements. Spectrographic analyses could have been made at the start of the investigation to at least define the presence of rare or exotic constituents.

Hydraulic Properties of the Aquifer

No pumping tests were conducted to define the water-bearing properties of the upper glacial aquifer, and most of the information developed in the study was obtained empirically through test drilling and sampling. It is likely that a better knowledge of the hydraulic behavior of the groundwater system at the site could have expedited the evaluation of the problem and saved some of the costs of drilling. Predictions of flow velocities, for example, might have been made at an early stage in order to estimate the probable length of the plume.

Contaminant Transport Model

Another useful step would have been to develop a mathematical model to predict rates of movement of contaminated water and changes in concentration, distribution, and dispersion of the contaminants. This would require information on such factors as porosity, hydrodynamic dispersion coefficient, hydraulic conductivity of the aquifer materials, and the thickness and hydraulic conductivity of the stream bed material in Massapequa Creek, which was the discharge point at the southern end of the plume. Such a model has been made by Pinder (1973), who predicted that contamination of Massapequa Creek would effectively cease about 7 years after cessation of disposal, or institution of complete treatment.

WHAT SHOULD HAVE BEEN DONE
Improved Treatment

As with many such projects, after the extent and composition of the plume had been reasonably well defined between 1949 and 1962, the public agencies decreased the detailed monitoring effort. In order to abate the contamination, a recommendation was made to the plant owners to improve the effectiveness of the methods of treating the hexavalent chromium, but little was done to eliminate other metals in the waste. A more complete treatment method should have been used to remove all of the toxic constituents.

Sampling of Wells and Streams

Although a reasonably good picture of the extent and movement of the plume was developed during the detailed field investigation, little has been done since to determine if the hydrologic situation has altered significantly. No new information has been obtained, for example, on the attenuation of the plume or whether the contamination has started to move downward into the underlying Magothy aquifer, which is the principal source of potable water in Nassau County. It is conceivable, with the growing stress being placed on the Magothy aquifer by public water supply systems, that a downward hydraulic gradient could develop eventually, with possible long-term implications to the quality of drinking water. It probably would be desirable to conduct follow-up studies involving periodic sampling of strategically located shallow wells and parts of Massapequa Creek, and determinations of changes of heads and directions of flow in both the shallow and deep aquifers.

Recovery of Contaminants

Another step that should have been given greater consideration is the feasibility of pumping out the contaminated groundwater for transport to a water-treatment plant or other disposal facility. The transmissivity of the upper glacial aquifer is reasonably high, so that even a single shallow pumping well could withdraw a fairly large quantity of the contaminated groundwater. This could have been tried, at least experimentally, to show how effective such a procedure would be.

LANDFILL LEACHATE CONTAMINATION IN MILFORD, CONNECTICUT

BACKGROUND

In 1973, the State of Connecticut authorized an investigation of a sanitary landfill site in Milford, Connecticut, that was under consideration as the location for a new State park. The main objectives of the study were to determine if contamination of the ground and surface waters would prevent the use of the area for this purpose, and whether or not the contamination problem was severe enough to warrant shutting down the landfill. Drilling and sampling were carried out to define the chemical quality and pattern of movement of the ground and surface water beneath and adjacent to the landfill, and an evaluation was made of gases being generated by the landfill materials (Geraghty and Miller, 1973).

The landfill area, part of which is an old fly ash disposal site formerly operated by the Devon Power Plant, covers approximately 90 acres. The refuse is derived from nearby communities, and consists largely of ordinary household wastes, construction rubble, brush and vegetative materials, and various types of solid and liquid wastes from local industries. A volume-reduction plant was constructed at the site to shred a portion of the refuse prior to its deposition in the landfill.

Hydrogeologic Setting

The landfill is located on an old tidal marsh, about one-half square mile in area, bordering on Long Island (Figure 6-6). Part of the original marsh is still visible around the landfill materials. The entire project site was at one time underlain by swamp deposits ranging from several feet to a few tens of feet in thickness. Directly beneath the landfill, the marsh deposits have been somewhat compressed and mixed with fill materials so that they do not show up as a distinct unit in drilling logs. The marsh is now mostly isolated from tidal effects, except for a small channel in the eastern section. The channel discharges freshwater during low tide, and contains some salty water part way upstream during high tide. About half of the original marsh area is covered by the artificial fill.

The marsh deposits are underlain by unconsolidated materials about 40 to 60 feet thick (Figure 6-7). These materials are essentially glacial till and some outwash sediments, consisting chiefly of layers of fine to medium sand, silt, and clay. The individual beds do not appear to be very extensive laterally.

The glacial materials are underlain by consolidated bedrock consisting primarily of schist and some gneiss. A bedrock valley extends from west to east across the northern portion of the landfill site. The bedrock slopes to the southeast, and an outcrop is present northwest of the landfill. The water table ranges in altitude from sea level to about 8 feet above sea level (Figure 6-7). The altitude has been raised above its normal level due to the construction of the landfill mounds.

Dimensions of Plume

Because the landfill is in a hydrologic system of limited areal extent, the contaminated fluid has not moved a great distance away from the site. A mound of highly contaminated water is within the landfill materials. The contaminated water moves out radially from the center of the mound, and most of it eventually discharges into Long Island Sound, only a few hundred feet to the south. Leachate moving to the west and

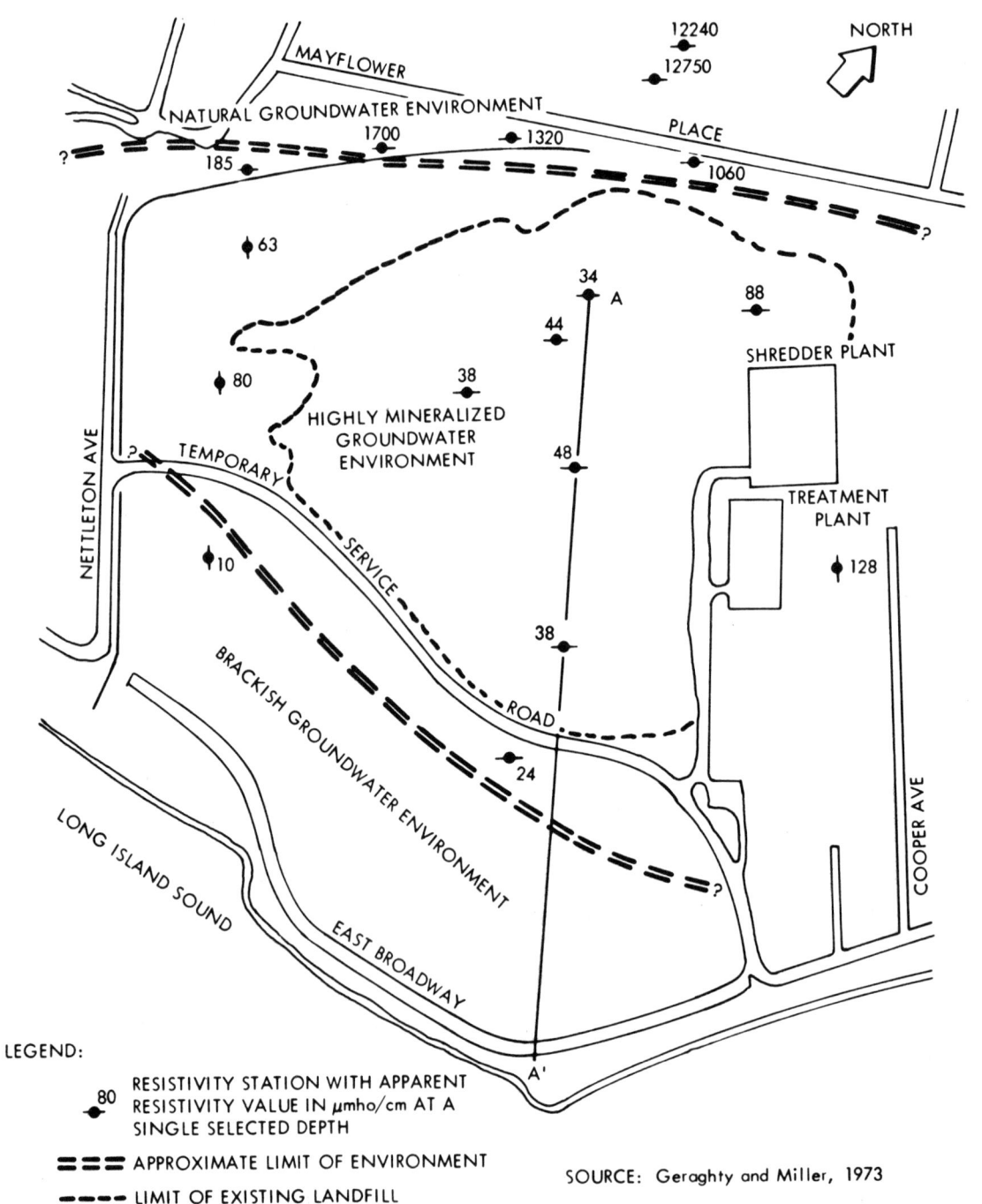

Figure 6-6. Three major groundwater environments as delineated by interpretation of resistivity data 15 to 20 feet below land surface, Milford, Connecticut.

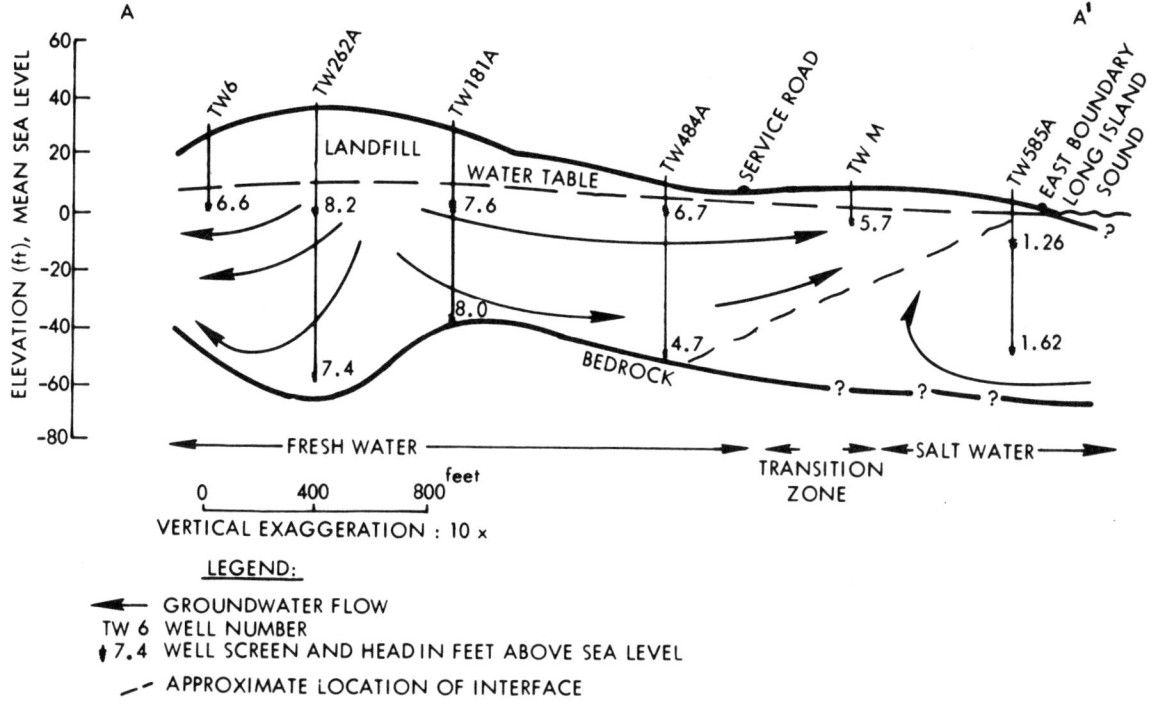

Figure 6-7. Schematic hydraulic profile along section A-A' of Figure 6-6, Milford, Connecticut.

southwest from the landfill discharges almost immediately into the surface waters in the marsh and the small streams that drain the area. In some parts of the marsh, the water table is above land surface and has formed ponds containing leachate.

The investigation showed that some of the groundwater has moved vertically downward to invade deeper zones of the unconsolidated material directly below the landfill. The quality of the water at these depths is much better than that of the water in the landfill materials, suggesting that the underlying fine-grained sediments have been at least partly effective in attenuating the contamination. Several hundred million gallons of groundwater have been slightly to heavily contaminated by leachate at the site. A water-budget calculation indicates that about 80,000 gallons per day of water derived from precipitation is recharged into the landfill materials, and that an equivalent amount of contaminated fluid discharges through the bottom and sides of the landfill.

Mapping of Plume

Initially, background information was reviewed on topography, vegetation, records of wells and borings, rainfall, tides, and surface drainage. Following this, seismic and electrical-resistivity geophysical methods were utilized to give a preliminary idea of the character and thickness of the materials at the site. In the next step, 36 test observation wells were drilled in and around the landfill. At some of the well sites, two wells (one deep and one shallow) were installed to define vertical head relationships. The wells ranged in depth from 12 to 96 feet.

Periodic measurements of water levels, referenced to sea level, were made in all wells. Water samples were collected from the wells for chemical analysis, and where the water levels were below suction lift, the samples were collected by first blowing water out of the well casings with compressed air, and then bailing out the water sample. A field laboratory was set up to make determinations of some chemical constituents, and samples were sent to a certified laboratory for more detailed analyses.

Water temperature profiles were made in the deeper wells by means of an electronic thermometer. Specific-conductance and dissolved oxygen determinations were made on water samples from principal surface water bodies. Complete chemical and bacteriological analyses were run for some of these samples.

A series of shallow gas sampling tubes was installed in the landfill at depths ranging from 3 to 5 feet. Analyses were made to determine the percentage of methane in the gas mixture and its explosive levels.

The vegetation was studied by biologists from a local research institute to define stresses on the vegetation and relationships with the groundwater system. Color, stereo, and multispectral photographs were taken of the landfill and the surrounding area and used to construct base maps, establish topographic contours, and define vegetative patterns. The abnormally high water table, contaminated groundwater, and insect infestation accounted for most of the stress on the vegetation.

WHAT COULD HAVE BEEN DONE

The information collected during the investigation provided adequate answers to the questions asked by the State, and led to the shutting down of the landfill. Funds were not allocated in the first stage for a more intensive monitoring effort, simply because it was not needed. However, a number of other monitoring steps could have been taken for general research purposes, or to provide data that might have proved useful, if an early decision could have been made on covering the landfill.

Gas Generation

The investigation provided some information on the generation of gases within the landfill. However, no detailed observations were made to define which types and concentrations of gases were being generated, where they were concentrated, and the pressure distribution. More studies could have been conducted along these lines, since, ultimately there would have to be some requirement for venting gases in order to permit multipurpose use of the landfill area.

Additional Chemical Determinations

Although numerous common chemical constituents and a few heavy metals such as iron, lead, copper, and manganese were determined in the water samples, little attention was given to the possible presence of other toxic heavy metals and trace elements. At some additional expense, these could have been determined through spectrographic analysis, which might have helped detect other toxic constituents possibly responsible for some of the ecological damage.

Bacteriological Studies

Coliform bacteria found in nearby surface-water bodies were believed to be largely, if not entirely, from contaminated materials in the landfill. However, because of the great difficulty in sterilizing pumps and wells and disinfecting the environment around the wells, it was not considered feasible to collect samples of groundwater for bacteriological analysis. With sufficient funds, time, and suitably constructed wells, it would have been possible to study this aspect of the problem.

Additional Test Wells

Samples taken from the limited number of wells drilled directly within the landfill materials showed a wide divergence in chemical composition, owing partly to differences in the types of materials placed at different locations within the landfill and partly to dilution of the contaminated water. It would have been useful, therefore, to have installed a denser grid of sampling wells in order to define particular hot spots of highly contaminated water. The results of such sampling might have helped locate the sources of particularly objectionable contaminants. However, problems in constructing wells in the landfill materials and the fact that the landfill was still in operation during the test program made it impractical to fully explore the entire landfill area.

Infiltration Rates

Another useful procedure would have been to prepare a more accurate and detailed water budget for the landfill area to determine the rate of leachate production. Field measurements of infiltration, runoff, and evaporation would have been useful in this regard. However, the estimates made were considered to be reasonably useful for the purpose of the investigation, since the landfill was still active at the time, and the rates of leachate production were probably variable.

WHAT SHOULD HAVE BEEN DONE

Because of the intention to convert the landfill area into a State park, more intensive monitoring of stresses on the vegetation, during the field study and following the investigation, should have been planned. This additional work was not authorized, consequently, although the landfill is covered over and converted into a recreational area, monitoring has ceased, although the original test wells are still in place. Additional monitoring of vegetative stress and leachate discharge would prove of considerable value in showing the effectiveness of the landfill cover and the slow changes anticipated as the production of leachate slowly diminishes.

POLLUTION POTENTIAL OF AN OXIDATION POND NEAR TUCSON, ARIZONA*

The Ina Road oxidation pond site is located about 10 miles northwest of Tucson, Arizona, in SE1/4, Section 1, T13S, R12E (see Figure 6-8). The site abuts the Santa Cruz River, the principal drainage channel of the Tucson Basin, and is immediately downstream of the confluences of Canada del Oro and Rillito Rivers. Discharge in these channels is primarily ephemeral; however, the Santa Cruz River drains the entire effluent discharge from the City of Tucson Treatment Plant about six miles upstream, as well as overflow from the Ina Road ponds.

The Ina Road ponds, serving as principal treatment facilities for sewage in northwest Tucson, are managed by The Metropolitan Utilities Management Authority. These ponds will be replaced in the near future by a regional treatment plant of standard design. A sanitary landfill is located along the Santa Cruz River immediately to the southwest of the ponds.

---

*The work reported in this subsection was supported in part by a grant from The Office of Water Research and Technology, U.S. Department of the Interior, Washington, D.C.

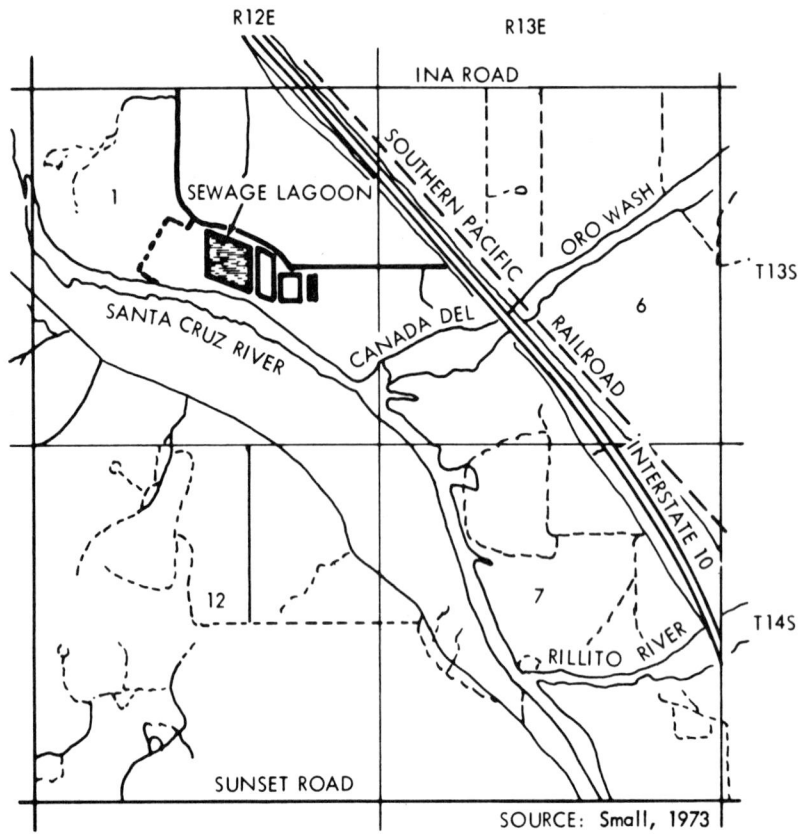

Figure 6-8. Location of pond near Tucson, Arizona.

Sediments underlying the pond site are typical of the basin and range physiographic province. Specific geologic units and their water-bearing properties are discussed by Davidson (1973). The source materials for these sediments are volcanic rocks from the Tucson Mountains, immediately to the west, and the granitic rocks of the Santa Catalina and Tanque Verde Mountains to the east. Principal aquifers in the region comprise surficial material, Fort Lowell formation and the Tinaja Beds. For the Ina Road site, Randall (1974) estimated the transmissivities of the aquifers comprising the surficial deposits to be between 150,000 - 300,000 gallons per day per foot. The corresponding transmissivity of the combined Fort Lowell and Tinaja Beds aquifers was estimated to be 35,000 gallons per day per foot. Depth to the water table at the time of the tests was about 70 feet.

Groundwater flow in the vicinity of the pond is predominantly in a northwesterly direction, corresponding to underflow in the Santa Cruz River, but is moderated by southwesterly flow in the Canada del Oro system. Similarly, groundwater quality reflects two distinct sources: underflow and recharge in the Canada del Oro and Rillito systems; and underflow and recharge in the Santa Cruz River, including the contribution of sewage effluent (Schmidt, 1972a).

An additional complicating factor is that shallow perching layers within the vadose zone may conduct water laterally at substantial rates (Wilson and DeCook, 1968). If such layers are hydraulically connected to the Santa Cruz River, sewage effluent may move laterally from the river into the pond area, eventually leaking into the water table. Leaching of the landfill deposits by river seepage is also a distinct possibility, and is the subject of a study (Wilson, 1974).

MAP OF NITRATE LEVELS

Nitrate was the principal ion considered in the pond study. In general, groundwater quality in the region downstream from the City of Tucson Treatment Plant is noted for localized, high concentrations of nitrate (Matlock et al. 1972; Schmidt, 1972a). Such nitrate may have originated from sewage effluent recharging in the Santa Cruz River, recharging of effluent applied during irrigation of cropland, leaching of nitrogenous fertilizers, leaching of indigenous nitrogen, or leaching of landfill deposits.

One of the major purposes of the study by Wilson et al. (1973) at the pond site was to monitor the movement of nitrate during deep seepage, particularly during the period immediately after the pond was placed into operation. Wilson et al. found no positive evidence that lagoon seepage had resulted in nitrate contamination of groundwater in the area.

The areal distribution of nitrate in wells within the area encompassing the pond in October 1971, three months after the pond was initiated, is shown in Figure 6-9. Data were obtained by the Department of Soils, Water and Engineering, University of Arizona. The two wells sampled in the southeast quarter, upstream of the pond, contained nitrate levels of 72 milligram/liter and 48 milligram/liter. Wells in the northwest quarter, downstream of the pond, generally contained nitrate concentrations of about 30 milligram/liter except for the furthermost northwest well, with a level of 45 milligram/liter.

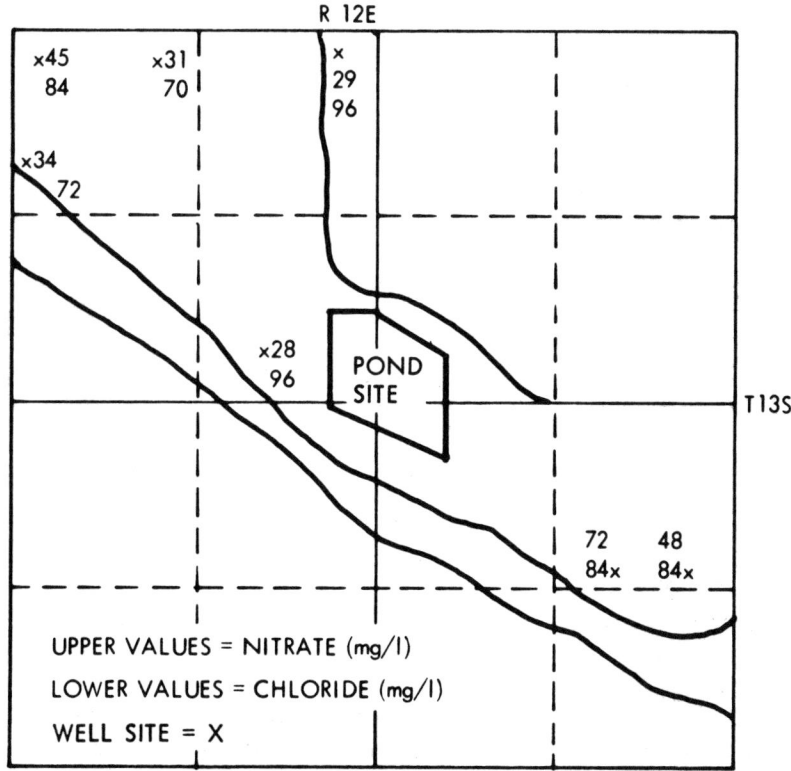

Figure 6-9. Nitrate and chloride distribution in wells near the pond site, October 1971.

An irrigation well 300 feet downstream of the pond (see Figure 6-10) contained 28 milligram/liter nitrate in October 1971. The vertical distribution of nitrate in the profile beneath the pond after several months of pond operation is reported by Wilson et al. (1973). Subsequent nitrate values on 19 April 1973, within the same profile reported by Wilson and Small (1973) were: 60-foot PVC well 27.28 milligram/liter, No. 1 access well 7.48 milligram/liter, and irrigation well 18.04 milligram/liter. In 1974, the PVC wells were dry and could not be sampled. Nitrate levels in the No. 1 and No. 2 access wells were reported by Wilson (1974) to be 1.2 milligram/liter and 1.5 milligram/liter respectively. Pump water from the irrigation well contained 11.7 milligram/liter nitrate.

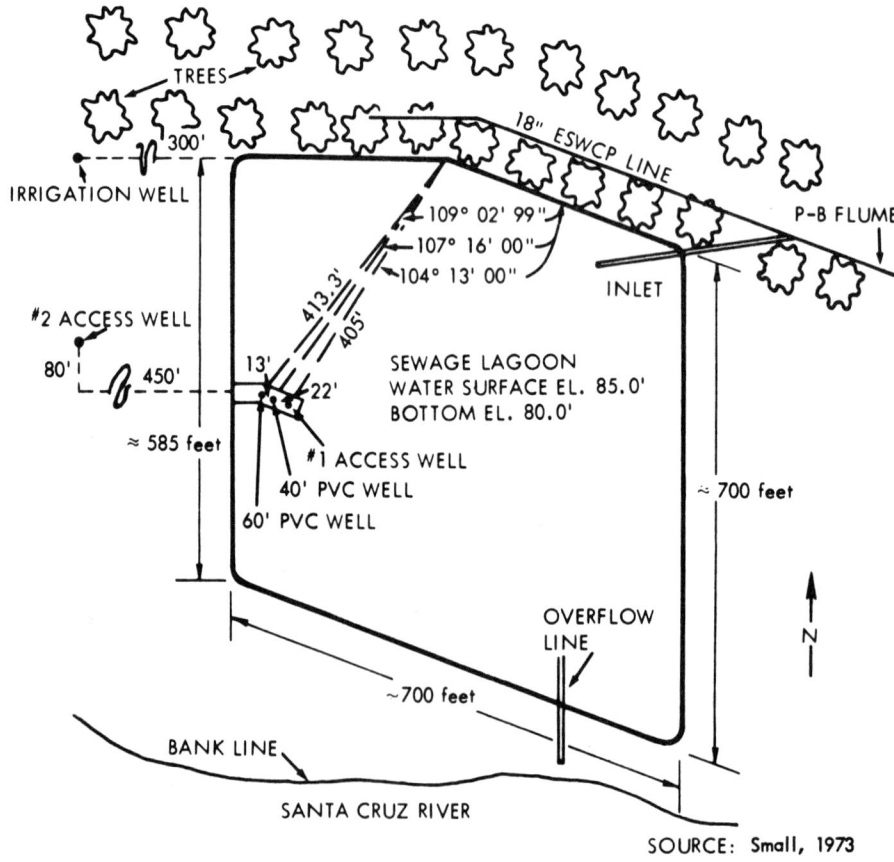

Figure 6-10. Location of monitoring facilities at the pond site.

RATIONALE OF PROJECT

The monitoring program at the pond site was based on experience from prior studies. One of these studies was conducted at the site before construction of the pond. Other studies were conducted at the University of Arizona Water Resources Research Center (WRRC) field laboratory about 6 miles south of the pond. The latter studies involved investigations of artificial and natural recharge, with particular references to mechanisms of water movement in the vadose zone and water quality changes during such movement. Results of some of these studies were reported by Wilson and DeCook (1968) and Wilson (1971). Monitoring facilities included observation wells, pumping wells, shallow piezometer-water sampling wells, and access wells. By means of neutron logs in the latter wells, the growth and dissipation of two perched water tables in the sediments overlying the principal aquifer were clearly observed. The existence of such perched tables in surficial deposits of the Tucson Basin has been known for many years. The advantage of neutron logging is that the location and behavior of the tables can be followed. Furthermore, knowing the location of regions in which tables develop, it is possible to terminate sampling wells in sediments which saturate during recharge. Based in part on such reasoning, two batteries of sampling wells were installed at the WRRC recharge site. Four wells in each battery terminated within the vadose zone, and a fifth terminated below the water table.

One observation of interest from the recharge study was that water moved very rapidly (up to 200 feet per day) in the perched layers. In fact, it appeared that the upper layer served, essentially, to spread water for a considerable lateral distance away from the recharge source, with leakage into the lower layers. Furthermore, samples from the wells in the vicinity of the water table showed a gradual displacement of native groundwater by recharge effluent, suggesting that recharge effluent flows along the top of the main, but slower moving, water body. Mixing then takes place by a variety of physical and chemical processes.

The experience gained during the installation and operation of the facilities at the WRRC site prompted the installation and operation of similar facilities at the oxidation pond site, during investigations of grass and soil filtration in 1967-1968. Results of these investigations were reported by Wilson and Lehman (1967) and Lehman (1968). Basically, the studies involved metering oxidation pond effluent onto three Bermuda grass plots, each 25 feet by 1,000 feet. Two 100-foot access wells were installed on the central plot. Neutron logging in these wells during preliminary experiments seemed to indicate the presence of two perched layers within the vadose zone, one at about 30 feet below land surface, and a deeper layer at about 50 feet. Two 4-inch-diameter PVC wells terminating at depths of 40 and 60 feet were installed by the cable-tool method. Each well contains a 4-foot-long plastic well screen. PVC was selected as material for these wells to minimize interference during studies involving the monitoring of heavy metals, during effluent irrigation. Drill cuttings were obtained during construction of the wells, but unfortunately, except for the upper 8 feet, these cuttings were not examined for physical or chemical makeup.

In addition to the PVC wells and access wells, the central plot was instrumented with three sets of four suction cup batteries, extending 2 feet below ground surface. These units were installed to permit soil solution sampling in the unsaturated state.

Results of flooding trials showed that grass filtration was not particularly effective as an overall, tertiary-treatment technique, compared with soil filtration. As expected, effluent arriving in the two PVC wells contained excessive levels of nitrate (in excess of 90 milligram/liter), but phosphate concentrations were lowered.

In 1970, the Pima County Department of Sanitation began plans for a new 10-acre stabilization lagoon, which would encompass several hundred feet of the grass plots. The County and City had received unfavorable publicity a short time before this period, due to a threatened lawsuit by a homeowner near the ponds. The homeowner claimed that he could not drill a well for fear of nitrate contamination. Although the matter was settled before the new pond was due to be constructed, the University approached the County Department of Sanitation with the suggestion that a joint study be undertaken to monitor seepage from the new pond, with particular emphasis on nitrate movement. Fortunately, the two PVC wells and one access well were close enough to the western dyke of the lagoon that it was possible to construct a platform to reach them. The resultant arrangement is shown in Figure 6-10. Not shown in Figure 6-10 are two batteries of ceramic suction cups, one located on the western side of the pond, and the second near the eastern side.

With the physical arrangement of wells shown in Figure 6-10, it was conjectured that a fair representation of the vertical changes in effluent quality could be obtained during deep seepage. Thus, the shallow suction cups would provide samples of soil solution, the two PVC wells would sample from mounds within the vadose zone, the access wells would sample just below the water table, and the irrigation well would provide an integrated sample from its perforated region of 80 feet to 278 feet. Also, the two access wells, one within the pond and one outside, offered the opportunity to detect the lateral movement of water in the vicinity of the main water table.

WHAT COULD HAVE BEEN DONE

Additional steps could have been taken to upgrade the monitoring program, but were not because of limited funds.

An interdisciplinary team could have been assembled to ensure that all parameters of significance would be monitored during the study. The following disciplines would have been desirable: soil chemistry, soil microbiology, soil physics, sanitary engineering, aquatic biology, and hydrology. An interdisciplinary team would have allowed relating changes in physical, chemical, and biological properties of the aqueous environment of the pond to corresponding changes in effluent, during flow across the benthic-soil interface and deep seepage. Some of the specific parameters, which could have been monitored by the team, are presented below.

Since the pond site is located in a region of complex hydrogeology at the confluences of the Canada del Oro and Santa Cruz Rivers, bulk groundwater flow from the two systems creates a complex effect on flow patterns and water quality. Also, the effect of flood recharge and inflow of sewage effluent and landfill leachate, through perching layers, should be taken into account. Therefore, a thorough hydrogeological study would have helped to delineate and separate the interrelated effects, and thereby assisted in interpreting results.

Standard techniques for hydrogeological studies (Walton, 1970) could have been employed to delineate sources and sinks, boundary conditions, and other important characteristics. In addition, test wells could have been constructed to provide drill cuttings for particle-size analyses and chemical composition. Resistivity and seismic surveys and down-hole gamma and neutron logging would have been included. Results would have been carefully examined for more precise delineation of possible perching layers in the vadose zone, as well as regions of varying permeability below the water table.

As part of the hydrogeologic study, the team could have attempted to trace inflow of sewage effluent from the Santa Cruz River by introducing suitable dyes upstream of the pond and collecting samples of well water. (Unfortunately, chloride levels are about the same in all sources.)

Based on the hydrogeological studies, additional monitoring facilities could have been installed around the pond. For example, from moisture logging (i.e., neutron probe) data, additional shallow wells could have been constructed down to perching layers. These wells would have allowed more accurate examination of the lateral spread of pond effluent through the layers. Similarly, better estimates could have been made of the mixing, in these layers, of inflowing sewage effluent (and possibly landfill leachate), natural recharge, and downward flowing pond effluent.

Several deeper observation wells could have been constructed around the pond. Additional information on aquifer transmissivity and storability could have been obtained by pumping these wells, and water samples could have been extracted from various zones beneath the water table before and after initiating the pond operation. Data from such a program would have provided a picture of the vertical distribution of quality (e.g., nitrate content) below the water table, and indicated the effects of dispersion and other important information.

A ring of access wells would provide moisture content data via neutron moisture-logging, to facilitate water balance studies.

The drill cuttings obtained as part of the hydrogeological study could have been examined for chemical composition. In particular, the concentration of indigenous nitrogen and phosphorous in saturated extracts from the cuttings would have indicated the vertical distribution of these constituents in the vadose zone.

The carbon to nitrogen (C:N) ratio of sludge within the benthic layer of the pond could have been evaluated periodically to determine the effects of changes in this ratio on mineralization of nitrogen (see Miller, in Sopper and Kardos, 1973). Similarly, soil cores from the soil-benthic interface could have been taken to determine changes in organic matter, cation exchange capacity (CEC), and exchangeable cations [particularly ammonium-nitrogen (NH_4-N)]. Wilson et al. (1973) hypothesize that increase of soil organic matter by penetration of sludge would increase the CEC. Changes in NH_4-N or nitrate-nitrogen (NO_3-N) levels in shallow tensiometer samples could have been examined for a relationship between organic matter content and CEC.

Vertical movement of heavy metals and organic toxins originating in pond effluent could have been monitored in shallow tensiometers, deeper PVC wells, and access wells.

An attempt could have been made to install a system of electrodes in the soil-benthic interface for monitoring redox potential. However, as pointed out by Ellis (in Sopper and Kardos, 1973), in-situ determination of redox has not proven to be successful.

Wells constructed for the sampling program could have been used to obtain data for the development and calibration of computer models. In particular, data could have been obtained to provide realistic estimates of dispersivity coefficients of aquifer materials.

At the time of the study (1971), a few finite difference models were available to simulate groundwater flow. Mass transport effects were handled by the method of characteristics. Today finite element models are being developed to simulate joint hydraulic-mass transport phenomena in aquifer systems. Such models could be adapted to the pond site.

WHAT SHOULD HAVE BEEN DONE

Some monitoring programs or techniques that should have been used in the pond study were not included because of lack of insight or lack of time.

A hydrochemical balance, albeit gross, of the groundwater system of the area should have been conducted before the pond was placed in operation. A fair amount of data was readily available for developing such a balance. For example, the Department of Soils, Water, and Engineering at the University of Arizona had been involved in obtaining hydraulic data of the Tucson aquifer system for a number of years. These data could have been examined to estimate flow trends in the vicinity of the pond. In the early 1970s, the same department was also actively involved in collecting chemical data from well-water samples. These data should have been examined and used to construct trilinear diagrams and other chemical analysis formats.

Much data could have been collected from wells in the area; a basic program to monitor water levels and quality in nearby domestic and irrigation wells should have been established. Water levels and chemical data were obtained in access wells and the irrigation well at the pond site for several months before the pond was put into operation, but this program should have been expanded.

The major oversight in the pond study, vis-à-vis the hydrochemical balance, was in not monitoring seepage of effluent in the Santa Cruz River. Later studies by Wilson and Small (1973) showed that intake rates of sewage effluent in the reach of the Santa Cruz River along the pond site are substantial, ranging from 1.5 feet per day to 7.7 feet per day. A program to monitor trends in the quality of river effluent, including, for example, total nitrate and boron, should have been implemented before and after the pond was placed in operation. Resultant data could have been used with groundwater data to construct trilinear diagrams.

One other possible source of subsurface inflow into the area was not examined, namely, deep seepage from irrigation across the river from the ponds. In particular, the movement of leached fertilizers should have been considered.

Pond overflow should have been metered or sampled. The rationale at the time was that the various transformations within the aqueous system of the pond were of interest only as the pond filled. When the groundwater monitoring program was continued, a metering device should have been installed on the pond overflow, and a sampling program established. A nitrogen balance of the aqueous system could have been conducted, subsequently, to relate to changes in the groundwater system. For example, since the pond frequently shifted from an aerobic to an anaerobic state, valuable data could have been obtained on nitrification-denitrification processes. Samples from the shallow suction cups might have reflected these processes. Also, a meter on the overflow line would have allowed calculation of long-term intake rates.

In addition to monitoring seepage rates in the pond by the gross inflow-change in storage technique, seepage measurements should also have been attempted at several locations via seepage meters, infiltrometers, and other means. The measurement results should have been related to soil core data on such items as bulk density and particle size distribution. (Consideration was given to mounting one or two seepage meters permanently near the platform in order to relate seepage rates to development of the benthic layer at precise locations.)

Although the primary purpose of the study was to monitor the movement of nitrogen species, the chemical data (see Table 2 of Wilson et al., 1973) showed that the total phosphate increased from 3.7 milligram/liter to 52 milligram/liter in the 40-foot PVC well, and from 6.7 milligram/liter to 24.5 milligram/liter in the 60-foot PVC well. Normally, migration of phosphate in soils and groundwater systems is not considered a problem, and soil filtration studies at the site showed a diminution of phosphorus during soil filtration. Consequently, the observed trends should have prompted additional sampling for phosphate beyond the period of the study. Furthermore, the forms of phosphorus, i.e., organic versus ortho, or condensed, phosphate, should have been determined. A soil microbiologist could perhaps have related the mobility of indigenous, or effluent, phosphate to soil transformations in the soil-benthos region and underlying zones. (As pointed out by Ellis in Sopper and Kardos (1973), a soil under reducing conditions will not adsorb as much phosphorus as the same soil would in well-aerated conditions.)

Although it is known that soils are capable of the chemical filtration of boron (Ellis, in Sopper and Kardos, 1973), there are cases reported where boron levels increase in the soil solution, during irrigation with sewage effluent (Bouwer, in Sopper and Kardos, 1973). Therefore, the movement of boron should have been monitored in the well system at the pond site.

Additional technical details relating to monitoring should have included:

- Checking the interaction, if any, of the ceramic materials used for suction cups with nitrogen, phosphorus, boron, or other elements
- Installing a system of tensiometers to measure soil moisture tension near ceramic cups. This would have ensured applying the proper suction so as not to affect moisture flow
- Taking samples periodically to other laboratories as a quality control measure
- Monitoring groundwater temperatures in the network of wells

MULTIPLE-SOURCE NITRATE POLLUTION IN THE FRESNO-CLOVIS, CALIFORNIA, METROPOLITAN AREA

BACKGROUND

The objective of the Fresno-Clovis study was to determine the extent of nitrate pollution in the groundwater, the sources of pollution, and time trends in nitrate content of water pumped by wells.

The Fresno-Clovis Metropolitan Area (FCMA) is a predominately urban area of 145 square miles in the central San Joaquin Valley of California (Figure 6-11). The surrounding lands are agricultural, and rainfall averages about 11 inches per year. Groundwater occurs in permeable alluvial deposits, and water levels average about 70 feet in depth. Wells average several hundred feet in depth, with yields exceeding 1,000 gallons per minute being common. Groundwater is the sole source of water supply in the urban area, whereas irrigation demand in the surrounding area is supplied by both canal water and groundwater. The primary means of liquid waste disposal, other than evapotranspiration, is by percolation, as there are no significant discharges to surface water.

Major sources of nitrogen include septic tank effluent in unsewered areas, sewage effluent, leakage from sewers, fertilizers, and meat-packing plant and winery wastes. Natural sources of nitrogen do not appear to be of major significance, and background levels of nitrate in the aquifer are less than 10 parts per million.

SUMMARY OF THE MONITORING PROGRAM

The two major constraints on the project were time and funding. The project was a doctoral dissertation, the time available was about 1 year, and no grant funds were available on such short notice. Maximum use had to be made of existing data and the cooperation of local individuals and agencies. Limited personal funds were used for research (mainly photocopying and kits for chemical quality determinations in the field).

The monitoring phase of the project encompassed the following steps:

1. Determination of the extent of data on groundwater and water quality in the proposed study area. Sufficient data were available to warrant proceeding with the program.

2. Completion of an exhaustive literature review on the pollutant of interest, in this case nitrogen or nitrate. Studies of sources of nitrogen, and the occurrence of nitrate in soils and groundwater were reviewed (American Water Works Association, 1967; Schroepfer and Polta, 1969; and Stout et al., 1965).

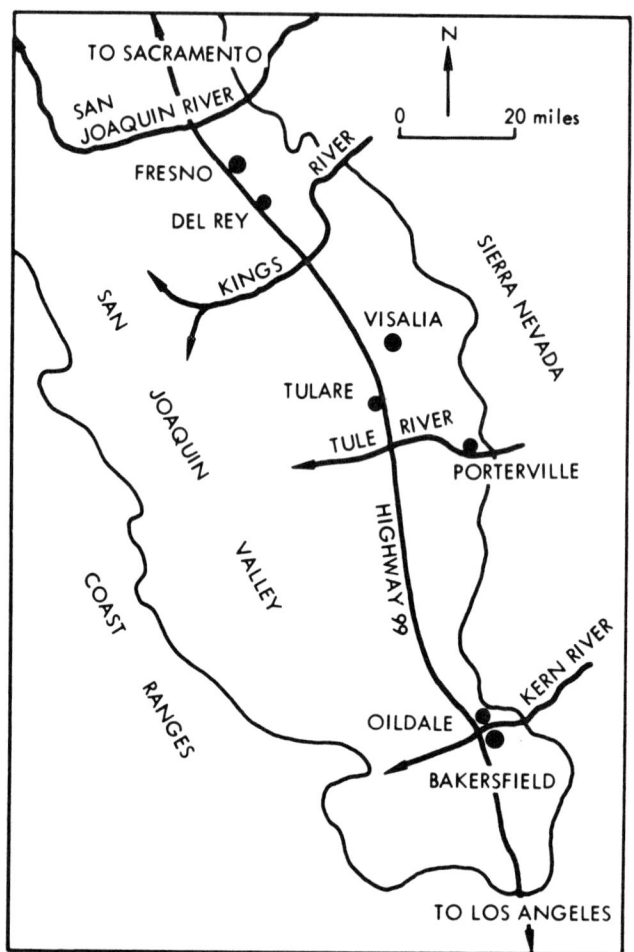

Figure 6-11. Map of part of the San Joaquin Valley, California.

3. Collection of all available reports and data on (a) groundwater, (b) soils, (c) well data, (d) pollution sources, and (e) chemical analyses of groundwater and pollution sources in the study area (Behnke and Haskell, 1968; California Department of Water Resources, 1965; Nightingale, 1970; and Page and LeBlanc, 1969).

4. Preliminary evaluation of the areal distribution of sources of nitrogen and nitrate in groundwater.

5. Collection of supplementary data to fill in gaps, such as water samples for more extensive areal coverage of groundwater quality, recent hydrogeologic records, historical development of nitrogen sources, and chemical analyses of waste waters.

The remainder of the project was interpretation and report preparation (Schmidt, 1972b and 1975).

Collection of Additional Water Samples

High capacity wells (500 to 2500 gallons per minute) were selected for water sampling at the discharge, after prolonged pumping. Localized situations, such as the effect of a septic tank or lawn fertilizer on groundwater beneath one lot, were not of concern in this study. Low capacity domestic wells (less than 50 gallons per minute), pumping for short time periods, reflect very localized conditions. Water samples taken from high capacity wells, after long periods of pumping, are much more indicative of regional conditions, which were of interest in this investigation. Figures 6-12 and 6-13 illustrate the areal distribution of chloride southwest of Fresno, as determined from analysis of water samples from high capacity wells. Chloride was evaluated in this case because of its use as a tracer in this area, it being present in waste waters, but almost absent in native groundwater. In the case of point or line sources, high capacity wells can be used for monitoring at a distance of several hundred or thousand feet from the source. This is due to the lateral extent of the cone of depression after prolonged pumping.

Because nitrate content varies vertically in groundwater, well construction is an important parameter in the selection of monitor wells. Figure 6-14 illustrates the vertical distribution of nitrate in a septic tank area of the FCMA, as determined from pumping of open-bottom (unperforated) wells. Highest nitrates occur in the shallowest part of the aquifer. Because of this vertical stratification of water quality, nitrate contents of well water often change with pumping time over short time periods. Short-term trends in some cases can be plotted as straignt-line relations on semilogarithmic graph paper (Figure 6-15). Seasonal fluctuations were also considered in evaluating chemical analyses of water samples from monitor wells (Figure 6-16). Short-term and seasonal trends must be established before long-term time trends can be evaluated.

As a number of individuals and agencies operate wells within the FCMA, and no uniform monitoring program for groundwater quality exists, the existing chemical analyses are often for water samples taken at different times. It was desirable to sample many wells over a period of several days to several weeks to establish the areal and vertical distribution of nitrate at a specific time. The optimum sampling time was during the warmest time of the year, when the maximum pumpage occurred from high capacity wells. This served two purposes: (1) sampling was much easier, with almost all wells already pumping, and (2) the most typical chemical quality of the regional groundwater could be monitored.

Several hundred municipal wells were sampled by one individual within two or three days. This was because only five or six agencies operated these wells. Sampling of irrigation wells southwest of the urban area took much longer because of access problems (poor roads), individual ownership in many cases, and lack of a faucet or open discharge. Because nitrate content can change with storage time, determinations were made immediately after collection. Electrical conductivity, water temperature, and chloride content were also measured.

Site-Specific Data and Interpretation

The FCMA was subdivided on the basis of the predominant nitrogen source and on hydrogeology and the areal distribution of nitrate in the aquifer. This subdivision (Figure 6-17) was probably the key aspect of the entire study. The Figarden-Bullard area and Fresno sewage treatment plant were selected because of the predominance of one nitrogen source in each case (septic tanks and sewage effluent, respectively). Evidence gained from studies in these areas was used in the Tarpey Village and Mayfair-Fresno Air Terminal areas, where both sources, as well as others, were present. The latter two areas had the highest nitrates in groundwater of the urban area. The Downtown Fresno area had no obvious source, but high nitrate contents were present in the aquifer. Each area had distinct hydrogeologic conditions with respect to the other areas.

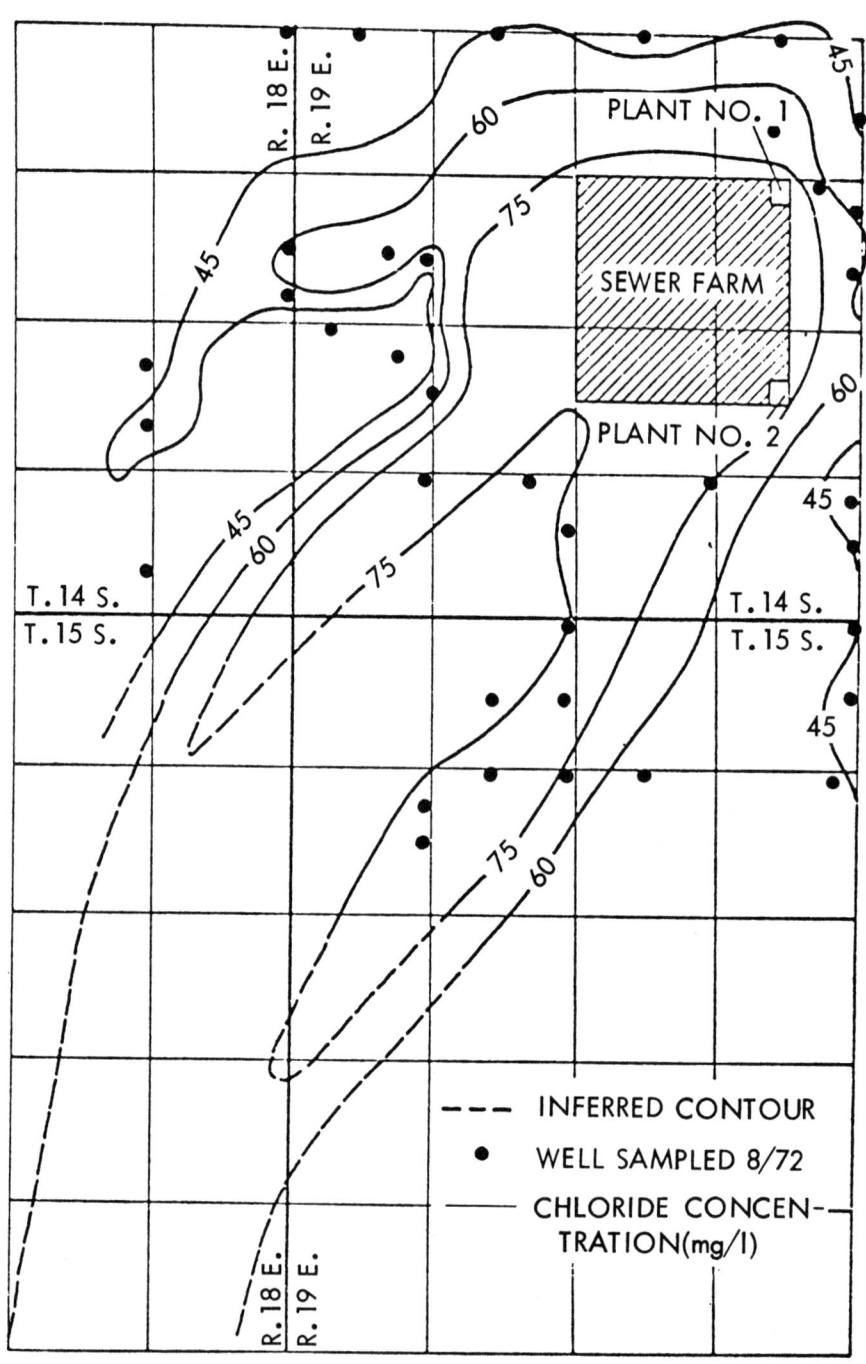

Figure 6-12. Chloride concentration contours (mg/ℓ) in groundwater at and downgradient of Fresno sewage treatment plant.

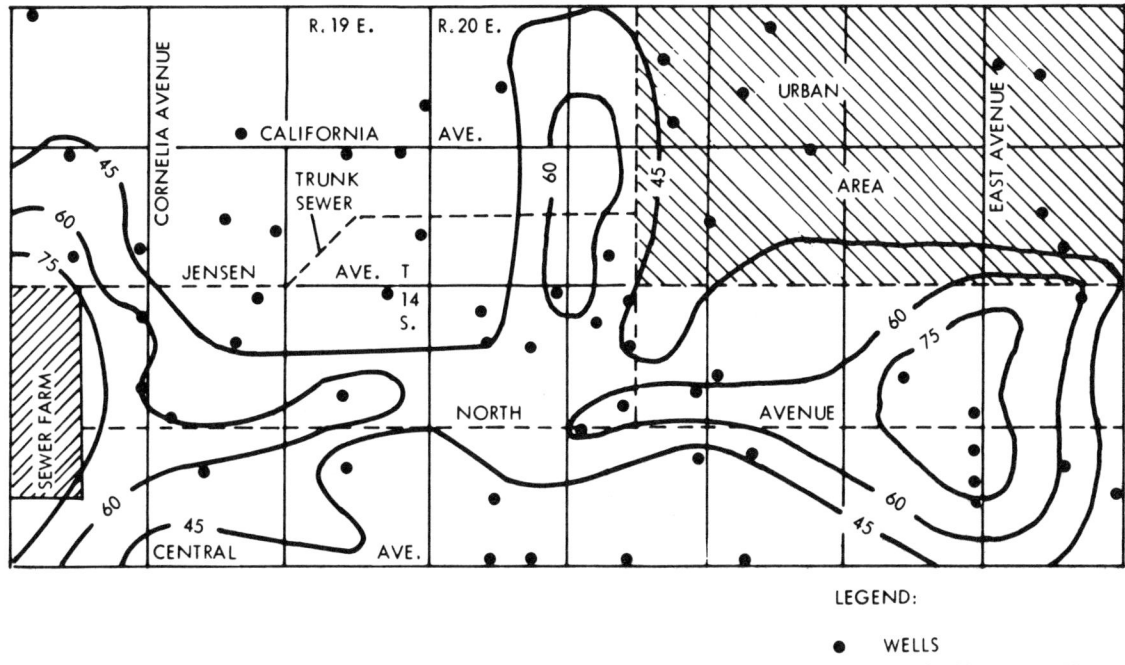

Figure 6-13. Chloride concentration contours (mg/ℓ) in groundwater east of the Fresno sewage treatment plant.

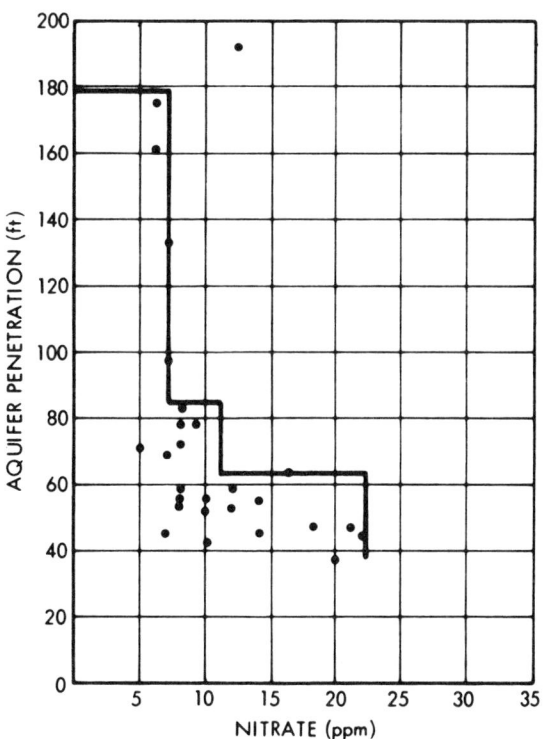

Figure 6-14. Relation between aquifer penetration and 1970 nitrate for wells in Figarden-Bullard area.

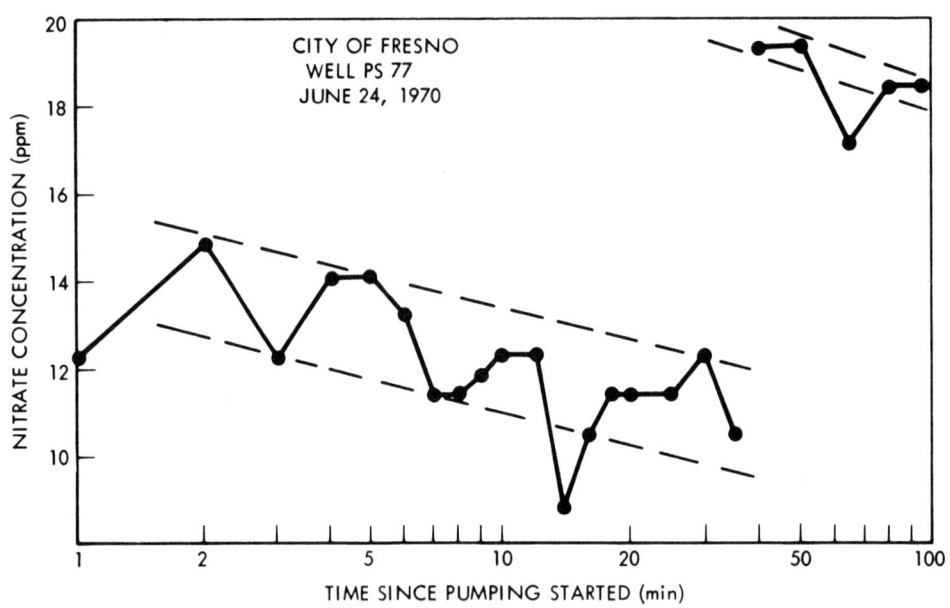

Figure 6-15. Short-term trends in nitrate during pump test on a large-capacity well in FCMA.

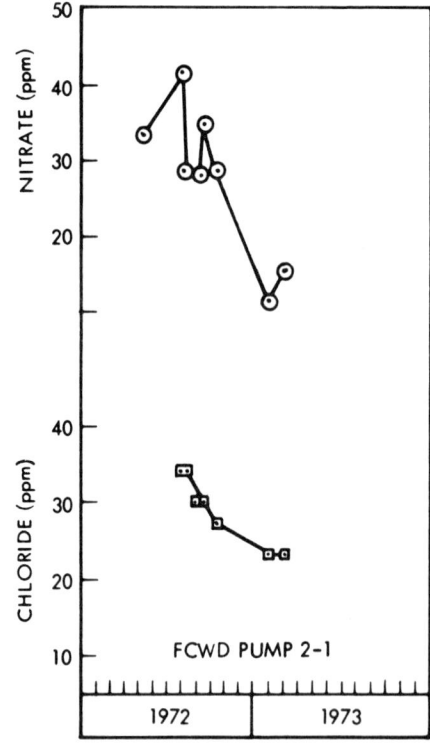

Figure 6-16. Seasonal trends in nitrate and chloride for a large-capacity well in a septic tank area.

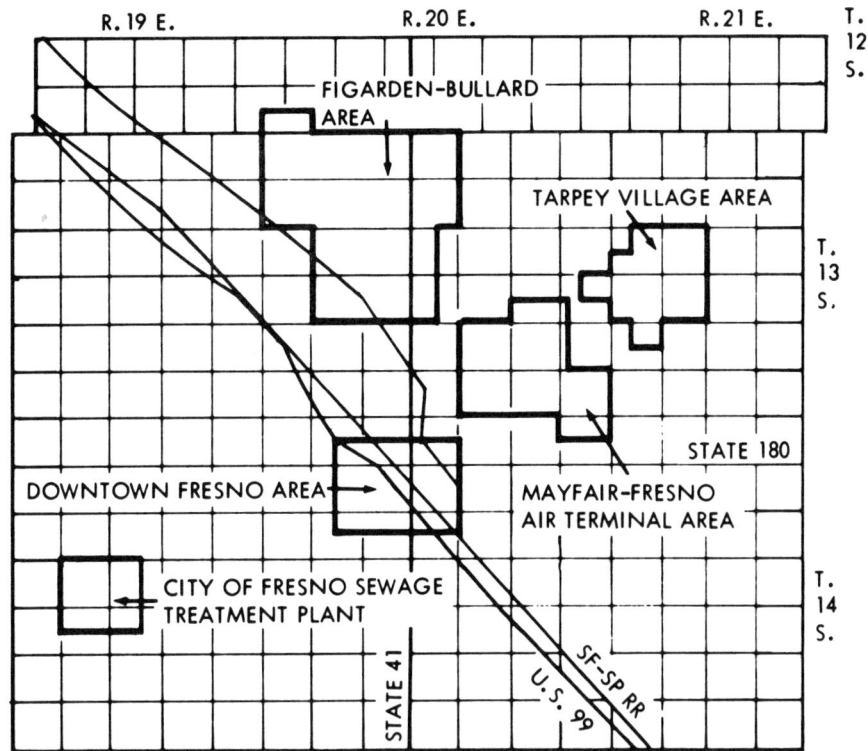

Figure 6-17. Study areas in the Fresno-Clovis Metropolitan area.

In areas of diffuse sources, such as septic tanks or fertilizers, semiannual analyses are often sufficient to detect seasonal trends. However, near point sources of large volume, such as the Fresno sewage treatment plant disposal ponds, weekly or monthly sampling of wells is necessary (Figure 6-18). The density of existing wells is usually sufficient to delineate the areal water quality distribution. Open-bottom wells drilled by the cable-tool method are especially valuable, as they tend to draw water from specific depth zones, and thus give an indication of vertical stratification of groundwater quality. Other constituents, such as chloride, potassium, ammonium, and calcium, are valuable in differentiating among various sources of nitrate. Trilinear diagrams can be prepared for waste waters and groundwater to illustrate similarities in chemical types of water. Historical chemical analyses in the FCMA have documented the buildup of nitrate in groundwater due to the development of nitrogen sources at certain times.

Special Cases and Assumptions

The travel time of recharged waste waters from the land surface through the vadose zone to the water table is in terms of weeks, months, or several years in the FCMA. Calculations on the water budget indicate that, near point sources in particular, wastes must move rather rapidly to the water table. If not, there would be no storage space in the vadose zone for storage of these large volumes of water. Hydrographs of water levels and water quality data, as related to land surface phenomena, confirm that movement of water through the vadose zone is relatively rapid.

In the particular case of nitrate, which originates usually as organic or ammonium nitrogen at the land surface, there is no gross uptake in the vadose zone, as might be the case for certain trace metals. The fate of most of the nitrogen applied at the land surface is: (1) plant uptake, (2) denitrification and loss to the atmosphere, or (3) nitrification and leaching to the groundwater. In this particular case, natural nitrate contents are low, making it easier to detect nitrate groundwater pollution.

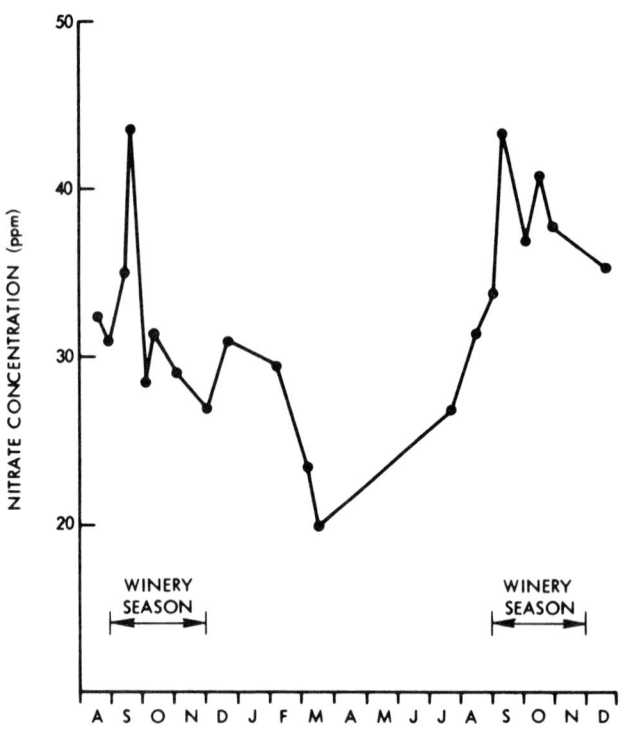

Figure 6-18. Nitrate concentrations in groundwater near Fresno sewage treatment plant.

Institutional Constraints

Little or no monitoring has been done by regulatory agencies, thus there was a lack of supplementary data, particularly on waste dischargers. Most polluters had an ingrained belief that they were not polluters, therefore, they tended to hesitate to sample or permit sampling. In addition, in the literature reviewed, the researchers often represent some interest, such as agriculture, and bias is sometimes evident. The literature, therefore, is confusing with regard to effects of specific pollutants on groundwater.

DESCRIPTION OF ALTERNATIVE MONITORING PROGRAMS

Additional monitoring would have been possible without the two major constraints of time and funding. If more money had been available, the following could have been done:

- Sampling of nitrate in soil moisture and the vadose zone in septic tank areas, as well as near some point sources

- Measurement of water movement in the vadose zone beneath point sources

- More detailed sampling of waste at the land surface, including determinations of viruses, stable organics, nitrogen forms, chloride, boron, and trace metals

- Test well drilling near some point sources, such as wineries and meat packing plants, where no nearby wells existed

- Possibly, use of stable nitrogen isotopes to differentiate among sources of groundwater nitrate

- More complete chemical analyses of groundwater, as many existing analyses were incomplete

If more time was available, the following could have been done:

- Establishment of seasonal trends in groundwater nitrate in more detail
- Evaluation of minor sources, such as lawn fertilizers
- Statistical correlation of septic tank density and groundwater nitrate content
- Calculation of water budgets and nitrogen budgets for various sources, including determination of evapotranspiration, percolation, denitrification, crop uptake, and other factors

STRENGTHS AND WEAKNESSES OF MONITORING PROGRAM

The major strengths of the investigation were the development of a detailed hydrogeologic framework in the area and the comprehensive use of water well sampling. The selection of high capacity wells for water sampling and chemical analyses permitted evaluation of regional groundwater conditions. Historical records compiled during the investigation permitted development of long-term time trends in groundwater nitrate related to nitrogen sources at the land surface. Other constituents, such as chloride, proved to be strong tools in determination of possible sources of nitrate. The thorough literature study preceding the field work was invaluable.

Many of the weaknesses of the investigation were due to the limited experience of the investigator, as well as time and funding constraints. Some sources were ruled out or considered negligible, without acquisition of data to support such a decision. Leaking sewers and lawn fertilizers should have been analyzed in more detail. Insufficient consideration was given to soils and the vadose zone. Too large an area was selected for study, and the Fresno sewage treatment plant and the agricultural area southwest of Fresno should have been studied separately. Reliable chemical analyses were unavailable for some pollution sources, and data on plant operation were lacking.

DESCRIPTION OF OPTIMAL MONITORING PROGRAM

An optimal monitoring program would have included the following:

- More land surface or source monitoring, specifically waste water sampling
- Test well drilling at selected point sources of pollution
- Study of the Fresno sewage treatment plant separately from the rest of the FCMA
- Sampling in the topsoil and the vadose zone, specifically measurement of the movement of percolating water and nitrate and chloride contents

Most of the additional types of monitoring would be costly. The least costly would be more monitoring at the land surface. Sampling in the vadose zone would probably have to be limited to a reconnaissance level, especially for diffuse sources. The existence of a monitoring program by regulatory agencies for waste waters and other potential sources of groundwater pollution would have greatly enhanced the study.

SECTION 3 – SITE-SPECIFIC GROUNDWATER QUALITY MONITORING EXAMPLES

Four potential categories of groundwater pollution have been selected to illustrate application of the major steps in development of the groundwater pollution monitoring methodology for site-specific conditions: agricultural return flow, septic tanks, percolation ponds, and landfills. Agricultural return flow represents one of the major potential sources of groundwater pollution in the western United States. Septic tanks are known to be a major source of groundwater pollution in some suburban areas. Both of these sources are generally diffuse, and thus monitoring programs for the two have certain similarities. Disposal and storage of various types of liquid wastes in ponds or pits subject to percolation represent another major potential source of groundwater pollution. Landfills for disposal of solid wastes can be major sources of groundwater pollution in humid areas. Percolation ponds and landfills are both point sources, and thus monitoring programs for the two have certain similarities.

After discussion of the salient aspects of monitoring each type of pollution source, an example illustrating the procedure for selecting site-specific monitoring alternatives and estimating associated costs is presented. Steps 1, 10, 14, and 15 of the methodology, which relate to areawide aspects in application of the methodology, are not included in the discussions.

AGRICULTURAL RETURN FLOW

The area selected to monitor return flow depends primarily on soil conditions, the type of crops irrigated, irrigation methods, farming practice, and groundwater characteristics. In most cases, large farms (hundreds to thousands of acres) and irrigation districts (tens to hundreds of thousands of acres) would be areas of suitable sizes for monitoring.

Nonquality parameters to be monitored include volumes of applied water, precipitation, recharge from other pollution sources in the area, evapotranspiration, and infiltration of excess applied irrigation water. Quality-related parameters include the quality of applied water, amounts of additives such as fertilizers and pesticides, concentration due to evapotranspiration, dissolution and precipitation reactions in the soil-groundwater system, crop uptake of some constituents from the irrigation water, and quality of percolated water. Sampling is usually necessary for the applied water, percolate in the vadose zone, and groundwater.

LAND SURFACE MONITORING

Land surface monitoring encompasses an inventory of volumes of water applied to the land surface and other sources of recharge, such as seepage from streams or canals. Evapotranspiration and rainfall rates must occasionally be measured in the field, but can often be extrapolated from nearby areas. Compilation of these data in conjunction with runoff determinations will allow calculation of infiltration rates. Amounts of additives must be inventoried, including fertilizers, soil amendments, and pesticides. Some of these are directly introduced to the irrigation water, whereas others are applied on the land. Estimates for application rates per unit area should be compiled. The chemical quality of applied water must be known, or sampling and analysis undertaken. The approximate volumes and quality of other sources of recharge must also be determined. In some irrigated areas of the western United States, preliminary inventories of most of these items have been made for large areas, such as irrigation districts.

VADOSE ZONE MONITORING

Monitoring in the vadose zone includes laboratory and field tests for determination of percolate quality and determination of storage capacity and travel times for water and specific pollutants in the vadose zone. The occurence of restricting layers in the topsoil and relatively impermeable strata above the water table should be ascertained. Native soil and geological materials may be sampled for determinations such as nitrogen and total dissolved solids. Because return flow is a diffuse source, detailed sampling of percolating water in the vadose zone is impractical over large areas. The major disadvantage of sampling in the vadose zone, in the case of a diffuse source of large areal extent, is the cost of obtaining a sample representative of the entire system. To compensate for this, selected target areas can be chosen as typical of the larger area. Neutron probe and tensiometer measurements are the most effective method for tracing water movement, and water samples can be effectively collected from soil-water samplers in the vadose zone, or from wells in the saturated zone. In most cases, analytical determination would be limited to the major inorganic chemical constituents, nitrogen forms, and boron. Pesticides, phosphorus, chloride, and potassium could be important in some areas. Sampling frequency, in part, depends on travel time in the vadose zone. Calculation or determination of travel time can be made based on infiltration rates and storage capacity of the vadose zone. If wells are drilled to penetrate the entire vadose zone and travel time of percolate to the water table is sufficiently slow, sampling during drilling and once every 5 or 10 years thereafter may be sufficient. Where travel times are less than 1 or 2 years, semiannual sampling may be necesssary. Where pressure-vacuum soil-water samplers are installed as permanent sampling points, monthly sampling may be done; however, this usually will be unnecessary.

SATURATED ZONE MONITORING

Monitoring beneath a diffuse source in the zone of saturation should usually focus on sampling existing wells after long-term pumping. Large-capacity wells should be selected in order to provide an integrated sample of the water quality in the area of the well. In areas with few wells, construction of monitor wells may be necessary. Open-bottom cased wells drilled by the cable-tool method in unconsolidated materials may provide the most suitable results. Such wells produce water from well-defined depth zones. Where large seasonal variations in groundwater quality occur, seasonal trends must be established, based on at least monthly measurements to represent the extremes of chemical quality. Thereafter, a semiannual or annual sampling program will usually suffice. Long-term chemical hydrographs can be plotted to illustrate groundwater quality changes due to return flow. Often, detection of meaningful changes requires chemical analyses for a period of a decade or longer.

An analysis for the major inorganic chemical constituents is advisable in most cases. Total dissolved solids, boron, sodium percentage, nitrate, and hardness are the major concern. Selected pesticides also should be periodically determined. The wells to be sampled should be chosen to reflect the quality of water in the upper part of the aquifer, where pollution from return flow will first become apparent.

AGRICULTURAL RETURN FLOW EXAMPLE

STEP 2 -- IDENTIFY POLLUTION SOURCES, CAUSES, AND METHODS OF WASTE DISPOSAL

For this example, a 50,000-acre irrigation district in Central California is the area in need of monitoring. The source is given as return flow. The area is rural, with two towns of population less than 500 each. The area has been intensively farmed for over 80 years. In this case, there is no specific method of waste disposal, as groundwater pollution can result from normal crop irrigation.

STEP 3 -- IDENTIFY POTENTIAL POLLUTANTS

The chemical quality of irrigation water is known from records of State water agencies. Farmers, farm advisors, manufacturers, and regulatory agencies provide information on application rates of fertilizers, soil amendments, and pesticides. The primary fertilizer elements are nitrogen and phosphorus. The major soil amendment is gypsum widely used in only two of the subareas. There are five major types of pesticides in use, including two chlorinated hydrocarbons. The use of pesticides is related primarily to cropping patterns. Two of these pesticides are applied entirely in one subarea because of cropping patterns.

STEP 4 -- DEFINE GROUNDWATER USAGE

In this district, 95 percent of the groundwater use is for irrigation of agricultural lands. The remainder is used for domestic purposes in small communities and rural areas. Despite the fact that the water used for domestic purposes is only a small portion of the total water use, groundwater provides the sole source of drinking water in the area.

STEP 5 -- DEFINE HYDROGEOLOGIC SITUATION

Subsurface materials are alluvial sediments comprised of interbedded sand, silt, and clay layers. The depth to groundwater ranges from 50 to 200 feet below land surface. The average annual rainfall is 10 inches. Irrigation water is supplied from an extensive system of canals, utilizing surface runoff from nearby areas, and groundwater. Soils range from highly permeable sandy soils, developed on sand dunes, to low permeability hardpans. The regional direction of groundwater movement is westward through the district toward pumping depressions. Aquifer transmissivities range from 100,000 to 300,000 gallons per day/per foot, and irrigation well yields range from 500 to 1,000 gallons per minute.

Preliminary investigation includes calculation of an approximate hydrologic water budget. Surface water inflow and outflow are large items, whereas groundwater inflow and outflow are small items. Data on precipitation and evapotranspiration can be combined with data on the foregoing parameters to compute the water budget. The budget indicates whether there is an imbalance between groundwater recharge and discharge. This, in turn, indicates whether groundwater levels are relatively constant, rising, or falling. This information is pertinent to the monitoring effort, as the thickness of the vadose zone may vary substantially seasonally and over a period of years or decades. Sources of recharge are seepage from streams and canals, return flow of excess applied irrigation water, and groundwater inflow. Groundwater discharge is primarily through pumping and natural groundwater outflow. Domestic waste disposal volumes are negligible compared to agricultural return flow volumes.

STEP 6 -- STUDY EXISTING GROUNDWATER QUALITY

Maps prepared based on existing chemical analyses indicate the areal distribution of total dissolved solids, nitrate, and sodium percentage in the groundwater. High values of these parameters are generally related to soil type, cropping and irrigation patterns, and duration of irrigation. Water quality records indicate high total dissolved solids and nitrate contents in the upper 50 feet of the groundwater body, with much lower contents at deeper intervals. Previous studies by the United States Geological Survey indicate the chemical quality of sources of recharge other than return flow.

At this point, available information indicates the advisability of subdividing the study area. As a result, the study area is divided into five subareas for the following reasons:

- Surface water in the district is supplied from two separate sources of differing quality
- Soils in the district can be categorized into several groups on the basis of permeability
- Irrigation methods vary from place to place, but sprinkler irrigation is predominant in some areas and furrow irrigation in others
- Fertilizers, soil amendments, and pesticides are applied at different rates in various portions of the district
- Characteristics of the groundwater basin vary, especially from east to west

Each subarea has a unique combination of these factors, and thus lends itself to a separate determination of water budgets and salt balance.

STEP 7 -- EVALUATE INFILTRATION POTENTIAL OF WASTES AT THE LAND SURFACE

Determination of the volume of return flow escaping the root zone is the objective of this step. Average applied water volumes per acre for the district are available from a State water agency. Knowlege of the soils, irrigation methods, and cropping patterns enables more accurate estimates to be made for each subarea. Estimates of canal seepage are available from the irrigation district, and streamflow seepage is known from streamflow records at various gaging stations operated by the United States Geological Survey. Crop surveys by District personnel and evapotranspiration rates from lysimeter tests at a local agricultural experiment station can be used to determine crop evapotranspiration. Precipitation is measured at several stations in the District. Average annual return flow is calculated for each subarea.

STEP 8 -- EVALUATE MOBILITY OF POLLUTANTS FROM THE LAND SURFACE TO THE WATER TABLE

Travel time of return flow to the water table is generally unknown. However, preliminary calculations indicate that in cases of shallow water tables, this travel time would generally range from 6 months to 5 or 10 years. In cases of deep water tables, the travel time could range from 5 to 50 years. Application rates of irrigation water are the primary controlling factor.

Because of the pollutant attenuation characteristics of soils and alluvial deposits in the District, forms of fertilizer such as phosphorus and potassium would be adsorbed by soils and geologic materials. The primary form of nitrogen fertilizer is anhydrous ammonia. Ammonia tends to be sorbed to materials in the vadose zone. However, oxidizing conditions in the vadose zone permit formation of nitrate, which is subsequently leached to the groundwater. The gypsum contains sulfate which is fairly mobile in the vadose zone. Although calcium may be adsorbed in the vadose zone, it may subsequently be replaced by other cations and reach the water table. Some precipitation of gypsum may occur in the vadose zone. Tests at agricultural experiment stations indicate that only one of the pesticides used is subject to significant leaching; however, its mobility in the vadose zone is unknown.

STEP 9 -- EVALUATE ATTENUATION OF POLLUTANTS IN THE SATURATED ZONE

Horizontal movement of pollutants in the aquifer is not of primary concern, in this example, as a diffuse source is being considered. However, the extent of return flow in the aquifer in a vertical sense needs to be approximated. This is necessary in order to effectively utilize existing wells for monitoring purposes. Wells that are perforated too deep may not indicate any effect of return flow. Return flow tends to occur in the upper 50 to 100 feet of the aquifer, as shown by existing well data and chemical analyses. This occurrence is largely due to the layered nature of the alluvium, which results in small vertical permeabilities, compared to horizontal permeabilities.

STEP 11 -- EVALUATE EXISTING MONITORING PROGRAMS

A brief investigation of the records of the local water resource agencies indicates that there are no existing monitoring programs for groundwater quality in the area, except for limited sampling of domestic wells in the two towns.

STEP 12 -- ESTABLISH ALTERNATIVE MONITORING APPROACHES

Analysis of the hydrogeologic framework, groundwater quality, and irrigation practice of the district indicates that monitoring at the land surface, in the vadose zone, and in the saturated zone is necessary. Monitoring in the vadose zone is where the test drilling and relatively expensive, time-consuming aspects of monitoring come into play. The cost of monitoring in the vadose zone depends highly on the density of sampling devices.

Land Surface Monitoring

This phase of the monitoring program was previously developed largely in Steps 3 and 7. This monitoring will be continuously updated approximately every five years. No additional sampling or analysis will be necessary.

Vadose Zone Monitoring

One site in the District is selected for detailed monitoring of percolate in the vadose zone. This monitoring is necessary due to the virtual absence of such data in the project area. The site is 40 acres in size. Soils, irrigation methods, and cropping patterns are judged typical of the larger area. Primary costs of this phase are for well construction, installation of sampling devices, sample retrieval, and chemical analyses and interpretation.

Three alternatives have been selected for monitoring in the vadose zone. The most effective means of sampling and analysis are derived from Everett et al. (1976). Alternative A has two access wells for neutron probes and three holes for pressure-vacuum soil-water sampler nests (Everett et al., 1976, Figure 15). Alternative B has five access wells for neutron probes and ten holes for pressure-vacuum soil-water sampler nests. Alternative C has 10 access wells for neutron probes and 20 holes for pressure-vacuum soil-water sampler nests. The access wells for neutron probes are 2 inches in diameter and 150 feet deep. Three pressure-vacuum soil-water samplers are placed at 10-, 25-, and 50-foot depths in a 6-inch diameter hole to comprise each nest.

Neutron probe analysis and lysimeter sampling are carried out on a monthly basis. The percolate is analyzed for the major chemical constituents and boron. For Alternatives A, B, and C, the number of percolate samples collected monthly are 9, 30, and 60, respectively. Vadose zone monitoring is envisioned to be unnecessary after the first two years, because this period is believed to be sufficient for determination of rates of water movement and pollutant attenuation.

Saturated Zone Monitoring

Due to their large number, existing wells are determined to be sufficient for sampling. A carefully conducted well-data collection procedure is necessary before wells are selected for monitoring. Seasonal variations in well-water quality in the district are generally unknown. A 2-year period is chosen for bimonthly sampling of 300 large-capacity wells. Thereafter, a semiannual sampling program is selected for 50 wells chosen from the 300. The major inorganic chemical constituents and boron are determined for the well-water samples. Primary costs of this phase are for sample collection, chemical analysis, and interpretation of results.

STEP 13 -- SELECT AND IMPLEMENT THE MONITORING PROGRAM

Given the alternatives from Step 12, costs are derived based on Everett et al., 1976.

Land Surface Monitoring

It is estimated that a person with a B.S. degree in hydrogeology or water resources engineering and a minimum of two years of work experience in groundwater (salary $12,000 per year) could collect most of the data required on surface water, groundwater, soils, climatology, and waste loads in about two months. Costs for the time of the junior-level individual are calculated by applying a multiplier of 2.5 times the salary. A senior-level hydrogeologist, with an M.S. degree in hydrogeology and a minimum of five years of work experience in groundwater, (salary $18,000 per year) could supervise the monitoring effort. About one month of his time would be necessary to review and interpret collected data, interpret hydrogeologic conditions, with respect to groundwater pollution, delineate subdivisions of the study area, and establish monitoring alternatives. Costs for the time of the senior-level individual are also calculated by applying a multiplier of 2.5 times the salary. This phase is primarily a one-time effort, but might have to be periodically updated, depending on future land use, irrigation methods, fertilizer application rates, and other factors. Costs for this phase of the program would be $5,000 for the junior-level individual's time and $3,750 for the senior-level individual's time, or about $8,750. It is estimated that $2,000 per year would cover periodic updating of this phase every 3 or 4 years if necessary.

Vadose Zone Monitoring

Costs for access wells for the neutron probe and for soil-water sampler wells are taken from Everett et al. (1976, Figure 18). However, in this case, slightly different well-construction techniques are necessary. Casing is not necessary, but special plugs must be installed to separate the three samplers in each hole. Reasonable estimates can be derived from Everett et al. (1976, Figure 18) Table 6-1 shows figures applicable to this study.

Table 6-1

WELL-CONSTRUCTION COSTS FOR VADOSE ZONE MONITORING

Type of Well	Alternatives (cost in dollars)		
	A	B	C
Access Well Construction (including casing and development)	500	1,100	2,100
Soil-water Sampler Well Construction (including sampler installation	500	1,750	3,380
TOTAL	1,000	2,850	5,480

In addition, logging test holes and supervision of sampling device installation by the junior-level individual would require one month for Case A, three months for Case B, and six months for Case C. The costs of his time for these alternatives are $2,500, $7,500, and $15,000, respectively.

One neutron probe device, with 200 feet of cable, is purchased for $3,000 (Everett et al., 1976, p. 36). Each soil-water sampler costs $20, and one hand-pump service kit costs $30. Including tubing, the samplers required for each hole cost $70. Only one service kit is needed for all of the sampler wells.

The moisture logging cost per run per well for Alternative A, with a density of one per 20 acres, is approximately $250, using the cost curve of one per acre density (Everett et al., 1976, Figure 9), totaling $500 for the 40-acre tract. Since this cost curve was prepared for point-source application, densities in this example are too low to use the graph directly, and costs must be estimated by extrapolation. The cost per run per well for Alternative B, with a density of one per 8 acres, would be approximately $225. The cost per run per well for Alternative C, with a density of one per 4 acres, would be approximately $200, or $2,000 total. The monthly time for the junior-level individual for sampling soil-water samplers is 0.5 day for Alternative A, 1.5 days for Alternative B, and 2.5 days for Alternative C.

The chemical analyses for percolate obtained from the soil-water samplers include the major inorganic chemical constituents (Everett et al., 1976, p. 114) and boron (Everett et al., 1976, Table 14). The cost of analysis of percolate for the major inorganic chemical constituents is $12 per sample, and the cost for boron (dissolved) is $10. For purposes of this example, a special group rate of $17 is assumed for a combination of the foregoing. Discounts of 10 percent are applied for total cost over $500, and 20 percent for total cost over $1,000 (Everett et al., 1976, p. 113). Analytical costs are $17 per water sample for Alternative A, $15 per sample for Alternative B, and $13.50 per sample for Alternative C. The costs for vadose zone monitoring are given in Table 6-2.

Table 6-2

COSTS FOR VADOSE ZONE MONITORING

	Alternatives (cost in dollars)		
	A	B	C
One-Time Costs			
Well Construction and Logging	3,500	10,350	20,480
Neutron Probe and Soil-water Samplers	3,240	3,730	4,430
TOTAL	6,740	14,080	24,910
Annual Costs (first two years only)			
Neutron Moisture Logging	6,000	13,500	24,000
Soil-water Sampling	700	2,100	3,500
Chemical Analyses	1,840	5,400	9,720
TOTAL	8,540	21,000	37,220

In order to select the most cost-effective alternative, consideration is given to impacts of pollution on groundwater use. Nitrate and pesticides pose a potential health effect on groundwater used for drinking purposes. No feasible alternative water-supply sources for drinking water are available. Pollution due to return flow also creates economic impacts, as eventually the degraded groundwater can result in decreased crop yields in the District. Assessment of long-term damages is not possible. Another consideration is the net worth of the farm produce. A final consideration is the money available for monitoring in the district and other districts in the region. Consideration of all these factors leads to selection of Alternative B.

Saturated Zone Monitoring

The junior-level individual would spend one month in collecting existing well data and groundwater quality data. Ten days would be spent collecting water samples from wells for each round during the first two years. Five additional days would be spent by this individual checking chemical analyses and tabulating the results for each round. First-year costs would be $13,750, and second-year costs would be $11,250. The senior-level individual would spend two weeks each year for supervision and review of the program. Costs would be about $1,880 each year. A routine irrigation water analysis would be $17 per sample, as calculated previously for analyses of percolate in the vadose zone. For groups of 300 samples, the analytical cost per sample is lowered to $13.50. For the first year, personnel costs would be $15,630, and chemical analyses $24,300. For the second year, personnel costs would be $13,130, and chemical analyses $24,300.

After the first two years, the junior-level individual would spend two days collecting samples for each round, and one additional day checking chemical analyses and tabulating results for each round. Costs for his time would be $630 per year. Supervision and review by the senior-level individual would be about one week each year after the first two years, at a cost of $940 per year. For groups of 50 water samples, the analytical cost per sample is $15. For each year after the first two, personnel costs are $1,570 and chemical analyses $1,500.

Summary

Table 6-3 summarizes costs for the entire program.

Table 6-3

TOTAL COSTS FOR AGRICULTURAL RETURN
FLOW MONITORING

Type of Monitoring	Annual Costs (dollars)		
	Year 1	Year 2	Subsequent Years
Land Surface	8,750	2,000	2,000
Vadose Zone	35,080	21,000	0
Saturated Zone	39,930	37,430	3,070
TOTAL	83,760	60,430	5,070

These costs reflect several factors. The greatest costs are incurred during the first two years of the monitoring program. One-time costs make the first-year costs almost 40 percent greater than second-year costs. Annual costs after the intensive 2-year monitoring period are only about 10-percent of the average annual cost during the first two years.

SEPTIC TANKS

The area selected for monitoring depends primarily on the location and configuration of unsewered areas. In general, septic tanks in sparsely populated rural areas are insignificant sources of groundwater pollution. In more densely populated areas, soil conditions, septic tank density, method of disposal, and groundwater characteristics influence the selection of the area to be monitored. The area should be chosen to insure uniformity of as many of these factors as possible. Downgradient areas within one mile of the unsewered area should also be included to monitor movement of recharged septic tank effluent. The area selected typically ranges from several hundred to several thousand acres in size.

Nonquality factors to be considered include volume of the septic tank effluent, method of effluent disposal, and soil hydraulic characteristics. Septic tank density, or lot size, is an important factor. Disposal methods range from shallow drainfields, where substantial losses due to evapotranspiration occur, to seepage pits, where this loss is insignificant. Usually, the disposal methods are selected on the basis of soil conditions. Percolation rates and the presence of restricting layers influence the impact of septic tank effluent on groundwater quality. Quality-related factors include the quality of septic tank effluent and percolate in the vadose zone. The major sampling required is for effluent, percolate, and shallow groundwater. Soils and geologic materials may occasionally be sampled to determine pollutant attenuation mechanisms, such as adsorption.

LAND SURFACE MONITORING

The most effective type of land-surface monitoring comprises inventorying septic tank densities, volumes of septic tank effluent, and water use, and possibly, determining a water budget for the area. Data on septic tank densities and volume of septic tank effluent can be determined, for residential areas, from lot sizes in unsewered areas. Special attention should be focused on schools, shopping centers, and other facilities, with effluent volumes significantly greater than those for individual households. The volume of effluent has been carefully documented in some areas. Figures for in-home water use are available for most areas. The water subject to septic tank disposal is used in the home for toilets, sinks, garbage disposers, bathtubs, showers, dishwashers, washing machines, and water softeners. In areas where little or no data have been developed, representative households can be chosen for detailed monitoring. The effluent should be characterized as to quality, especially for total dissolved solids and total nitrogen. Sampling of representative effluent may be necessary. The inventory should include pertinent data on the types of detergents used and the extent of the use of water-softening devices.

In the case of shallow disposal, such as seepage trenches, water budget analyses may be necessary in order to determine infiltration of septic effluent. Precipitation and evapotranspiraiton can be estimated from records in the area. However, judgment is necessary to calculate infiltration, as some seepage trenches may be below the root zones of most plants. In other cases, plant uptake of nutrients, such as nitrogen, from the effluent could be significant.

VADOSE ZONE MONITORING

Determination of storage capacity and travel time for water and specific pollutants in the vadose zone is an important component of the monitoring program. Delineation of restricting layers in the topsoil and relatively impermeable geologic materials above the water table is important. Laboratory and field tests can be conducted to determine the quality changes of effluent during percolation through native soil and geologic materials. Sampling of soil and geologic materials for nitrogen determinations may be necessary where natural sources of nitrogen are present. Because septic tank effluent is a diffuse source over a monitoring area, detailed sampling of percolating water in the vadose zone is impractical in most cases. However, monitoring areas are generally of a size such that several holes could be drilled for sampling in each area.

Neutron probes and tensiometer measurements can be effectively used to trace water movement. Percolate samples can be collected from soil-water samplers in the vadose zone, or from wells in the saturated zone. In most cases, analyses of percolate samples are limited to the major inorganic chemical constituents and nitrogen forms. Boron, detergents, stable organics, and bacteriological constituents could be important in some cases. The frequency of sampling depends on travel time in the vadose zone. For wells penetrating the entire vadose zone, and when travel times are very slow, sampling may be necessary only once every few years. However, percolate should be collected from the soil-water samplers generally on a monthly basis. Percolation rates of septic tank effluent often are in the range of one to two feet per acre per year.

SATURATED ZONE MONITORING

As septic tanks represent a diffuse source, effective monitoring in the saturated zone usually entails sampling of existing wells after long-term pumping. These wells should be large-capacity wells, if possible, in order to provide an integrated sample. Such wells are often present in urban or suburban areas. Shallow wells should be selected which tap the upper part of the aquifer; an occasional deep well should be included. Seasonal trends should be established in some areas, and a semiannual sampling program will usually suffice thereafter. The establishment of seasonal trends may require monthly sampling for several years. Wells to be sampled include not only those beneath septic tank areas, but also upgradient and downgradient wells. Upgradient sampling can provide an indication of water quality unaffected by septic tank effluent. Downgradient sampling indicates the downgradient movement of septic tank effluent in the aquifer. Usually a distance of 1/2 mile or so from the downgradient boundary of the unsewered area will be adequate. This limit is due to pollutant attenuation mechanisms in the saturated zone. Chemical hydrographs can then be plotted to illustrate long-term trends in groundwater quality. Many years of records may be necessary for proper interpretation.

Analysis of well water for the major inorganic chemical constituents is advisable in most cases. Total dissolved solids, nitrate, chloride, and possibly other nitrogen forms are of chief concern. Detergents, boron, hardness, stable organics, and bacteriological constituents may also be important.

SEPTIC TANK EXAMPLE

STEP 2 - IDENTIFY POLLUTION SOURCES AND CAUSES AND METHODS OF WASTE DISPOSAL

In this example, a 2,000-acre suburban area in an alluvial basin of Central California is given as the area for monitoring of septic tank effluent. Septic tank treatment and disposal has been practiced in the area for about 30 years. The area to be monitored includes upgradient and downgradient areas within one mile of the unsewered area.

A portion of the water pumped in the area is returned to the groundwater by lawn irrigation return flow and septic tank effluent disposal systems. The method of septic tank effluent disposal used is rather uniform over the area; seepage fields are used that are usually 8 to 10 feet below the land surface. Several schools and small shopping centers are points of heavy effluent discharge, whereas the remainder of the area is residential. Fertilizers used for lawns, gardens, trees, and shrubs may contribute to groundwater pollution. Urban runoff is diverted from the area by storm drains and disposed of elsewhere.

STEP 3 - IDENTIFY POTENTIAL POLLUTANTS

Well water in the area averages about 250 parts per million total dissolved solids. Based on analyses of domestic sewage effluent from nearby sewered areas, the dissolved solids content of the septic tank effluent is estimated at 500 parts per million. There is no water softening in the area. Chlorides averaging 80 parts per million are present in sewage effluent discharged from nearby areas. Boron is introduced through the use of detergents, and fluoride is added for health reasons at the well sites. Boron, fluoride, and total nitrogen concentrations in the septic tank effluent are estimated at 0.5, 1.2, and 25 parts per million, respectively, based on data in the literature.

STEP 4 - DEFINE GROUNDWATER USAGE

Groundwater is pumped for municipal use, which includes domestic use and lawn irrigation. One hundred percent is for municipal use, of which about 25 percent is for in-house use (including drinking water), and about 50 percent is for yard irrigation. The remainder is used for cooling purposes (15 percent) and commercial use (10 percent). Additional information on water use is presented in Step 5.

STEP 5 - DEFINE HYDROGEOLOGIC SITUATION

The depth to groundwater ranges from 50 to 70 feet, and no perched water is present in the alluvium. The soils are uniform over the area, and no restricting layers are present. Water is supplied entirely by groundwater. The aquifer transmissivity is 100,000 gallons per day per foot, and the water level slopes uniformly to the south. Well yields range from 500 to 1,000 gallons per minute and well depths range from 100 to 200 feet. Most wells have been drilled by the cable tool method, and casings are perforated over short intervals, or are unperforated (open-bottomed).

Groundwater recharge is primarily from groundwater inflow, and the quality of this water is determined primarily by natural factors. Available studies indicate that this water is of the calcium-sodium bicarbonate type, with total dissolved solids content less than 200 parts per million. Groundwater discharge is by well-pumping and groundwater outflow. Water levels are relatively stable from year-to-year, although they fluctuate seasonally.

STEP 6 - STUDY EXISTING GROUNDWATER QUALITY

Maps have been prepared by County agencies, based on existing chemical analyses. These maps indicate the areal distribution of total dissolved solids, chloride, nitrate, and hardness in the groundwater. High values are most often found in the central portion of the study area. Chemical analyses of water from shallow wells in the area indicate higher total dissolved solids, nitrate, and chloride contents than groundwater beneath surrounding areas. Deep wells produce water of a chemical quality similar to that in the surrounding area.

STEP 7 - EVALUATE INFILTRATION POTENTIAL OF WASTES AT THE LAND SURFACE

Calculation of the volume of septic tank effluent percolating below septic system leach lines is the objective of this step. Daily water consumption is available from the water purveyor, since well pumpage and household use are metered. Lawn irrigation is a major use in the summer, but minor in the winter. Based on literature studies and a comparison of water usage in the summer and winter, domestic use subject to septic tank treatment and disposal is calculated. The number and size of lots are available from the County public works department. There are 3,900 lots in the residential portion of the study area (1,850 acres), and the population is 8,900.

The average water use in the residential area is 300 gallons per capita per day, and the average effluent volume is 75 gallons per capita per day. The total volume of septic tank effluent in the residential area is about 670,000 gallons per day or 750 acre-feet per year. This averages about 14 inches annually over the part of the residential area not occupied by streets or structures (650 acres). Two schools and two shopping centers are located on a total of about 150 acres within the study area. Annual water use for the schools and shopping centers averages about 90,000 gallons per day. The effluent volume from these four sources averages 100,000 gallons per day during the school year, and 20,000 gallons per day during the rest of the year. As the septic tank effluent is disposed of below the root zone of most plants, no loss of effluent to evapotranspiration is assumed; thus no water budget analysis is necessary.

On the basis of winter and summer use, evapotranspiration rates, and irrigation methods, it is estimated that return flow from irrigation averages 1.0 million gallons per day, or about 21 inches per year over the irrigated part of the residential area (650 acres).

STEP 8 - EVALUATE MOBILITY OF POLLUTANTS FROM THE LAND SURFACE TO THE WATER TABLE

Previous studies have been undertaken at a nearby Agricultural Research Service site and the local university. These studies have documented travel times of septic tank effluent to the water table and evaluated the movement of nitrate in the vadose zone. Travel times, where effluent application rates are 14 inches per year and the water table is 50 feet deep, are about 5 years. Little denitrification occurs because of the aerobic conditions prevailing in the vadose zone and the lack of organic matter. Hence, nitrogen forms in the effluent are oxidized to nitrate and leached to the water table. Bacteriological constituents, including viruses, are removed within several feet of travel in the alluvium above the water table. Phosphorus is strongly retained in the vadose zone due to sorption and chemical precipitation.

For return flow from irrigation, travel times to the water table are comparable to that discussed above. Nitrogen behaves similarly as in the case of septic tank effluent.

STEP 9 - EVALUATE ATTENUATION OF POLLUTANTS IN THE SATURATED ZONE

This step is necessary in order to determine the extent of downgradient monitoring needed. In this case, inspection of the maps prepared for well-water quality indicates no detectable effects occur more than 1/2 mile downgradient of the unsewered area after 30 years of operation. The extent of septic tank effluent in the aquifer, in a vertical sense, also needs to be estimated. This is necessary in order to effectively utilize existing wells for monitoring purposes. Septic tank effluent tends to occur in the upper 50 feet of the aquifer, as shown by existing well data and chemical analyses.

STEP 11 - EVALUATE EXISTING MONITORING PROGRAMS

Fairly comprehensive monitoring programs are in effect for all wells for the major chemical constituents (including nitrate, fluoride, and total dissolved solids). One laboratory provides all of the analytical services, and the chemical analyses appear to be adequate for monitoring groundwater pollution. No determination for detergents or stable organics has been made. Bacteriological sampling is routinely performed by the water purveyor, and analyses for fecal coliform are consistently negative.

STEP 12 - ESTABLISH ALTERNATIVE MONITORING APPROACHES

Analysis of alternative monitoring programs in relation to the hydrogeologic framework, groundwater quality, and waste disposal practice indicates that no vadose zone monitoring is necessary. Routine land surface monitoring and sampling of large-capacity wells can be used to effectively monitor the source. The most cost-effective alternative is selected after consideration of the impact of septic tank effluent on water use in the area and downgradient areas. Nitrate and stable organics in groundwater used for drinking purposes pose a potential health effect. Pollution due to septic tank effluent can also degrade municipal supplies in a monetary sense, especially for parameters such as hardness. Water treatment in the home, or by the water purveyor, may be necessary to enable the use of the degraded water. A final consideration is the funds available for monitoring in the subdivision and other unsewered areas in the region. Consideration of all these factors leads to selection of the most cost-effective method.

Land Surface Monitoring

This phase of the monitoring program was largely developed in Steps 3 and 7. No annual updating is deemed necessary.

Saturated Zone Monitoring

Because the source to be monitored is normally diffuse and travel times of percolate to the water table are in terms of about 5 years, long-term monitoring of large-capacity pumping wells can be effectively used. Previous monthly sampling has established that peak nitrate concentrations occur in late summer, and the lowest values occur in early spring. Thus, about 25 wells are selected for continuous monitoring, and chemical analyses are performed on water samples taken semiannually. Analyses include the major inorganic chemical constituents, boron, and fluoride. Samples are collected annually to determine the stable organic, detergent, and ammonia contents in water from six selected wells.

STEP 13 - SELECT AND IMPLEMENT THE MONITORING PROGRAM

The monitoring program was already selected in Step 12, based on experience and hydrogeologic judgment. The following costs are derived from Everett et al. (1976).

Land Surface Monitoring

A person of minimum qualifications (i.e., a B.S. degree in hydrogeology or water resources engineering and 2 years working experience) could collect most of the required data on water use, lot size, septic tank effluent disposal methods, quality of source water and septic tank effluent, and lawn irrigation and fertilizers in about one month. Data on groundwater conditions, well data, and historical chemical analyses could also be collected during this period. A one-half week review by a senior individual costs $470. This phase is primarily a one-time effort. No costs for updating have been calculated for this analysis. Total cost during the first year of the program would be $2,970.

Vadose Zone Monitoring

Vadose zone monitoring encompasses the review of existing reports and consultations with researchers in the area. The junior-level individual collects these data in one week and a senior-level individual reviews and interprets them in one-half week. This phase is a one-time effort during the first year, and costs $1,160.

Saturated Zone Monitoring

Sampling of water wells is conducted by local agencies at no additional cost. Costs for the major inorganic chemical constituents are given in Everett et al. (1976, p. 114) as $12 per sample. Costs for boron and fluoride determinations are given in Table 14 of the same reference as $10 each. A special group rate is assumed for determination of all of these constituents. For the samples collected semiannually, determinations of the major inorganic chemical constituents, including boron and fluoride, cost $22 per sample, which is discounted to $20 per sample for groups of 25 samples. The annual analytical cost is $1,000 per year. For the samples collected annually, costs per sample are $5 for ammonia (Everett et al., 1976, Table 14), $10 for methylene blue active substances (Everett et al., 1976, Table 15), and $20 for biochemical oxygen demand (Everett et al., 1976, Table 15). A special group rate of $30 per sample is assumed for these annual determinations, or a total of $180. Total analytical costs are thus about $1,180 per year.

Checking chemical analyses and tabulating results requires one week each year for the junior-level individual at a cost of $630. Supervision of the program and annual review by the senior-level individual requires one week at a cost of $940. Total annual personnel costs are $1,570.

Summary

Table 6-4 summarizes costs for the entire program.

Table 6-4

COST SUMMARY FOR MONITORING SEPTIC TANK POLLUTION

Type of Monitoring	Annual Cost (dollars)	
	Year 1	Subsequent Years
Land Surface	2,970	0
Vadose Zone	1,160	0
Saturated Zone	2,750	2,750
TOTAL	6,880	2,750

These costs reflect several factors. The annual cost is relatively low due to the use of existing programs for collecting samples. The overall program cost is relatively low due to the lack of well drilling and vadose zone sampling, neither of which are necessary in this example.

PERCOLATION PONDS AND LINED PONDS

Ponds for containment of liquid wastes are potential point sources of groundwater pollution. Two major categories of ponds presented in this discussion are percolation ponds and lined ponds. The first category is represented by large-scale percolation of sewage effluent to remove bacteria and possibly other constituents. In this case large volumes of water are recharged per acre; for example, in the range of 20 to 100 feet per year. The second category is represented by disposal or storage of some oilfield wastes, industrial wastes, and certain toxic materials in lined ponds. Artificial liners have come into wide use to limit infiltration, and small amounts of seepage or leakage per acre usually occur (less than 1 inch to several inches or feet per year). The type of monitoring effective for each category is basically different.

In the case of percolation ponds, water budget evaluation can be effectively used to determine infiltration. Thus nonquality parameters to be measured include rates of waste discharge, precipitation, and evaporation. The infiltration can be calculated by measuring inflow to the pond, storage changes, precipitation, and evaporation.

Artificial liners range from compacted soil or clay to impermeable plastic and rubber liners. *Seepage* may be termed slow-flow through a liner over the entire lined area, whereas *leakage* is flow through breaks or perforations in the liner. Monitoring may be necessary for almost all ponds despite the lining material. Liners can be pierced or joints not carefully sealed during installation. Chemical deterioration is common for some types of liners exposed to toxic wastes. The water budget approach usually is not applicable in this case, as the infiltration rate is less than the error inherent in calculating infiltration. Rather, monitoring focuses on the physical integrity of the liner, reactions between wastes and the liner, and detection of leaks.

Factors determining percolate quality in both cases include the waste discharge quality, concentration of pond water by evaporation, dilution of pond water by precipitation, chemical changes in pond water, and dissolution and precipitation reactions in the topsoil and vadose zone. Retrieval of water samples from the waste discharge stream and ponds, percolate in the vadose zone, and shallow groundwater are the most effective sampling methods. Topsoil and geologic materials are sampled where significant retention of some constituents is important.

LAND SURFACE MONITORING

Nonquality monitoring for percolation ponds involves accumulating data for calculation of the water budget. The waste discharge flow can usually be measured at the point of entry to the pond. In some ponds recycling occurs, and outputs must also be measured. Flow meters can often be installed in discharge pipes, and weirs and flumes can be installed to measure flow in other cases. Thus a continuous record of waste discharge is available. For large ponds, evaporation and precipitation may have to be measured onsite. To a large degree this depends on the climatological homogeneity of the area. Land pans and floating pans can be used to calculate evaporation from a free-water surface. Extrapolations from nearby areas can be made over climatologically homogenous areas. Daily measurements enable calculation of infiltration. Radioactive isotopes, such as tritium, have been used to directly measure seepage rates. Also, stable hydrogen isotopes are fractionated during evaporation, and thus the deuterium content of pond water can indicate the relative percentages of evaporation and infiltration.

Water samples can be collected from the discharge stream or the open pond. Sampling in open ponds may be greatly hampered by netting or other features designed for wildlife protection. Large fluctuations in discharge stream quality often occur due to variations in plan operational characteristics; therefore compositing of samples from the discharge stream is necessary. Continuous recording devices may be used for some parameters, such as electrical conductivity. Sampling open pond-water quality to determine percolate may be more representative than sampling the waste discharge stream. Boats may be used or, in some cases, special walkways constructed for sample retrieval. Of importance is the collection of a sample that would be representative of water that would eventually percolate. Certain parts of large ponds may be much more favorable for infiltration than others, and consideration should be given to this factor when selecting sampling sites. Sample frequency often must be established on a trial and error basis. For composite samples, weekly composites of samples collected at four to eight hour intervals is ordinarily sufficient. For pond samples, weekly, biweekly, or monthly sampling is usually sufficient. The constituents to be analyzed depend on the type and characteristics of the waste, as well as water use in the area.

VADOSE ZONE MONITORING

Monitoring in the vadose zone beneath percolation ponds is especially important when percolate travel time from the land surface to the water table is so long (20 to 30 years or greater) as to render saturated zone monitoring ineffective. Where substantial retention of toxic pollutants occurs in the vadose zone, it may also be necessary to obtain soil samples at various depths.

Neutron probes and tensiometers can be used to trace water movement, and pressure-vacuum soil-water samplers can be used for water sample collection. Laboratory and field tests can be performed to evaluate the reaction of percolated waste water with soils and geologic materials. For example, the sorptive capacity for various trace metals can be determined. Also, the effect of extreme pH values in percolated waste water on the dissolution of minerals in geologic materials can be evaluated.

Appropriate sampling devices can be installed for artificial liners that may leak. One method is to use two liners separated by a layer of soil and tile drain pipes. The lower liner is graded toward a central point for sample collection. Any leakage through the upper liner tends to accumulate on the lower liner, and can be collected. Obviously, the lower linear must be relatively impermeable and carefully constructed for this method to be effective. Sampling of nearby wells can also provide information on leakage.

SATURATED ZONE MONITORING

Resistivity methods can be used where high salinity wastes are ponded and existing wells are not suitable for monitoring. Specially designed monitor wells, tapping the shallow portion of the saturated zone, are often necessary. Cable-tool drilled wells that are either unperforated, or are perforated over short intervals, may be effective in many cases. A number of small-diameter observation wells can be effectively sampled periodically by use of a portable submersible pump. In some cases, bailing or air-jetting may be used for sample retrieval. Consideration should also be given to sampling one or more large-capacity wells typical of those used in the area. Such wells may provide data on regional groundwater quality. Once seasonal trends are established, the frequency of sample collection can be determined. As a rule, monthly or bimonthly sampling is sufficient. The specific determinations depend on the composition of the waste water and the water use in the area.

PERCOLATION POND EXAMPLE

STEP 2 - IDENTIFY POLLUTION SOURCES, CAUSES, AND METHODS OF WASTE DISPOSAL

In this example, a percolation pond for the disposal of toxic industrial wastes in the eastern United States is the source to be monitored. The pond has been in operation for 5 years, is 20 acres in size, and about 5 feet deep. The topsoil has been removed to achieve higher infiltration rates.

STEP 3 - IDENTIFY POTENTIAL POLLUTANTS

The discharged waste is a low pH sodium chloride brine containing some trace metals. Total dissolved solids commonly exceed 10,000 parts per million and there are high concentrations of arsenic, cadmium, chromium, barium, and silver. The salinity and trace element content of the waste discharge fluctuates considerably, due to plant operation. Pond-water quality shows less fluctuation due to damping by the storage in the pond. Dilution by precipitation falling on the water surface also occurs.

STEP 4 - DEFINE GROUNDWATER USAGE

There is no use of groundwater in the immediate area. However, County-wide plans indicate the probability of urbanization within 20 years. The future water supply would likely be provided by wells.

STEP 5 - DEFINE HYDROGEOLOGIC SITUATION

The average annual rainfall is about 50 inches per year, and glacial till materials about 100 feet thick comprise the aquifer. These aquifer materials are highly permeable, and well yields in the region range from 500 to 1,000 gallons per minute. The aquifer is underlain by relatively impermeable igneous and metamorphic rocks. Groundwater flow is to the north at a uniform gradient of about 20 feet per mile. Groundwater flow rates are about 1 foot per day. The regional depth to water is about 20 feet; however, a mound is present beneath the pond, and groundwater is believed to be less than 10 feet deep. No wells are in the immediate area. Recharge is from precipitation, and groundwater discharge is to streams in the area.

STEP 6 - STUDY EXISTING GROUNDWATER QUALITY

Native groundwater in the region is of excellent chemical quality, with total dissolved solids less than 50 parts per million. Groundwater quality beneath the percolation pond is unknown.

STEP 7 - EVALUATE INFILTRATION POTENTIAL OF WASTES AT THE LAND SURFACE

Since no previous measurements are available on which to accurately calculate infiltration rates, a monitoring program will be established for this purpose. However, a preliminary evaluation can be made. Regional data are gathered on precipitation and evaporation from free-water surfaces. It is estimated from the amount of water used by the industry that about 2,000 acre-feet per year of waste water is discharged to the pond. A water budget analysis is used to estimate infiltration. Infiltration is estimated to be about 1,950 acre-feet per year, or almost 100 feet per year over each acre of the pond.

STEP 8 - EVALUATE MOBILITY OF POLLUTANTS FROM THE LAND SURFACE TO THE WATER TABLE

There are no significant restricting layers present above the water table to obstruct downward percolation of recharged waste water. There are no field data in the region on the mobility of pollutants in this type of waste. However, sodium and chloride are ordinarily highly mobile in such situations. Studies in similar areas also reveal that arsenic, cadmium, and chromium may be mobile. On the other hand, barium and silver are not expected to be mobile, due to chemical precipitation above the water table.

STEP 9 - EVALUATE ATTENUATION OF POLLUTANTS IN THE SATURATED ZONE

This step is necessary in order to properly determine the location of monitor wells to be drilled near the percolation pond. Physical factors important in the analysis for this step are the water budget of the area, slope of the water table, transmissivity of aquifer materials, and dynamics of groundwater flow. The most valid data are derived from previous pollution studies in other areas of similar hydrogeology. Waste plumes for comparable hydrogeologic situations, and for these waste percolation rates, indicate detectable salinity increases in the aquifer for a distance of only about 1/2 mile downgradient of the source after 20 years of operation.

STEP 11 - EVALUATE EXISTING MONITORING PROGRAMS

There are no existing monitoring programs near this site.

STEP 12 - ESTABLISH ALTERNATIVE MONITORING APPROACHES

Analysis of the hydrogeologic framework, waste characterisitics, and disposal method indicates that monitoring at the land surface, in the vadose zone, and in the saturated zone is necessary. Alternatives are involved primarily with monitoring in the vadose zone and the saturated zone. The most effective methods are chosen from Everett et al., 1976.

Land Surface Monitoring

Daily precipitation and evaporation rates are measured onsite by use of a standard U.S. Weather Bureau raingage, a Class A land pan, and a floating pan in the pond. In order to estimate evaporation from the free-water surface, corrections for salinity are necessary. The waste flow is continually measured at the inlet to the pond by a propeller type meter installed in the discharge pipe.

Because of the large fluctuations in chemical quality of the discharge stream, an electrical conductivity recorder is installed on the discharge pipe at the pond inlet. This record allows continuous monitoring of the electrical conductivity of the waste discharge. Composite samples of the waste discharge are collected for determination of trace metal content. Weekly composites of the waste discharge samples collected at four hour intervals are sufficient to characterize seasonal variations in chemical quality. Monthly samples of pond water are collected in order to evaluate significant differences in chemical quality between the waste discharge and pond water. The primary differences are due to dilution by precipitation and chemical reactions caused by exposure of the waste water to the atmosphere and sunlight in the open pond.

Vadose Zone Monitoring

As some trace metals in water precolating from the pond would likely be removed by sorption, chemical precipitation, and other processes, the percolate and geologic materials should be sampled. Soil-water samplers are placed in access holes drilled by power auger beneath the pond. Three alternatives, consisting of 5, 10, and 20 holes, are considered for soil-water sampler emplacement. Three samplers are emplaced in each hole at depths of 5, 10, and 15 feet beneath the land surface, depending on the actual depth to the water table. Group seals are used to separate the samplers in each hole. During drilling, the access holes are logged by the junior-level individual, and samples of subsurface materials are taken at 2-foot intervals. Exchangeable cations, cation exchange capacity, electrical conductivity, and pH of the saturation extract are determined on selected samples. Percolate samples are taken from the soil-water samplers on a monthly basis, and electrical conductivity, pH, aresenic, cadmium, chromium, barium, and silver contents are determined.

Saturated Zone Monitoring

As there are no existing wells near the pond, a number of monitor wells are necessary. Prior to the selection of drilling sites, a surface resistivity survey is conducted to delineate the zone of polluted groundwater. Since much of the salinity is due to chloride, and chloride is mobile in this soil-groundwater system, the high salinity zone should delineate the maximum extent of polluted groundwater. Other pollutants are generally less mobile and occupy smaller zones in the groundwater. It is estimated that the zone of polluted groundwater extends over 1/2 mile downgradient from the ponds.

From 5 to 10 monitor wells are considered for installation by the cable-tool method. Eight-inch diameter holes are to be drilled and six-inch diameter steel casing installed.

Under Alternative A, two wells would be located immediately adjacent to the pond in the downgradient direction. One well would be drilled 500 feet upgradient of the pond. One well would be drilled 500 feet and another 2,000 feet downgradient of the pond. Both of the downgradient wells are within the plume delineated by the geophysical survey. Under Alternative B, two upgradient wells would be drilled, at 500 feet distance, three wells near the pond, and five wells from 500 to 2,000 feet downgradient. Wells near the ponds are 50 feet deep and perforated from 20 to 50 feet. Upgradient and downgradient wells are 100 feet deep and perforated from 20 to 100 - feet. After development and pump testing for 24 hours, the monitor wells are sampled each month after 24 hours of continuous pumping. A portable submersible pump is purchased for use in all of the monitor wells. The water samples are analyzed for the major inorganic chemical constituents and the five trace metals.

STEP 13 - SELECT AND IMPLEMENT THE MONITORING PROGRAM

Given the alternatives from Step 12, costs are derived from Everett et al. (1976). A preliminary investigation of hydrogeology, existing water quality, and characteristics of industrial wastes is conducted to meet the requirements of Steps 3 through 6. The junior-level individual spends two weeks at $1,250, and the senior-level individual one week at $940. The preliminary study thus costs $2,190.

Land Surface Monitoring

The costs of a recording precipitation gage, land pan, floating pan, and flow meter are $200, $150, $200, and $100 respectively, including installation. The electrical conductivity instrument costs $700. The annual cost for maintaining these devices is $1,000. Sample retrieval from the waste discharge by the junior-level individual requires one day per month, and costs $1,200 per year. Analytical costs for electrical conductivity (Everett et al., 1976, Table 14) are $3 per sample. Everett et al. (1976, p. 114) list a group rate for analysis of drinking water trace elements. Five of the 12 trace constituents listed are analyzed at a group rate of $25. Analytical costs for the waste discharge are $1,460 per year.

Collection of monthly pond samples is by small boat, requires one-half day per month by the junior-level individual, and costs $600 per year. Analyses of the monthly pond samples for the same constituents as in the waste discharge costs $340 per year. Supervision by a senior-level professional requires one week per year or $940. Checking chemical analyses and tabulating results by a junior-level individual requires two weeks at $1,250 per year.

Total one-time costs are $1,350 for equipment. Annual costs are $1,000 for maintaining equipment, $1,800 for sample collection, $1,800 for chemical analyses, and $2,190 for other personnel costs. Total annual costs are $6,790.

Vadose Zone Monitoring

Costs for drilling and pressure-vacuum soil-water sampler installation, logging, sampling of geologic materials, chemical analyses of geologic materials, sampling of percolate, and chemical analyses of percolate for the three alternatives (A, 5 holes; B, 10 holes; and C, 20 holes) are given in Table 6-5. Costs of 8-inch diameter augered holes are determined from Everett et al. (1976, Figure 6). Five holes are $50 per hole, 10 holes are $45 per hole, and 20 holes are $40 per hole. Due to the low density of holes in this example, these values are extrapolated from the data on Figure 6 of Everett et al. (1976). Soil-water samplers cost $70 for each hole, and one service kit costs $30. Geologic logging and sampling require 1 day for 5 holes, 2 days for 10 holes, and 4 days for 20 holes, all by the junior-level individual. Chemical analyses of subsurface materials are performed on samples at 5-foot depth intervals, including one at the land surface. Cation exchange capacity is $11 per sample (from Everett et al., 1976, p. 112). Electrical conductivity of the saturation extract is $5.50 per sample (from Everett et al., 1976, p. 112). The pH of the saturation extract is $2.50 per sample (from Everett et al., 1976 p. 112). The exchangeable cations are $12 per sample (from Everett et al., 1976, p. 113). In this case, a special group rate of $25 per sample is used. The determinations are discounted to $22 per sample for 25 samples, $20 per sample for 50 samples, and $18 per sample for 100 samples.

Table 6-5

COSTS FOR MONITORING THE VADOSE ZONE

COST COMPONENTS	Alternatives (costs in dollars)		
	A	B	C
One-time Costs			
Drilling and Sampler Installation (including sampler)	630	1,180	2,230
Geologic Logging and Sampling Materials	130	250	380
Chemical Analyses of Materials	500	880	1,600
TOTAL	1,260	2,310	4,210
Annual Costs			
Sampling of Percolate	130	250	380
Chemical Analyses of Percolate	5,400	9,970	17,280
TOTAL	5,530	9,970	17,660

Sampling of percolate requires 1 day per month for 5 holes, 2 days for 10 holes, and 3 days for 20 holes, all by the junior-level individual. Costs for analytical determinations of the five trace metals in the percolate are $25 per sample. The trace metals were selected from the 12 listed by Everett et al. (1976, p. 114), and a special group rate was applied. Electrical conductivity and pH (Everett et al., 1976, Table 14) are $3 each. A special group rate of $30 for all of these determinations is applied. Chemical analyses for monthly percolate samples are $30 per sample up to 30 samples, $27 per sample for 30 samples, and $24 per sample for 60 samples.

Saturated Zone Monitoring

The cost of the surface resistivity survey is determined from Everett et al. (1976, Table 6). The depth to top of the plume is 20 feet, and the plume is estimated to be 50 feet thick. The areal extent of the plume is estimated at 50 acres. Based on an electrode spread of 100 feet, 6 surveys totaling about 18 hours would be required. At $80 per hour, the total cost is about $1,400.

For Alternative A, 2 monitor wells are drilled near the pond to a depth of 50 feet and the remaining 3 wells are 100 feet deep. For Alternative B, 3 monitor wells are 50 feet deep and the remaining 7 are 100 feet. Monitor-well drilling costs are calculated from Everett et al. (1976, Figure 19). A 6-inch casing is necessary to permit installation of a 50 gallons per minute capacity pump at the estimated lift of 60 feet (Everett et al., 1976, Table 11). Well drilling costs, including casings and well development, are about $2,600 each for the 100-foot deep monitor wells, and $1,700 each for the 50-foot deep monitor wells. Logging requires 1 day of the junior-level individual's time for 5 wells, and 2 days for 10 wells. The logging cost is $125 for 5 wells and $250 for 10 wells.

Twenty-four-hour pump tests are run on each monitor well at a cost of $20 per hour, exclusive of the junior-level individual's time. His time for pump tests, tabulating and plotting results, and interpretation is four days per well. Costs for pump tests including the junior-level individual's time are $980 per well. Chemical analyses are made for pumped water during the pump test. The major inorganic chemical constituents are determined at $12 per sample, and five trace metals are run at $25 per sample (Everett et al., 1976, p. 114). A group rate is applied for five of the 12 drinking water trace element determinations. Chemical analyses are thus $37 per sample, or $185 for Alternative A and $370 for Alternative B.

A portable 2-hp submersible pump costs $500 (Everett et al., 1976, Table 11). Eighty feet of cable are an additional $70 (Everett et al., 1976, Figure 26). Sample retrieval requires 5 days of pumping per month for 5 monitor wells, and 10 days for 10 monitor wells at $50 per day. Annual pumping costs are thus $3,000 for Alternative A, and $6,000 for Alternative B. The monthly samples of water from the monitor wells are analyzed for the major inorganic chemical constituents and the five trace elements. Chemical analyses for the monitor well water are $37 per sample, as determined previously.

Time required of the junior-level individual for sampling, checking chemical analyses, and tabulating results is one month for Alternative A, and four months for Alternative B. Supervision by the senior-level individual is one week for Alternative A and two weeks for Alternative B. Table 6-6 summarizes costs for monitoring the saturated zone.

Table 6-6

COSTS FOR SATURATED ZONE MONITORING

COST COMPONENTS	Alternatives (cost in dollars)	
	A	B
One-time Costs		
Surface Resistivity Survey	1,400	1,400
Drilling, Casing, and Development for Monitor Wells	11,200	23,300
Logging, Pump Testing, and Chemical Analyses	5,210	10,420
Portable Submersible Pump	570	570
TOTAL	18,380	35,690
Annual Costs		
Pumping for Monthly Sample Retrieval	3,000	6,000
Chemical Analyses of Monthly Samples	2,220	4,440
Personnel	3,440	6,880
TOTAL	8,660	17,320

In order to select the most cost-effective alternatives, consideration is given to impacts of groundwater pollution on subsequent groundwater use. In this case, no groundwater is presently used, but it is projected that groundwater will be used for municipal purposes in 20 years. Arsenic, cadmium, and hexavalent chromium pose a distinct health threat to groundwater used for drinking purposes. The high salinity, chloride, and sodium contents could easily render groundwater near the disposal site unusable. Secondly, the net worth of the industrial product has to be considered. A final consideration is the money available for monitoring the site. Consideration of all these factors, in conjunction with experience in monitoring groundwater pollution, leads to selection of Alternative B for the vadose zone, and Alternative A for the saturated zone.

Summary

Total costs for the program selected are given in Table 6-7.

SOLID WASTE LANDFILLS

The area selected for monitoring above the water table includes the landfill itself. In many areas tens of feet of soil and geologic materials have been excavated and replaced by solid wastes, comprising the landfill. Sampling and analysis are done to determine the composition of the solid materials present in the landfill, the leachate, and percolate in the vadose zone. If impermeable layers underlie the landfill, vadose zone monitoring may be necessary for distances of several hundred feet laterally from

Table 6-7

TOTAL COSTS FOR MONITORING
INDUSTRIAL WASTE PERCOLATION POND

Cost Components	Monitoring Costs (dollars)	
	One-Time	Annual
Preliminary Investigation	2,190	----
Surface Monitoring	1,350	6,790
Vadose Zone Monitoring	2,310	9,970
Saturated Zone Monitoring	18,380	8,660
TOTAL	24,230	25,420

the landfill. Sampling of leachate and shallow groundwater is almost always necessary. Groundwater should generally be monitored for a distance of at least several hundred or several thousand feet from the landfill. Sampling of topsoil and geologic materials for toxic constituents may be necessary.

LAND SURFACE MONITORING

Land surface monitoring encompasses an inventory of the volumes, or weights, of solid materials present in the landfill. Rainfall and evapotranspiration rates must be known, and can often be extrapolated from nearby meteorological stations. Compilation of these data allows calculation of the potential leachate production. The moisture characteristics of the solid wastes should be monitored, and usually, shallow holes drilled by augering will suffice. It is important to determine if aerobic or anaerobic decomposition is occurring at specific locations. This depends on the moisture content of the landfill, depth to groundwater, groundwater inflow, and other factors. Detection of leakage is a prime concern for landfills with liners installed to limit percolation.

The chemical composition of solid materials in the landfill and the leachate must be determined. The composition of landfill materials often can be broadly characterized, whereas detailed sampling of leachate is often necessary to determine its composition. The frequency of leachate sample collection depends highly on the frequency and rate of rainfall. Rather than a uniform sampling frequency, samples should be collected at intervals based on the rate of leachate production. The approximate volumes and quality of other sources of recharge and groundwater inflow must also be determined.

Leachate analyses should include the major inorganic chemical species, nitrogen forms, total dissolved solids, pH, and oxidation potential. The primary metals in the landfill that may be leached should be determined. Examples are iron, manganese, barium, chromium, lead, selenium, and zinc.

VADOSE ZONE MONITORING

Monitoring in the vadose zone includes determination of percolate quality, flow rate, attenuation characteristics of soil and geologic materials, and chemical analyses of solid materials. In areas of low rainfall, little or no leachate will be produced, and sampling in the vadose zone may be unnecessary. Test drilling is often necessary in or near landfills because of a lack of information on materials comprising the vadose zone. Drilling of test holes can provide geologic information on the vadose zone and permit the installation of devices for water sample retrieval. Neutron probes can be effectively used to trace water movement above the water table when necessary. Analytical determinations for percolate quality include the major inorganic chemical constituents; nitrogen forms; selected trace metals, particularly iron and manganese; and oxidation potential. Cation exchange capacity, electrical conductivity, pH, and exchangeable cations are important analyses for soil and geologic materials.

SATURATED ZONE MONITORING

Wells have often been successfully used in monitoring groundwater pollution beneath or near solid waste disposal sites. In areas of moderate to heavy rainfall, significant amounts of leachate are produced. Most or all of the leachate can subsequently percolate to the groundwater. Monitor wells may be installed and sampled; monthly sampling appears to be sufficient in most cases. Several existing large-capacity wells in the area should also be periodically sampled. A portable submersible pump can be used to pump water for sampling from a number of test wells in one area, thus avoiding the cost of equipping each well with a permanent pump. Analyses of water from wells are similar to those for percolate quality, but can be modified to reflect the water usage in the area. As leachate is generally high in total dissolved solids, surface resistivity surveys can be used in some cases to delineate the extent of polluted groundwater. The water table must be shallow and groundwater conditions fairly well understood. Remote sensing can provide information in areas where the water table is shallow, and/or the leachate is forced to the land surface.

SOLID WASTE LANDFILL EXAMPLE

STEP 2 - IDENTIFY POLLUTION SOURCES, CAUSES, AND METHODS OF WASTE DISPOSAL

In this example a 20-acre landfill in the southeastern United States, which has been in operation for 10 years, is to be monitored. The landfill receives solid refuse from a medium-sized city. A plastic liner has been installed to limit percolation. The other potential sources of groundwater pollution include small amounts of fertilizers, scattered septic tanks and polluted rainfall. Evaluation of these other sources indicates that they are insignificant.

STEP 3 - IDENTIFY POTENTIAL POLLUTANTS

The liner has been graded toward a central point so that leachate samples can be collected. Previous analyses of grab samples obtained from a sump drain have indicated that the quality of the leachate is similar to that to be expected from such a source. The leachate contains high total dissolved solids concentrations, some organic chemicals, and selected trace elements. Materials placed in the landfill indicate that chloride, nitrogen forms, potassium, calcium, sulfate, iron, manganese, lead, silver, chromium, cadmium, and zinc are of concern.

STEP 4 - DEFINE GROUNDWATER USAGE

Wells in this rural area are used primarily for domestic purposes, however, small amounts of groundwater are used for crop irrigation. Of all groundwater pumpage, about 80 percent is for domestic use, and the remainder for irrigation.

STEP 5 - DEFINE HYDROGEOLOGIC SITUATION

The average annual rainfall in the area is 40 inches. Alluvial deposits about 50 feet thick overlie hundreds of feet of permeable limestone. Both the alluvium and limestone are developed aquifers in the region, and beneath the landfill, the hydraulic head is lower in the limestone than in the alluvium. A number of water level measurements in wells in the two aquifers are available, and indicate the regional direction of groundwater movement. The water table in the alluvium is about 20 feet beneath the landfill, and water in the limestone is under artesian pressure. Groundwater in the alluvium tends to move toward a stream about one mile downgradient. There is a tendency for downward movement of shallow groundwater into the confined groundwater in the limestone. The extent of the polluted zone in the alluvium has been previously delineated, but the extent of pollution in the lower aquifer is unknown. There are several wells within the polluted zone in the alluvium.

Annual recharge to both aquifers was previously calculated by the U.S. Geological Survey. Annual pumpage from both aquifers in the region is about 20,000 acre-feet, whereas annual recharge to both aquifers is about 100,000 acre-feet.

STEP 6 - STUDY EXISTING GROUNDWATER QUALITY

Previous studies by the U.S. Geological Survey have delineated the regional groundwater quality. Groundwater in the limestone is calcium bicarbonate in type, with total dissolved solids less than 100 parts per million. Groundwater in the alluvium has a greater variation in chemical quality than groundwater in the limestone. Alluvial groundwater is a sodium-calcium bicarbonate type, and total dissolved solids range from about 70 to 500 parts per million. Beneath the landfill, groundwater is a sodium bicarbonate-chloride type, with high nitrate, chromium, iron, and manganese contents. Maps are available indicating the regional distribution of total dissolved solids, chloride, and nitrate in both aquifers.

STEP 7 - EVALUATE INFILTRATION POTENTIAL OF WASTES AT THE LAND SURFACE

Water budget analysis is unnecessary due to the presence of the liner, which greatly limits percolation of leachate. The presence of a polluted zone of groundwater beneath the site is ample indication that limited seepage and/or leakage has occurred. There was no inspection of the liner during installation, and there has been no monitoring of the liner integrity.

STEP 8 - EVALUATE MOBILITY OF POLLUTANTS FROM THE LAND SURFACE TO THE WATER TABLE

Neutron probe moisture logging and soil-water sampling were done in a similar hydrogeologic situation by researchers at a nearby university. This research indicated travel times of leachate to the shallow water table of several weeks to months. Attenuation of bacteriological constituents and some organic chemicals and trace elements was noted. These results could be directly extrapolated to the study area.

STEP 9 - EVALUATE ATTENUATION OF POLLUTANTS IN THE SATURATED ZONE

A surface resistivity survey was previously conducted to delineate the extent of the polluted zone in the alluvium. Several existing wells tap the alluvium, both in and outside of the polluted zone. However, although several nearby wells tap the limestone, no such wells are in the anticipated polluted zone. The extent of the polluted zone, determined by resistivity measurements, corresponds to high total dissolved solids content. Previous well sampling indicates that high calcium, bicarbonate, and chloride contents also occur in this zone. Several trace metals occur in high concentration, however, over a smaller zone. This is believed to be due primarily to precipitation and adsorption.

STEP 11 - EVALUATE EXISTING MONITORING PROGRAMS

Existing programs of most direct value in the area are related to monitoring in the vadose zone. There is no routine program for leachate sampling at the landfill, nor is there any for routine well sampling in the immediate vicinity of the landfill.

STEP 12 - ESTABLISH ALTERNATIVE MONITORING APPROACHES

Consideration of information developed in the previous steps indicates that no additional monitoring in the vadose zone is necessary. Routine sampling of leachate, well drilling, and water sampling from wells are deemed necessary. Hydrogeologic judgment is used to select the most effective monitoring program in each case.

To determine cost-effectiveness, consideration is given to water use in areas that could be impacted by the waste disposal. Most of the water use is for domestic purposes, and, due to the rural nature of the area, groundwater is considered the sole source of the drinking water supply. Nitrate, chromium, and cadmium pose a potential threat to health for persons drinking water from wells in the polluted zone. The maximum extent of the polluted zone is believed to not exceed 1,000 acres for the alluvium. For the limestone the maximum extent is believed to not exceed 200 acres. Within these areas at present, only three wells are used for drinking water purposes, serving a total of 10 people. Given the rural nature of the area and projected land use, this situation should not greatly change in the next 30 years.

Inorganic chemical constituents may also pose a health threat. Chloride, sulfate, calcium, iron, and manganese may degrade drinking water. Increased total dissolved solids from pollution may decrease crop yields in a small area near the pond.

Land Surface Monitoring

As part of the recommended monitoring program, leachate samples will be collected daily for electrical conductivity determinations. In turn, these samples will be composited weekly for analysis. The major inorganic chemical constituents, total nitrogen, COD, dissolved oxygen, iron, manganese, lead, silver, chromium, cadmium, and zinc will be determined.

Saturated Zone Monitoring

Monitoring includes routine sampling of water from existing wells and drilling of additional wells in the limestone for water sampling. Three monitor wells are to be drilled into the limestone by the cable-tool method. These wells are 300 feet deep and perforated from 100 to 300 feet. The water level in the wells is only 60 feet below the landfill, due to the artesian condition existing in the limestone aquifer. Hole diameter is 8 inches and casing diameter is 6 inches. After development, 1-week pump tests are conducted, and permanent submersible pumps of 25 gallons per minute capacity at 75-foot lift are installed. The pump tests provide information on the hydraulic connection of the upper and lower aquifer, as well as vertical permeability of the confining bed, for the confined aquifer. The electrical conductivity and temperature of the water discharged are frequently measured. On this basis, five water samples are chosen for routine chemical analyses for each well, including the seven trace metals listed previously.

For the first year, monthly samples are collected from these three wells and five previously existing wells. The major inorganic chemicals are determined, including iron, manganese, chromium, and cadmium. Lead, silver, and zinc are deleted from the program because of attenuation characteristics of materials above the water table. After the first year, quarterly samples from five wells are sufficient.

STEP 13 - SELECT AND IMPLEMENT THE MONITORING PROGRAM

As the approach has already been determined in Step 12, this step primarily involves determination of costs. A preliminary investigation is conducted to meet the requirements of Steps 3 through 6, including collection and interpretation of records on groundwater conditions, water quality, landfills, and leachate. The junior-level individual spends one month at a cost of $2,500. The senior-level individual spends one week at a cost of $940. The cost of this phase is thus $3,440.

Land Surface Monitoring

A portable meter for daily electrical conductivity determinations costs $350. Leachate samples are taken by the caretaker, at no additional expense. Major inorganic chemical constituents determinations are $12 per sample (Everett et al., 1976, p. 114). The 7 trace elements determinations are $30 per sample, based on a group rate derived from the 12 drinking water trace elements from Everett et al. (1976, p. 114). Dissolved oxygen determinations are $5 per sample (Everett et al., 1976, Table 14). Chemical oxygen demand determinations are $10 per sample (Everett et al., 1976, Table 15). Total nitrogen determinations are $10 per sample (Everett et al., 1976, Table 14).

Leachate analyses on a weekly basis are $67, or $3,480 during the first year. Checking chemical analyses and plotting results during the first year requires one month of time by the junior professional, at a cost of $2,500. Supervision by a senior level professional during the first year totals one week, at a cost of $940. After the first year, weekly samples are composited and analyzed monthly. Thus, analytical costs after the first year are $804 per year. Personnel time after the first year includes two weeks for the junior-level individual at $1,250 and one-half week by the senior-level individual at $470.

Vadose Zone Monitoring

Review of existing studies requires two weeks by a junior professional at $1,250, and one week by a senior professional at $940. Total one-time costs are $2,190.

Saturated Zone Monitoring

Drilling costs are about $3,800 each for 8-inch diameter holes, with 6-inch steel casing (Everett et al., 1976, Figure 20 - for consolidated formations). The pump tests of one week duration for each well are conducted at a cost of $3,500 per well, exclusive of geologist time. The junior professional spends one day logging cuttings, and two weeks for pump tests (including interpretation) on each well. Personnel costs for well-logging, pump testing, and aquifer analysis are $1,460 per well.

Major inorganic chemical constituents are determined at $12 per sample (Everett et al., 1976, p. 114). Seven trace elements are determined at $30 per sample (Everett et al., 1976, p. 114), applying a special group rate. Chemical analyses are $42 each for the five water samples collected during the pump test for each well. Three 1-hp submersible pumps, with 25-gallons per minute capacity at the projected 75-foot lift, cost $300 each. Three hundred feet of cable for the pumps costs $240 (Everett et al., 1976, Figure 26). Supervision by the senior-level individual for this phase is three weeks at a cost of $2,830. Total one-time cost for this phase of the program is $29,070.

Water samples are collected at no pumping cost, as the monitor wells are also used for water supply in the area. Sampling can be done during normal operation of the pump by the user. Major inorganic chemical constituents are determined for $12 per sample (Everett et al., 1976, p. 114). The 4 trace metals are analyzed at a cost of $17 per sample, applying a discount rate to that given for the 12 drinking water trace constituents (Everett et al., 1976, p. 114). Costs for the water analyses are $29 per sample. Monthly analyses during the first year total $2,784. Sample collection, checking chemical analyses, and tabulating results require one month of the junior-level individual's time, at a cost of $2,500. Supervision by the senior-level individual requires one week at $940.

After the first year, analytical costs are $145 per quarterly sampling round, of $580 per year. The junior-level individual spends two weeks annually after the first year at a cost of $1,250. The senior professional spends one-half week at a cost of $470.

Summary

Total monitoring costs are summarized in Table 6-8.

Table 6-8

MONITORING COSTS FOR SOLID WASTE LANDFILL

Cost Component	Cost (dollars)		
	One-Time	Annual First Year	Annual Subsequent Years
Preliminary Investigation	3,440	---	---
Land Surface Monitoring	350	6,920	1,720
Monitoring Vadose Zone	2,190	---	---
Monitoring Saturated Zone	30,870	6,224	2,300
TOTAL	36,850	13,144	4,020

SELECTED BIBLIOGRAPHY

American Water Works Association, "Sources of Nitrogen and Phosphorus in Water Supplies," *Journal of American Water Works Association, Task Group Report*, pp. 344-366, March 1967.

J.J. Behnke and E.E. Haskell, "Ground Water Nitrate Distribution Beneath Fresno, California," *Journal of American Water Works Association*, Vol. 60, No. 4, pp. 477-480, 1968.

H. Bouwer, "Renovating Secondary Effluent by Groundwater Recharge with Infiltration Basins," *Recycling Treated Municipal Wastewater and Sludge through Forest and Cropland*, W.E. Sopper and L.T. Kardos (eds), Pennsylvania State University Press, 1973.

California Department of Water Resources, *Fresno-Clovis Metropolitan Area Water Quality Investigation*, Bulletin 143-3, 1965.

E.S. Davidson, *Geohydrology and Water Resources of the Tucson Basin, Arizona*, U.S. Geological Survey, Water Supply Paper 1939-E, 1973.

B.G. Ellis, "The Soil as a Chemical Filter," *Recycling Treated Municipal Wastewater and Sludge through Forest and Cropland*, W.E. Sopper and L.T. Kardos (eds), Pennsylvania State University Press, 1973.

L.G. Everett, K.D. Schmidt, R.M. Tinlin, and D.K. Todd, *Monitoring Groundwater Quality: Methods and Costs*, EPA-600/4-76-023, U.S. Environmental Protection Agency, Las Vegas, Nevada, May 1976.

J.S. Fryberger, *Rehabilitation of a Brine-Polluted Aquifer*, EPA-R2-72-014, Environmental Protection Technology Series, 1972.

Geraghty and Miller, Inc., *Geologic and Hydrologic Investigation of the Proposed Silver Sands State Park, Milford, Connecticut*, for the Connecticut Departments of Public Works and Environmental Protection, 1973.

G.S. Lehman, *Soil and Grass Filtration of Domestic Sewage Effluent for the Removal of Trace Elements*, unpublished Ph. D. Dissertation, University of Arizona, 1968.

A.H. Ludwig, *Water Resources of Hempstead, Lafayette, Little Rock, Miller, and Nevada Counties, Arkansas*, Geological Survey Water Supply Paper 1998, 1973.

W.G. Matlock, P.R. Davis, and R.L. Roth, *Sewage Pollution of a Groundwater Aquifer*, Paper presented at the 1972 Winter Meeting of American Society of Agricultural Engineers, Chicago, Illinois, 1972.

R.H. Miller, "The Soil as a Biological Filter," *Recycling Treated Municipal Wastewater and Sludge through Forest and Cropland*, W.E. Sopper and L.T. Kardos (eds), Pennsylvania State University Press, 1973.

H.I. Nightingale, "Nitrates in Soil and Ground Water Beneath Irrigated and Fertilized Crops," *Soil Science*, Vol. 114, No. 4, pp. 300-311, 1972.

H.I. Nightingale, "Statistical Evaluation of Salinity and Nitrate Content and Trends Beneath Urban and Agricultural Areas – Fresno, California," *Ground Water*, Vol. 8, No. 1, pp. 22-28, 1970.

Oklahoma Water Resources Board, *Salt Water Detection in the Cimarron Terrace, Oklahoma*, EPA-660/3-74-033, 1974.

R.W. Page and R.A. LeBlanc, *Geology, Hydrology, and Water Quality in the Fresno Area, California*, U.S. Geological Survey Open-File Report, Menlo Park, California, 1969. (189 pp.)

N.M. Perlmutter and J.J. Geraghty, *Geology and Groundwater Conditions in Southern Nassau and Souteastern Queens Counties, Long Island, New York*, U.S. Geological Survey Water-Supply Paper 1613-A, 1963. (205 pp.)

N.M. Perlmutter and M. Lieber, *Disposal of Plating Wastes and Sewage Contaminants in Ground Water and Surface Water, South Farmingdale – Massapequa Area, Nassau County, New York*, U.S. Geological Survey Water-Supply Paper 1879G, 1970.

N.M. Perlmutter, M. Lieber, and H.L. Frauenthal, "Movement of Waterborne Cadmium and Hexavalent Chromium Wastes in South Farmingdale, Nassau County, Long Island, New York," *Short Papers in Geology and Hydrology*, U.S. Geological Survey Professional Paper 475-C, pp. C179-C184, 1963.

G.F. Pinder, "A Galerkin-Finite Element Simulation of Ground-Water Contamination on Long Island, N.Y.," *Water Resources Research*, Vol. 9, No. 6, pp. 1657-1669, 1973.

J.H. Randall, *Hydrogeology and Water Quality, Pima County Landfill Project*, Unpublished Report, University of Arizona, 1974.

K.D. Schmidt, "Regional Sewering and Groundwater Quality in the Southern San Joaquin Valley," Water Resources Bulletin, Vol. 11, No. 3, pp. 514-525, 1975.

K.D. Schmidt, "Groundwater Contamination in the Cortaro Area, Pima County, Arizona," *Hydrology and Water Resources in Arizona and the Southwest*, Vol. 2, Proceedings of 1972 Meetings of Arizona Section, American Water Resources Association, Prescott, Arizona, May 5-6, 1972a.

K.D. Schmidt, "Nitrate in Ground Water of the Fresno-Clovis Metropolitan Area," *Ground Water*, Vol. 10, No. 1, pp. 50-64, 1972b.

G.J. Schroepfer and R.C. Polta, *Travel of Nitrogen Compounds in Soils*, University of Minnesota, Sanitary Engineering Report 172-5, 1969.

G.G. Small, *Groundwater Recharge and Quality Transformations During the Initiation and Management of a New Stabilization Lagoon*, Unpublished M.S. Thesis, University of Arizona, 1973.

P.R. Stout, R.G. Burau, and W.R. Allardice, *A Study of the Vertical Movement of Nitrogenous Matter from the Ground Surface to the Water Table in the Vicinity of Grover City and Arroyo Grande, San Luis Obispo County*, Report to Central Coastal Regional Water Pollution Control Board, 1965. (51 pp.)

R. Suter, W. deLaguna, and N.M. Perlmutter, *Mapping of Geologic Formations and Aquifers of Long Island, New York*, New York Water Power and Control Commission Bulletin GW-18, 1949. (212 pp.)

D.K. Todd, R.M. Tinlin, K.D. Schmidt, and L.G. Everett, *Monitoring Groundwater Quality: Monitoring Methodology*, U.S. Environmental Protection Agency, Las Vegas, Nevada, 1976.

W.C. Walton, *Groundwater Resource Evaluation*, McGraw-Hill Book Co., 1970.

L.G. Wilson, *Quality Transformation in Recharged River Water During Possible Interactions with Landfill Deposits Along the Santa Cruz River*, Annual Report to Pima County Department of Sanitation, Water Resources Research Center, University of Arizona, 1974.

L.G. Wilson, "Observations on Water Content Changes in Stratified Sediments During Pit Recharge," *Ground Water*, Vol. 9, No. 3, pp. 29-40, 1971.

L.G. Wilson, W.L. Clark III, and G.G. Small, "Subsurface Transformations During the Initiation of a New Stabilization Lagoon, *Water Resources Bulletin*, Vol. 9, No. 2, pp. 243-257, 1973.

L.G. Wilson and K.J. DeCook, "Field Observations on Changes in the Subsurface Water Regime During Influent Seepage in the Santa Cruz River, *Water Resources Research*, Vol. 4, No. 6, pp. 1219-1234, 1968.

L.G. Wilson and G.S. Lehman, "Reclaiming Sewage Effluent," *Progressive Agriculture in Arizona*, Vol. 19, No. 4, pp. 22-24, 1967.

L.G. Wilson and C.C. Small, "Population Potential of a Sanitary Landfill Near Tucson," *Hydraulic Engineering and the Environment*, Proceedings of 21st Annual Hydrology Division, 1973.